T0213830

Lecture Notes in Artificial Intelligence 9355

Subseries of Lecture Notes in Computer Science

LNAI Series Editors

Randy Goebel
University of Alberta, Edmonton, Canada
Yuzuru Tanaka
Hokkaido University, Sapporo, Japan
Wolfgang Wahlster
DFKI and Saarland University, Saarbrücken, Germany

LNAI Founding Series Editor

Joerg Siekmann
DFKI and Saarland University, Saarbrücken, Germany

More information about this series at http://www.springer.com/series/1244

Kamalika Chaudhuri · Claudio Gentile
Sandra Zilles (Eds.)

Algorithmic
Learning Theory

26th International Conference, ALT 2015
Banff, AB, Canada, October 4–6, 2015
Proceedings

 Springer

Editors
Kamalika Chaudhuri
University of California
La Jolla, CA
USA

Claudio Gentile
Università dell'Insubria
21100 Varese
Italy

Sandra Zilles
Department of Computer Science
University of Regina
Regina, SK
Canada

ISSN 0302-9743 ISSN 1611-3349 (electronic)
Lecture Notes in Artificial Intelligence
ISBN 978-3-319-24485-3 ISBN 978-3-319-24486-0 (eBook)
DOI 10.1007/978-3-319-24486-0

Library of Congress Control Number: 2015949454

LNCS Sublibrary: SL7 – Artificial Intelligence

Springer Cham Heidelberg New York Dordrecht London
© Springer International Publishing Switzerland 2015
This work is subject to copyright. All rights are reserved by the Publisher, whether the whole or part of the material is concerned, specifically the rights of translation, reprinting, reuse of illustrations, recitation, broadcasting, reproduction on microfilms or in any other physical way, and transmission or information storage and retrieval, electronic adaptation, computer software, or by similar or dissimilar methodology now known or hereafter developed.
The use of general descriptive names, registered names, trademarks, service marks, etc. in this publication does not imply, even in the absence of a specific statement, that such names are exempt from the relevant protective laws and regulations and therefore free for general use.
The publisher, the authors and the editors are safe to assume that the advice and information in this book are believed to be true and accurate at the date of publication. Neither the publisher nor the authors or the editors give a warranty, express or implied, with respect to the material contained herein or for any errors or omissions that may have been made.

Printed on acid-free paper

Springer International Publishing AG Switzerland is part of Springer Science+Business Media
(www.springer.com)

Preface

This volume contains the papers presented at the 26th International Conference on Algorithmic Learning Theory (ALT 2015), which was held in Banff, Alberta, Canada during October 4–6, 2015. ALT 2015 was co-located with the 18th International Conference on Discovery Science (DS 2015). The technical program of ALT 2015 had four invited talks (presented jointly to both ALT 2015 and DS 2015) and 23 papers selected from 46 submissions by the ALT Program Committee. ALT 2015 took place in the Banff Park Lodge in the heart of Banff. It provided a stimulating interdisciplinary forum to discuss the theoretical foundations of machine learning as well as their relevance to practical applications.

ALT is dedicated to the theoretical foundations of machine learning and provides a forum for high-quality talks and scientific interaction in areas such as reinforcement learning, inductive inference and grammatical inference, learning from queries, active learning, probably approximate correct learning, online learning, bandit theory, statistical learning theory, Bayesian and stochastic learning, unsupervised or semi-supervised learning, clustering, universal prediction, stochastic optimization, high dimensional and non-parametric inference, information-based methods, decision tree methods, kernel-based methods, graph methods and/or manifold-based methods, sample complexity, complexity of learning, privacy preserving learning, learning based on Kolmogorov complexity, new learning models, and applications of algorithmic learning theory.

The present LNAI volume contains the 23 papers presented at ALT 2015 and the following invited papers, either as full papers or as abstracts:

- Inderjit Dhillon (University of Texas at Austin), "Efficient Matrix Sensing Using Rank-1 Gaussian Measurements" (Invited Speaker for ALT 2015);
- Sham Kakade (Microsoft Research and University of Washington), "Tensor Decompositions for Learning Latent Variable Models" (Invited Speaker for ALT 2015);
- Cynthia Rudin (Massachusetts Institute of Technology), "Turning Prediction Tools Into Decision Tools" (Invited Speaker for DS 2015);
- Kiri Wagstaff (Jet Propulsion Laboratory, Los Angeles, CA), "Overcoming Obstacles to the Adoption of Machine Learning by Domain Experts" (Invited Speaker for DS 2015).

Since 1999, ALT has been awarding the E.M. Gold Award for the most outstanding student contribution. This year, the award was given to Ana Ozaki for her paper "Exact Learning of Multivalued Dependencies" coauthored with Montserrat Hermo.

ALT 2015 was the 26th meeting in the ALT conference series, established in Japan in 1990. The ALT series is supervised by its Steering Committee: Shai Ben-David (University of Waterloo, Canada), Nader H. Bshouty (Technion - Israel Institute of Technology, Israel), Kamalika Chaudhuri (University of California, San Diego, USA),

Claudio Gentile (Università dell'Insubria, Varese, Italy), Marcus Hutter (Australian National University, Canberra, Australia), Sanjay Jain (National University of Singapore, Republic of Singapore), Frank Stephan (Co-chair, National University of Singapore, Republic of Singapore), Gilles Stoltz (École normale supérieure, Paris, France), Csaba Szepesvári (University of Alberta, Edmonton, Canada), Eiji Takimoto (Kyushu University, Fukuoka, Japan), György Turán (University of Illinois at Chicago, USA, and University of Szeged, Hungary), Akihiro Yamamoto (Kyoto University, Japan), and Sandra Zilles (Chair, University of Regina, Saskatchewan, Canada).

We thank various people and institutions who contributed to the success of the conference. Most importantly, we would like to thank the authors for contributing and presenting their work at the conference. Without their contribution this conference would not have been possible. We would like to thank ISM Canada, an IBM company, and Alberta Innovates Technology Futures (AITF) for the generous financial support for the conference.

ALT 2015 and DS 2015 were organized by the University of Alberta and the University of Regina. We thank the local arrangement team for their efforts in organizing the two conferences. In particular, we thank Csaba Szepesvári, who co-chaired the local arrangements team together with Sandra Zilles. Special thanks to Deon Nicholas and Reka Szepesvári from the University of Alberta for their help in setting up the registration page for the two conferences. We are grateful for the collaboration with the conference series Discovery Science. In particular, we would like to thank the general chair of DS 2015 and ALT 2015, Randy Goebel, and the DS 2015 Program Committee chairs, Nathalie Japkowicz and Stan Matwin.

We are also grateful to EasyChair, the excellent conference management system, which was used for putting together the program for ALT 2015. EasyChair was developed mainly by Andrei Voronkov and is hosted at the University of Manchester. The system is cost-free.

We are grateful to the members of the Program Committee for ALT 2015 and the subreferees for their hard work in selecting a good program for ALT 2015. Last but not least, we thank Springer for their support in preparing and publishing this volume in the Lecture Notes in Artificial Intelligence series.

July 2015

Kamalika Chaudhuri
Claudio Gentile
Sandra Zilles

Organization

Program Committee

Dana Angluin	Yale University, USA
András Antos	Budapest University of Technology and Economics, Hungary
Shai Ben-David	University of Waterloo, Canada
Nader Bshouty	Technion Israel Institute of Technology, Israel
Nicolò Cesa-Bianchi	Università degli Studi di Milano, Italy
Kamalika Chaudhuri	University of California San Diego, USA
Vitaly Feldman	IBM, Almaden Research Center, San Jose, USA
Claudio Gentile	Università dell'Insubria, Varese, Italy
Marcus Hutter	Australian National University, Australia
Sanjay Jain	National University of Singapore, Singapore
Wouter Koolen	Queensland University of Technology, Australia and University of California, Berkeley, USA
Timo Kötzing	Universität Jena, Germany
Lihong Li	Microsoft Research, USA
Odalric-Ambrym Maillard	Technion Israel Institute of Technology, Israel
Yishay Mansour	Microsoft Research and Tel Aviv University, Israel
Claire Monteleoni	George Washington University, USA
Gergely Neu	Inria, France
Francesco Orabona	Yahoo Labs New York, USA
Mark Reid	Australian National University, Australia
Lev Reyzin	University of Illinois, Chicago, USA
Philippe Rigollet	MIT, USA
Sivan Sabato	Ben Gurion University of the Negev, Israel
Hans Simon	Ruhr Universität Bochum, Germany
Karthik Sridharan	Cornell University, USA
Liwei Wang	Peking University, China

Additional Reviewers

Abbasi-Yadkori, Yasin	Choromanski, Krzysztof
Arias, Marta	Daniely, Amit
Balcazar, Jose	Daswani, Mayank
Blanchard, Gilles	Everitt, Tom
Brautbar, Michael	Fetaya, Ethan
Case, John	Frongillo, Rafael
Choromanska, Anna	Helmbold, David

Huang, Ruitong
Jackson, Jeffrey
Jiang, Nan
Kanade, Varun
Karampatziakis, Nikos
Kloft, Marius
Kontorovich, Aryeh
Koren, Tomer
Kun, Jeremy
Kuzborskij, Ilja
Leike, Jan
Lelkes, Ádám D.
Levy, Kfir
Liberty, Edo
Menon, Aditya
Mineiro, Paul

Moseley, Benjamin
Ortner, Ronald
Pentina, Anastasya
Perchet, Vianney
Servedio, Rocco
Spall, James
Stephan, Frank
Tamon, Christino
Tang, Cheng
Theocharous, Georgios
Tolstikhin, Ilya
van Rooyen, Brendan
Zeugmann, Thomas
Zhang, Chicheng
Zhao, Peilin

Abstracts of Invited Talks

Efficient Matrix Sensing
Using Rank-1 Gaussian Measurements

Kai Zhong[1], Prateek Jain[2], and Inderjit S. Dhillon[1]

[1] University of Texas at Austin, USA
[2] Microsoft Research, India
zhongkai@ices.utexas.edu, prajain@microsoft.com,
inderjit@cs.utexas.edu

Abstract. In this paper, we study the problem of low-rank matrix sensing where the goal is to reconstruct a matrix *exactly* using a small number of linear measurements. Existing methods for the problem either rely on measurement operators such as random element-wise sampling which cannot recover arbitrary low-rank matrices or require the measurement operator to satisfy the Restricted Isometry Property (RIP). However, RIP based linear operators are generally full rank and require large computation/storage cost for both measurement (encoding) as well as reconstruction (decoding).

In this paper, we propose simple rank-one Gaussian measurement operators for matrix sensing that are significantly less expensive in terms of memory and computation for both encoding and decoding. Moreover, we show that the matrix can be reconstructed *exactly* using a simple alternating minimization method as well as a nuclear-norm minimization method. Finally, we demonstrate the effectiveness of the measurement scheme vis-a-vis existing RIP based methods.

Tensor Decompositions for Learning Latent Variable Models
(a survey for ALT)

Anima Anandkumar[1], Rong Ge[2], Daniel Hsu[3],
Sham M. Kakade[4], and Matus Telgarsky[5]

University of California, Irvine
Microsoft Research, New England
Columbia University
Rutgers University
University of Michigan

Abstract. This note is a short version of that in [1]. It is intended as a survey for the 2015 Algorithmic Learning Theory (ALT) conference.

This work considers a computationally and statistically efficient parameter estimation method for a wide class of latent variable models— including Gaussian mixture models, hidden Markov models, and latent Dirichlet allocation—which exploits a certain tensor structure in their low-order observable moments (typically, of second- and third-order). Specifically, parameter estimation is reduced to the problem of extracting a certain (orthogonal) decomposition of a symmetric tensor derived from the moments; this decomposition can be viewed as a natural generalization of the singular value decomposition for matrices. Although tensor decompositions are generally intractable to compute, the decomposition of these specially structured tensors can be efficiently obtained by a variety of approaches, including power iterations and maximization approaches (similar to the case of matrices). A detailed analysis of a robust tensor power method is provided, establishing an analogue of Wedin's perturbation theorem for the singular vectors of matrices. This implies a robust and computationally tractable estimation approach for several popular latent variable models.

Turning Prediction Tools Into Decision Tools

Cynthia Rudin

MIT CSAIL and Sloan School of Management
Building E62-576, 100 Main Street, Cambridge, MA 02142, USA
rudin@mit.edu

Arguably, the main stumbling block in getting machine learning algorithms used in practice is the fact that people do not trust them. There could be many reasons for this, for instance, perhaps the models are not sparse or transparent, or perhaps the models are not able to be customized to the user's specifications as to what a decision tool should look like. I will discuss some recent work from the Prediction Analysis Lab on how to build machine learning models that have helpful decision-making properties. I will show how these models are applied to problems in healthcare and criminology.

Overcoming Obstacles to the Adoption
of Machine Learning by Domain Experts

Kiri L. Wagstaff

Jet Propulsion Laboratory, California Institute of Technology,
4800 Oak Grove Drive, Pasadena, CA 91109, USA
kiri.l.wagstaff@jpl.nasa.gov

The ever-increasing volumes of scientific data being collected by fields such as astronomy, biology, planetary science, medicine, etc., present a need for automated data analysis methods to assist investigators in understanding and deriving new knowledge from their data. Partnerships between domain experts and computer scientists can open that door. However, there are obstacles that sometimes prevent the successful adoption of machine learning by those who stand to benefit most.

We devote a lot of effort to solving technological challenges (e.g., scalability, performance), but less effort to overcoming psychological and logistical barriers. Domain experts may fail to be persuaded to adopt a tool based on performance results that are otherwise compelling to those in machine learning, which can be frustrating and perplexing. Algorithm aversion is the phenomenon in which people place more trust in human predictions than those generated by an algorithm, even when the algorithm demonstrably performs better. Media hype about the dangers of artificial intelligence and fears about the loss of jobs or the loss of control create additional obstacles.

I will describe two case studies in which we have developed and delivered machine learning systems to solve problems from radio astronomy and planetary science domains. While we cannot claim to have a magic wand that ensures the adoption of machine learning systems, we can share lessons learned from our experience. Key elements include progressive integration, enthusiasm on the part of the domain experts, and a system that visibly learns or adapts to user feedback to correct any mistakes.

This work was performed at the Jet Propulsion Laboratory, California Institute of Technology, under a contract with NASA. Government sponsorship acknowledged.

Contents

Statistical Learning Theory and Sample Complexity

Online Learning, Stochastic Optimization

Kolmogorov Complexity, Algorithmic Information Theory

Invited Papers

Efficient Matrix Sensing
Using Rank-1 Gaussian Measurements

Kai Zhong[1], Prateek Jain[2], and Inderjit S. Dhillon[1][(✉)]

[1] University of Texas at Austin, Austin, USA
[2] Microsoft Research, Bangalore, India
zhongkai@ices.utexas.edu, prajain@microsoft.com, inderjit@cs.utexas.edu

Abstract. In this paper, we study the problem of low-rank matrix sensing where the goal is to reconstruct a matrix *exactly* using a small number of linear measurements. Existing methods for the problem either rely on measurement operators such as random element-wise sampling which cannot recover arbitrary low-rank matrices or require the measurement operator to satisfy the Restricted Isometry Property (RIP). However, RIP based linear operators are generally full rank and require large computation/storage cost for both measurement (encoding) as well as reconstruction (decoding).

In this paper, we propose simple rank-one Gaussian measurement operators for matrix sensing that are significantly less expensive in terms of memory and computation for both encoding and decoding. Moreover, we show that the matrix can be reconstructed *exactly* using a simple alternating minimization method as well as a nuclear-norm minimization method. Finally, we demonstrate the effectiveness of the measurement scheme vis-a-vis existing RIP based methods.

Keywords: Matrix sensing · Matrix completion · Inductive learning · Alternating minimization

1 Introduction

In this paper, we consider the matrix sensing problem, where the goal is to recover a low-rank matrix using a small number of linear measurements. The matrix sensing process contains two phases: a) compression phase (encoding), and b) reconstruction phase (decoding).

In the compression phase, a sketch/measurement of the given low-rank matrix is obtained by applying a linear operator $\mathcal{A} : \mathbb{R}^{d_1 \times d_2} \rightarrow \mathbb{R}^m$. That is, given a rank-$k$ matrix, $W_* \in \mathbb{R}^{d_1 \times d_2}$, its linear measurements are computed by:

$$b = \mathcal{A}(W_*) = [\text{Tr}(A_1^T W_*) \ \text{Tr}(A_2^T W_*) \ \ldots \ \text{Tr}(A_m^T W_*)]^T , \qquad (1)$$

Electronic supplementary material The online version of this chapter (doi:10.1007/978-3-319-24486-0_1) contains supplementary material, which is available to authorized users.

© Springer International Publishing Switzerland 2015
K. Chaudhuri et al. (Eds.): ALT 2015, LNAI 9355, pp. 3–18, 2015.
DOI: 10.1007/978-3-319-24486-0_1

where $\{A_l \in \mathbb{R}^{d_1 \times d_2}\}_{l=1,2,\ldots,m}$ parameterize the linear operator $\mathcal{A}$ and Tr denotes the trace operator. Then, in the reconstruction phase, the underlying low-rank matrix is reconstructed using the given measurements $\boldsymbol{b}$. That is, W_* is obtained by solving the following optimization problem:

$$\min_{W} \ \text{rank}(W) \quad \text{s.t.} \quad \mathcal{A}(W) = \boldsymbol{b}. \tag{2}$$

The matrix sensing problem is a matrix generalization of the popular compressive sensing problem and has several real-world applications in the areas of system identification and control, Euclidean embedding, and computer vision (see [20] for a detailed list of references).

Naturally, the design of the measurement operator $\mathcal{A}$ is critical for the success of matrix sensing as it dictates cost of both the compression as well as the reconstruction phase. Most popular operators for this task come from a family of operators that satisfy a certain Restricted Isometry Property (RIP). However, these operators require each A_l, that parameterizes $\mathcal{A}$, to be a full rank matrix. That is, cost of compression as well as storage of $\mathcal{A}$ is $O(md_1d_2)$, which is infeasible for large matrices. Reconstruction of the low-rank matrix W_* is also expensive, requiring $O(md_1d_2 + d_1^2 d_2)$ computation steps. Moreover, m is typically at least $O(k \cdot d \log(d))$ where $d = d_1 + d_2$. But, these operators are universal, i.e., every rank-k W_* can be compressed and recovered using such RIP based operators.

In this paper, we seek to reduce the computational/storage cost of such operators but at the cost of the universality property. That is, we propose to use simple rank-one operators, i.e., where each A_l is a rank one matrix. We show that using similar number of measurements as the RIP operators, i.e., $m = O(k \cdot d \log(d))$, we can recover a fixed rank W_* exactly.

In particular, we propose two measurement schemes: a) *rank-one independent* measurements, b) *rank-one dependent* measurements. In rank-one independent measurement, we use $A_l = \boldsymbol{x}_l \boldsymbol{y}_l^T$, where $\boldsymbol{x}_l \in \mathbb{R}^{d_1}$, $\boldsymbol{y}_l \in \mathbb{R}^{d_2}$ are both sampled from zero mean sub-Gaussian product distributions, i.e., each element of $\boldsymbol{x}_l$ and $\boldsymbol{y}_l$ is sampled from a fixed zero-mean univariate sub-Gaussian distribution. Rank-one dependent measurements combine the above rank-one measurements with element-wise sampling, i.e., $A_l = \boldsymbol{x}_{i_l} \boldsymbol{y}_{j_l}^T$ where $\boldsymbol{x}_{i_l}, \boldsymbol{y}_{j_l}$ are sampled as above. Also, $(i_l, j_l) \in [n_1] \times [n_2]$ is a randomly sampled index, where $n_1 \geq d_1$, $n_2 \geq d_2$. These measurements can also be viewed as the inductive version of the matrix completion problem (see Section 3), where $\boldsymbol{x}_i$ represents features of the i-th user (row) and $\boldsymbol{y}_j$ represents features of the j-th movie (column). In fact, an additional contribution of our work is that we can show that the inductive matrix completion problem can also be solved using only $O(k(d_1+d_2) \log(d_1+d_2) \log(n_1+n_2))$ samples, as long as X, Y, W_* satisfy certain incoherence style assumptions (see Section 5 for more details)[1].

Next, we provide two recovery algorithms for both of the above measurement operators: a) alternating minimization, b) nuclear-norm minimization.

---

[1] A preliminary version of this work appeared in [11]. Since the above mentioned work, several similar rank-one measurement operators have been studied [2,15,25].

Table 1. Comparison of sample complexity and computational complexity for different approaches and different measurements

METHODS		SAMPLE COMPLEXITY	COMPUTATIONAL COMPLEXITY
ALS	RANK-1 INDEP.	$O(k^4\beta^2 d\log^2 d)$	$O(kdm)$
	RANK-1 DEP.	$O(k^4\beta^2 d\log d)$	$O(dm + knd)$
	RIP	$O(k^4 d\log d)$	$O(d^2 m)$
NUCLEAR	RANK-1 INDEP.	$O(kd)$	$O(\hat{k}dm)$
	RANK-1 DEP.	$O(kd\log n\log d)$	$O(\hat{k}^2 m + \hat{k}nd)$
	RIP	$O(kd\log d)$	$O(d^2 m)$

Note that, in general, the recovery problem (2) is NP-hard to solve. However, for the RIP based operators, both alternating minimization as well as nuclear norm minimization methods are known to solve the problem exactly in polynomial time [13]. Note that the existing analysis of both the methods crucially uses RIP and hence does not extend to the proposed operators.

We show that if $m = O(k^4 \cdot \beta^2 \cdot (d_1 + d_2) \log^2(d_1 + d_2))$, where β is the condition number of W_* then alternating minimization for the rank-one independent measurements recovers W_* in time $O(kdm)$, where $d = d_1 + d_2$. Similarly, if $m = O(k \cdot (d_1 + d_2) \cdot \log^2(d_1 + d_2))$ then the nuclear norm minimization based method recovers W_* in time $O(d^2 m)$ in the worst case. Note that alternating minimization has slightly higher sample complexity requirement but is significantly faster than the nuclear norm minimization method. Due to this, most practical low-rank estimation systems in fact use alternating minimization method for recovery. We obtain similar results for the rank-one dependent measurements.

We summarize the sample complexity and computational complexity for different approaches and different measurements in Table 1. In the table, ALS refers to alternating least squares, i.e., alternating minimization. For the symbols, $d = d_1 + d_2$, $n = n_1 + n_2$, and $\hat{k}$ is the maximum of rank estimate used as an input in the nuclear-norm solver [10][9]. This number can be as large as $\min\{d_1, d_2\}$ when we have no prior knowledge of the rank.

Paper Organization: We summarize related work in Section 2. In Section 3 we formally introduce the matrix sensing problem and our proposed rank-one measurement operators. In Section 4, we present the alternating minimization method for matrix reconstruction. We then present a *generic* analysis for alternating minimization when applied to the proposed rank-one measurement operators. We present the nuclear-norm minimization based method in Section 5 and present its analysis for the rank-one dependent measurements. Finally, we provide empirical validation of our methods in Section 6.

2   Related Work

Matrix Sensing: Matrix sensing[20][12][16] is a generalization of the popular compressive sensing problem for the sparse vectors and has applications in several domains such as control, vision etc. [20] introduced measurement operators that satisfy RIP and showed that using only $O(kd \log d)$ measurements, a rank-k $W_* \in \mathbb{R}^{d_1 \times d_2}$ can be recovered. Recently, a set of universal Pauli measurements, used in quantum state tomography, have been shown to satisfy the RIP condition [17]. These measurement operators are Kronecker products of 2×2 matrices, thus, they have appealing computation and memory efficiency. In concurrent work, [15][2] also proposed an independent rank-one measurement using nuclear-norm minimization. In contrast, we use two different measurement operators and show that the popular alternating minimization method also solves the problem exactly.

Matrix Completion and Inductive Matrix Completion: Matrix completion [3][14][13] is a special case of rank-one matrix sensing problem when the operator takes a subset of the entries. However, to guarantee exact recovery, the target matrix has to satisfy the incoherence condition. Using our rank-one Gaussian operators, we don't require any condition on the target matrix. For inductive matrix completion (IMC), which is a generalization of matrix completion utilizing movies' and users' features, the authors of [22] provided the theoretical recovery guarantee for nuclear-norm minimization. In this paper, we show that IMC is equivalent to the matrix sensing problem using dependent rank-one measurements, and provide a similar result for nuclear-norm based methods for IMC but eliminate a "joint" incoherent condition on the rank-one measurements and an upper bound condition on the sample complexity. Moreover, we give a theoretical guarantee using alternating minimization methods.

Alternating Minimization: Although nuclear-norm minimization enjoys nice recovery guarantees, it usually doesn't scale well. In practice, alternating minimization is employed to solve problem (2) by assuming the rank is known. Alternating minimization solves two least square problems alternatively in each iteration, thus is very computationally efficient[23]. Although widely used in practice, its theoretical guarantees are relatively less understood due to non-convexity. [13] first showed optimality of alternating minimization in the matrix sensing/low-rank estimation setting under the RIP setting. Subsequently, several other papers have also shown global convergence guarantees for alternating minimization, e.g. matrix completion [8][7], robust PCA [18] and dictionary learning [1]. In this paper, we provide a *generic* analysis for alternating minimization applied to the proposed rank-one measurement operators. Our results distill out certain key problem specific properties that would imply global optimality of alternating minimization. We then show that the rank-one Gaussian measurements satisfy those properties.

3   Problem Formulation

The goal of matrix sensing is to design a linear operator $\mathcal{A} : \mathbb{R}^{d_1 \times d_2} \rightarrow \mathbb{R}^m$ and a recovery algorithm so that a low-rank matrix $W_* \in \mathbb{R}^{d_1 \times d_2}$ can be recovered exactly using $\mathcal{A}(W_*)$. In this paper, we focus on rank-one measurement operators, $A_l = \boldsymbol{x}_l \boldsymbol{y}_l^T$, and call such problems as *Low-Rank matrix estimation using*

Rank One Measurements (**LRROM**): recover the rank-k matrix $W_* \in \mathbb{R}^{d_1 \times d_2}$ by using rank-1 measurements of the form:

$$b = [x_1^T W_* y_1 \quad x_2^T W_* y_2 \quad \ldots \quad x_m^T W_* y_m]^T, \tag{3}$$

where x_l, y_l are "feature" vectors and are provided along with the measurements b.

We propose two different kinds of rank-one measurement operators based on Gaussian distribution.

3.1 Rank-one Independent Gaussian Operator

Our first measurement operator is a simple rank-one Gaussian operator, $\mathcal{A}_{GI} = [A_1 \ldots A_m]$, where, $\{A_l = x_l y_l^T\}_{l=1,2,\ldots,m}$, and x_l, y_l are sampled i.i.d. from spherical Gaussian distribution.

3.2 Rank-one Dependent Gaussian Operator

Our second operator can introduce certain "dependencies" in our measurement and has in fact interesting connections to the matrix completion problem. We provide the operator as well as the connection to matrix completion in this sub-section. To generate the rank-one dependent Gaussian operator, we first sample two Gaussian matrices $X \in \mathbb{R}^{n_1 \times d_1}$ and $Y \in \mathbb{R}^{n_2 \times d_2}$, where each entry of both X, Y is sampled independently from Gaussian distribution and $n_1 \geq C d_1$, $n_2 \geq C d_2$ for a global constant $C \geq 1$. Then, the Gaussian dependent operator $\mathcal{A}_{GD} = [A_1, \ldots A_m]$ where $\{A_l = x_{i_l} y_{i_l}^T\}_{(i_l, j_l) \in \Omega}$. Here x_i^T is the i-th row of X and y_j^T is the j-th row of Y. Ω is a uniformly random subset of $[n_1] \times [n_2]$ such that $E[|\Omega|] = m$. For simplicity, we assume that each entry $(i_l, j_l) \in [n_1] \times [n_2]$ is sampled i.i.d. with probability $p = m/(n_1 \times n_2)$. Therefore, the measurements using the above operator are given by: $b_l = x_{i_l}^T W y_{j_l}, (i_l, j_l) \in \Omega$.

Connections to Inductive Matrix Completion (IMC): Note that the above measurements are inspired by matrix completion style sampling operator. However, here we first multiply W with X, Y and then sample the obtained matrix XWY^T. In the domain of recommendation systems (say user-movie system), the corresponding reconstruction problem can also be thought as the inductive matrix completion problem. That is, let there be n_1 users, n_2 movies, X represents user features, and Y represents the movie features. Then, the true ratings matrix is given by $R = XWY^T \in \mathbb{R}^{n_1 \times n_2}$.

That is, given the user/movie feature vectors $x_i \in \mathbb{R}^{d_1}$ for $i = 1, 2, \ldots, n_1$ and $y_j \in \mathbb{R}^{d_2}$ for $j = 1, 2, \ldots, n_2$, our goal is to recover a rank-k matrix W_* of size $d_1 \times d_2$ from a few observed entries $R_{ij} = x_i^T W_* y_j$, for $(i, j) \in \Omega \subset [n_1] \times [n_2]$. Because of the equivalence between the dependent rank-one measurements and the entries of the rating matrix, in the rest of the paper, we will use $\{R_{ij}\}_{(i,j) \in \Omega}$ as the dependent rank-one measurements.

Now, if we can reconstruct W_* from the above measurements, we can predict ratings *inductively* for new users/movies, provided their feature vectors are given.

Hence, our reconstruction procedure also solves the IMC problem. However, there is a key difference: in matrix sensing, we can select X, Y according to our convenience, while in IMC, X and Y are provided a priori. But, for general X, Y one cannot solve the problem because if say $R = XW_*Y^T$ is a 1-sparse matrix, then W_* cannot be reconstructed even with a large number of samples.

Interestingly, our proof for reconstruction using nuclear-norm based method does not require Gaussian X, Y. Instead, we can distill out two key properties of R, X, Y ensuring that using only $O(k(d_1+d_2)\log(d_1+d_2)\log(n_1+n_2))$ samples, we can reconstruct W_*. Note that a direct application of matrix completion results [3][4] would require $O(k(n_1 + n_2)\log(n_1 + n_2))$ samples which can be much large if $n_1 \gg d_1$ or $n_2 \gg d_2$. See Section 5 for more details on the assumptions that we require for the nuclear-norm minimization method to solve the IMC problem exactly.

4   Rank-one Matrix Sensing via Alternating Minimization

We now present the alternating minimization approach for solving the reconstruction problem (2) with rank-one measurements (3). Since W to be recovered is restricted to have at most rank-k, (2) can be reformulated as the following non-convex optimization problem:

$$\min_{U\in\mathbb{R}^{d_1\times k}, V\in\mathbb{R}^{d_2\times k}} \sum_{l=1}^{m}(b_l - x_l^T UV^T y_l)^2 . \tag{4}$$

Alternating minimization is an iterative procedure that alternately optimizes for U and V while keeping the other factor fixed. As optimizing for U (or V) involves solving just a least squares problem, so each individual iteration of the algorithm is linear in matrix dimensions. For the rank-one measurement operator, we use a particular initialization method to initialize U (see line 3 of Algorithm 1). See Algorithm 1 for a pseudo-code of the algorithm.

4.1   General Theoretical Guarantee for Alternating Minimization

As mentioned above, (4) is non-convex in U, V and hence standard analysis would only ensure convergence to a local minima. However, [13] recently showed that the alternating minimization method in fact converges to the global minima of two low-rank estimation problems: matrix sensing with RIP matrices and matrix completion.

The rank-one operator given above does not satisfy RIP (see Definition 1), even when the vectors x_l, y_l are sampled from the normal distribution (see Claim 4.2). Furthermore, each measurement need not reveal exactly one entry of W_* as in the case of matrix completion. Hence, the proof of [13] does not apply directly. However, inspired by the proof of [13], we distill out three key properties that the operator should satisfy, so that alternating minimization would converge to the global optimum.

Algorithm 1. AltMin-LRROM : Alternating Minimization for LRROM

1: **Input:** Measurements: $\boldsymbol{b}_{all}$, Measurement matrices: $\mathcal{A}_{all}$, Number of iterations: H
2: Divide $(\mathcal{A}_{all}, \boldsymbol{b}_{all})$ into $2H + 1$ sets (each of size m) with h-th set being $\mathcal{A}^h = \{A_1^h, A_2^h, \ldots, A_m^h\}$ and $\boldsymbol{b}^h = [b_1^h\ b_2^h\ \ldots\ b_m^h]^T$
3: **Initialization:** U_0 =top-k left singular vectors of $\frac{1}{m} \sum_{l=1}^m b_l^0 A_l^0$
4: **for** $h = 0$ to $H - 1$ **do**
5:    $b \leftarrow b^{2h+1}, \mathcal{A} \leftarrow \mathcal{A}^{2h+1}$
6:    $\widehat{V}_{h+1} \leftarrow \operatorname{argmin}_{V \in \mathbb{R}^{d_2 \times k}} \sum_l (b_l - \boldsymbol{x}_l^T U_h V^T \boldsymbol{y}_l)^2$
7:    $V_{h+1} = QR(\widehat{V}_{h+1})$ //orthonormalization of $\widehat{V}_{h+1}$
8:    $b \leftarrow b^{2h+2}, \mathcal{A} \leftarrow \mathcal{A}^{2h+2}$
9:    $\widehat{U}_{h+1} \leftarrow \operatorname{argmin}_{U \in \mathbb{R}^{d_1 \times k}} \sum_l (b_l - \boldsymbol{x}_l^T U V_{h+1}^T \boldsymbol{y}_l)^2$
10:   $U_{h+1} = QR(\widehat{U}_{h+1})$ //orthonormalization of $\widehat{U}_{h+1}$
11: **end for**
12: **Output:** $W_H = U_H (\widehat{V}_H)^T$

Theorem 1. *Let $W_* = U_* \Sigma_* V_*^T \in \mathbb{R}^{d_1 \times d_2}$ be a rank-k matrix with k-singular values $\sigma_*^1 \geq \sigma_*^2 \cdots \geq \sigma_*^k$. Also, let $\mathcal{A} : \mathbb{R}^{d_1 \times d_2} \to \mathbb{R}^m$ be a linear measurement operator parameterized by m matrices, i.e., $\mathcal{A} = \{A_1, A_2, \ldots, A_m\}$ where $A_l = \boldsymbol{x}_l \boldsymbol{y}_l^T$. Let $\mathcal{A}(W)$ be as given by (1).*

Now, let $\mathcal{A}$ satisfy the following properties with parameter $\delta = \frac{1}{k^{3/2} \cdot \beta \cdot 100}$ ($\beta = \sigma_*^1 / \sigma_*^k$):

1. **Initialization:** $\|\frac{1}{m} \sum_l b_l A_l - W_*\|_2 \leq \|W_*\|_2 \cdot \delta$.
2. **Concentration of operators B_x, B_y:** *Let $B_x = \frac{1}{m} \sum_{l=1}^m (\boldsymbol{y}_l^T \boldsymbol{v})^2 \boldsymbol{x}_l \boldsymbol{x}_l^T$ and $B_y = \frac{1}{m} \sum_{l=1}^m (\boldsymbol{x}_l^T \boldsymbol{u})^2 \boldsymbol{y}_l \boldsymbol{y}_l^T$, where $\boldsymbol{u} \in \mathbb{R}^{d_1}, \boldsymbol{v} \in \mathbb{R}^{d_2}$ are two unit vectors that are independent of randomness in $\boldsymbol{x}_l, \boldsymbol{y}_l, \forall i$. Then the following holds: $\|B_x - I\|_2 \leq \delta$ and $\|B_y - I\|_2 \leq \delta$.*
3. **Concentration of operators G_x, G_y:** *Let $G_x = \frac{1}{m} \sum_l (\boldsymbol{y}_l^T \boldsymbol{v})(\boldsymbol{y}_l \boldsymbol{v}_\perp) \boldsymbol{x}_l \boldsymbol{x}_l^T$, $G_y = \frac{1}{m} \sum_l (\boldsymbol{x}_l^T \boldsymbol{u})(\boldsymbol{u}_\perp^T \boldsymbol{x}_l) \boldsymbol{y}_l \boldsymbol{y}_l^T$, where $\boldsymbol{u}, \boldsymbol{u}_\perp \in \mathbb{R}^{d_1}$, $\boldsymbol{v}, \boldsymbol{v}_\perp \in \mathbb{R}^{d_2}$ are unit vectors, s.t., $\boldsymbol{u}^T \boldsymbol{u}_\perp = 0$ and $\boldsymbol{v}^T \boldsymbol{v}_\perp = 0$. Furthermore, let $\boldsymbol{u}, \boldsymbol{u}_\perp, \boldsymbol{v}, \boldsymbol{v}_\perp$ be independent of randomness in $\boldsymbol{x}_l, \boldsymbol{y}_l, \forall i$. Then, $\|G_x\|_2 \leq \delta$ and $\|G_y\|_2 \leq \delta$.*

Then, after H-iterations of the alternating minimization method (Algorithm 1), we obtain $W_H = U_H V_H^T$ s.t., $\|W_H - W_*\|_2 \leq \epsilon$, where $H \leq 100 \log(\|W_*\|_F / \epsilon)$.

See Supplementary A for a detailed proof. Note that we require intermediate vectors $\boldsymbol{u}, \boldsymbol{v}, \boldsymbol{u}_\perp, \boldsymbol{v}_\perp$ to be independent of randomness in A_l's. Hence, we partition $\mathcal{A}_{all}$ into $2H + 1$ partitions and at each step $(\mathcal{A}^h, \boldsymbol{b}^h)$ and $(\mathcal{A}^{h+1}, \boldsymbol{b}^{h+1})$ are supplied to the algorithm. This implies that the measurement complexity of the algorithm is given by $m \cdot H = m \log(\|W_*\|_F / \epsilon)$. That is, given $O(m \log(\|(d_1 + d_2) W_*\|_F))$ samples, we can estimate matrix W_H, s.t., $\|W_H - W_*\|_2 \leq \frac{1}{(d_1 + d_2)^c}$, where $c > 0$ is any constant.

4.2   Independent Gaussian Measurements

In this subsection, we consider the rank-one independent measurement operator $\mathcal{A}_{GI}$ specified in Section 3. Now, for this operator $\mathcal{A}_{GI}$, we show that if

$m = O(k^4\beta^2 \cdot (d_1 + d_2) \cdot \log^2(d_1 + d_2))$, then w.p. $\geq 1 - 1/(d_1 + d_2)^{100}$, any fixed rank-$k$ matrix W_* can be recovered by AltMin-LRROM (Algorithm 1). Here $\beta = \sigma_*^1/\sigma_*^k$ is the condition number of W_*. That is, using nearly linear number of measurements in d_1, d_2, one can exactly recover the $d_1 \times d_2$ rank-k matrix W_*.

As mentioned in the previous section, the existing matrix sensing results typically assume that the measurement operator $\mathcal{A}$ satisfies the Restricted Isometry Property (RIP) defined below:

Definition 1. *A linear operator $\mathcal{A} : \mathbb{R}^{d_1 \times d_2} \to \mathbb{R}^m$ satisfies RIP iff, for $\forall W$ s.t. $rank(W) \leq k$, the following holds:*

$$(1 - \delta_k)\|W\|_F^2 \leq \|\mathcal{A}(W)\|_F^2 \leq (1 + \delta_k)\|W\|_F^2 ,$$

where $\delta_k > 0$ is a constant dependent only on k.

Naturally, this begs the question whether we can show that our rank-1 measurement operator $\mathcal{A}_{GI}$ satisfies RIP, so that the existing analysis for RIP based low-rank matrix sensing can be used [13]. We answer this question in the negative, i.e., for $m = O((d_1 + d_2)\log(d_1 + d_2))$, $\mathcal{A}_{GI}$ does not satisfy RIP even for rank-1 matrices (with high probability):

Claim. 4.2. Let $\mathcal{A}_{GI} = \{A_1, A_2, \ldots A_m\}$ be a measurement operator with each $A_l = \boldsymbol{x}_l \boldsymbol{y}_l^T$, where $\boldsymbol{x}_l \in \mathbb{R}^{d_1} \sim \mathcal{N}(0, I)$, $\boldsymbol{y}_l \in \mathbb{R}^{d_2} \sim \mathcal{N}(0, I), 1 \leq l \leq m$. Let $m = O((d_1 + d_2)\log^c(d_1 + d_2))$, for any constant $c > 0$. Then, with probability at least $1 - 1/m^{10}$, $\mathcal{A}_{GI}$ does not satisfy RIP for rank-1 matrices with a constant δ.

See Supplementary B for a detailed proof of the above claim. Now, even though $\mathcal{A}_{GI}$ does not satisfy RIP, we can still show that $\mathcal{A}_{GI}$ satisfies the three properties mentioned in Theorem 1. and hence we can use Theorem 1 to obtain the exact recovery result.

Theorem 2 (Rank-One Independent Gaussian Measurements using ALS). *Let $\mathcal{A}_{GI} = \{A_1, A_2, \ldots A_m\}$ be a measurement operator with each $A_l = \boldsymbol{x}_l \boldsymbol{y}_l^T$, where $\boldsymbol{x}_l \in \mathbb{R}^{d_1} \sim \mathcal{N}(0, I)$, $\boldsymbol{y}_l \in \mathbb{R}^{d_2} \sim \mathcal{N}(0, I), 1 \leq l \leq m$. Let $m = O(k^4\beta^2(d_1 + d_2)\log^2(d_1 + d_2))$. Then, Property 1, 2, 3 required by Theorem 1 are satisfied with probability at least $1 - 1/(d_1 + d_2)^{100}$.*

Proof. Here, we provide a brief proof sketch. See Supplementary B for a detailed proof.

Initialization: Note that,

$$\frac{1}{m}\sum_{l=1}^m b_l \boldsymbol{x}_l \boldsymbol{y}_l^T = \frac{1}{m}\sum_{l=1}^m \boldsymbol{x}_l \boldsymbol{x}_l^T U_* \Sigma_* V_*^T \boldsymbol{y}_l \boldsymbol{y}_l^T = \frac{1}{m}\sum_{l=1}^m Z_l ,$$

where $Z_l = \boldsymbol{x}_l \boldsymbol{x}_l^T U_* \Sigma_* V_*^T \boldsymbol{y}_l \boldsymbol{y}_l^T$. Note that $\mathbb{E}[Z_l] = U_* \Sigma_* V_*^T$. Hence, to prove the initialization result, we need a tail bound for sums of random matrices. To this end, we use Theorem 1.6 in [21]. However, Theorem 1.6 in [21] requires a bounded random variable while Z_l is an unbounded variable. We handle this issue by clipping Z_l to ensure that its spectral norm is always bounded. Furthermore, by using properties of normal distribution, we can ensure that w.p. $\geq 1 - 1/m^3$, Z_l's do not require clipping and the new "clipped" variables converge to nearly the same quantity as the original "non-clipped" Z_l's. See Supplementary B for more details.

Concentration of B_x, B_y, G_x, G_y: Consider $G_x = \frac{1}{m} \sum_{l=1}^{m} x_l x_l^T y_l^T v v_\perp^T y_l$. As, $v, v_\perp$ are unit-norm vectors, $y_l^T v \sim \mathcal{N}(0,1)$ and $v_\perp^T x_l \sim \mathcal{N}(0,1)$. Also, since v and $v_\perp$ are orthogonal, $y_l^T v$ and $v_\perp^T y_l$ are independent variables. Hence, $G_x = \frac{1}{m} \sum_{l=1}^{m} Z_l$ where $\mathbb{E}[Z_l] = 0$. Here again, we apply Theorem 1.6 in [21] after using a clipping argument. We can obtain the required bounds for B_x, B_y, G_y also in a similar manner.

Note that the clipping procedure ensures that Z_l's don't need to be clipped with probability $\geq 1 - 1/m^3$ only. That is, we cannot apply the *union* bound to ensure that the concentration result holds *for all* $v, v_\perp$. Hence, we need a fresh set of measurements after each iteration to ensure concentration.

Global optimality of the rate of convergence of the Alternating Minimization procedure for this problem now follows directly by using Theorem 1. We would like to note that while the above result shows that the $\mathcal{A}_{GI}$ operator is almost as powerful as the RIP based operators for matrix sensing, there is one critical drawback: while RIP based operators are universal that is they can be used to recover any rank-k W_*, $\mathcal{A}_{GI}$ needs to be resampled for each W_*. We believe that the two operators are at two extreme ends of randomness vs universality trade-off and intermediate operators with higher success probability but using larger number of random bits should be possible.

4.3 Dependent Gaussian Measurements

For the dependent Gaussian measurements, the alternating minimization formulation is given by:

$$\min_{U \in \mathbb{R}^{d_1 \times k}, V \in \mathbb{R}^{d_2 \times k}} \sum_{(i,j) \in \Omega} (x_i^T U V^T y_j - R_{ij})^2 . \tag{5}$$

Here again, we can solve the problem by alternatively optimizing for U and V. Later in Section 4.4, we show that using such dependent measurements leads to a faster recovery algorithm when compared to the recovery algorithm for independent measurements.

Note that both the measurement matrices X, Y can be thought of as orthonormal matrices. The reason being, $X W_* Y^T = U_X \Sigma_X V_X^T W_* V_Y \Sigma_Y U_Y^T$, where $X = U_X \Sigma_X V_X^T$ and $Y = U_Y \Sigma_Y V_Y^T$ is the SVD of X, Y respectively. Hence, $R = X W_* Y^T = U_X (\Sigma_X V_X^T W_* V_Y \Sigma_Y) U_Y^T$. Now U_X, U_Y can be treated as the true "X", "Y" matrices and $W_* \leftarrow (\Sigma_X V_X^T W_* V_Y \Sigma_Y)$ can be thought of as W_*. Then the "true" W_* can be recovered using the obtained W_H as: $W_H \leftarrow V_X \Sigma_X^{-1} W_H \Sigma_Y^{-1} V_Y^T$. We also note that such a transformation implies that the condition number of R and that of $W_* \leftarrow (\Sigma_X V_X^T W_* V_Y \Sigma_Y)$ are exactly the same.

Similar to the previous section, we utilize our general theorem for optimality of the LRROM problem to provide a convergence analysis of rank-one Gaussian dependent operators $\mathcal{A}_{GD}$. We prove if X and Y are random orthogonal matrices, defined in [3], the above mentioned dependent measurement operator $\mathcal{A}_{GD}$ generated from X, Y also satisfies Properties 1, 2, 3 in Theorem 1. Hence, AltMin-LRROM (Algorithm 1) converges to the global optimum in $O(\log(\|W_*\|_F/\epsilon))$ iterations.

Theorem 3 (Rank-One Dependent Gaussian Measurements using ALS). *Let $X_0 \in \mathbb{R}^{n_1 \times d_1}$ and $Y_0 \in \mathbb{R}^{n_2 \times d_2}$ be Gaussian matrices, i.e. every entry is sampled i.i.d from $\mathcal{N}(0, 1)$. Let $X_0 = X\Sigma_X V_X^T$ and $Y_0 = Y\Sigma_Y V_Y^T$ be the thin SVD of X_0 and Y_0 respectively. Then the rank-one dependent operator $\mathcal{A}_{GD}$ formed by X, Y with $m \geq O(k^4\beta^2(d_1 + d_2)\log(d_1 + d_2))$ satisfy Property 1,2,3 required by Theorem 1 with high probability.*

See Supplementary C for a detailed proof. Interestingly, our proof does not require X, Y to be Gaussian. It instead utilizes only two key properties about X, Y which are given by:

1. **Incoherence:** For some constant μ, c,

$$\max_{i\in[n]} \|\boldsymbol{x}_i\|_2^2 \leq \frac{\mu\bar{d}}{n}, \tag{6}$$

   where $\bar{d} = \max(d, \log n)$
2. **Averaging Property:** For H different orthogonal matrices $U_h \in \mathbb{R}^{d\times k}$, $h = 1, 2, \ldots, H$, the following hold for these U_h's,

$$\max_{i\in[n]} \|U_h^T \boldsymbol{x}_i\|_2^2 \leq \frac{\mu_0\bar{k}}{n}, \tag{7}$$

   where μ_0, c are some constants and $\bar{k} = \max(k, \log n)$.

Hence, the above theorem can be easily generalized to solve the inductive matrix completion problem (IMC), i.e., solve (5) for arbitrary X, Y. Moreover, the sample complexity of the analysis would be nearly in $(d_1 + d_2)$, instead of $(n_1 + n_2)$ samples required by the standard matrix completion methods.

The following lemma shows that the above two properties hold w.h.p. for random orthogonal matrices .

Lemma 1. *If $X \in \mathbb{R}^{n\times d}$ is a random orthogonal matrix, then both Incoherence and Averaging properties are satisfied with probability $\geq 1 - (c/n^3)\log n$, where c is a constant.*

The proof of Lemma 1 can be found in Supplementary C.

4.4   Computational Complexity for Alternating Minimization

In this subsection, we briefly discuss the computational complexity for Algorithm 1. For simplicity, we set $d = d_1 + d_2$ and $n = n_1 + n_2$, and in practical implementation, we don't divide the measurements and use the whole measurement operator $\mathcal{A}$ for every iteration. The most time-consuming part of Algorithm 1 is the step for solving the least square problem. Given $U = U_h$, V can be obtained by solving the following linear system,

$$\sum_{l=1}^{m} \langle V, A_l^T U_h \rangle A_l U_h = \sum_{l=1}^{m} b_l A_l^T U_h \ . \tag{8}$$

The dimension of this linear system is kd, which could be large, thus we use conjugate gradient (CG) method to solve it. In each CG iteration, different measurement operators have different computational complexity. For RIP-based full-rank operators, the computational complexity for each CG step is $O(d^2 m)$ while it is $O(kdm)$ for rank-one independent operators. However, for rank-one dependent operators, using techniques introduced in [24], we can reduce the per iteration complexity to be $O(kdn + md)$. Furthermore, if $n = d$, the computational complexity of dependent operators is only $O(kd^2 + md)$, which is better than the complexity of rank-one independent operators in an order of k.

5   Rank-one Matrix Sensing via Nuclear Norm Minimization

In this section, we consider solving LRROM by nuclear norm relaxation. We first note that using nuclear norm relaxation, [15] provided the analysis for independent rank-one measurement operators when the underlying matrix is Hermitian. It can be shown that non-Hermitian matrices problem can be transformed to Hermitian cases. Their proof uses the bowling scheme and only requires $O(k(d_1+d_2))$ measurements for Gaussian case and $O(k(d_1+d_2)\log(d_1+d_2)\log(n_1+n_2))$ measurements for 4-designs case. In this paper, we consider dependent measurement operators which have a similar sample complexity as the independent operators, but less computational complexity and memory footprint than those of independent ones.

The nuclear norm minimization using rank-one dependent Gaussian operator is of form,

$$\min \|W\|_* $$
$$\text{s.t. } \boldsymbol{x}_i^T W \boldsymbol{y}_j = R_{ij}, \ (i,j) \in \Omega \ . \tag{9}$$

(9) can be solved exactly by semi-definite programming or approximated by

$$\min_W \sum_{(i,j)\in\Omega} (\boldsymbol{x}_i^T W \boldsymbol{y}_j - R_{ij})^2 + \lambda \|W\|_* \ , \tag{10}$$

where λ is a constant and can be viewed as a Lagrange multiplier.

5.1   Recovery Guarantee for Nuclear-norm Minimization

In this subsection, we show that using rank-one dependent Gaussian operators, the nuclear-norm minimization can recover *any* low-rank matrix exactly with $O(k(d_1 + d_2) \log(d_1 + d_2) \log(n_1 + n_2))$ measurements. We also generalize the theorem to the IMC problem where the feature matrices X and Y can be arbitrary instead of being Gaussian. We show that as long as X, W_*, Y satisfy certain incoherence style properties, the nuclear norm minimization can guarantee exact recovery using only $O(k(d_1 + d_2) \log(d_1 + d_2))$ samples.

We first provide recovery guarantees for our rank-one dependent operator, i.e., when X, Y are sampled from the Gaussian distribution.

Theorem 4 (Rank-one Dependent Gaussian Measurements using Nuclear-norm Minimization). *Let $W_* = U_* \Sigma_* V_*^T \in \mathbb{R}^{d_1 \times d_2}$ be a rank-k matrix. Let $X \in \mathbb{R}^{n_1 \times d_1}$ and $Y \in \mathbb{R}^{n_2 \times d_2}$ be random orthogonal matrices. Assume each $(i, j) \in \Omega$ is sampled from $[n_1] \times [n_2]$ i.i.d.. Then if $m = |\Omega| \geq O(k(d_1 + d_2) \log(d_1 + d_2) \log(n_1 + n_2))$, the minimizer to the problem (9) is unique and equal to W_* with probability at least $1 - c_1(d_1 + d_2)^{-c_2}$, where c_1 and c_2 are universal constants.*

The above theorem is a directly corollary of Theorem 5 combined with Lemma 1. Lemma 1 shows that random orthonormal matrices X, Y (can be generated using Gaussian matrices as stated in Theorem 2) satisfy the requirements of Theorem 5.

Nuclear-norm minimization approach for inductive matrix completion (9) has also been studied by [22]. However, their recovery guarantee holds under a much more restrictive set of assumptions on X, W_*, Y and in fact requires that the number of samples is not only lower bounded by certain quantity but also upper bounded by some other quantity. Our general analysis below doesn't rely on this upper bound. Moreover, their proof also requires a "joint" incoherent condition, i.e., an upper bound on $\max_{i,j} |\boldsymbol{x}_i^T U_* V_*^T \boldsymbol{y}_j|$ which is not required by our method; to this end, we use a technique introduced by [5] to bound an $\ell_{\infty,2}$-norm.

Theorem 5 (Inductive Matrix Completion using Nuclear-norm Minimization). *Let $W_* = U_* \Sigma_* V_*^T \in \mathbb{R}^{d_1 \times d_2}$ be a rank-k matrix. Assume X, Y are orthogonal matrices, and satisfy the following conditions with respect to W_* for some constant μ and μ_0,*

C1. $\displaystyle \max_{i \in [n_1]} \|\boldsymbol{x}_i\|_2^2 \leq \frac{\mu d_1}{n_1}, \quad \max_{j \in [n_2]} \|\boldsymbol{y}_j\|_2^2 \leq \frac{\mu d_2}{n_2},$

C2. $\displaystyle \max_{i \in [n_1]} \|U_*^T \boldsymbol{x}_i\|_2^2 \leq \frac{\mu_0 k}{n_1}, \quad \max_{j \in [n_2]} \|V_*^T \boldsymbol{y}_j\|_2^2 \leq \frac{\mu_0 k}{n_2}.$

Then if each observed entry $(i, j) \in \Omega$ is sampled from $[n_1] \times [n_2]$ i.i.d. with probability p,

$$p \geq \max \left\{ \frac{c_0 \mu_0 \mu k d \log(d) \log(n)}{n_1 n_2}, \frac{1}{\min\{n_1, n_2\}^{10}} \right\}, \tag{11}$$

the minimizer to the problem (9) is unique and equal to W_* with probability at least $1 - c_1 d^{-c_2}$, where c_0, c_1 and c_2 are universal constants, $d = d_1 + d_2$ and $n = n_1 + n_2$.

Note that the first condition **C1** is actually the incoherence condition on X, Y, while the second one **C2** is the incoherence of XU_*, YV_*. Additionally, **C2** is weaker than the *Averaging* property in Lemma 1, as it only asks for one U_* rather than H different U_h's to satisfy the property.

Proof. We follow the popular proof ideas used by [3][19], that is, finding a dual feasible solution for (9) to certify the uniqueness of the minimizer of (9). Unlike the analysis in [22], we build our dual certificate in the $\mathbb{R}^{d_1 \times d_2}$ matrix space rather than the $\mathbb{R}^{n_1 \times n_2}$ space. This choice makes it easy to follow the analysis in standard matrix completion problem. In Proposition 1 in Supplementary D, we give certain conditions the dual certificate should satisfy for the uniqueness. Then we apply golfing scheme [6] to find such a certificate. When building the dual certificate, we use an $\ell_{\infty,2}$-norm adapted from [5]. This enables us to discard the assumption of "joint" incoherence. The details can be found in Supplementary D.

5.2   Computational Complexity for Nuclear-norm Minimization

The optimization for nuclear-norm formulation is much more complex. Recently [10] proposed an active subspace method to solve Problem (10). The computational bottleneck is the approxSVD step and the inner problem step, both of which involve calculating a similar equation as shown on the left hand side of Eq (8). However, the rank of U or V is not fixed in each iteration as that of ALS, and in the worst case, it can be as large as $\min\{d_1, d_2\}$. The computational complexity for this basic operation is shown in Table 1.

6   Experiments

In this section, we demonstrate empirically that our Gaussian rank-one linear operators are significantly more efficient for matrix sensing than the existing RIP based measurement operators. In particular, we apply the two recovery methods namely alternating minimization (ALS) and nuclear norm minimization (Nuclear) to the measurements obtained using three different operators: rank-one independent (Rank1 Indep), rank-one dependent (Rank1 Dep), and a RIP based operator generated using random Gaussian matrices (RIP).

The experiments are conducted on Matlab and the nuclear-norm solver is adapted from the code by [10]. We first generated a random rank-5 signal $W_* \in \mathbb{R}^{50 \times 50}$, and compute $m = 1000$ measurements using different measurement operators. Here, we fix a small $\lambda = 10^{-6}$ for solving Eq (10) in order to exactly recover the matrix. And we set the maximum possible rank $\hat{k} = k$ as the input of the nuclear-norm solver. Figure 1a plots the relative error in recovery, $err = \|W - W^*\|_F^2 / \|W^*\|_F^2$, against computational time required by each method. Clearly, recovery using rank-one measurements requires significantly

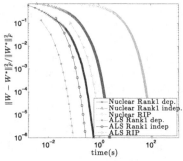
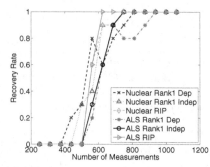

(a) Relative error in recovery v.s. computation time

(b) Recovery rate v.s. number of measurements

Fig. 1. Comparison of computational complexity and measurement complexity for different approaches and different operators

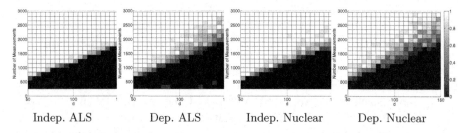

Indep. ALS   Dep. ALS   Indep. Nuclear   Dep. Nuclear

Fig. 2. Recovery rate for different matrix dimension d (x-axis) and different number of measurements m (y-axis). The color reflects the recovery rate scaled from 0 to 1. The white color indicates perfect recovery, while the black color denotes failure in all the experiments.

less time compared to the RIP based operator. Moreover, ALS in general seems to be significantly faster than Nuclear methods.

Next, we compare the measurement complexity (m) of each method. Here again, we first generate a random rank-5 signal $W_* \in \mathbb{R}^{50 \times 50}$ and its measurements using different operators. We then measure error in recovery by each of the method and consider success if the relative error $err \leq 0.05$. We repeat the experiment 10 times to obtain the recovery rate (number of success/10) for each value of m (number of measurements). Figure 1b plots the recovery rate of different approaches for different m. Clearly, the rank-one based measurements have similar recovery rate and measurement complexity as the RIP based operators. However, our rank-one operator based methods are much faster than the corresponding methods for the RIP-based measurement scheme.

Finally, in Figure 2, we validate our theoretical analysis on the measurement complexity by showing the recovery rate for different d and m. We fix the rank

$k = 5$, set $d = d_1 = d_2$ and $n_1 = d_1$, $n_2 = d_2$ for dependent operators. Figure 3 plots the recovery rate for various d and m. As shown in Figure 2, both independent and dependent operators using alternating minimization or nuclear-norm minimization require a number of measurements proportional to the dimension of d. We also see that dependent operators require a slight larger number of measurements than that of independent ones. Another interesting observation is that although our theoretical analysis requires a higher measurement complexity of ALS than that of Nuclear methods, the empirical results show their measurement complexities are almost identical for the same measurement operator.

References

1. Agarwal, A., Anandkumar, A., Jain, P., Netrapalli, P., Tandon, R.: Learning sparsely used overcomplete dictionaries via alternating minimization. COLT (2014)
2. Cai, T.T., Zhang, A., et al.: Rop: Matrix recovery via rank-one projections. The Annals of Statistics **43**(1), 102–138 (2015)
3. Candès, E.J., Recht, B.: Exact matrix completion via convex optimization. Foundations of Computational Mathematics **9**(6), 717–772 (2009)
4. Candès, E.J., Tao, T.: The power of convex relaxation: Near-optimal matrix completion. IEEE Trans. Inform. Theory **56**(5), 2053–2080 (2009)
5. Chen, Y.: Incoherence-optimal matrix completion. arXiv preprint arXiv:1310.0154 (2013)
6. Gross, D.: Recovering low-rank matrices from few coefficients in any basis. IEEE Transactions on Information Theory **57**(3), 1548–1566 (2011)
7. Hardt, M.: Understanding alternating minimization for matrix completion. In: Foundations 2014 IEEE 55th Annual Symposium on of Computer Science (FOCS), pp. 651–660. IEEE (2014)
8. Hardt, M., Wootters, M.: Fast matrix completion without the condition number. In: Proceedings of The 27th Conference on Learning Theory, pp. 638–678 (2014)
9. Hsieh, C.J., Dhillon, I.S., Ravikumar, P.K., Becker, S., Olsen, P.A.: Quic & dirty: A quadratic approximation approach for dirty statistical models. In: Advances in Neural Information Processing Systems, pp. 2006–2014 (2014)
10. Hsieh, C.J., Olsen, P.: Nuclear norm minimization via active subspace selection. In: Proceedings of The 31st International Conference on Machine Learning, pp. 575–583 (2014)
11. Jain, P., Dhillon, I.S.: Provable inductive matrix completion (2013). CoRR. http://arxiv.org/abs/1306.0626
12. Jain, P., Meka, R., Dhillon, I.S.: Guaranteed rank minimization via singular value projection. In: NIPS, pp. 937–945 (2010)
13. Jain, P., Netrapalli, P., Sanghavi, S.: Low-rank matrix completion using alternating minimization. In: STOC (2013)
14. Keshavan, R.H., Montanari, A., Oh, S.: Matrix completion from a few entries. IEEE Transactions on Information Theory **56**(6), 2980–2998 (2010)
15. Kueng, R., Rauhut, H., Terstiege, U.: Low rank matrix recovery from rank one measurements. arXiv preprint arXiv:1410.6913 (2014)
16. Lee, K., Bresler, Y.: Guaranteed minimum rank approximation from linear observations by nuclear norm minimization with an ellipsoidal constraint. arXiv preprint arXiv:0903.4742 (2009)

17. Liu, Y.K.: Universal low-rank matrix recovery from pauli measurements. In: Advances in Neural Information Processing Systems, pp. 1638–1646 (2011)
18. Netrapalli, P., Niranjan, U., Sanghavi, S., Anandkumar, A., Jain, P.: Nonconvex robust PCA. In: Advances in Neural Information Processing Systems, pp. 1107–1115 (2014)
19. Recht, B.: A simpler approach to matrix completion. The Journal of Machine Learning Research **12**, 3413–3430 (2011)
20. Recht, B., Fazel, M., Parrilo, P.A.: Guaranteed minimum-rank solutions of linear matrix equations via nuclear norm minimization. SIAM Review **52**(3), 471–501 (2010)
21. Tropp, J.A.: User-friendly tail bounds for sums of random matrices. Foundations of Computational Mathematics **12**(4), 389–434 (2012)
22. Xu, M., Jin, R., Zhou, Z.H.: Speedup matrix completion with side information: application to multi-label learning. In: Advances in Neural Information Processing Systems, pp. 2301–2309 (2013)
23. Yu, H.F., Hsieh, C.J., Si, S., Dhillon, I.S.: Scalable coordinate descent approaches to parallel matrix factorization for recommender systems. In: ICDM, pp. 765–774 (2012)
24. Yu, H.F., Jain, P., Kar, P., Dhillon, I.S.: Large-scale multi-label learning with missing labels. In: Proceedings of The 31st International Conference on Machine Learning, pp. 593–601 (2014)
25. Zuk, O., Wagner, A.: Low-rank matrix recovery from row-and-column affine measurements. In: Proceedings of the 32nd International Conference on Machine Learning, ICML 2015, Lille, France, 6–11 July 2015, pp. 2012–2020 (2015)

Tensor Decompositions for Learning Latent Variable Models (A Survey for ALT)

Anima Anandkumar[1], Rong Ge[2], Daniel Hsu[3],
Sham M. Kakade[4(✉)], and Matus Telgarsky[5]

[1] University of California, Irvine, USA
[2] Microsoft Research, New England, USA
[3] Columbia University, New York, USA
[4] Rutgers University, New Brunswick, USA
skakade@microsoft.cm
[5] University of Michigan, Ann Arbor, USA

Abstract. This note is a short version of that in [1]. It is intended as a survey for the 2015 Algorithmic Learning Theory (ALT) conference.

This work considers a computationally and statistically efficient parameter estimation method for a wide class of latent variable models— including Gaussian mixture models, hidden Markov models, and latent Dirichlet allocation—which exploits a certain tensor structure in their low-order observable moments (typically, of second- and third-order). Specifically, parameter estimation is reduced to the problem of extracting a certain (orthogonal) decomposition of a symmetric tensor derived from the moments; this decomposition can be viewed as a natural generalization of the singular value decomposition for matrices. Although tensor decompositions are generally intractable to compute, the decomposition of these specially structured tensors can be efficiently obtained by a variety of approaches, including power iterations and maximization approaches (similar to the case of matrices). A detailed analysis of a robust tensor power method is provided, establishing an analogue of Wedin's perturbation theorem for the singular vectors of matrices. This implies a robust and computationally tractable estimation approach for several popular latent variable models.

1 Introduction

The method of moments is a classical parameter estimation technique [29] from statistics which has proved invaluable in a number of application domains. The basic paradigm is simple and intuitive: (i) compute certain statistics of the data — often empirical moments such as means and correlations — and (ii) find model parameters that give rise to (nearly) the same corresponding population quantities. In a number of cases, the method of moments leads to consistent estimators which can be efficiently computed; this is especially relevant in the context of latent variable models, where standard maximum likelihood approaches are typically computationally prohibitive, and heuristic methods can be unreliable and difficult to validate with high-dimensional data. Furthermore, the method

© Springer International Publishing Switzerland 2015
K. Chaudhuri et al. (Eds.): ALT 2015, LNAI 9355, pp. 19–38, 2015.
DOI: 10.1007/978-3-319-24486-0_2

of moments can be viewed as complementary to the maximum likelihood approach; simply taking a single step of Newton-Ralphson on the likelihood function starting from the moment based estimator [22] often leads to the best of both worlds: a computationally efficient estimator that is (asymptotically) statistically optimal.

The primary difficulty in learning latent variable models is that the latent (hidden) state of the data is not directly observed; rather only observed variables correlated with the hidden state are observed. As such, it is not evident the method of moments should fare any better than maximum likelihood in terms of computational performance: matching the model parameters to the observed moments may involve solving computationally intractable systems of multivariate polynomial equations. Fortunately, for many classes of latent variable models, there is rich structure in low-order moments (typically second- and third-order) which allow for this inverse moment problem to be solved efficiently [2,4,6,8,9,16,18,27]. What is more is that these decomposition problems are often amenable to simple and efficient iterative methods, such as gradient descent and the power iteration method.

This survey observes that a number of important and well-studied latent variable models—including Gaussian mixture models, hidden Markov models, and Latent Dirichlet allocation—share a certain structure in their low-order moments, and this permits certain tensor decomposition approaches to parameter estimation. In particular, this decomposition can be viewed as a natural generalization of the singular value decomposition for matrices.

While much of this (or similar) structure was implicit in several previous works [2,4,9,16,18,27], here we make the decomposition explicit under a unified framework. Specifically, we express the observable moments as sums of rank-one terms, and reduce the parameter estimation task to the problem of extracting a symmetric orthogonal decomposition of a symmetric tensor derived from these observable moments. The problem can then be solved by a variety of approaches, including fixed-point and variational methods.

One approach for obtaining the orthogonal decomposition is the tensor power method of [21, Remark3]. We provide a convergence analysis of this method for orthogonally decomposable symmetric tensors, as well as a robust (and computationally tractable) variant. The perturbation analysis in [1] can be viewed as an analogue of Wedin's perturbation theorem for singular vectors of matrices [32], providing a bound on the error of the recovered decomposition in terms of the operator norm of the tensor perturbation.

1.1   Related Work

See [1] for a discussion of related work.

1.2   Organization

The rest of the survey is organized as follows. Section 2 reviews some basic definitions of tensors. Section 3 provides examples of a number of latent variable models which, after appropriate manipulations of their low order moments, share a certain natural tensor structure. Section 4 reduces the problem of parameter estimation to that of extracting a certain (symmetric orthogonal) decomposition of a tensor. See [1] which states establishes an analogue of Wedin's perturbation theorem for the singular vectors of matrices.

2 Preliminaries

We introduce some tensor notations borrowed from [23]. A real *p-th order tensor* $A \in \bigotimes_{i=1}^{p} \mathbb{R}^{n_i}$ is a member of the tensor product of Euclidean spaces $\mathbb{R}^{n_i}$, $i \in [p]$. We generally restrict to the case where $n_1 = n_2 = \cdots = n_p = n$, and simply write $A \in \bigotimes^p \mathbb{R}^n$. For a vector $v \in \mathbb{R}^n$, we use $v^{\otimes p} := v \otimes v \otimes \cdots \otimes v \in \bigotimes^p \mathbb{R}^n$ to denote its p-th tensor power. As is the case for vectors (where $p = 1$) and matrices (where $p = 2$), we may identify a p-th order tensor with the p-way array of real numbers $[A_{i_1,i_2,\ldots,i_p} : i_1, i_2, \ldots, i_p \in [n]]$, where $A_{i_1,i_2,\ldots,i_p}$ is the $(i_1, i_2, \ldots, i_p)$-th coordinate of A (with respect to a canonical basis).

We can consider A to be a multilinear map in the following sense: for a set of matrices $\{V_i \in \mathbb{R}^{n \times m_i} : i \in [p]\}$, the $(i_1, i_2, \ldots, i_p)$-th entry in the p-way array representation of $A(V_1, V_2, \ldots, V_p) \in \mathbb{R}^{m_1 \times m_2 \times \cdots \times m_p}$ is

$$[A(V_1, V_2, \ldots, V_p)]_{i_1,i_2,\ldots,i_p} := \sum_{j_1,j_2,\ldots,j_p \in [n]} A_{j_1,j_2,\ldots,j_p} [V_1]_{j_1,i_1} [V_2]_{j_2,i_2} \cdots [V_p]_{j_p,i_p}.$$

Note that if A is a matrix ($p = 2$), then

$$A(V_1, V_2) = V_1^\top A V_2.$$

Similarly, for a matrix A and vector $v \in \mathbb{R}^n$, we can express Av as

$$A(I, v) = Av \in \mathbb{R}^n,$$

where I is the $n \times n$ identity matrix. As a final example of this notation, observe

$$A(e_{i_1}, e_{i_2}, \ldots, e_{i_p}) = A_{i_1,i_2,\ldots,i_p},$$

where $\{e_1, e_2, \ldots, e_n\}$ is the canonical basis for $\mathbb{R}^n$.

Most tensors $A \in \bigotimes^p \mathbb{R}^n$ considered in this work will be *symmetric* (sometimes called *supersymmetric*), which means that their p-way array representations are invariant to permutations of the array indices: i.e., for all indices $i_1, i_2, \ldots, i_p \in [n]$, $A_{i_1,i_2,\ldots,i_p} = A_{i_{\pi(1)},i_{\pi(2)},\ldots,i_{\pi(p)}}$ for any permutation π on $[p]$. It can be checked that this reduces to the usual definition of a symmetric matrix for $p = 2$.

The *rank* of a p-th order tensor $A \in \bigotimes^p \mathbb{R}^n$ is the smallest non-negative integer k such that $A = \sum_{j=1}^{k} u_{1,j} \otimes u_{2,j} \otimes \cdots \otimes u_{p,j}$ for some $u_{i,j} \in \mathbb{R}^n, i \in [p], j \in [k]$, and the *symmetric rank* of a symmetric p-th order tensor A is the smallest non-negative integer k such that $A = \sum_{j=1}^{k} u_j^{\otimes p}$ for some $u_j \in \mathbb{R}^n, j \in [k]$ (for even p, the definition is slightly different [11]). The notion of rank readily reduces to the usual definition of matrix rank when $p = 2$, as revealed by the singular value decomposition. Similarly, for symmetric matrices, the symmetric rank is equivalent to the matrix rank as given by the spectral theorem.

The notion of tensor (symmetric) rank is considerably more delicate than matrix (symmetric) rank. For instance, it is not clear *a priori* that the symmetric rank of a tensor should even be finite [11]. In addition, removal of the best rank-1 approximation of a (general) tensor may increase the tensor rank of the residual [31].

Throughout, we use $\|v\| = (\sum_i v_i^2)^{1/2}$ to denote the Euclidean norm of a vector v, and $\|M\|$ to denote the spectral (operator) norm of a matrix. We also use $\|T\|$ to denote the operator norm of a tensor, which we define later.

3   Tensor Structure in Latent Variable Models

In this section, we give several examples of latent variable models whose low-order moments can be written as symmetric tensors of low symmetric rank; many of these examples can be deduced using the techniques developed in [25]. The basic form is demonstrated in Theorem 1 for the first example, and the general pattern will emerge from subsequent examples.

3.1   Exchangeable Single Topic Models

We first consider a simple bag-of-words model for documents in which the words in the document are assumed to be *exchangeable*. Recall that a collection of random variables $x_1, x_2, \ldots, x_\ell$ are exchangeable if their joint probability distribution is invariant to permutation of the indices. The well-known De Finetti's theorem [5] implies that such exchangeable models can be viewed as mixture models in which there is a latent variable h such that $x_1, x_2, \ldots, x_\ell$ are conditionally i.i.d. given h (see Figure 1(a) for the corresponding graphical model) and the conditional distributions are identical at all the nodes.

In our simplified topic model for documents, the latent variable h is interpreted as the (sole) topic of a given document, and it is assumed to take only a finite number of distinct values. Let k be the number of distinct topics in the corpus, d be the number of distinct words in the vocabulary, and $\ell \geq 3$ be the number of words in each document. The generative process for a document is as follows: the document's topic is drawn according to the discrete distribution specified by the probability vector $w := (w_1, w_2, \ldots, w_k) \in \Delta^{k-1}$. This is modeled as a discrete random variable h such that

$$\Pr[h = j] = w_j, \quad j \in [k].$$

Given the topic h, the document's ℓ words are drawn independently according to the discrete distribution specified by the probability vector $\mu_h \in \Delta^{d-1}$. It will be convenient to represent the ℓ words in the document by d-dimensional random *vectors* $x_1, x_2, \ldots, x_\ell \in \mathbb{R}^d$. Specifically, we set

$$x_t = e_i \quad \text{if and only if} \quad \text{the t-th word in the document is } i, \quad t \in [\ell],$$

where $e_1, e_2, \ldots e_d$ is the standard coordinate basis for $\mathbb{R}^d$.

One advantage of this encoding of words is that the (cross) moments of these random vectors correspond to joint probabilities over words. For instance, observe that

$$
\begin{aligned}
\mathbb{E}[x_1 \otimes x_2] &= \sum_{1 \le i,j \le d} \Pr[x_1 = e_i, x_2 = e_j] \, e_i \otimes e_j \\
&= \sum_{1 \le i,j \le d} \Pr[\text{1st word} = i, \text{2nd word} = j] \, e_i \otimes e_j,
\end{aligned}
$$

so the (i,j)-the entry of the matrix $\mathbb{E}[x_1 \otimes x_2]$ is $\Pr[\text{1st word} = i, \text{2nd word} = j]$. More generally, the $(i_1, i_2, \ldots, i_\ell)$-th entry in the tensor $\mathbb{E}[x_1 \otimes x_2 \otimes \cdots \otimes x_\ell]$ is $\Pr[\text{1st word} = i_1, \text{2nd word} = i_2, \ldots, \ell\text{-th word} = i_\ell]$. This means that estimating cross moments, say, of $x_1 \otimes x_2 \otimes x_3$, is the same as estimating joint probabilities of the first three words over all documents. (Recall that we assume that each document has at least three words.)

The second advantage of the vector encoding of words is that the conditional expectation of x_t given $h = j$ is simply μ_j, the vector of word probabilities for topic j:

$$\mathbb{E}[x_t|h = j] = \sum_{i=1}^{d} \Pr[t\text{-th word} = i|h = j] \, e_i = \sum_{i=1}^{d}[\mu_j]_i \, e_i = \mu_j, \quad j \in [k]$$

(where $[\mu_j]_i$ is the i-th entry in the vector μ_j). Because the words are conditionally independent given the topic, we can use this same property with conditional cross moments, say, of x_1 and x_2:

$$\mathbb{E}[x_1 \otimes x_2|h = j] = \mathbb{E}[x_1|h = j] \otimes \mathbb{E}[x_2|h = j] = \mu_j \otimes \mu_j, \quad j \in [k].$$

This and similar calculations lead one to the following theorem.

Theorem 1 ([4]). *If*

$$
\begin{aligned}
M_2 &:= \mathbb{E}[x_1 \otimes x_2] \\
M_3 &:= \mathbb{E}[x_1 \otimes x_2 \otimes x_3],
\end{aligned}
$$

then

$$M_2 = \sum_{i=1}^{k} w_i \, \mu_i \otimes \mu_i$$

$$M_3 = \sum_{i=1}^{k} w_i \, \mu_i \otimes \mu_i \otimes \mu_i.$$

As we will see in Section 4.3, the structure of M_2 and M_3 revealed in Theorem 1 implies that the topic vectors $\mu_1, \mu_2, \ldots, \mu_k$ can be estimated by computing a certain symmetric tensor decomposition. Moreover, due to exchangeability, all triples (resp., pairs) of words in a document—and not just the first three (resp., two) words—can be used in forming M_3 (resp., M_2).

3.2   Beyond Raw Moments

In the single topic model above, the raw (cross) moments of the observed words directly yield the desired symmetric tensor structure. In some other models, the raw moments do not explicitly have this form. Here, we show that the desired tensor structure can be found through various manipulations of different moments.

Spherical Gaussian Mixtures. We now consider a mixture of k Gaussian distributions with spherical covariances. We start with the simpler case where all of the covariances are identical; this probabilistic model is closely related to the (non-probabilistic) k-means clustering problem [24]. We then consider the case where the spherical variances may differ.

Common Covariance. Let w_i be the probability of choosing component $i \in [k]$, $\{\mu_1, \mu_2, \ldots, \mu_k\} \subset \mathbb{R}^d$ be the component mean vectors, and $\sigma^2 I$ be the common covariance matrix. An observation in this model is given by

$$x := \mu_h + z,$$

where h is the discrete random variable with $\Pr[h = i] = w_i$ for $i \in [k]$ (similar to the exchangeable single topic model), and $z \sim \mathcal{N}(0, \sigma^2 I)$ is an independent multivariate Gaussian random vector in $\mathbb{R}^d$ with zero mean and spherical covariance $\sigma^2 I$.

The Gaussian mixture model differs from the exchangeable single topic model in the way observations are generated. In the single topic model, we observe multiple draws (words in a particular document) $x_1, x_2, \ldots, x_\ell$ given the same fixed h (the topic of the document). In contrast, for the Gaussian mixture model, every realization of x corresponds to a different realization of h.

Theorem 2 ([16]). *Assume $d \geq k$. The variance σ^2 is the smallest eigenvalue of the covariance matrix $\mathbb{E}[x \otimes x] - \mathbb{E}[x] \otimes \mathbb{E}[x]$. Furthermore, if*

$$M_2 := \mathbb{E}[x \otimes x] - \sigma^2 I$$

$$M_3 := \mathbb{E}[x \otimes x \otimes x] - \sigma^2 \sum_{i=1}^{d} \left(\mathbb{E}[x] \otimes e_i \otimes e_i + e_i \otimes \mathbb{E}[x] \otimes e_i + e_i \otimes e_i \otimes \mathbb{E}[x] \right),$$

then

$$M_2 = \sum_{i=1}^{k} w_i \, \mu_i \otimes \mu_i$$

$$M_3 = \sum_{i=1}^{k} w_i \, \mu_i \otimes \mu_i \otimes \mu_i.$$

Differing Covariances. See [1] for the case is where each component may have a *different* spherical covariance.

Independent Component Analysis (ICA). The standard model for ICA [7,10,12,19], in which independent signals are linearly mixed and corrupted with Gaussian noise before being observed, is specified as follows. Let $h \in \mathbb{R}^k$ be a latent random *vector* with independent coordinates, $A \in \mathbb{R}^{d \times k}$ the mixing matrix, and z be a multivariate Gaussian random vector. The random vectors h and z are assumed to be independent. The observed random vector is

$$x := Ah + z.$$

Let μ_i denote the i-th column of the mixing matrix A.

Theorem 3 ([12]). *Define*

$$M_4 := \mathbb{E}[x \otimes x \otimes x \otimes x] - T$$

where T is the fourth-order tensor with

$$[T]_{i_1,i_2,i_3,i_4} := \mathbb{E}[x_{i_1} x_{i_2}]\mathbb{E}[x_{i_3} x_{i_4}] + \mathbb{E}[x_{i_1} x_{i_3}]\mathbb{E}[x_{i_2} x_{i_4}] + \mathbb{E}[x_{i_1} x_{i_4}]\mathbb{E}[x_{i_2} x_{i_3}],$$

where $1 \leq i_1, i_2, i_3, i_4 \leq k$ (i.e., T is the fourth derivative tensor of the function $v \mapsto 8^{-1}\mathbb{E}[(v^\top x)^2]^2$, so M_4 is the fourth cumulant tensor). Let $\kappa_i := \mathbb{E}[h_i^4] - 3$ for each $i \in [k]$. Then

$$M_4 = \sum_{i=1}^{k} \kappa_i \, \mu_i \otimes \mu_i \otimes \mu_i \otimes \mu_i.$$

Note that κ_i corresponds to the excess kurtosis, a measure of non-Gaussianity as $\kappa_i = 0$ if h_i is a standard normal random variable. Furthermore, note that A is not identifiable if h is a multivariate Gaussian.

We may derive forms similar to that of M_2 and M_3 from Theorem 1 using M_4 by observing that

$$M_4(I, I, u, v) = \sum_{i=1}^{k} \kappa_i(\mu_i^\top u)(\mu_i^\top v) \, \mu_i \otimes \mu_i,$$

$$M_4(I, I, I, v) = \sum_{i=1}^{k} \kappa_i(\mu_i^\top v) \, \mu_i \otimes \mu_i \otimes \mu_i$$

for any vectors $u, v \in \mathbb{R}^d$.

Latent Dirichlet Allocation (LDA). An increasingly popular class of latent variable models are *mixed membership models*, where each datum may belong to several different latent classes simultaneously. LDA is one such model for the case of document modeling; here, each document corresponds to a mixture over topics (as opposed to just a single topic). The distribution over such topic mixtures is a Dirichlet distribution $\mathrm{Dir}(\alpha)$ with parameter vector $\alpha \in \mathbb{R}_{++}^k$ with strictly positive entries; its density over the probability simplex $\Delta^{k-1} := \{v \in \mathbb{R}^k : v_i \in [0,1] \forall i \in [k], \sum_{i=1}^k v_i = 1\}$ is given by

$$p_\alpha(h) = \frac{\Gamma(\alpha_0)}{\prod_{i=1}^k \Gamma(\alpha_i)} \prod_{i=1}^k h_i^{\alpha_i-1}, \quad h \in \Delta^{k-1}$$

where

$$\alpha_0 := \alpha_1 + \alpha_2 + \cdots + \alpha_k.$$

As before, the k topics are specified by probability vectors $\mu_1, \mu_2, \ldots, \mu_k \in \Delta^{d-1}$. To generate a document, first draw the topic mixture $h = (h_1, h_2, \ldots, h_k) \sim \mathrm{Dir}(\alpha)$, and then conditioned on h, we draw ℓ words $x_1, x_2, \ldots, x_\ell$ independently from the discrete distribution specified by the probability vector $\sum_{i=1}^k h_i \mu_i$ (i.e., for each x_t, we independently sample a topic j according to h and then sample x_t according to μ_j). Again, we encode a word x_t by setting $x_t = e_i$ iff the t-th word in the document is i.

The parameter α_0 (the sum of the "pseudo-counts") characterizes the concentration of the distribution. As $\alpha_0 \to 0$, the distribution degenerates to a single topic model (i.e., the limiting density has, with probability 1, exactly one entry of h being 1 and the rest are 0). At the other extreme, if $\alpha = (c, c, \ldots, c)$ for some scalar $c > 0$, then as $\alpha_0 = ck \to \infty$, the distribution of h becomes peaked around the uniform vector $(1/k, 1/k, \ldots, 1/k)$ (furthermore, the distribution behaves like a product distribution). We are typically interested in the case where α_0 is small (e.g., a constant independent of k), whereupon h typically has only a few large entries. This corresponds to the setting where the documents are mainly comprised of just a few topics.

Theorem 4 ([2]). *Define*

$$M_1 := \mathbb{E}[x_1]$$

$$M_2 := \mathbb{E}[x_1 \otimes x_2] - \frac{\alpha_0}{\alpha_0 + 1} M_1 \otimes M_1$$

$$M_3 := \mathbb{E}[x_1 \otimes x_2 \otimes x_3]$$
$$- \frac{\alpha_0}{\alpha_0 + 2}\Big(\mathbb{E}[x_1 \otimes x_2 \otimes M_1] + \mathbb{E}[x_1 \otimes M_1 \otimes x_2] + \mathbb{E}[M_1 \otimes x_1 \otimes x_2]\Big)$$
$$+ \frac{2\alpha_0^2}{(\alpha_0 + 2)(\alpha_0 + 1)} M_1 \otimes M_1 \otimes M_1.$$

Then

$$M_2 = \sum_{i=1}^k \frac{\alpha_i}{(\alpha_0 + 1)\alpha_0} \mu_i \otimes \mu_i$$

$$M_3 = \sum_{i=1}^k \frac{2\alpha_i}{(\alpha_0 + 2)(\alpha_0 + 1)\alpha_0} \mu_i \otimes \mu_i \otimes \mu_i.$$

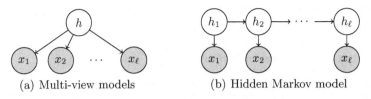

(a) Multi-view models   (b) Hidden Markov model

Fig. 1. Examples of latent variable models.

Note that α_0 needs to be known to form M_2 and M_3 from the raw moments. This, however, is a much weaker than assuming that the entire distribution of h is known (*i.e.*, knowledge of the whole parameter vector α).

3.3 Multi-view Models

Multi-view models (also sometimes called naïve Bayes models) are a special class of Bayesian networks in which observed variables $x_1, x_2, \ldots, x_\ell$ are conditionally independent given a latent variable h. This is similar to the exchangeable single topic model, but here we do not require the conditional distributions of the $x_t, t \in [\ell]$ to be identical. Techniques developed for this class can be used to handle a number of widely used models including hidden Markov models (HMMs) [4,27], phylogenetic tree models [9,27], certain tree mixtures [3], and certain probabilistic grammar models [17].

As before, we let $h \in [k]$ be a discrete random variable with $\Pr[h = j] = w_j$ for all $j \in [k]$. Now consider random vectors $x_1 \in \mathbb{R}^{d_1}$, $x_2 \in \mathbb{R}^{d_2}$, and $x_3 \in \mathbb{R}^{d_3}$ which are conditionally independent given h, and

$$\mathbb{E}[x_t | h = j] = \mu_{t,j}, \quad j \in [k], \ t \in \{1, 2, 3\}$$

where the $\mu_{t,j} \in \mathbb{R}^{d_t}$ are the conditional means of the x_t given $h = j$. Thus, we allow the observations $x_1, x_2, \ldots, x_\ell$ to be random vectors, parameterized only by their conditional means. Importantly, these conditional distributions may be discrete, continuous, or even a mix of both.

We first note the form for the raw (cross) moments.

Proposition 1. *We have that:*

$$\mathbb{E}[x_t \otimes x_{t'}] = \sum_{i=1}^{k} w_i \, \mu_{t,i} \otimes \mu_{t',i}, \quad \{t, t'\} \subset \{1, 2, 3\}, t \neq t'$$

$$\mathbb{E}[x_1 \otimes x_2 \otimes x_3] = \sum_{i=1}^{k} w_i \, \mu_{1,i} \otimes \mu_{2,i} \otimes \mu_{3,i}.$$

The cross moments do not possess a symmetric tensor form when the conditional distributions are different. Nevertheless, the moments can be "symmetrized" via a simple linear transformation of x_1 and x_2 (roughly speaking, this relates x_1 and x_2 to x_3); this leads to an expression from which the conditional means of x_3 ($i.e.$, $\mu_{3,1}, \mu_{3,2}, \ldots, \mu_{3,k}$) can be recovered. For simplicity, we assume $d_1 = d_2 = d_3 = k$; the general case (with $d_t \geq k$) is easily handled using low-rank singular value decompositions.

Theorem 5 ([2]). *Assume that the vectors $\{\mu_{v,1}, \mu_{v,2}, \ldots, \mu_{v,k}\}$ are linearly independent for each $v \in \{1, 2, 3\}$. Define*

$$\tilde{x}_1 := \mathbb{E}[x_3 \otimes x_2]\mathbb{E}[x_1 \otimes x_2]^{-1}x_1$$
$$\tilde{x}_2 := \mathbb{E}[x_3 \otimes x_1]\mathbb{E}[x_2 \otimes x_1]^{-1}x_2$$
$$M_2 := \mathbb{E}[\tilde{x}_1 \otimes \tilde{x}_2]$$
$$M_3 := \mathbb{E}[\tilde{x}_1 \otimes \tilde{x}_2 \otimes x_3].$$

Then

$$M_2 = \sum_{i=1}^{k} w_i \, \mu_{3,i} \otimes \mu_{3,i}$$

$$M_3 = \sum_{i=1}^{k} w_i \, \mu_{3,i} \otimes \mu_{3,i} \otimes \mu_{3,i}.$$

Hidden Markov Models. Our last example is the time-homogeneous HMM for sequences of vector-valued observations $x_1, x_2, \ldots \in \mathbb{R}^d$. Consider a Markov chain of discrete hidden states $y_1 \to y_2 \to y_3 \to \cdots$ over k possible states $[k]$; given a state y_t at time t, the observation x_t at time t (a random vector taking values in $\mathbb{R}^d$) is independent of all other observations and hidden states. See Figure 1(b).

Let $\pi \in \Delta^{k-1}$ be the initial state distribution ($i.e.$, the distribution of y_1), and $T \in \mathbb{R}^{k \times k}$ be the stochastic transition matrix for the hidden state Markov chain: for all times t,

$$\Pr[y_{t+1} = i | y_t = j] = T_{i,j}, \quad i, j \in [k].$$

Finally, let $O \in \mathbb{R}^{d \times k}$ be the matrix whose j-th column is the conditional expectation of x_t given $y_t = j$: for all times t,

$$\mathbb{E}[x_t | y_t = j] = Oe_j, \quad j \in [k].$$

Proposition 2 ([4]). *Define $h := y_2$, where y_2 is the second hidden state in the Markov chain. Then*

- *x_1, x_2, x_3 are conditionally independent given h;*
- *the distribution of h is given by the vector $w := T\pi \in \Delta^{k-1}$;*
- *for all $j \in [k]$,*

$$\mathbb{E}[x_1 | h = j] = O \operatorname{diag}(\pi) T^\top \operatorname{diag}(w)^{-1} e_j$$
$$\mathbb{E}[x_2 | h = j] = Oe_j$$
$$\mathbb{E}[x_3 | h = j] = OTe_j.$$

Note the matrix of conditional means of x_t has full column rank, for each $t \in \{1, 2, 3\}$, provided that: (i) O has full column rank, (ii) T is invertible, and (iii) π and $T\pi$ have positive entries.

4   Orthogonal Tensor Decompositions

We now show how recovering the μ_i's in our aforementioned problems reduces to the problem of finding a certain orthogonal tensor decomposition of a symmetric tensor. We start by reviewing the spectral decomposition of symmetric matrices, and then discuss a generalization to the higher-order tensor case. Finally, we show how orthogonal tensor decompositions can be used for estimating the latent variable models from the previous section.

4.1   Review: The Matrix Case

We first build intuition by reviewing the matrix setting, where the desired decomposition is the eigendecomposition of a symmetric rank-k matrix $M = V \Lambda V^\top$, where $V = [v_1|v_2|\cdots|v_k] \in \mathbb{R}^{n \times k}$ is the matrix with orthonormal eigenvectors as columns, and $\Lambda = \mathrm{diag}(\lambda_1, \lambda_2, \ldots, \lambda_k) \in \mathbb{R}^{k \times k}$ is diagonal matrix of non-zero eigenvalues. In other words,

$$M = \sum_{i=1}^{k} \lambda_i\, v_i v_i^\top = \sum_{i=1}^{k} \lambda_i\, v_i^{\otimes 2}. \tag{1}$$

Such a decomposition is guaranteed to exist for every symmetric matrix.

Recovery of the v_i's and λ_i's can be viewed at least two ways. First, each v_i is fixed under the mapping $u \mapsto Mu$, up to a scaling factor λ_i:

$$M v_i = \sum_{j=1}^{k} \lambda_j (v_j^\top v_i) v_j = \lambda_i v_i$$

as $v_j^\top v_i = 0$ for all $j \neq i$ by orthogonality. The v_i's are not necessarily the only such fixed points. For instance, with the multiplicity $\lambda_1 = \lambda_2 = \lambda$, then any linear combination of v_1 and v_2 is similarly fixed under M. However, in this case, the decomposition in (1) is not unique, as $\lambda_1 v_1 v_1^\top + \lambda_2 v_2 v_2^\top$ is equal to $\lambda(u_1 u_1^\top + u_2 u_2^\top)$ for any pair of orthonormal vectors, u_1 and u_2 spanning the same subspace as v_1 and v_2. Nevertheless, the decomposition is unique when $\lambda_1, \lambda_2, \ldots, \lambda_k$ are distinct, whereupon the v_j's are the only directions fixed under $u \mapsto Mu$ up to non-trivial scaling.

The second view of recovery is via the variational characterization of the eigenvalues. Assume $\lambda_1 > \lambda_2 > \cdots > \lambda_k$; the case of repeated eigenvalues again leads to similar non-uniqueness as discussed above. Then the *Rayleigh quotient*

$$u \mapsto \frac{u^\top M u}{u^\top u}$$

is maximized over non-zero vectors by v_1. Furthermore, for any $s \in [k]$, the maximizer of the Rayleigh quotient, subject to being orthogonal to $v_1, v_2, \ldots, v_{s-1}$, is v_s. Another way of obtaining this second statement is to consider the *deflated* Rayleigh quotient

$$u \mapsto \frac{u^\top \left(M - \sum_{j=1}^{s-1} \lambda_j v_j v_j^\top \right) u}{u^\top u}$$

and observe that v_s is the maximizer.

Efficient algorithms for finding these matrix decompositions are well studied [15, Section 8.2.3], and iterative power methods are one effective class of algorithms.

We remark that in our multilinear tensor notation, we may write the maps $u \mapsto Mu$ and $u \mapsto u^\top Mu / \|u\|_2^2$ as

$$u \mapsto Mu \ \equiv \ u \mapsto M(I, u), \tag{2}$$

$$u \mapsto \frac{u^\top Mu}{u^\top u} \ \equiv \ u \mapsto \frac{M(u, u)}{u^\top u}. \tag{3}$$

4.2   The Tensor Case

Decomposing general tensors is a delicate issue; tensors may not even have unique decompositions. Fortunately, the orthogonal tensors that arise in the aforementioned models have a structure which permits a unique decomposition under a mild non-degeneracy condition. We focus our attention to the case $p = 3$, *i.e.*, a third order tensor; the ideas extend to general p with minor modifications.

An *orthogonal decomposition* of a symmetric tensor $T \in \bigotimes^3 \mathbb{R}^n$ is a collection of orthonormal (unit) vectors $\{v_1, v_2, \ldots, v_k\}$ together with corresponding positive scalars $\lambda_i > 0$ such that

$$T = \sum_{i=1}^k \lambda_i v_i^{\otimes 3}. \tag{4}$$

Note that since we are focusing on odd-order tensors ($p = 3$), we have added the requirement that the λ_i be positive. This convention can be followed without loss of generality since $-\lambda_i v_i^{\otimes p} = \lambda_i (-v_i)^{\otimes p}$ whenever p is odd. Also, it should be noted that orthogonal decompositions do not necessarily exist for every symmetric tensor.

In analogy to the matrix setting, we consider two ways to view this decomposition: a fixed-point characterization and a variational characterization. Related characterizations based on optimal rank-1 approximations can be found in [33].

Fixed-Point Characterization. For a tensor T, consider the vector-valued map

$$u \mapsto T(I, u, u) \tag{5}$$

which is the third-order generalization of (2). This can be explicitly written as

$$T(I, u, u) = \sum_{i=1}^{d} \sum_{1 \le j, l \le d} T_{i,j,l}(e_j^\top u)(e_l^\top u)e_i.$$

Observe that (5) is *not* a linear map, which is a key difference compared to the matrix case.

An eigenvector u for a matrix M satisfies $M(I, u) = \lambda u$, for some scalar λ. We say a unit vector $u \in \mathbb{R}^n$ is an *eigenvector* of T, with corresponding *eigenvalue* $\lambda \in \mathbb{R}$, if

$$T(I, u, u) = \lambda u.$$

(To simplify the discussion, we assume throughout that eigenvectors have unit norm; otherwise, for scaling reasons, we replace the above equation with $T(I, u, u) = \lambda \|u\| u$.) This concept was originally introduced in [23, 30]. For orthogonally decomposable tensors $T = \sum_{i=1}^{k} \lambda_i v_i^{\otimes 3}$, ,

$$T(I, u, u) = \sum_{i=1}^{k} \lambda_i (u^\top v_i)^2 v_i \ .$$

By the orthogonality of the v_i, it is clear that $T(I, v_i, v_i) = \lambda_i v_i$ for all $i \in [k]$. Therefore each (v_i, λ_i) is an eigenvector/eigenvalue pair.

There are a number of subtle differences compared to the matrix case that arise as a result of the non-linearity of (5). First, even with the multiplicity $\lambda_1 = \lambda_2 = \lambda$, a linear combination $u := c_1 v_1 + c_2 v_2$ may *not* be an eigenvector. In particular,

$$T(I, u, u) = \lambda_1 c_1^2 v_1 + \lambda_2 c_2^2 v_2 = \lambda(c_1^2 v_1 + c_2^2 v_2)$$

may not be a multiple of $c_1 v_1 + c_2 v_2$. This indicates that the issue of repeated eigenvalues does not have the same status as in the matrix case. Second, even if all the eigenvalues are distinct, it turns out that the v_i's are not the only eigenvectors. For example, set $u := (1/\lambda_1)v_1 + (1/\lambda_2)v_2$. Then,

$$T(I, u, u) = \lambda_1 (1/\lambda_1)^2 v_1 + \lambda_2 (1/\lambda_2)^2 v_2 = u,$$

so $u/\|u\|$ is an eigenvector. More generally, for any subset $S \subseteq [k]$, we have that $\sum_{i \in S}(1/\lambda_i)v_i$ is (proportional to) an eigenvector.

As we now see, these additional eigenvectors can be viewed as spurious. We say a unit vector u is a *robust eigenvector* of T if there exists an $\epsilon > 0$ such that for all $\theta \in \{u' \in \mathbb{R}^n : \|u' - u\| \le \epsilon\}$, repeated iteration of the map

$$\bar{\theta} \mapsto \frac{T(I, \bar{\theta}, \bar{\theta})}{\|T(I, \bar{\theta}, \bar{\theta})\|} \ , \tag{6}$$

starting from θ converges to u. Note that the map (6) rescales the output to have unit Euclidean norm. Robust eigenvectors are also called attracting fixed points of (6) (see, *e.g.*, [20]).

The following theorem implies that if T has an orthogonal decomposition as given in (4), then the set of robust eigenvectors of T are precisely the set $\{v_1, v_2, \ldots v_k\}$, implying that the orthogonal decomposition is unique. (For even order tensors, the uniqueness is true up to sign-flips of the v_i.)

Theorem 6. *Let T have an orthogonal decomposition as given in* (4).

1. *The set of $\theta \in \mathbb{R}^n$ which do not converge to some v_i under repeated iteration of (6) has measure zero.*
2. *The set of robust eigenvectors of T is equal to $\{v_1, v_2, \ldots, v_k\}$.*

The proof of Theorem 6 is given in [1] and follows readily from simple orthogonality considerations. Note that *every* v_i in the orthogonal tensor decomposition is robust, whereas for a symmetric matrix M, for almost all initial points, the map $\bar\theta \mapsto \frac{M\bar\theta}{\|M\bar\theta\|}$ converges only to an eigenvector corresponding to the largest magnitude eigenvalue. Also, since the tensor order is odd, the signs of the robust eigenvectors are fixed, as each $-v_i$ is mapped to v_i under (6).

Variational Characterization. We now discuss a variational characterization of the orthogonal decomposition. The *generalized Rayleigh quotient* [33] for a third-order tensor is

$$u \mapsto \frac{T(u, u, u)}{(u^\top u)^{3/2}},$$

which can be compared to (3). For an orthogonally decomposable tensor, the following theorem shows that a non-zero vector $u \in \mathbb{R}^n$ is an *isolated local maximizer* [28] of the generalized Rayleigh quotient if and only if $u = v_i$ for some $i \in [k]$.

Theorem 7. *Assume $n \geq 2$. Let T have an orthogonal decomposition as given in* (4), *and consider the optimization problem*

$$\max_{u \in \mathbb{R}^n} T(u, u, u) \ s.t. \ \|u\| = 1.$$

1. *The stationary points are eigenvectors of T.*
2. *A stationary point u is an isolated local maximizer if and only if $u = v_i$ for some $i \in [k]$.*

The proof of Theorem 7 is given in [1]. It is similar to local optimality analysis for ICA methods using fourth-order cumulants (*e.g.*, [13,14]).

Again, we see similar distinctions to the matrix case. In the matrix case, the only local maximizers of the Rayleigh quotient are the eigenvectors with the largest eigenvalue (and these maximizers take on the globally optimal value). For the case of orthogonal tensor forms, the robust eigenvectors are precisely the isolated local maximizers.

An important implication of the two characterizations is that, for orthogonally decomposable tensors T, (i) the local maximizers of the objective function

$u \mapsto T(u, u, u)/(u^\top u)^{3/2}$ correspond precisely to the vectors v_i in the decomposition, and (ii) these local maximizers can be reliably identified using a simple fixed-point iteration (*i.e.*, the tensor analogue of the matrix power method). Moreover, a second-derivative test based on $T(I, I, u)$ can be employed to test for local optimality and rule out other stationary points.

4.3   Estimation via Orthogonal Tensor Decompositions

We now demonstrate how the moment tensors obtained for various latent variable models in Section 3 can be reduced to an orthogonal form. For concreteness, we take the specific form from the exchangeable single topic model (Theorem 1):

$$M_2 = \sum_{i=1}^{k} w_i \, \mu_i \otimes \mu_i,$$

$$M_3 = \sum_{i=1}^{k} w_i \, \mu_i \otimes \mu_i \otimes \mu_i.$$

(The more general case allows the weights w_i in M_2 to differ in M_3, but for simplicity we keep them the same in the following discussion.) We now show how to reduce these forms to an orthogonally decomposable tensor from which the w_i and μ_i can be recovered. See [1] for a discussion as to how previous approaches [2,4,16,27] achieved this decomposition through a certain simultaneous diagonalization method.

Throughout, we assume the following non-degeneracy condition.

Condition 41 (Non-degeneracy). *The vectors $\mu_1, \mu_2, \ldots, \mu_k \in \mathbb{R}^d$ are linearly independent, and the scalars $w_1, w_2, \ldots, w_k > 0$ are strictly positive.*

Observe that Condition 41 implies that $M_2 \succeq 0$ is positive semidefinite and has rank-k. This is a mild condition; furthermore, when this condition is not met, learning is conjectured to be hard for both computational [27] and information-theoretic reasons [26].

The Reduction. First, let $W \in \mathbb{R}^{d \times k}$ be a linear transformation such that

$$M_2(W, W) \; = \; W^\top M_2 W \; = \; I$$

where I is the $k \times k$ identity matrix (*i.e.*, W whitens M_2). Since $M_2 \succeq 0$, we may for concreteness take $W := UD^{-1/2}$, where $U \in \mathbb{R}^{d \times k}$ is the matrix of orthonormal eigenvectors of M_2, and $D \in \mathbb{R}^{k \times k}$ is the diagonal matrix of positive eigenvalues of M_2. Let

$$\tilde{\mu}_i := \sqrt{w_i} \, W^\top \mu_i.$$

Observe that

$$M_2(W, W) \; = \; \sum_{i=1}^{k} W^\top (\sqrt{w_i} \mu_i)(\sqrt{w_i} \mu_i)^\top W \; = \; \sum_{i=1}^{k} \tilde{\mu}_i \tilde{\mu}_i^\top \; = \; I,$$

so the $\tilde{\mu}_i \in \mathbb{R}^k$ are orthonormal vectors.

Now define $\widetilde{M}_3 := M_3(W, W, W) \in \mathbb{R}^{k \times k \times k}$, so that

$$\widetilde{M}_3 = \sum_{i=1}^{k} w_i \, (W^\top \mu_i)^{\otimes 3} = \sum_{i=1}^{k} \frac{1}{\sqrt{w_i}} \, \tilde{\mu}_i^{\otimes 3}.$$

As the following theorem shows, the orthogonal decomposition of $\widetilde{M}_3$ can be obtained by identifying its robust eigenvectors, upon which the original parameters w_i and μ_i can be recovered. For simplicity, we only state the result in terms of robust eigenvector/eigenvalue pairs; one may also easily state everything in variational form using Theorem 7.

Theorem 8. *Assume Condition 41 and take $\widetilde{M}_3$ as defined above.*

1. *The set of robust eigenvectors of $\widetilde{M}_3$ is equal to $\{\tilde{\mu}_1, \tilde{\mu}_2, \ldots, \tilde{\mu}_k\}$.*
2. *The eigenvalue corresponding to the robust eigenvector $\tilde{\mu}_i$ of $\widetilde{M}_3$ is equal to $1/\sqrt{w_i}$, for all $i \in [k]$.*
3. *If $B \in \mathbb{R}^{d \times k}$ is the Moore-Penrose pseudoinverse of $W^\top$, and (v, λ) is a robust eigenvector/eigenvalue pair of $\widetilde{M}_3$, then $\lambda B v = \mu_i$ for some $i \in [k]$.*

The theorem follows by combining the above discussion with the robust eigenvector characterization of Theorem 6. Recall that we have taken as convention that eigenvectors have unit norm, so the μ_i are exactly determined from the robust eigenvector/eigenvalue pairs of $\widetilde{M}_3$ (together with the pseudoinverse of $W^\top$); in particular, the scale of each μ_i is correctly identified (along with the corresponding w_i). Relative to previous works on moment-based estimators for latent variable models (*e.g.*, [2,4,16]), Theorem 8 emphasizes the role of the special tensor structure, which in turn makes transparent the applicability of methods for orthogonal tensor decomposition.

5   Tensor Power Method

In this section, we consider the tensor power method of [21, Remark 3] for orthogonal tensor decomposition. We first state a simple convergence analysis for an orthogonally decomposable tensor T.

When only an approximation $\hat{T}$ to an orthogonally decomposable tensor T is available (*e.g.*, when empirical moments are used to estimate population moments), an orthogonal decomposition need not exist for this perturbed tensor (unlike for the case of matrices), and a more robust approach is required to extract the approximate decomposition. Here, we propose such a variant in Algorithm 1 and provide a detailed perturbation analysis. We note that alternative approaches such as simultaneous diagonalization can also be employed (see [1]).

5.1   Convergence Analysis for Orthogonally Decomposable Tensors

The following lemma establishes the quadratic convergence of the tensor power method (*i.e.*, repeated iteration of (6)) for extracting a single component of the orthogonal decomposition. Note that the initial vector θ_0 determines which robust eigenvector will be the convergent point. Computation of subsequent eigenvectors can be computed with deflation, *i.e.*, by subtracting appropriate terms from T.

Lemma 1. *Let* $T \in \bigotimes^3 \mathbb{R}^n$ *have an orthogonal decomposition as given in* (4). *For a vector* $\theta_0 \in \mathbb{R}^n$, *suppose that the set of numbers* $|\lambda_1 v_1^\top \theta_0|, |\lambda_2 v_2^\top \theta_0|, \ldots, |\lambda_k v_k^\top \theta_0|$ *has a unique largest element. Without loss of generality, say* $|\lambda_1 v_1^\top \theta_0|$ *is this largest value and* $|\lambda_2 v_2^\top \theta_0|$ *is the second largest value. For* $t = 1, 2, \ldots,$ *let*

$$\theta_t := \frac{T(I, \theta_{t-1}, \theta_{t-1})}{\|T(I, \theta_{t-1}, \theta_{t-1})\|}.$$

Then

$$\|v_1 - \theta_t\|^2 \leq \left(2\lambda_1^2 \sum_{i=2}^{k} \lambda_i^{-2} \right) \cdot \left| \frac{\lambda_2 v_2^\top \theta_0}{\lambda_1 v_1^\top \theta_0} \right|^{2^{t+1}}.$$

That is, repeated iteration of (6) *starting from* θ_0 *converges to* v_1 *at a quadratic rate.*

To obtain all eigenvectors, we may simply proceed iteratively using deflation, executing the power method on $T - \sum_j \lambda_j v_j^{\otimes 3}$ after having obtained robust eigenvector / eigenvalue pairs $\{(v_j, \lambda_j)\}$.

Proof. Let $\bar{\theta}_0, \bar{\theta}_1, \bar{\theta}_2, \ldots$ be the sequence given by $\bar{\theta}_0 := \theta_0$ and $\bar{\theta}_t := T(I, \theta_{t-1}, \theta_{t-1})$ for $t \geq 1$. Let $c_i := v_i^\top \theta_0$ for all $i \in [k]$. It is easy to check that (i) $\theta_t = \bar{\theta}_t / \|\bar{\theta}_t\|$, and (ii) $\bar{\theta}_t = \sum_{i=1}^{k} \lambda_i^{2^t - 1} c_i^{2^t} v_i$. (Indeed, $\bar{\theta}_{t+1} = \sum_{i=1}^{k} \lambda_i (v_i^\top \bar{\theta}_t)^2 v_i = \sum_{i=1}^{k} \lambda_i (\lambda_i^{2^t - 1} c_i^{2^t})^2 v_i = \sum_{i=1}^{k} \lambda_i^{2^{t+1} - 1} c_i^{2^{t+1}} v_i$.) Then

$$1 - (v_1^\top \theta_t)^2 = 1 - \frac{\lambda_1^{2^{t+1} - 2} c_1^{2^{t+1}}}{\sum_{i=1}^{k} \lambda_i^{2^{t+1} - 2} c_i^{2^{t+1}}} \leq \frac{\sum_{i=2}^{k} \lambda_i^{2^{t+1} - 2} c_i^{2^{t+1}}}{\sum_{i=1}^{k} \lambda_i^{2^{t+1} - 2} c_i^{2^{t+1}}} \leq \lambda_1^2 \sum_{i=2}^{k} \lambda_i^{-2} \cdot \left| \frac{\lambda_2 c_2}{\lambda_1 c_1} \right|^{2^{t+1}}.$$

Since $\lambda_1 > 0$, we have $v_1^\top \theta_t > 0$ and hence $\|v_1 - \theta_t\|^2 = 2(1 - v_1^\top \theta_t) \leq 2(1 - (v_1^\top \theta_t)^2)$ as required.

5.2 Perturbation Analysis of a Robust Tensor Power Method

Now we summarize the case where we have an approximation $\hat{T}$ to an orthogonally decomposable tensor T. Here, a more robust approach is required to extract an approximate decomposition. We propose such an algorithm in Algorithm 1, and provide a detailed perturbation analysis. For simplicity, we assume the tensor $\hat{T}$ is of size $k \times k \times k$ as per the reduction from Section 4.3. In some applications, it may be preferable to work directly with a $n \times n \times n$ tensor of rank $k \leq n$ (as in Lemma 1); our results apply in that setting with little modification.

Algorithm 1. Robust tensor power method

---

input symmetric tensor $\tilde{T} \in \mathbb{R}^{k \times k \times k}$, number of iterations L, N.
output the estimated eigenvector/eigenvalue pair; the deflated tensor.
1: **for** $\tau = 1$ to L **do**
2:     Draw $\theta_0^{(\tau)}$ uniformly at random from the unit sphere in $\mathbb{R}^k$.
3:     **for** $t = 1$ to N **do**
4:         Compute power iteration update

$$\theta_t^{(\tau)} := \frac{\tilde{T}(I, \theta_{t-1}^{(\tau)}, \theta_{t-1}^{(\tau)})}{\|\tilde{T}(I, \theta_{t-1}^{(\tau)}, \theta_{t-1}^{(\tau)})\|} \tag{7}$$

5:     **end for**
6: **end for**
7: Let $\tau^* := \arg\max_{\tau \in [L]} \{\tilde{T}(\theta_N^{(\tau)}, \theta_N^{(\tau)}, \theta_N^{(\tau)})\}$.
8: Do N power iteration updates (7) starting from $\theta_N^{(\tau^*)}$ to obtain $\hat{\theta}$, and set $\hat{\lambda} := \tilde{T}(\hat{\theta}, \hat{\theta}, \hat{\theta})$.
9: **return** the estimated eigenvector/eigenvalue pair $(\hat{\theta}, \hat{\lambda})$; the deflated tensor $\tilde{T} - \hat{\lambda} \hat{\theta}^{\otimes 3}$.

Assume that the symmetric tensor $T \in \mathbb{R}^{k \times k \times k}$ is orthogonally decomposable, and that $\hat{T} = T + E$, where the perturbation $E \in \mathbb{R}^{k \times k \times k}$ is a symmetric tensor with small operator norm:

$$\|E\| := \sup_{\|\theta\|=1} |E(\theta, \theta, \theta)|.$$

In our latent variable model applications, $\hat{T}$ is the tensor formed by using empirical moments, while T is the orthogonally decomposable tensor derived from the population moments for the given model. In the context of parameter estimation (as in Section 4.3), E must account for any error amplification throughout the reduction, such as in the whitening step.

[1] provides a perturbation analysis which is similar to Wedin's perturbation theorem for singular vectors of matrices [32] in that it bounds the error of the (approximate) decomposition returned by Algorithm 1 on input $\hat{T}$ in terms of the size of the perturbation, provided that the perturbation is small enough.

Acknowledgments. We thank Boaz Barak, Dean Foster, Jon Kelner, and Greg Valiant for helpful discussions. This work was completed while DH was a postdoctoral researcher at Microsoft Research New England, and partly while AA, RG, and MT were visiting the same lab. AA is supported in part by the NSF Award CCF-1219234, AFOSR Award FA9550-10-1-0310 and the ARO Award W911NF-12-1-0404.

References

1. Anandkumar, A., Ge, R., Hsu, D., Kakade, S., Telgarsky, M.: Tensor decompositions for learning latent variable models. Journal of Machine Learning Research **15**, (2014)

2. Anandkumar, A., Foster, D.P., Hsu, D., Kakade, S.M., Liu, Y.-K.: A spectral algorithm for latent Dirichlet allocation. In: Advances in Neural Information Processing Systems 25, (2012)
3. Anandkumar, A., Hsu, D., Huang, F., Kakade, S.M.: Learning mixtures of tree graphical models. In: Advances in Neural Information Processing Systems 25 (2012)
4. Anandkumar, A., Hsu, D., Kakade, S.M.: A method of moments for mixture models and hidden Markov models. In: Twenty-Fifth Annual Conference on Learning Theory, vol. 23, pp. 33.1–33.34 (2012)
5. Austin, T.: On exchangeable random variables and the statistics of large graphs and hypergraphs. Probab. Survey **5**, 80–145 (2008)
6. Cardoso, J.-F.: Super-symmetric decomposition of the fourth-order cumulant tensor. Blind identification of more sources than sensors. In: ICASSP-91, 1991 International Conference on Acoustics, Speech, and Signal Processing, pp. 3109–3112. IEEE (1991)
7. Cardoso, J.-F., Comon, P.: Independent component analysis, a survey of some algebraic methods. In: IEEE International Symposium on Circuits and Systems, pp. 93–96 (1996)
8. Cattell, R.B.: Parallel proportional profiles and other principles for determining the choice of factors by rotation. Psychometrika **9**(4), 267–283 (1944)
9. Chang, J.T.: Full reconstruction of Markov models on evolutionary trees: Identifiability and consistency. Mathematical Biosciences **137**, 51–73 (1996)
10. Comon, P.: Independent component analysis, a new concept? Signal Processing **36**(3), 287–314 (1994)
11. Comon, P., Golub, G., Lim, L.-H., Mourrain, B.: Symmetric tensors and symmetric tensor rank. SIAM Journal on Matrix Analysis Appl. **30**(3), 1254–1279 (2008)
12. Comon, P., Jutten, C.: Handbook of Blind Source Separation: Independent Component Analysis and Applications. Academic Press, Elsevier (2010)
13. Delfosse, N., Loubaton, P.: Adaptive blind separation of independent sources: a deflation approach. Signal Processing **45**(1), 59–83 (1995)
14. Alan, M., Frieze, M.J., Kannan, R.: Learning linear transformations. In: Thirty-Seventh Annual Symposium on Foundations of Computer Science, pp. 359–368 (1996)
15. Golub, G.H., van Loan, C.F.: Matrix Computations. Johns Hopkins University Press (1996)
16. Hsu, D., Kakade, S.M.: Learning mixtures of spherical Gaussians: moment methods and spectral decompositions. In: Fourth Innovations in Theoretical Computer Science (2013)
17. Hsu, D., Kakade, S.M., Liang, P.: Identifiability and unmixing of latent parse trees. In: Advances in Neural Information Processing Systems 25 (2012)
18. Hsu, D., Kakade, S.M., Zhang, T.: A spectral algorithm for learning hidden Markov models. Journal of Computer and System Sciences **78**(5), 1460–1480 (2012)
19. Hyvärinen, A., Oja, E.: Independent component analysis: algorithms and applications. Neural Networks **13**(4–5), 411–430 (2000)
20. Kolda, T.G., Mayo, J.R.: Shifted power method for computing tensor eigenpairs. SIAM Journal on Matrix Analysis and Applications **32**(4), 1095–1124 (2011)
21. De Lathauwer, L., De Moor, B., Vandewalle, J.: On the best rank-1 and rank-$(R_1, R_2, ., R_n)$ approximation and applications of higher-order tensors. SIAM J. Matrix Anal. Appl. **21**(4), 1324–1342 (2000)
22. Le Cam, L.: Asymptotic Methods in Statistical Decision Theory. Springer (1986)

23. Lim, L.-H.: Singular values and eigenvalues of tensors: a variational approach. In: Proceedings of the IEEE International Workshop on Computational Advances in Multi-Sensor Adaptive Processing vol. 1, pp. 129–132 (2005)
24. MacQueen, J.B.: Some methods for classification and analysis of multivariate observations. In: Proceedings of the fifth Berkeley Symposium on Mathematical Statistics and Probability, vol. 1, pp. 281–297. University of California Press (1967)
25. McCullagh, P.: Tensor Methods in Statistics. Chapman and Hall (1987)
26. Moitra, A., Valiant, G.: Settling the polynomial learnability of mixtures of Gaussians. In: Fifty-First Annual IEEE Symposium on Foundations of Computer Science, pp. 93–102 (2010)
27. Mossel, E., Roch, S.: Learning nonsingular phylogenies and hidden Markov models. Annals of Applied Probability 16(2), 583–614 (2006)
28. Nocedal, J., Wright, S.J.: Numerical Optimization. Springer, 1999
29. Pearson, K.: Contributions to the mathematical theory of evolution. In: Philosophical Transactions of the Royal Society, London, A., p. 71 (1894)
30. Qi, L.: Eigenvalues of a real supersymmetric tensor. Journal of Symbolic Computation 40(6), 1302–1324 (2005)
31. Stegeman, A., Comon, P.: Subtracting a best rank-1 approximation may increase tensor rank. Linear Algebra and Its Applications 433, 1276–1300 (2010)
32. Wedin, P.: Perturbation bounds in connection with singular value decomposition. BIT Numerical Mathematics 12(1), 99–111 (1972)
33. Zhang, T., Golub, G.: Rank-one approximation to high order tensors. SIAM Journal on Matrix Analysis and Applications 23, 534–550 (2001)

Inductive Inference

Priced Learning

Sanjay Jain[1]([✉]), Junqi Ma[2], and Frank Stephan[3]

[1] School of Computing, National University of Singapore,
Singapore 117417, Singapore
sanjay@comp.nus.edu.sg
[2] School of Computing, National University of Singapore,
Singapore 117417, Singapore
ma.junqi@nus.edu.sg
[3] Department of Mathematics and Department of Computer Science,
National University of Singapore, Singapore 119076, Singapore
fstephan@comp.nus.edu.sg

Abstract. In iterative learning the memory of the learner can only be updated when the hypothesis changes; this results in only finitely many updates of memory during the overall learning history. Priced learning relaxes this constraint on the update of memory by imposing some price on the updates of the memory – depending on the current datum – and requiring that the overall sum of the costs incurred has to be finite. There are priced-learnable classes which are not iteratively learnable. The current work introduces the basic definitions and results for priced learning. This work also introduces various variants of priced learning.

1 Introduction

Learning from positive data in Gold's model [6] has the following basic scenario: Given a class of r.e. languages and an unknown language L from this class, the learner observes an infinite list containing all and only the elements of L. The order and multiplicity of the elements of L in the list may be arbitrary. As the learner is observing the members of the list, it outputs a sequence of hypotheses about what the input language might be. For learning the language, this sequence of hypotheses is required to converge to a grammar for L (for every input list of elements of L as above).

In general, in the above learning process, the learner can memorise all data observed so far and do comprehensive calculations without restrictions to the memory amount and usage. Several approaches have been formalised in order to restrict the amount of memory used. One of them is the strict separation between short-term and long-term memory where, whenever a datum is read, some computations are done in an unlimited short-term memory and then the data for the next round of learning is archived in a size-constrained long-term memory [4,7]. The other approaches to limit the memory usage do not count

---

Research for this work is supported in part by NUS grants C252-000-087-001 (S. Jain) and R146-000-181-112 (S. Jain and F. Stephan).

© Springer International Publishing Switzerland 2015
K. Chaudhuri et al. (Eds.): ALT 2015, LNAI 9355, pp. 41–55, 2015.
DOI: 10.1007/978-3-319-24486-0_3

bits and bytes, but rather allow only fixed number of elements of the input to be memorized. In such models, if the learner does not update its memory/hypothesis on a receipt of a datum, then at later stages it is not able to know whether the corresponding datum was observed or not. In its most restrictive form, this type of learning is called incremental or iterative learning [3,8,13]. An iterative learner can memorise data only when it revises the hypothesis and not at any other point of time; thus it can, through the overall learning history, only finitely often revise its long-term memory. This restriction is quite severe and, for example, the class consisting of the set $\{1, 2, \ldots\}$ and all sets of the form $\{0, 1, \ldots, x\}$, where x is a natural number, is not iteratively learnable. To see this, note that if the learner first sees data from the infinite set $\{1, 2, \ldots\}$, then, after some time, the learner will need to converge to a grammar for $\{1, 2, \ldots\}$. Thereafter the learner cannot archive any data, in particular it cannot archive the maximum datum x seen so far. If later the datum 0 shows up, the learner knows that the input set must be of the form $\{0, 1, \ldots, x\}$; however, the learner can no longer recover the exact value of x from its memory.

The idea of priced learning is to relax the severe constraints placed on iterative learning. In priced learning, a price is charged for each update of the memory and it is required that during the learning process, the overall costs incurred is finite. In the case that this price charged is always the same constant c for each memory update, the corresponding notion would be exactly that of iterative learning. However, by allowing the price to depend on the datum x read when the memory is updated, one can give discounts for various types of data and permit, under certain circumstances, to update the memory infinitely often during the learning process. This concept of priced learning, however, still does not permit, in general, to do every possible update of the memory. Priced learning is therefore between the two extremes of iterative learning and the original unrestricted model of learning by Gold where the memory could be updated as often as desired. For the priced learning model $\mathbf{Priced}_f\mathbf{Ex}$ introduced in this paper, the costs incurred are guided by a price function f which is a parameter of the learning criterion. A $\mathbf{Priced}_f\mathbf{Ex}$-learner incurs, at every update of the memory on input datum x, the cost $1/(f(x) + 1)$ (where f is a recursive function). We refer the reader to Section 3 for the formal definitions. On reading a datum x, the learner has to decide whether it wants to update the memory so that on one hand, all important information are preserved in the long term memory and, on the other hand, the overall price of these updates does not go to infinity.

The present work investigates when, for different price functions f and g, $\mathbf{Priced}_f\mathbf{Ex}$-learnability implies $\mathbf{Priced}_g\mathbf{Ex}$-learnability. This depends on the existence of sets S such that $\sum_{x \in S} \frac{1}{f(x)+1}$ is finite but $\sum_{x \in S} \frac{1}{g(x)+1}$ is infinite (see Theorem 9 and Theorem 11). Theorem 15 gives a characterisation of classes which are $\mathbf{Priced}_f\mathbf{Ex}$-learnable for some recursive function f: these are exactly the classes which are learnable by a set-driven learner. Here a set-driven learner [9] is a learner whose output conjecture depends only on the set of elements it has observed in the input and not on their order or amount of repetitions.

Further results are based on whether a class can be learnt when the price functions are from a natural class $\mathcal{C}$ of price functions, irrespective of which price function from the class is actually chosen. Here, one may allow the learner to be chosen depending on the price function, uniformly or non-uniformly based on the program for the price function, or may even require it to be the same learner for all the price functions from the class $\mathcal{C}$. Section 6 onwards explore such questions. Let $\mathbf{MF} = \{f : f$ is recursive and unbounded and $\forall x\,[f(x) \leq f(x+1)]\}$ and $\mathbf{FF} = \{f : f$ is recursive and $\forall y\,[\mathrm{card}(f^{-1}(y))$ is finite$]\}$. Note that $\mathbf{MF}$ is a proper subset of $\mathbf{FF}$. It is shown in Theorem 17 that there exists a class of languages which can be $\mathbf{Priced}_f$-learnt for every $f \in \mathbf{FF}$, but which cannot be iteratively learnt. On the other hand, Theorem 18 shows advantages of price function being from $\mathbf{MF}$ compared to it being from $\mathbf{FF}$. Results in Section 7 deal with the question of when and what kind of uniformity constraints can be placed on learners which learn a class with respect to a price function from a class of price functions such as $\mathbf{MF}$ or $\mathbf{FF}$. Due to space constraints, some proofs are omitted.

2   Notations and Preliminaries

Any unexplained recursion theoretic notation is from Rogers' book [11]. The symbol $\mathbb{N}$ denotes the set of natural numbers $\{0, 1, 2, \ldots\}$. Subsets of natural numbers are referred to as languages. We let $\emptyset, \subseteq, \subset, \supseteq$ and $\supset$ respectively denote empty set, subset, proper subset, superset and proper superset. Cardinality of a set S is denoted by $\mathrm{card}(S)$. The maximum and minimum of a set S are denoted by $\max(S)$ and $\min(S)$, respectively, where $\max(\emptyset) = 0$ and $\min(\emptyset) = \infty$.

Let $\langle \cdot, \cdot \rangle$ denote a recursive pairing function which is a bijection from $\mathbb{N} \times \mathbb{N}$ to $\mathbb{N}$. Without loss of generality, we assume that pairing function is monotonic in both its arguments; in particular $\langle 0, 0 \rangle = 0$. The pairing function can be extended to pairing of n-tuples by using $\langle x_1, x_2, \ldots, x_n \rangle = \langle x_1, \langle x_2, \ldots, x_n \rangle \rangle$. We define projection functions π_1, π_2 as $\pi_1(\langle x_1, x_2 \rangle) = x_1$ and $\pi_2(\langle x_1, x_2 \rangle) = x_2$. Similarly, let $\pi_i^n(\langle x_1, x_2, \ldots, x_n \rangle) = x_i$. For a set L, let $\mathrm{cyl}(L) = \{\langle x, i \rangle : x \in L, i \in \mathbb{N}\}$.

By φ we denote a fixed *acceptable* programming system for the partial computable functions mapping $\mathbb{N}$ to $\mathbb{N}$ [11]. By φ_i we denote the partial function computed by the i-th program in the φ-system. Thus, i is also called a program/index for φ_i. Let $\mathcal{R}$ denote the class of all total recursive functions. For any partial function η, we let $\eta(x)\!\downarrow$ denote that $\eta(x)$ is defined and $\eta(x)\!\uparrow$ denote that $\eta(x)$ is undefined. By Φ we denote an arbitrary fixed Blum complexity measure [2] for the φ-system. Let

$$\varphi_{i,s}(x) = \begin{cases} \varphi_i(x), & \text{if } x < s \text{ and } \Phi_i(x) < s; \\ \uparrow, & \text{otherwise.} \end{cases}$$

Let $W_i = \mathrm{domain}(\varphi_i)$. W_i can be considered as the language accepted by the i-the φ-program φ_i. We also say that i is a grammar/index for W_i. Thus, $W_0, W_1, \ldots,$ is an acceptable programming system for recursively enumerable languages. The symbol $\mathcal{E}$ denotes the set of all recursively enumerable (r.e.) languages. The symbol L ranges over $\mathcal{E}$; the symbol $\mathcal{L}$ ranges over subsets of $\mathcal{E}$.

By $\overline{L}$, we denote the complement of L, that is $\overline{L} = \mathbb{N} - L$. By $L(x)$ we denote the value of the characteristic function of L at x; that is, $L(x) = 1$ if $x \in L$ and $L(x) = 0$ if $x \notin L$. By $W_{i,s}$ we denote the set $\{x < s : \Phi_i(x) < s\}$.

3   Models of Learning

We now present some concepts from language learning theory. A *text* is a mapping from $\mathbb{N}$ to $\mathbb{N} \cup \{\#\}$. The *content* of a text T, denoted content(T) is the set of natural numbers in the range of T, that is, $\{T(m) : m \in \mathbb{N}\} - \{\#\}$. We say that T is a *text for* L if content$(T) = L$. Intuitively, $\#$ can be considered as pauses in the presentation of data to a learner. We let T range over texts. $T[n]$ denotes the initial segment of a text T with length n, that is, $T(0), T(1), \ldots, T(n-1)$.

Initial segments of texts are called finite *sequences* (or just sequences). We let σ and τ range over finite sequences. We let content(σ) denote the set of natural numbers in the range of σ. Let SEQ denote the set of all finite sequences. We assume some fixed computable ordering of members SEQ (so that we can talk about the least sequence, etc.). We let $|\sigma|$ denote the length of σ. We let Λ denote the empty sequence and $\sigma[n]$ denote the initial segment of σ of length n (where $\sigma[n] = \sigma$, if $n \geq |\sigma|$). Let $\sigma \diamond \tau$ denote the concatenation of two finite sequence σ and τ. Similarly, let $\sigma \diamond T$ denote the concatenation of finite sequence σ and text T. For $x \in \mathbb{N} \cup \{\#\}$, we let $\sigma \diamond x$ denote the concatenation of σ with the finite sequence containing just one element x. Let $\sigma \preceq T$ and $\sigma \preceq \tau$ denote that σ is an initial segment of T and τ respectively.

Definition 1 (Based on Gold [6]).

(a) A *learner* **M** is a (possibly partial) recursive mapping from $(\mathbb{N} \cup \{?\}) \times (\mathbb{N} \cup \{\#\})$ to $(\mathbb{N} \cup \{?\}) \times (\mathbb{N} \cup \{?\})$. A learner has initial memory mem_0 and initial hypothesis hyp_0.

(b) Suppose a learner **M** with initial memory mem_0 and initial hypothesis hyp_0 is given. Let $mem_0^{\mathbf{M},T} = mem_0$ and $hyp_0^{\mathbf{M},T} = hyp_0$. For $k \geq 0$, let $(mem_{k+1}^{\mathbf{M},T}, hyp_{k+1}^{\mathbf{M},T}) = \mathbf{M}(mem_k^{\mathbf{M},T}, T(k))$. Extend the definition of **M** to initial segments of texts by letting $\mathbf{M}(T[k]) = (mem_k^{\mathbf{M},T}, hyp_k^{\mathbf{M},T})$.

Intuitively, $mem_k^{\mathbf{M},T}$ is considered as the memory of the learner after having seen data $T[k]$ and $hyp_k^{\mathbf{M},T}$ is considered as the hypothesis of the learner after having seen data $T[k]$. Without loss of generality we assume that the memory is revised whenever the hypothesis is revised. Sometimes, we just say $\mathbf{M}(T[k]) = hyp_k^{\mathbf{M},T}$, where the memory of the learner is implicit.

(c) We say that **M** *converges* on a text T to a hypothesis hyp if the sequence $hyp_0^{\mathbf{M},T}, hyp_1^{\mathbf{M},T}, hyp_2^{\mathbf{M},T}, \ldots$ converges syntactically to hyp.

(d) We say that **M** **Ex**-learns a language L on text T if **M** is defined on all initial segments of the text T and **M** converges on T to a hypothesis hyp such that $W_{hyp} = L$.

(e) We say that **M** **Ex**-learns a language L (written: $L \in \mathbf{Ex}(\mathbf{M})$), if **M** **Ex**-learns L from each text T for L.

(f) We say that **M** **Ex**-learns a class $\mathcal{L}$ of languages (written: $\mathcal{L} \subseteq \mathbf{Ex}(\mathbf{M})$) iff **M** **Ex**-learns each $L \in \mathcal{L}$.

(g) $\mathbf{Ex} = \{\mathcal{L} : \exists \mathbf{M} [\mathcal{L} \subseteq \mathbf{Ex}(\mathbf{M})]\}$.

We say that **M** changes its mind (hypothesis, conjecture) at $T[n+1]$, if $hyp_n^{\mathbf{M},T} \neq hyp_{n+1}^{\mathbf{M},T}$. We say that **M** changes its memory at $T[n+1]$, if $mem_n^{\mathbf{M},T} \neq mem_{n+1}^{\mathbf{M},T}$ or $hyp_n^{\mathbf{M},T} \neq hyp_{n+1}^{\mathbf{M},T}$; that is, a learner cannot bring down the costs of mind changes by only changing the hypothesis and not doing adequate bookkeeping in the memory. Blum and Blum [1] gave a useful lemma for learnability of languages by learners based on topological considerations.

Definition 2 (Fulk [5]). Suppose $\mathbf{M}(\sigma) = (mem, hyp)$. Sequence σ is said to be a *stabilising sequence* for **M** on a language L if

(i) content$(\sigma) \subseteq L$ and
(ii) for all τ such that $\sigma \subseteq \tau$ and content$(\tau) \subseteq L$, $\mathbf{M}(\tau) = (\cdot, hyp)$. That is, **M** does not change its hypothesis beyond σ on any text T for L.

Definition 3 (Blum and Blum [1]). A sequence σ is said to be a *locking sequence* for **M** on a language L if

(i) σ is a stabilising sequence for **M** on L and
(ii) $W_{hyp} = L$, where $\mathbf{M}(\sigma) = (mem, hyp)$.

Lemma 4 (Blum and Blum [1]). *Suppose* **M** **Ex**-*learns L. Then the following statements hold:*

(i) *There exists a locking sequence for* **M** *on L;*
(ii) *Every stabilising sequence for* **M** *on L is a locking sequence for* **M** *on L.*

A lemma similar to above can be shown for all of the criteria of learning considered in this paper.

For some of the results in this paper it is useful to consider set-driven learning, where the output of the learner depends just on the set of inputs seen.

Definition 5 (Osherson, Stob and Weinstein [9]).

(a) A learner **M** is said to be *set-driven* iff for all σ, τ, if content$(\sigma) =$ content(τ), and $\mathbf{M}(\sigma) = (mem, hyp)$, then $\mathbf{M}(\tau) = (mem', hyp)$, for some memory mem'. That is, the hypothesis of the learner depends just on the content of the input and not on the exact order or the number of repetitions of the elements presented.

(b) $\mathbf{SD} = \{\mathcal{L} : \exists \mathbf{M} [\mathbf{M} \text{ is set-driven and } \mathcal{L} \subseteq \mathbf{Ex}(\mathbf{M})]\}$.

A related notion of rearrangement independence, where the learner is able to base its conjectures on the content and length of the input was considered by [12]. We now consider iterative learning.

Definition 6 (Based on Wexler and Culicover [13] and Wiehagen [14]).

(a) A learner $\mathbf{M}$ is said to be *iterative* if for any text T and any k $mem_k^{\mathbf{M},T} = hyp_k^{\mathbf{M},T}$ (where it is possible that both are undefined).

(b) $\mathbf{M}$ **ItEx**-learns a language L (class of languages $\mathcal{L}$) iff $\mathbf{M}$ is iterative and $\mathbf{M}$ **Ex**-learns L (class of languages $\mathcal{L}$).

(c) $\mathbf{ItEx} = \{\mathcal{L} : \exists \mathbf{M} \, [\mathbf{M} \text{ is iterative and } \mathcal{L} \subseteq \mathbf{Ex}(\mathbf{M})]\}$.

It can be shown that iterative learning is restrictive: $\mathbf{ItEx} \subset \mathbf{Ex}$ (see [14]).

There are several other variations of iterative learning considered in the literature where the memory is constrained in some way. For example, a learner $\mathbf{M}$ has k-bounded example memory [8,10] iff for all texts T and n, $mem_n^{\mathbf{M},T}$ represents a set of size at most k, where $mem_0^{\mathbf{M},T} = \emptyset$ and $mem_{n+1}^{\mathbf{M},T} \subseteq mem_n^{\mathbf{M},T} \cup \{T(n)\}$.

As iterative learning (and most of its variations considered until now) put severe constraints on what can be memorised by a learner, we consider a variation where we allow the memory of grow arbitrarily large. However, there is a cost associated with each change of memory: if the learner changes its memory (or hypothesis) after seeing datum x, then it is charged some cost $f(x)$. Now the learner is said to identify the input language (for any given text) iff besides **Ex**-learning the language, the total cost charged to the learner on the input text is finite. Formally, this is defined as follows.

A *price function* is a recursive function mapping $\mathbb{N} \cup \{\#\}$ to $\mathbb{N}$.

Definition 7. Suppose f is a price function and $\mathcal{C}$ is a class of price functions.

(a) $\mathbf{M}$ $\mathbf{Priced}_f\mathbf{Ex}$-learns a language L (written: $L \in \mathbf{Priced}_f\mathbf{Ex}(\mathbf{M})$) iff $\mathbf{M}$ **Ex**-learns L and for all texts T for L, $\sum_{n:\mathbf{M}(T[n+1]) \neq \mathbf{M}(T[n])} \frac{1}{f(T(n))+1} < \infty$.

(b) $\mathbf{Priced}_f\mathbf{Ex} = \{\mathcal{L} : \exists \mathbf{M} \, [\mathcal{L} \subseteq \mathbf{Priced}_f\mathbf{Ex}(\mathbf{M})]\}$.

(c) $\mathbf{Priced}_{\mathcal{C}}\mathbf{Ex} = \{\mathcal{L} : \forall f \in \mathcal{C} \, [\mathcal{L} \in \mathbf{Priced}_f\mathbf{Ex}]\}$.

Note that in the above definition $\mathbf{M}(T[n])$ is taken as the pair $(mem_n^{\mathbf{M},T}, hyp_n^{\mathbf{M},T})$. For a learner $\mathbf{M}$ and price function g, $\mathbf{cost}_{\mathbf{M},g}(\sigma)$ is defined as follows. Let $S = \{m < |\sigma| : \mathbf{M} \text{ makes a memory change or a mind change at } \sigma[m+1]\}$. Let $\mathbf{cost}_{\mathbf{M},g}(\sigma) = \sum_{m \in S} 1/(g(\sigma(m))+1)$ and $\mathbf{cost}_{\mathbf{M},g}(T) = \sup_{\sigma \preceq T} \mathbf{cost}_{\mathbf{M},g}(\sigma)$. Here $\mathbf{cost}_{\mathbf{M},g}(\sigma)$ ($\mathbf{cost}_{\mathbf{M},g}(T)$) is called the cost of $\mathbf{M}$ on input σ (input T) with respect to the cost function g.

Let $price_f(S) = \sum_{x \in S} \frac{1}{f(x)+1}$. Intuitively, $price_f(S)$ denotes the cost with respect to the price function f if a mind change is made by the learner exactly once with respect to each member of S.

Let $\mathbf{M}_0, \mathbf{M}_1, \ldots$ denote a recursive enumeration of all the learning machines.

Definition 8. Suppose $\mathcal{C} \subseteq \mathcal{R}$. $\mathbf{EffPriced}_{\mathcal{C}}\mathbf{Ex} = \{\mathcal{L} : \exists f \in \mathcal{R} \, \forall i \in \mathbb{N} \, [\text{if } \varphi_i \in \mathcal{C}$ then $\mathbf{M}_{f(i)} \, \mathbf{Priced}_{\varphi_i}\mathbf{Ex}$-learns $\mathcal{L}]\}$.

4 General Cost Functions

The theorems in this section explore when $\mathbf{Priced}_f\mathbf{Ex} \subseteq \mathbf{Priced}_g\mathbf{Ex}$. Theorem 9 shows when $\mathcal{L} \in \mathbf{Priced}_f\mathbf{Ex}$ implies $\mathcal{L} \in \mathbf{Priced}_g\mathbf{Ex}$, and Theorem 11 shows when such an implication does not hold.

Theorem 9. *Suppose f and g are price functions for which no set S satisfies: $price_f(S)$ is finite and $price_g(S)$ is infinite. Then, $\mathbf{Priced}_f\mathbf{Ex} \subseteq \mathbf{Priced}_g\mathbf{Ex}$.*

Proposition 10. *Let S be an infinite r.e. set, g be a price function such that $price_g(S)$ is infinite. Then, $\mathcal{L} = \{S - \{x\} : x \in S - \min(S)\}$ is not $\mathbf{Priced}_g\mathbf{Ex}$-learnable.*

Proof. Suppose by way of contradiction that $\mathbf{M}$ $\mathbf{Priced}_g\mathbf{Ex}$-learns $\mathcal{L}$.

Suppose there are distinct $x, y \in S - \{\min(S)\}$ such that for some σ satisfying $\{x, y\} \subseteq \mathrm{content}(\sigma) \subseteq S$, $\mathbf{M}$ did not change its memory and hypothesis whenever it received x, y in the input σ. Let T be a text for $S - \{x, y\}$. Let σ' be obtained from σ by deleting the occurrences of x and σ'' be obtained from σ by deleting the occurrences of y. Now, $\mathbf{M}$ converges to the same hypothesis on both $\sigma' \diamond T$ and $\sigma'' \diamond T$, however $\sigma' \diamond T$ and $\sigma'' \diamond T$ are respectively texts for two different languages $S - \{x\}$ and $S - \{y\}$ in $\mathcal{L}$. Thus, $\mathbf{M}$ does not $\mathbf{Priced}_g\mathbf{Ex}$-learn at least one of these languages.

On the other hand, if such distinct x, y and corresponding σ as above do not exist, then on any text T for a language $L \in \mathcal{L}$, the cost incurred by $\mathbf{M}$ on T is infinite. Thus, $\mathbf{M}$ does not $\mathbf{Priced}_g\mathbf{Ex}$-learn $\mathcal{L}$. □

Theorem 11. *Suppose f and g are price functions such that there exists a set S satisfying that $price_f(S)$ is finite but $price_g(S)$ is infinite. Then $\mathbf{Priced}_f\mathbf{Ex} \not\subseteq \mathbf{Priced}_g\mathbf{Ex}$.*

Proof. Suppose there is a set S such that $price_f(S)$ is finite and $price_g(S)$ is infinite. Then, for each c, there exists a set S_c such that $price_f(S_c) < 1/(c+1)$ and $price_g(S_c) = \infty$. Such sets can be found by considering tails of the sum of the members in S. Hence one can, by effective search, find disjoint finite sets $\tilde{S}_0, \tilde{S}_1, \ldots$ such that $price_f(\tilde{S}_c) < 1/(c+1)$ and $price_g(\tilde{S}_c) \geq 1$. To see this, define $\tilde{S}_c$ by induction as the first finite set found (in some ordering of finite sets) such that $price_f(\tilde{S}_c) < 1/(w + 1)$ and $price_g(\tilde{S}_c) \geq 1$, where $w > c + \max(\{f(x) : x \in \bigcup_{c' < c} \tilde{S}_{c'}\})$; note that the constraint on w implies that $\tilde{S}_c \cap (\bigcup_{c' < c} \tilde{S}_{c'}) = \emptyset$. Now let $\tilde{S}$ be the union of all sets $\tilde{S}_{2^k}$ for $k \in \mathbb{N}$. The set $\tilde{S}$ is an r.e. set.

One can let $\mathcal{L} = \{\tilde{S} - \{x\} : x \in \tilde{S}\}$. It is easy to see that $\mathcal{L} \in \mathbf{Priced}_f\mathbf{Ex}$, as on any input text T for $L \in \mathcal{L}$, the learner can memorise all data seen, and, in the limit, output a grammar for $\tilde{S} - \{x\}$, where x is the minimal element, if any, in $\tilde{S} - \mathrm{content}(T)$. Note that such a learner makes a memory / mind change on any x at most once.

By Proposition 10, $\mathcal{L} \notin \mathbf{Priced}_g\mathbf{Ex}$. The theorem follows. □

Theorem 9 and Theorem 11 imply that $\mathbf{Priced}_f\mathbf{Ex} = \mathbf{ItEx}$ iff $price_f(S)$ is infinite for every infinite subset S of $\mathbb{N}$. Theorem 11 can be generalised to find classes which have one learner working with all price functions which permit to learn this class. These classes can be chosen such that they are priced learnable for some but not all price functions and hence they do not coincide with iteratively learnable classes.

Theorem 12. *Given any price function f, there is a class $\mathcal{C}_f$ and a learner **M** such that for all price functions g, the following conditions are equivalent:*

- **Priced$_f$Ex** $\subseteq$ **Priced$_g$Ex;**
- **M Priced$_g$Ex**-*learns* $\mathcal{C}_f$;
- *There is no set S such that $price_f(S)$ is finite but $price_g(S)$ is infinite.*

5   Priced Learning and Set-Drivenness

In this section we show that classes which are **Priced$_f$Ex**-learnable with respect to some price function f are learnable by a set-driven learner and vice versa, classes which are set-driven-learnable are **Priced$_g$Ex**-learnable with respect to some price function g. To this end we first give a modification of the definition of stabilising/locking sequences with respect to a cost function.

Definition 13. *A sequence σ is a g-cost-stabilising sequence (g-cost-locking sequence) for **M** on L, if the following conditions hold:*

(a) σ *is a stabilising sequence (locking sequence) for **M** on L and*
(b) *for all τ such that $\sigma \preceq \tau$ and content$(\tau) \subseteq L$, $\mathbf{cost}_{\mathbf{M},g}(\tau) \leq \mathbf{cost}_{\mathbf{M},g}(\sigma) + 1$.*

The following can be proved in a way similar to locking sequence lemma in [1].

Lemma 14 (Based on Blum and Blum [1]). *If **M Priced$_g$Ex**-learns L then there exists a g-cost-locking sequence for **M** on L; Furthermore, all g-cost-stabilising sequences for **M** on L are g-cost-locking sequences for **M** on L.*

Furthermore, note that for any finite set S contained in some set learned by **M**, and any price function g one can check if σ is a g-cost-stabilising sequence for **M** on S. This holds as the number of memory / mind changes of **M** on any text extending σ for S will be at most $\max(\{g(x) + 1 : x \in S \cup \{\#\}\})$ if σ is indeed a g-cost-stabilising sequence for **M** on S. Thus, we can check if the tree of memory/mind changes for **M** on S above σ is bounded as above.

Theorem 15. $\bigcup_{g \in \mathcal{R}}$ **Priced$_g$Ex** $=$ **SD**.

Proof. Suppose $\mathcal{L} \in$ **SD** as witnessed by **M**. Let $g(x) = 1/2^x$. Let $\mathbf{N}(\sigma) =$ (content$(\sigma), hyp)$, where hyp is the hypothesis of **M** on input σ. Note that, as **M** is set-driven, **N** can easily obtain the hypothesis of **M** on input σ, using its memory content(σ), by running **M** on any τ such that content$(\tau) =$ content(σ). It is easy to verify that **N Priced$_g$Ex**-learns $\mathcal{L}$.

Now suppose $\mathcal{L} \in$ **Priced$_g$Ex** as witnessed by **M**. Then, consider the following set-driven learner **N**. The aim of the set-driven learner **N** is to find the least g-cost-stabilising sequence for **M** on the input language. **N** on input σ searches for the least g-cost-stabilising sequence τ for **M** on content(σ), if any, of length at most $|$content$(\sigma)|$. Note that one can check if such a stabilising sequence exists. Now, $\mathbf{N}(\sigma) =$ (content$(\sigma), hyp)$, where hyp is the hypothesis of **M** on input τ if a τ as above exists; otherwise hyp is a canonical grammar for content(σ).

For a finite language $L \in \mathcal{L}$, whether there exists a g-cost-stabilising sequence for **M** on L of length at most card(L) or not, the learner **N** will output a grammar for L after it has received all the elements of L. Thus, **N** would **Ex**-learn L.

For an infinite language $L \in \mathcal{L}$, there exists a g-cost-stabilising sequence for **M** on L of finite length, and thus for large enough initial segment of any text T for L, **N** will find the least g-cost-stabilising sequence τ. Then, **N** will output the hypothesis of **M** on τ as its hypothesis. Thus, **N** would **Ex**-learn L.

As **N** defined above is set-driven, we have that $\mathcal{L} \in$ **SD**.   □

Theorem 9 and Theorem 11 along with **SD** = **Priced$_g$Ex**, for the g used in the proof of Theorem 15, imply that **Priced$_f$Ex** = **SD** iff $price_f(\mathbb{N})$ is finite.

6   Classes of Cost Functions

In this section we will consider priced learning when the price function might be any of the functions in a class of price functions. In particular we look at the classes of monotonic functions (**MF**) and finite-one functions (**FF**) which map only finitely many inputs to the same output:

$$\mathbf{MF} = \{f \in \mathcal{R} : \forall x \, [f(x) \leq f(x+1)] \text{ and } f \text{ is unbounded}\};$$
$$\mathbf{FF} = \{f \in \mathcal{R} : \forall y \, [\text{card}(f^{-1}(y)) \text{ is finite}]\}.$$

The following lemma is useful for proving some results in this paper. It shows that if $\mathcal{L}$ is **Ex**-learnable, then the cylindrification of $\mathcal{L}$ is **Priced$_{FF}$Ex**-learnable.

Lemma 16. *If $\mathcal{L} \in$ **Ex** and $\mathcal{L}' = \{\text{cyl}(L) : L \in \mathcal{L}\}$ then $\mathcal{L}' \in$ **EffPriced$_{FF}$Ex**.*

The next theorem gives a class $\mathcal{L}$ that can be **Priced$_f$Ex**-learnt with respect to every price function f in **FF**, even though $\mathcal{L}$ is not iteratively learnable. Moreover, the **Priced$_f$Ex**-learner can be obtained effectively from an index for f.

Theorem 17. EffPriced$_{FF}$Ex $\not\subseteq$ ItEx.

Proof. The class $\mathcal{L}_1 \cup \mathcal{L}_2$ with

$$\mathcal{L}_1 = \{W_e : e = \min(W_e) \text{ and } \forall x \, [\{2x, 2x+1\} \not\subseteq W_e]\} \text{ and}$$
$$\mathcal{L}_2 = \{L : \text{card}(L) < \infty \text{ and } \exists x \, [\{2x, 2x+1\} \subseteq L]\}$$

is **Ex**-learnable: The learner, on input σ, checks whether there is an x with $2x, 2x+1 \in \text{content}(\sigma)$. If so then the learner conjectures a canonical index for content(σ) else the learner conjectures $\min(\text{content}(\sigma))$. So, on a text T with both $2x, 2x+1 \in \text{content}(T)$, the learner converges to a canonical index of content(T) whenever the latter is finite and thus the learner **Ex**-learns $\mathcal{L}_2$; on a text T which contains at most one of $2x, 2x+1$ for any x, the learner converges to the minimum of content(T) and thus the learner **Ex**-learns $\mathcal{L}_1$. Let

$$\mathcal{L} = \{\text{cyl}(L) : L \in \mathcal{L}_1 \cup \mathcal{L}_2\}.$$

By Lemma 16, $\mathcal{L} \in \textbf{EffPriced}_{\textbf{FF}}\textbf{Ex}$. Now it remains to show that $\mathcal{L} \notin \textbf{ItEx}$. Suppose by way of contradiction that $\textbf{M}$ $\textbf{ItEx}$-learns $\mathcal{L}$. Then by Kleene's recursion theorem there exists an e such that W_e is the set of all x for which $\langle x, 0 \rangle$ occurs in some σ_s which are defined below. Here, initially, $\sigma_0 = \langle e, 0 \rangle$ and in stage s below, σ_{s+1} is constructed.

Stage s.
1.   Search for an extension τ of σ_s satisfying the following conditions:
      (i) $\textbf{M}(\sigma_s) \neq \textbf{M}(\tau)$;
      (ii) for all x and i, if $\langle 2x, i \rangle \in \text{content}(\tau)$, then for all j, $\langle 2x + 1, j \rangle \notin \text{content}(\tau)$;
      (iii) for all i, for all $x < e$, $\langle x, i \rangle \notin \text{content}(\tau)$.
2.   If and when such a τ is found, let σ_{s+1} be an extension of τ satisfying the following conditions:
      (i) $\forall x, i \, [[\langle x, i \rangle \in \text{content}(\tau)] \Rightarrow [\forall j \leq s \, [\langle x, j \rangle \in \text{content}(\sigma_{s+1})]]]$;
      (ii) $\forall x, i \, [[\langle x, i \rangle \in \text{content}(\sigma_{s+1})] \Rightarrow [\exists j \, [\langle x, j \rangle \in \text{content}(\tau)]]]$.
3.   Go to stage $s + 1$.
End Stage s.

Now if there are infinitely many stages in the above construction, then $T = \bigcup_{s \in \mathbb{N}} \sigma_s$ is a text for $\text{cyl}(W_e)$, where $W_e \in \mathcal{L}_1$. However, $\textbf{M}$ makes infinitely many mind changes on T. On the other hand if stage s starts but does not finish, then, for $y > \max(\{x : \langle x, 0 \rangle \in \text{content}(\sigma_s)\})$, let $L_y = \{x : \langle x, 0 \rangle \in \text{content}(\sigma_s)\} \cup \{2y, 2y + 1\}$. Then for all texts T extending σ_s for any $\text{cyl}(L_y)$ with $y > \max(\{x : \langle x, 0 \rangle \in \text{content}(\sigma_s)\})$, $\textbf{M}(T)$ converges to $\textbf{M}(\sigma_s)$, and thus it fails to $\textbf{ItEx}$-learn $\mathcal{L}$, as all such L_y belong to $\mathcal{L}_2$. $\qquad\square$

The next theorem shows that some class $\mathcal{L}$ can be $\textbf{Priced}_f\textbf{Ex}$-learnt with respect to every price function in $\textbf{MF}$, but not with respect to some price function in $\textbf{FF}$. Furthermore, the $\textbf{Priced}_f\textbf{Ex}$ learner for $f \in \textbf{MF}$ can be obtained effectively from an index for f.

Theorem 18. $\textbf{EffPriced}_{\textbf{MF}}\textbf{Ex} \not\subseteq \textbf{Priced}_{\textbf{FF}}\textbf{Ex}$.

Proof. Let $F_0, F_1, \ldots$ be an enumeration of disjoint finite sets such that $F_0 = \{0\}$, $\bigcup_x F_x = \mathbb{N}$ and for all i, if φ_i is total, then for all $x > i$, there exists a $y > \varphi_i(x)$ in F_x. Note that such an enumeration can be easily constructed.
    Let $L_0 = \bigcup_{x>0} F_x$. For $e \in \mathbb{N}$, let $L_{e+1} = \bigcup_{x \in D_e \cup \{0\}} F_x$. Let $\mathcal{L} = \{L_e : e \in \mathbb{N}\}$.

Claim. $\mathcal{L} \in \textbf{EffPriced}_{\textbf{MF}}\textbf{Ex}$.

Let h be a recursive function such that for all i, $\varphi_{h(i)}(x)$ is the first z found in a search such that $\varphi_i(z) > 2^x$. Now, if $\varphi_i \in \textbf{MF}$, then for all $x > h(i)$, there exists a $y \in F_x$ such that $y > \varphi_{h(i)}(x)$ and thus $\varphi_i(y) > 2^x$.
    Given an index i for a cost function $g \in \textbf{MF}$, consider the following learner $\textbf{M}$ obtained effectively from i. Learner $\textbf{M}$ will have a finite set as its memory. Initially, $\textbf{M}$'s memory is $\emptyset$. Whenever $\textbf{M}$ receives $y \in F_x$ as input, it adds x

to the memory if x is not already in the memory and $[x \leq h(i)$ or $g(y) > 2^x]$; otherwise the memory is unchanged. If the memory does not contain 0, then $\mathbf{M}$ outputs a canonical grammar for L_0. Otherwise it outputs a canonical grammar for L_{e+1}, where $D_e = S - \{0\}$, where S is the set in $\mathbf{M}$'s memory. It is easy to verify that for any text T, $\mathbf{cost}_{\mathrm{M},g}(T)$ is finite and $\mathbf{M}$ $\mathbf{Priced}_g\mathbf{Ex}$-learns $\mathcal{L}$.

Claim. $\mathcal{L} \notin \mathbf{Priced}_{\mathbf{FF}}\mathbf{Ex}$.

To see this, let g be such that $g(y) = x$, for $y \in F_x$. Suppose by way of contradiction that $\mathbf{M}$ $\mathbf{Priced}_g\mathbf{Ex}$-learns $\mathcal{L}$. Now, consider $\mathbf{M}$'s behaviour on text for L_0, where the elements are presented to it in order of elements from $F_1, F_2, \ldots$. If for every x, $\mathbf{M}$ makes mind change on some input from F_x, then clearly the cost is infinite. Otherwise, for some x, $\mathbf{M}$ did not make any mind change on the above text when presented with elements from F_x. Now, if after elements of F_x, $\mathbf{M}$ only gets elements from F_0, then $\mathbf{M}$ cannot distinguish between the input being L_e or $L_{e'}$, where $D_e = \{1, 2, \ldots, x\}$ and $D_{e'} = \{1, 2, \ldots, x-1\}$.

The above two claims prove the theorem.   □

The following theorem shows that some class of languages $\mathcal{L}$ can be learnt with respect to every price function in $\mathbf{FF}$, but one cannot effectively obtain a learner for even monotonic price functions.

Theorem 19. $\mathbf{Priced}_{\mathbf{FF}}\mathbf{Ex} \nsubseteq \mathbf{EffPriced}_{\mathbf{MF}}\mathbf{Ex}$.

Proof. We will define the class $\mathcal{L}$ in dependence of a set $A \subseteq \{\langle x, r \rangle : x \geq 1, r \in \mathbb{N}\}$ to be constructed below by letting

$$L_m = \{0\} \cup (A \cap \{\langle x, r \rangle : x \leq m, r \in \mathbb{N}\});$$
$$\mathcal{L} = \{A\} \cup \{L_m : m \in \mathbb{N}\}.$$

We will have the property that for all $g \in \mathbf{FF}$, for all but finitely many x, there exists an r such that $\langle x, r \rangle \in A$ and $g(\langle x, r \rangle) > 2^x$.

The above property allows for easy learning of $\mathcal{L}$ in the model $\mathbf{Priced}_{\mathbf{FF}}\mathbf{Ex}$. To see this, for $g \in \mathbf{FF}$, let R_g be the set of finitely many x such that for all r, $g(\langle x, r \rangle) \leq 2^x$. Then, the $\mathbf{Priced}_g\mathbf{Ex}$-learner $\mathbf{M}$ is defined as follows. The memory of $\mathbf{M}$ is a finite set. The memory mem_σ of $\mathbf{M}$ after having seen input σ consists of the following elements: (i) 0 if $0 \in \mathrm{content}(\sigma)$, (ii) all $x \in R_g$ such that for some $r \in \mathbb{N}$, $\langle x, r \rangle \in \mathrm{content}(\sigma)$, and (iii) all $x \geq 1$ such that for some $r \in \mathbb{N}$, $\langle x, r \rangle \in \mathrm{content}(\sigma)$ and $g(\langle x, r \rangle) > 2^x$. Furthermore, the hypothesis of the learner $\mathbf{M}$ after having seen input σ is: (a) a canonical index for A if $0 \notin mem_\sigma$, and (b) a canonical index for L_m, if $0 \in mem_\sigma$ and $m = \max(mem_\sigma)$. It is easy to verify that $\mathbf{M}$ is a $\mathbf{Priced}_g\mathbf{Ex}$-learner for $\mathcal{L}$.

We will now construct A. For this we will define functions $h_{i,j}$ along with sets $B_{i,j}$, for $j < 2^{i+1}$. The functions $h_{i,j}$, if total, will be in $\mathbf{MF}$. $B_{i,j}$ will be finite, but may change over time. Let $B_{i,j}^t$ denote its value at time t, where we will only start with $B_{i,j}$ at time $t \geq i$ (so it is assumed to be empty before that).

Intuitively, aim of $B_{i,j}$ and $h_{i,j}$, $j < 2^{i+1}$ is to make sure that φ_i does not witness that $\mathcal{L} \in \mathbf{EffPriced_{MF}Ex}$. For $j < 2^{i+1}$, $X_{i,j}$ denotes the j-th subset of $\{0, 1, 2, \ldots, i\}$.

In the definitions below, we assume some steps to be atomic so that there is no interference between different parts. For ease of notation we will write $\varphi_i(h_{i,j})$ below rather than $\varphi_i(\text{index for } h_{i,j})$.

Definition of A.

    Stage s

        Each stage s is supposed to be atomic.

    1.   Enumerate $\langle s, 0 \rangle$ into A.

    2.   For each $i < s$,

        2.1. Let $x < s$ be least, if any, such that

$$x \notin \bigcup_{j < i, k < 2^{j+1}} B_{j,k}^s \text{ and}$$

                  for all r such that $\langle x, r \rangle \in A$ enumerated up to now

$$[\varphi_{i,s}(\langle x, r \rangle)\downarrow \leq 2^x \text{ or } \varphi_{i,s}(\langle x, r \rangle)\uparrow].$$

        2.2. If there is such an x as above and $\exists r \leq s\,[\varphi_{i,s}(\langle x, r \rangle)\downarrow > 2^x]$

             Then enumerate one such $\langle x, r \rangle$ into A.

    End For

    Go to stage $s + 1$

    End Stage s

Note that the construction of different $h_{i,j}$ is run in parallel (in dovetailing way, along with procedure for A above) for all $i \in \mathbb{N}$, $j < 2^{i+1}$.

Definition of $h_{i,j}$.

    Let $h_{i,j}(0) = 1$.

    Initially $\sigma_{i,j} = \Lambda$.

    Loop

    0.   Let $\sigma_{i,j}$ be extension of previous $\sigma_{i,j}$ so that it contains all of A enumerated up to now.

        (Above step is assumed to be atomic.)

    1.   Suppose $h_{i,j}$ has already been defined on inputs $\leq w$ up to now, and the maximum value in the range of $h_{i,j}$ up to now is z.

    2.   Pick $B_{i,j}$ of size $z + 2$ containing elements $> i + w$, such that none of the elements of $B_{i,j}$ have been used in any $B_{i',j'}$ at any earlier point in the construction up to now and all the elements of $B_{i,j}$ are larger than any element enumerated by A up to now.

        (We assume the above step to be atomic.)

    3.   Wait until for each $y \in B_{i,j}$,

        (i) $\langle y, 0 \rangle$ is enumerated into A and

        (ii) for each $i' \in X_{i,j}$, a pair $\langle y, r \rangle$ is enumerated into A with

$$\varphi_{i'}(\langle y, r \rangle) > 2^y.$$

    4.   Extend $h_{i,j}$ monotonically (non-decreasing) so that $h_{i,j}(\langle y, r \rangle) = z + 1$, for each $\langle y, r \rangle \in A$ enumerated up to now such that $y \in B_{i,j}$.

        (We assume above step to be atomic.)

5. Keep extending $h_{i,j}$ strictly monotonically increasing, while searching for an extension τ of $\sigma_{i,j}$ such that
    (i) content(τ) is contained in A, and
    (ii) for each $y \in B_{i,j}$, there exists a $\langle y, r \rangle$ in content(τ) such that $\mathbf{M}_{\varphi_i(h_{i,j})}$ made a mind change / memory change when it received $\langle y, r \rangle$ as input.
6. If such an extension τ is found, update $\sigma_{i,j}$ to τ and go to the next iteration of the loop. If no extension as above is found, then this iteration of the loop never ends.
End Loop

For each j and $k < 2^{j+1}$, let $B_{i,j}$ denote the eventual value of $B_{i,j}$ (if this value does not exist, then let $B_{i,j}$ be $\emptyset$).

Now consider any i such that $\varphi_i \in \mathbf{FF}$. Let $C_i = \bigcup_{i' < i, j' < 2^{i'+1}} B_{i',j'}$. Note that for $x \notin C_i$, eventually some element of the form $\langle x, r \rangle$, with $\varphi_i(\langle x, r \rangle) > 2^x$ will be enumerated into A. Thus, we have that $\mathcal{L} \in \mathbf{Priced_{FF}Ex}$.

Now we show that $\mathcal{L} \notin \mathbf{EffPriced_{MF}Ex}$. Suppose by way of contradiction that φ_i witnesses $\mathcal{L} \in \mathbf{EffPriced_{MF}Ex}$, that is, for all k, if $\varphi_k \in \mathbf{MF}$, then $\mathbf{M}_{\varphi_i(k)}$ is a $\mathbf{Priced}_{\varphi_k}\mathbf{Ex}$-learner for $\mathcal{L}$. Let j be such that $X_{i,j}$ is the set of $i' \leq i$ satisfying for all $y \notin \bigcup_{i'' < i, j'' < 2^{i''+1}} B_{i'',j''}$, some $\langle y, r \rangle$ with $\varphi_{i'}(\langle y, r \rangle) > 2^y$ is enumerated into A. Now consider the construction of $h_{i,j}$. Now consider the following cases.

Case 1: Some iteration of the Loop for $h_{i,j}$ does not terminate.

In this case we have that $\mathbf{M}_{\varphi_i(h_{i,j})}$ fails to $\mathbf{Priced}_{h_{i,j}}\mathbf{Ex}$ learn $\mathcal{L}$. To see this let $\sigma_{i,j}$ be as in the iteration of the loop (for $h_{i,j}$) which starts but does not terminate. Consider τ which extends $\sigma_{i,j}$, and contains elements exactly from content($\sigma_{i,j}$) $\cup$ $(A \cap \{\langle y, r \rangle : y \in B_{i,j}\})$, where these $\langle y, r \rangle$ appear in τ in increasing order of y. Then by step 5 not succeeding in the loop, there exists a $y \in B_{i,j}$, for which $\mathbf{M}_{\varphi_i(h_{i,j})}$ did not make a mind change / memory change when receiving $\langle y, r \rangle$ as input, for all $\langle y, r \rangle \in A$. Thus, we can fool the learner $\mathbf{M}_{\varphi_i(h_{i,j})}$ by giving 0 just after giving the elements $\langle y, r \rangle$ corresponding to the y above, and then extending it using all elements from L_{y-1}. Then, the learner $\mathbf{M}_{\varphi_i(h_{i,j})}$ is not able to distinguish between input being L_{y-1} or L_y.

Case 2: All iterations of the loop for $h_{i,j}$ terminate.

In this case consider the loop executions in which for each $y \in B_{i,j}$ (as defined in step 2 of that iteration of loop) none of the $i' \leq i$, $i' \notin X_{i,j}$ have a $\langle y, r \rangle \in A$ with $\varphi_{i'}(\langle y, r \rangle) > 2^y$. Note that all but finitely many loop executions have this property. In each of such executions, the construction would have found a mind change / memory change which costs in total at least 1 to the learner (as in each iteration, size of $B_{i,j}$ is $z + 2$, and the cost of memory / mind change on $\langle y, r \rangle$, $y \in B_{i,j}$ is $\frac{1}{z+2}$).

Thus, $\mathbf{M}_{\varphi_i(h_{i,j})}$ does not $\mathbf{Priced_{MF}Ex}$-learn $\mathcal{L}$. $\square$

7   Other Uniformity Criteria

In this section we consider some uniformity criteria for priced learning.

Definition 20. Suppose $\mathcal{C} \subseteq \mathcal{R}$.

(a) $\mathbf{UniPriced}_{\mathcal{C}}\mathbf{Ex} = \{\mathcal{L} : \exists \mathbf{M}\, \forall f \in \mathcal{C}\, [\mathbf{M}\, \mathbf{Priced}_f\mathbf{Ex}\text{-learns } \mathcal{L}]\}$.

(b) $\mathbf{EffBPriced}_{\mathcal{C}}\mathbf{Ex} = \{\mathcal{L} : \exists g \in \mathcal{R}\, \forall i\, [\text{if } \varphi_i \in \mathcal{C} \text{ then } \forall b \geq i\, [\mathbf{M}_{g(b)}$ $\mathbf{Priced}_{\varphi_i}\mathbf{Ex}\text{-learns } \mathcal{L}]]\}$.

(c) $\mathbf{UniFPriced}_{\mathcal{C}}\mathbf{Ex} = \{\mathcal{L} : \exists \mathbf{M}\, [\forall f \in \mathcal{C}\, [\mathrm{card}(\mathcal{L} - \mathbf{Priced}_f\mathbf{Ex}(\mathbf{M})) \text{ is finite }]$ and $\mathbf{M}\, \mathbf{Ex}\text{-learns } \mathcal{L}\,]\}$.

Intuitively, for $\mathbf{UniPriced}_{\mathcal{C}}\mathbf{Ex}$, the same learner $\mathbf{M}$ succeeds for all price functions in $\mathcal{C}$. For $\mathbf{EffBPriced}_{\mathcal{C}}\mathbf{Ex}$-learning the learner can be obtained effectively from a bound on the index for the price function.

For $\mathbf{UniFPriced}_{\mathcal{C}}\mathbf{Ex}$-learning, the same learner $\mathbf{M}\, \mathbf{Ex}$-learns the class $\mathcal{L}$ and for all price functions f in $\mathcal{C}$, $\mathbf{M}\, \mathbf{Priced}_f\mathbf{Ex}$-learns almost all the members of the class $\mathcal{L}$. Thus, the cost of learning can go infinite only for finitely many languages in the class for each price function $f \in \mathcal{C}$.

In addition we consider the case where the learner gets the cost function f as input rather than via index for it. In $\mathbf{ValPriced}_{\mathcal{C}}\mathbf{Ex}$ learning of $\mathcal{L}$, for all $f \in \mathcal{C}$, the same learner $\mathbf{M}\, \mathbf{Priced}_f\mathbf{Ex}$-learn each member of $\mathcal{L}$ when, in each iteration, instead of a single datum x the pair $(x, f(x))$ is presented to the learner in the iteration, so that the learner knows how expensive it is to process a datum.

Theorem 21. (a) *Let $\mathcal{C}$ be a class of price functions.*
   $\mathbf{ValPriced}_{\mathcal{C}}\mathbf{Ex} \subseteq \mathbf{EffPriced}_{\mathcal{C}}\mathbf{Ex} \subseteq \mathbf{Priced}_{\mathcal{C}}\mathbf{Ex}$.
(b) $\mathbf{EffPriced}_{\mathbf{FF}}\mathbf{Ex} \not\subseteq \mathbf{ValPriced}_{\mathbf{MF}}\mathbf{Ex}$.

Theorem 22. $\mathbf{ValPriced}_{\mathbf{FF}}\mathbf{Ex} \not\subseteq \mathbf{EffBPriced}_{\mathbf{MF}}\mathbf{Ex}$.

Theorem 23. $\mathbf{UniFPriced}_{\mathbf{FF}}\mathbf{Ex} \not\subseteq \mathbf{Priced}_{\mathbf{MF}}\mathbf{Ex}$.

Corollary 24. $\mathbf{UniFPriced}_{\mathbf{FF}}\mathbf{Ex} \not\subseteq \mathbf{EffPriced}_{\mathbf{MF}}\mathbf{Ex} \cup \mathbf{ValPriced}_{\mathbf{MF}}\mathbf{Ex}$.

Theorem 25. $\mathbf{UniFPriced}_{\mathbf{MF}}\mathbf{Ex} \not\subseteq \mathbf{Priced}_{\mathbf{FF}}\mathbf{Ex} \cup \mathbf{UniFPriced}_{\mathbf{FF}}\mathbf{Ex}$.

Theorem 26. $\mathbf{ValPriced}_{\mathbf{FF}}\mathbf{Ex} \not\subseteq \mathbf{UniFPriced}_{\mathbf{MF}}\mathbf{Ex}$.

It is open whether $\mathbf{UniPriced}_{\mathbf{MF}}\mathbf{Ex}$ or $\mathbf{UniPriced}_{\mathbf{FF}}\mathbf{Ex}$ contain a class which cannot be $\mathbf{ItEx}$-learnt. Similarly, it is open whether $\mathbf{EffBPriced}_{\mathbf{MF}}\mathbf{Ex}$ or $\mathbf{EffBPriced}_{\mathbf{FF}}\mathbf{Ex}$ contain a class which cannot be $\mathbf{ItEx}$-learnt.

8   Conclusions

In this paper we considered a generalization of memory limited learning, called priced learning, where the learner can update its memory based on any datum it receives, but it has a cost $\frac{1}{f(x)+1}$ associated with it, where f is the price function.

The learning is said to be successful if the overall cost in the learning process is finite. We gave a characterization that the classes learnable as above with respect to some cost function are exactly the classes learnable by a set-driven learner. We also gave complete picture of when different price functions lead to different learnable classes. In addition we considered when priced learning is possible for all price functions from a class of functions, and when a learner can be effectively found from an index for a price function. It is open at present whether there exists a single learner which learns a non iteratively learnable class from all price functions which are monotonically non-decreasing and unbounded. Similar question is also open for the price functions having each number in the range only finitely often. These questions are open even for the case when the learner is given a bound on an index for the price function.

Acknowledgments. We thank the referees for several helpful comments.

References

1. Blum, L., Blum, M.: Toward a mathematical theory of inductive inference. Information and Control **28**, 125–155 (1975)
2. Blum, M.: A machine independent theory of the complexity of recursive functions. Journal of the Association of Computing Machinery **14**, 322–336 (1967)
3. Case, J., Jain, S., Lange, S., Zeugmann, T.: Incremental concept learning for bounded data mining. Information and Computation **152**, 74–110 (1999)
4. Freivalds, R., Kinber, E., Smith, C.H.: On the impact of forgetting on learning machines. Journal of the ACM **42**, 1146–1168 (1995)
5. Fulk, M.: Prudence and other conditions on formal language learning. Information and Computation **85**, 1–11 (1990)
6. Mark, E.: Gold. Language identification in the limit. Information and Control **10**, 447–474 (1967)
7. Kinber, E., Stephan, F.: Language learning from texts: mind changes, limited memory and monotonicity. Information and Computation **123**, 224–241 (1995)
8. Lange, S., Zeugmann, T.: Incremental learning from positive data. Journal of Computer and System Sciences **53**, 88–103 (1996)
9. Osherson, D., Stob, M., Weinstein, S.: Learning strategies. Information and Control **53**, 32–51 (1982)
10. Osherson, D., Stob, M., Weinstein, S.: Systems That Learn, An Introduction to Learning Theory for Cognitive and Computer Scientists. Bradford - The MIT Press, Cambridge (1986)
11. Hartley Rogers. Theory of Recursive Functions and Effective Computability. McGraw-Hill, 1967. Reprinted by MIT Press in 1987
12. Gisela Schäfer-Richter. Uber Eingabeabhängigkeit und Komplexität von Inferenzstrategien. Ph.D. Thesis, RWTH Aachen (1984)
13. Wexler, K., Culicover, P.W.: Formal Principles of Language Acquisition. The MIT Press, Cambridge (1980)
14. Wiehagen, R.: Limes-Erkennung rekursiver Funktionen durch spezielle Strategien. Elektronische Informationsverarbeitung und Kybernetik (EIK) **12**, 93–99 (1976)

Combining Models of Approximation with Partial Learning

Ziyuan Gao[1]([⊠]) , Frank Stephan[2], and Sandra Zilles[3]

[1] Department of Computer Science, University of Regina,
Regina, SK S4S 0A2, Canada
gao257@cs.uregina.ca
[2] Department of Mathematics and Department of Computer Science,
National University of Singapore, Singapore 119076, Singapore
fstephan@comp.nus.edu.sg
[3] Department of Computer Science, University of Regina,
Regina, SK S4S 0A2, Canada
zilles@cs.uregina.ca

Abstract. In Gold's framework of inductive inference, the model of partial learning requires the learner to output exactly one correct index for the target object and only the target object infinitely often. Since infinitely many of the learner's hypotheses may be incorrect, it is not obvious whether a partial learner can be modified to "approximate" the target object.

Fulk and Jain (Approximate inference and scientific method. *Information and Computation* 114(2):179–191, 1994) introduced a model of approximate learning of recursive functions. The present work extends their research and solves an open problem of Fulk and Jain by showing that there is a learner which approximates and partially identifies every recursive function by outputting a sequence of hypotheses which, in addition, are also almost all finite variants of the target function.

The subsequent study is dedicated to the question how these findings generalise to the learning of r.e. languages from positive data. Here three variants of approximate learning will be introduced and investigated with respect to the question whether they can be combined with partial learning. Following the line of Fulk and Jain's research, further investigations provide conditions under which partial language learners can eventually output only finite variants of the target language.

1 Introduction

Gold [8] considered a learning scenario where the learner is fed with piecewise increasing amounts of finite data about a given target language L; at every stage where a new input datum is given, the learner makes a conjecture about L.

F. Stephan was partially supported by NUS grants R146-000-181-112 and R146-000-184-112; S. Zilles was partially supported by the Natural Sciences and Engineering Research Council of Canada (NSERC).

© Springer International Publishing Switzerland 2015
K. Chaudhuri et al. (Eds.): ALT 2015, LNAI 9355, pp. 56–70, 2015.
DOI: 10.1007/978-3-319-24486-0_4

If there is exactly one correct representation of L that the learner always outputs after some finite time (assuming that it never stops receiving data about L), then the learner is said to have "identified L in the limit." In this paper, it is assumed that all target languages are encoded as recursively enumerable (r.e.) sets of natural numbers, and that the learner uses Gödel numbers as its hypotheses.

Gold's learning paradigm has been used as a basis for a variety of theoretical models in subjects such as human language acquisition [12] and the theory of scientific inquiry in the philosophy of science [4,11]. This paper is mainly concerned with the *partial learning* model [13], which retains several features of Gold's original framework – the modelling of learners as recursive functions, the use of texts as the mode of data presentation and the restriction of target classes to the family of all r.e. sets – while liberalising the learning criterion by only requiring the learner to output exactly one hypothesis of the target set infinitely often while it must output any other hypothesis only finitely often. It is known that partial learning is so powerful that the class of all r.e. languages can be partially learnt [13].

However, the model of partial learning puts no further constraints on those hypotheses that are output only finitely often. In particular, it offers no notion of "eventually being correct" or even "approximating" the target object. From a philosophical point of view, if partial learning is to be taken seriously as a model of language acquisition, then it is quite plausible that learners are capable of gradually improving the quality of their hypotheses over time. For instance, if the learner M sees a sentence S in the text at some point, then it is conceivable that after some finite time, M will only conjecture grammars that generate S. This leads one to consider a notion of the learner "approximating" the target language.

The central question in this paper is whether any partial learner can be redefined in a way that it approximates the target object and still partially learns it. The first results, in the context of partial learning, deal with Fulk and Jain's [5] notion of approximating recursive functions. Fulk and Jain proved the existence of a learner that "approximates" every recursive function. This result is generalised as follows: partial learners can always be made to approximate recursive functions according to their model and, in addition, eventually output only finite variants of the target function, that is, they can be designed as BC^* learners[1]. This result solves an open question posed by Fulk and Jain, namely whether recursive functions can be approximated by BC^* learners. Note that BC^* learning can also, in some sense, be considered a form of approximation, as it requires that eventually all of the hypotheses (including those output only finitely often) differ from the target object in only finitely many values. It thus is interesting to see that partial learning can be combined not only with Fulk and Jain's model of approximation, but also with BC^* learning *at the same time*. Note that in this paper, when two learning criteria A and B are said to be combinable, it is generally not assumed that the new learner is effectively constructed from the A-learner and the B-learner.

---

[1] BC^* is mnemonic for "behaviourally correct with finitely many anomalies" [4].

This raises the question whether partial learners can also be turned into approximate learners in the more general case of learning r.e. languages. Unfortunately, Fulk and Jain's model applies only to learning recursive functions. The second contribution is the design of three notions of approximate learning of r.e. languages, two of which are directly inspired by Fulk and Jain's model. It is then investigated under which conditions partial learners can be modified to fulfill the corresponding constraints of approximate learning. These investigations are also extended to partial learners with additional constraints, such as consistency and conservativeness. It will be shown that while partial learners can always be constructed in a way so that for any given finite set D, their hypotheses will almost always agree with the target language on D, the same does not hold if D must be a finite variant of a fixed infinite set. Thus trade-offs between certain approximate learning constraints and partial learning are sometimes unavoidable – an observation that perhaps has a broader implication in the philosophy of language learning.

Following the line of Fulk and Jain's research, conditions are investigated under which partial language learners can eventually output only finite variants of the target function. While it remains open whether or not partial learners for a given BC^*-learnable class can be made BC^*-learners for this class without losing identification power, some natural conditions on a BC^* learner M are provided under which all classes learnable by M can be learnt by some BC^* learner N that is at the same time a partial learner.

Figure 1 summarises the main results of this paper. *RECPart* and *RECApproxBC*Part* refer respectively to partial learning of recursive functions and approximate BC^* partial learning of recursive functions. The remaining learning criteria are abbreviated (see Definitions 1, 2 and 6), and denote learning of classes of r.e. languages. An arrow from criterion A to criterion B means that the collection of classes learnable under model A is contained in that learnable under model B. Each arrow is labelled with the Corollary/Example/Remark/Theorem number(s) that proves (prove) the relationship represented by the arrow. If there is no path from A to B, then the collection of classes learnable under model A is not contained in that learnable under model B.

2   Preliminaries

The notation and terminology from recursion theory adopted in this paper follows in general the book of Rogers [14]. Background on inductive inference can be found in [9]. The symbol $\mathbb{N}$ denotes the set of natural numbers, $\{0, 1, 2, \ldots\}$. Let $\varphi_0, \varphi_1, \varphi_2, \ldots$ denote a fixed acceptable numbering [14] of all partial-recursive functions over $\mathbb{N}$. Given a set S, S^* denotes the set of all finite sequences in S. Wherever no confusion may arise, S will also denote its own characteristic function, that is, for all $x \in \mathbb{N}$, $S(x) = 1$ if $x \in S$ and $S(x) = 0$ otherwise. One defines the e-th r.e. set W_e as $dom(\varphi_e)$ and the e-th canonical finite set by choosing D_e such that $\sum_{x \in D_e} 2^x = e$. This paper fixes a one-one padding function pad with $W_{pad(e,d)} = W_e$ for all e, d. Furthermore, $\langle x, y \rangle$ denotes Cantor's

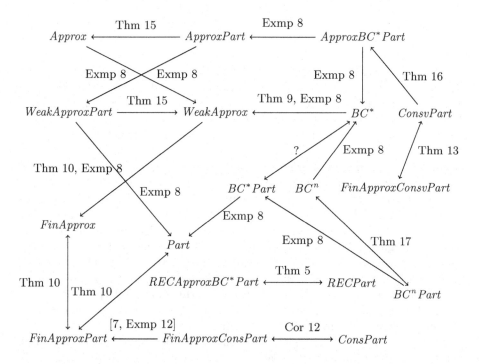

Fig. 1. Learning hierarchy

pairing function, given by $\langle x, y \rangle = \frac{1}{2}(x + y)(x + y + 1) + y$. A triple $\langle x, y, z \rangle$ denotes $\langle\langle x, y \rangle, z \rangle$.

For any $\sigma, \tau \in (\mathbb{N} \cup \{\#\})^*, \sigma \preceq \tau$ if and only if σ is a prefix of τ, $\sigma \prec \tau$ if and only if σ is a proper prefix of τ, and $\sigma(n)$ denotes the element in the nth position of σ, starting from $n = 0$. The concatenation of two strings σ and τ shall be denoted by $\sigma \circ \tau$; for convenience, and whenever there is no possibility of confusion, this is occasionally denoted by $\sigma\tau$. Let $\sigma[n]$ denote the sequence $\sigma(0) \circ \sigma(1) \circ \ldots \circ \sigma(n - 1)$. The length of σ is denoted by $|\sigma|$.

3   Learning

The basic learning paradigms studied in the present paper are *behaviourally correct learning* [2,3] and *partial learning* [13]. These learning models assume that the learner is presented with just positive examples of the target language, and that the learner is fed with a finite amount of data at every stage. They are modifications of the model of explanatory learning (or "learning in the limit"), first introduced by Gold [8], in which the learner must output in the limit a single correct representation h of the target language L; if L is an r.e. set, then h is usually an r.e. index of L with respect to the standard numbering $W_0, W_1, W_2, \ldots$ of all r.e. sets. Bārzdiņš [2] and Case [3] considered the more powerful model of behaviourally correct learning, whereby the learner must almost always output a correct hypothesis of the input set, but some of the correct hypotheses may be syntactically distinct. Case and Smith [4] also introduced a less stringent variant of BC learning of recursive functions, BC^* learning, which only requires the learner to output in the limit finite variants of the target recursive function. Still more general is the criterion of partial learning that Osherson, Stob and Weinstein [13] defined; in this model, the learner must output exactly one correct index of the input set infinitely often and output any other conjecture only finitely often.

One can also impose constraints on the quality of a learner's hypotheses. For example, Angluin [1] introduced the notion of *consistency*, which is the requirement that the learner's hypotheses must enumerate at least all the data seen up to the current stage. This seems to be a fairly natural demand on the learner, for it only requires that the learner's conjectures never contradict the available data on the target language. Angluin [1] also introduced the learning constraint of *conservativeness*; intuitively, a conservative learner never makes a mind change unless its prior conjecture does not enumerate all the current data. These two learning criteria have since been adapted to the partial learning model [6, 7].

The learning criteria discussed so far (and, where applicable, their partial learning analogues) are formally introduced below.

Let $\mathcal{C}$ be a class of r.e. sets. Throughout this paper, the mode of data presentation is that of a *text*, by which is meant an infinite sequence of natural numbers and the # symbol. Formally, a *text* T_L for some L in $\mathcal{C}$ is a map $T_L : \mathbb{N} \to \mathbb{N} \cup \{\#\}$ such that $L = range(T_L)$; here, $T_L[n]$ denotes the sequence $T_L(0) \circ T_L(1) \circ \ldots \circ T_L(n-1)$ and the range of a text T, denoted $range(T)$, is the set of numbers occurring in T. Analogously, for a finite sequence σ, $range(\sigma)$ is the set of numbers occurring in σ. A text, in other words, is a presentation of positive data from the target set. A *learner*, denoted by M in the following definitions, is a recursive function mapping $(\mathbb{N} \cup \{\#\})^*$ into $\mathbb{N}$.

Definition 1.   (I) [13] M *partially* (*Part*) learns $\mathcal{C}$ if, for every L in $\mathcal{C}$ and each text T_L for L, there is exactly one index e such that $M(T_L[k]) = e$ for infinitely many k; furthermore, if M outputs e infinitely often on T_L, then $L = W_e$.

(II) [3] M *behaviourally correctly* (*BC*) learns $\mathcal{C}$ if, for every L in $\mathcal{C}$ and each text T_L for L, there is a number n for which $L = W_{M(T_L[j])}$ whenever $j \geq n$.

(III) [1] M is *consistent* (*Cons*) if for all $\sigma \in (\mathbb{N} \cup \{\#\})^*$, $range(\sigma) \subseteq W_{M(\sigma)}$.

(IV) [1] For any text T, M is *consistent on* T if $range(T[n]) \subseteq W_{M(T[n])}$ for all $n > 0$.

(V) [7] M is said to *consistently partially* (*ConsPart*) learn $\mathcal{C}$ if it partially learns $\mathcal{C}$ from text and is consistent.

(VI) [6] M is said to *conservatively partially* (*ConsvPart*) learn $\mathcal{C}$ if it partially learns $\mathcal{C}$ and outputs on each text for every L in $\mathcal{C}$ exactly one index e with $L \subseteq W_e$.

(VII) [4] M is said to *behaviourally correctly* learn $\mathcal{C}$ *with at most a anomalies* (BC^a) iff for every $L \in \mathcal{C}$ and each text T_L for L, there is a number n for which $|(W_{M(T_L[j])} - L) \cup (L - W_{M(T_L[j])})| \leq a$ whenever $j \geq n$.

(VIII) [4] M is said to *behaviourally correctly* learn $\mathcal{C}$ *with finitely many anomalies* (BC^*) iff for every $L \in \mathcal{C}$ and each text T_L for L, there is a number n for which $|(W_{M(T_L[j])} - L) \cup (L - W_{M(T_L[j])})| < \infty$ whenever $j \geq n$.

This paper will also consider combinations of different learning criteria; for learning criteria $A_1, \ldots, A_n$, a class $\mathcal{C}$ is said to be $A_1 \ldots A_n$-learnable iff there is a learner M such that M A_i-learns $\mathcal{C}$ for all $i \in \{1, \ldots, n\}$. Due to space constraints, some proofs of formal statements are omitted throughout this paper. For the full version of the paper, see http://arxiv.org/abs/1507.01215.

4 Approximate Learning of Functions

Fulk and Jain [5] proposed a mathematically rigorous definition of *approximate inference*, a notion originally motivated by studies in the philosophy of science.

Definition 2. [5] An approximate (*Approx*) learner outputs on the graph of a function f a sequence of hypotheses such that there is a sequence $S_0, S_1, \ldots$ of sets satisfying the following conditions:
(a) The S_n form an ascending sequence of sets such that their union is the set of all natural numbers;
(b) There are infinitely many n such that $S_{n+1} - S_n$ is infinite;
(c) The n-th hypothesis is correct on all $x \in S_n$ but nothing is said about the $x \notin S_n$.

The next proposition simplifies this set of conditions. The proof is omitted.

Proposition 3. *M Approx learns a recursive function f iff the following conditions hold:*
(d) For all x and almost all n, M's n-th hypothesis is correct at x;
(e) There is an infinite set S such that for almost all n and all $x \in S$, M's n-th hypothesis is correct at x.

Fulk and Jain interpreted their notion of approximation as a process in scientific inference whereby physicists take the limit of the average result of a sequence of experiments. Their result that the class of recursive functions is approximately learnable seems to justify this view.

Theorem 4 (Fulk and Jain [5]). *There is a learner M that Approx learns every recursive function.*

The following theorem answers an open question posed by Fulk and Jain [5] on whether the class of recursive functions has a learner which outputs a sequence of hypotheses that approximates the function to be learnt and almost always differs from the target only on finitely many places.

Theorem 5. *There is a learner M which learns the class of all recursive functions such that (i) M is a BC^* learner, (ii) M is a partial learner and (iii) M is an approximate learner.*

Proof. Let $\psi_0, \psi_1, \ldots$ be an enumeration of all recursive functions and some partial ones such that in every step s there is exactly one pair (e, x) for which $\psi_e(x)$ becomes defined at step s and this pair satisfies in addition that $\psi_e(y)$ is already defined by step s for all $y < x$. Furthermore, a function ψ_e is said to make progress on σ at step s iff $\psi_e(x)$ becomes defined at step s and $x \in dom(\sigma)$ and $\psi_e(y) = \sigma(y)$ for all $y \leq x$.

Now one defines for every σ a partial-recursive function $\vartheta_{e,\sigma}$ as follows:

- $\vartheta_{e,\sigma}(x) = \sigma(x)$ for all $x \in dom(\sigma)$;
- Let $e_t = e$;
- Inductively for all $s \geq t$, if some index $d < e_s$ makes progress on σ at step $s + 1$ then let $e_{s+1} = d$ else let $e_{s+1} = e_s$;
- For each value $x \notin dom(\sigma)$, if there is a step $s \geq t + x$ for which $\psi_{e_s,s}(x)$ is defined then $\vartheta_{e,\sigma}(x)$ takes this value for the least such step s, else $\vartheta_{e,\sigma}(x)$ remains undefined.

The learner M, now to be constructed, uses these functions as hypothesis space; on input τ, M outputs the index of $\vartheta_{e,\sigma}$ for the unique e and shortest prefix σ of τ such that the following three conditions are satisfied at some time t:

- t is the first time such that $t \geq |\tau|$ and some function makes progress on τ;
- ψ_e is that function which makes progress at τ;
- for every $d < e$, ψ_d did not make progress on τ at any $s \in \{|\sigma|, \ldots, t\}$ and either $\psi_{d,|\sigma|}$ is inconsistent with σ or $\psi_{d,|\sigma|}(x)$ is undefined for at least one $x \in dom(\sigma)$.

For finitely many strings τ there might not be any such function $\vartheta_{e,\sigma}$, as τ is required to be longer than the largest value up to which some function has made progress at time $|\tau|$, which can be guaranteed only for almost all τ. For these finitely many exceptions, M outputs a default hypothesis, e.g., for the everywhere undefined function. Now the three conditions (i), (ii) and (iii) of M are verified. For this, let ψ_d be the function to be learnt, note that ψ_d is total.

Condition (i): M is a BC^* learner. Let d be the least index of the function ψ_d to be learnt and let u be the last step where some ψ_e with $e < d$ makes progress on ψ_d. Then every $\tau \preceq \psi_d$ with $|\tau| \geq u + 1$ satisfies that first $M(\tau)$ conjectures a function $\vartheta_{e,\sigma}$ with $e \geq d$ and $|\sigma| \geq u + 1$ and $\sigma \preceq \psi_d$ and second that almost all e_s used in the definition of $\vartheta_{e,\sigma}$ are equal to d; thus the function computed is a finite variant of ψ_d and M is a BC^* learner.

Condition (ii): M is a partial learner. Let $t_0, t_1, \ldots$ be the list of all times where ψ_d makes progress on itself with $u < t_0 < t_1 < \ldots$; note that whenever $\tau \preceq \psi_d$ and $|\tau| = t_k$ for some k then the conjecture $\vartheta_{e,\sigma}$ made by $M(\tau)$ satisfies $e = d$ and $|\sigma| = u + 1$. As none of these conjectures make progress from step $u + 1$ onwards on ψ_d, they also do not make progress on σ after step $|\sigma|$ and $\vartheta_{e,\sigma} = \psi_d$; hence the learner outputs some index for ψ_d infinitely often. Furthermore, all other indices $\vartheta_{e,\sigma}$ are output only finitely often: if $e < d$ then ψ_e makes no progress on the target function ψ_d after step u; if $e > d$ then the length of σ depends on the prior progress of ψ_d on itself, and if $|\tau| > t_k$ then $|\sigma| > t_k$.

Condition (iii): M is an approximate learner. Conditions (d) and (e) in Proposition 3 are used. Now it is shown that, for all $\tau \preceq \psi_d$ with $t_k \leq |\tau| < t_{k+1}$, the hypothesis $\vartheta_{e,\sigma}$ issued by $M(\tau)$ is correct on the set $\{t_0, t_1, \ldots\}$. If $|\tau| = t_k$ then the hypothesis is correct everywhere as shown under condition (ii). So assume that $e > d$. Then $|\tau| > t_k$ and $|\sigma| > t_k$, hence $\vartheta_{e,\sigma}(x) = \psi_d(x)$ for all $x \leq t_k$. Furthermore, as ψ_d makes progress on σ in step t_{k+1} and as no ψ_c with $c < d$ makes progress on σ beyond step $|\sigma|$, it follows that the e_s defined in the algorithm of $\vartheta_{e,\sigma}$ all satisfy $e_s = d$ for $s \geq t_{k+1}$; hence $\vartheta_{e,\sigma}(x) = \psi_d(x)$ for all $x \geq t_{k+1}$.   ∎

5   Approximate Learning of Languages

This section proposes three notions of approximation in language learning. The first two notions, *approximate* learning and *weak approximate* learning, are adaptations of the set of conditions for approximately learning recursive functions given in Proposition 3. Recall that a set V is a finite variant of a set W iff there is an x such that for all $y > x$ it holds that $V(y) = W(y)$.

Definition 6. Let S be a class of languages. S is *approximately (Approx) learnable* iff there is a learner M such that for every language $L \in S$ there is an infinite set W such that for all texts T and all finite variants V of W and almost all hypotheses H of M on T, $H \cap V = L \cap V$. S is *weakly approximately (WeakApprox) learnable* iff there is a learner M such that for every language $L \in S$ and for every text T for L there is an infinite set W such that for all finite variants V of W and almost all hypotheses H of M on T, $H \cap V = L \cap V$. S is *finitely approximately (FinApprox) learnable* iff there is a learner M such that for every language $L \in S$, all texts T for L, and any finite set D, it holds that for almost all hypotheses H of M on T, $H \cap D = L \cap D$.

Remark 7. Jain, Martin and Stephan [10] defined a partial-recursive function C to be an *In-classifier* for a class S of languages if, roughly speaking, for every $L \in S$, every text T for L, every finite set D and almost all n, C on $T[n]$ will correctly "classify" all $x \in D$ as either belonging to L or not belonging to L. A learner M that *FinApprox* learns a class S may be translated into a total In-classifier for S, and vice versa.

Approximate learning requires, for each target language, the existence of a set W suitable for all texts, while in weakly approximate learning the set W may depend on T. In the weakest notion, finitely approximate learning, on any text T for a target language L the learner is only required to be almost always correct on any finite set. As will be seen later, this model is so powerful that the whole class of r.e. sets can be finitely approximated by a partial learner. The following examples illustrate the models of approximate and weakly approximate learning.

Example 8. – If there is an infinite r.e. set W such that all members of $\mathcal{C}$ contain W then $\mathcal{C}$ is *Approx* learnable: the learner simply conjectures $range(\sigma) \cup W$ on any input σ. Such $\mathcal{C}$ is not necessarily BC^* learnable.
- If $\mathcal{C}$ consists only of coinfinite r.e. sets then $\mathcal{C}$ is *Approx* learnable.
- The class of all cofinite sets is BC^* learnable and *WeakApproxBC***Part* learnable but neither *Approx* learnable nor BC^n learnable for any n.
- The class of all infinite sets is *WeakApprox* learnable.
- Gold's class consisting of the set of natural numbers and all sets $\{0, 1, \ldots, m\}$ is not *WeakApprox* learnable.

The proofs are omitted. These examples establish that, in contrast to the function learning case, approximate language learnability does not imply BC^* learnability. BC^* learnability does not imply approximate learnability either, but weakly approximate learning is powerful enough to cover all BC^* learnable classes.

Theorem 9. *If $\mathcal{C}$ is BC^* learnable then $\mathcal{C}$ is WeakApprox learnable.*

Proof. By Example 8, there is a learner M that weakly approximates the class of all infinite sets. Let O be a BC^* learner for $\mathcal{C}$. Now the new learner N is given as follows: On input σ, $N(\sigma)$ outputs an index of the following set which first enumerates $range(\sigma)$ and then searches for some τ that satisfies the following conditions: (1) $range(\tau) = range(\sigma)$; (2) $|\tau| = 2 * |range(\sigma)|$; (3) $W_{O(\tau\#^s)}$ enumerates at least $|\sigma|$ many elements for all $s \leq |\sigma|$. If all three conditions are met then the set contains also all elements of $W_{M(\sigma)}$. Further details are omitted. ∎

6 Combining Partial Language Learning With Variants of Approximate Learning

This section is concerned with the question whether partial learners can always be modified to approximate the target language in the models introduced above.

6.1 Finitely Approximate Learning

The first results demonstrate the power of the model of finitely approximate learning: there is a partial learner that finitely approximates every r.e. language.

Theorem 10. *The class of all r.e. sets is FinApproxPart learnable.*

Proof. Let M_1 be a partial learner of all r.e. sets. Define a learner M_2 as follows. Given a text T, let $e_n = M_1(T[n+1])$ for all n. On input $T[n+1]$, M_2 determines the finite set $D = range(T[n+1]) \cap \{0, \ldots, m\}$, where m is the minimum $m \leq n$ with $e_m = e_n$. M_2 then outputs a canonical index for $D \cup (W_{e_n} \cap \{x : x > m\})$.

Suppose T is a text for some r.e. set L. Then there is a least l such that M_1 on T outputs e_l infinitely often and $W_{e_l} = L$. Furthermore, there is a least l' such that for all $l'' > l'$, $D_L = range(T[l''+1]) \cap \{0,\ldots,l\} = L \cap \{0,\ldots,l\}$. Hence M_2 will output a canonical index for $L = D_L \cup (W_{e_l} \cap \{x : x > l\})$ infinitely often. On the other hand, since, for every h with $e_h \neq e_l$ and $e_h \neq e_{h'}$ for all $h' < h$, M_1 outputs e_h only finitely often, M_2 will conjecture sets of the form $D' \cup (W_{e_h} \cap \{x : x > h\})$ only finitely often. Thus M_2 partially learns L.

To see that M_2 is also a finitely approximate learner, consider any number x. Suppose that M_1 on T outputs exactly one index e infinitely often; further, $W_e = L$ and j is the least index such that $e_j = e$. Let s be sufficiently large so that for all $s' > s$, $range(T[s'+1]) \cap \{0,\ldots,\max(\{x,j\})\} = L \cap \{0,\ldots,\max(\{x,j\})\}$. First, assume that M_1 outputs only finitely many distinct indices on T. It follows that M_1 on T converges to e. Thus M_2 almost always outputs a canonical index for $(L \cap \{0,\ldots,j\}) \cup (W_{e_j} \cap \{y : y > j\})$, and so it approximately learns L. Second, assume that M_1 outputs infinitely many distinct indices on T. Let $d_1,\ldots,d_x$ be the first x conjectures of M_1 that are pairwise distinct and are not equal to e. There is a stage $t > s$ large enough so that $e_{t'} \notin \{d_1,\ldots,d_x\}$ for all $t' > t$. Consequently, whenever $t' > t$, M_2 on $T[t'+1]$ will conjecture a set W such that $W \cap \{0,\ldots,x\} = L \cap \{0,\ldots,x\}$. This establishes that M_2 finitely approximately learns any r.e. set. ∎

Gao, Jain and Stephan [6] showed that consistently partial learners exist for all and only the subclasses of uniformly recursive families; the next theorem shows that such learners can even be finitely approximate at the same time, in addition to being *prudent*. A learner M is *prudent* if it learns the class $\{W_{M(\sigma)} : \sigma \in (\mathbb{N} \cup \{\#\})^*, M(\sigma) \neq ?\}$, that is, if M learns every set it conjectures [12]; the ? symbol allows M to abstain from conjecturing at any stage..

Theorem 11. *If $\mathcal{C}$ is a uniformly recursive family, then $\mathcal{C}$ is FinApproxCons-Part learnable by a prudent learner.*

Proof. Let $\mathcal{C} = \{L_0, L_1, L_2, \ldots, \}$ be a uniformly recursive family. On text T, define M at each stage s as follows:

> If there are $x \in \mathbb{N}$ and $i \in \{0, 1, \ldots, s\}$ such that
> - $range(T[s+1]) - range(T[s]) = \{x\}$,
> - $range(T[s+1]) \subseteq L_i \cup \{\#\}$ and
> - $range(T[s+1]) \cap \{0,\ldots,x\} = L_i \cap \{0,\ldots,x\}$
>
> Then M outputs the least such i
> Else M outputs a canonical index for $range(T[s+1]) - \{\#\}$.

The consistency of M follows directly by construction. If T is a text for a finite set then the "Else-Case" will apply almost always and M converges to a canonical index for $range(T)$. Now consider that T is a text for some infinite set $L_m \in \mathcal{C}$ and m is the least index of itself. Let t be large enough so that for all $t' > t$, all $x \in L - range(T[t+1]) - \{\#\}$ and all $j < m$, $L_j \cap \{0,\ldots,x\} \neq range(T[t'+1]) \cap \{0,\ldots,x\}$. There are infinitely many stages $s > \max(\{t,m\})$ at which $T(s) \notin range(T[s]) \cup \{\#\}$ and $range(T[s+1]) \cap \{0,\ldots,T(s)\} = L \cap \{0,\ldots,T(s)\}$. At each of these stages, M will conjecture L_m. Thus M conjectures L_m infinitely often. Furthermore, for every x there is some s_x such that for all $y \in L - $

$range(T[s_x + 1])$, it holds that $y > x$. Thus whenever $s' > s_x$, M's conjecture on $T[s' + 1]$ agrees with L on $\{0, \ldots, x\}$. M is therefore a finitely approximate learner, implying that it never conjectures any incorrect index infinitely often. ∎

Corollary 12. *If C is ConsPart learnable, then C is FinApproxConsPart learnable by a prudent learner.*

The following result shows that also conservative partial learning may always be combined with finitely approximate learning.

Theorem 13. *If C is ConsvPart learnable, then C is FinApproxConsvPart learnable.*

Proof. Let M_1 be a *ConsvPart* learner for C, and suppose that M_1 outputs the sequence of conjectures $e_0, e_1, \ldots$ on some given text T. The construction of a new learner M_2 is similar to that in Theorem 10; however, one has to ensure that M_2 does not output more than one index that is either equal to or a proper superset of the target language. On input $T[s+1]$, define $M_2(T[s+1])$ as follows.

1. If $range(T[s + 1]) \subseteq \{\#\}$ then output a canonical index for $\emptyset$ else go to 2.
2. Let $m \leq s$ be the least number such that $e_m = e_s$. If $W_{e_s,s} \cap \{0, \ldots, m\} = range(T[s+1]) \cap \{0, \ldots, m\} = D$ then output a canonical index for $D \cup (W_{e_m} \cap \{x : x > m\})$ else go to 3.
3. If $s \geq 1$ then output $M_2(T[s])$ else output a canonical index for $\emptyset$.

The details for verifying that M_2 is a *ConsvPart* learner for C are omitted. ∎

6.2   Weakly Approximate, Approximate and BC^* Learning

The next proposition shows that Theorem 11 cannot be improved and gives a negative answer to the question whether partial or consistent partial learning can be combined with weakly approximate learning.

Proposition 14. *The uniformly recursive class $\{A : A = \mathbb{N}$ or A contains all even and finitely many odd numbers or A contains finitely many even and all odd numbers$\}$ is WeakApprox learnable and ConsPart learnable, but not WeakApproxPart learnable.*

The next theorem shows that neither partial learning nor consistent partial learning can be combined with approximate learning. In fact, it establishes a stronger result: consistent partial learnability and approximate learnability are insufficient to guarantee both partial and weakly approximate learnability simultaneously.

Theorem 15. *There is a class of r.e. sets with the following properties:*
(i) The class is not BC^* learnable;
(ii) The class is not WeakApproxPart learnable;
(iii) The class is Approx learnable.

Proof. The key idea is to diagonalise against a list $M_0, M_1, \ldots$ of learners which are all total and which contains for every learner to be considered a delayed version. This permits to ignore the case that some learner is undefined on some input.

The class witnessing the claim consists of all sets L_d such that for each d, either L_d is $\{d, d+1, \ldots\}$ or L_d is a subset built by the following diagonalisation procedure: One assigns to each number $x \geq d$ a level $\ell(x)$.

- If some set $L_{d,e} = \{x \geq d : \ell(x) \leq e\}$ is infinite then
- let $L_d = L_{d,e}$ for the least such e and M_d does not partially learn L_d
- else let $L_d = \{d, d+1, \ldots\}$ and M_d does not weakly approximate L_d.

The construction of the sets is inductive over stages. For each stage $s = 0, 1, 2, \ldots$:

- Let τ_e be a sequence of all $x \in \{d, d+1, \ldots, d+s-1\}$ with $\ell(x) = e$ in ascending order;
- If there is an $e < s$ such that e has not been cancelled in any previous step and for each $\eta \preceq \tau_e$ the intersection $W_{M_d(\tau_0\tau_1\ldots\tau_{e-1}\eta),s} \cap \{y : d \leq y < d+s \wedge \ell(y) > e\}$ contains at least $|\tau_e|$ elements
  - Then choose the least such e and let $\ell(d+s) = e$ and cancel all e' with $e < e' \leq s$
  - Else let $\ell(d+s) = s$.

A text $T = \lim_e \sigma_e$ is defined as follows (where σ_0 is the empty sequence):

- Let τ_e be the sequence of all x with $\ell(x) = e$ in ascending order;
- If σ_e is finite then let $\sigma_{e+1} = \sigma_e \tau_e$ else let $\sigma_{e+1} = \sigma_e$.

In case some σ_e are infinite, let e be smallest such that σ_e is infinite. Then $T = \sigma_e$ and $L_d = L_{d,e}$ and T is a text for L_d. As $L_{d,e}$ is infinite, one can conclude that

$$\forall \eta \preceq \sigma_e \, \forall c \, [|W_{M_d(\tau_0\tau_1\ldots\tau_{e-1}\eta)} \cap \{y : \ell(y) > e\}| \geq c]$$

and thus M_d outputs on T almost always a set containing infinitely many elements outside L_d; so M_d does neither partially learn L_d nor BC^* learn L_d.

In case all σ_e are finite and therefore all $L_{d,e}$ are finite there must be infinitely many e that never get cancelled. Each such e satisfies

$$\exists \eta \preceq \tau_e \, [W_{M_d(\tau_0\tau_1\ldots\tau_{e-1}\eta)} \cap \{y : \ell(y) > e\} \text{ is finite}]$$

and therefore e also satisfies $\exists \eta \preceq \tau_e \, [W_{M_d(\tau_0\tau_1\ldots\tau_{e-1}\eta)}$ is finite]. Thus M_d outputs on the text T for the cofinite set $L_d = \{d, d+1, \ldots\}$ infinitely often a finite set and M_d is neither weakly approximately learning L_d (as there is no infinite set on which almost all conjectures are correct) nor BC^*-learning L_d. Thus claims (i) and (ii) are true.

Next it is shown that the class of all L_d is approximately learnable by some learner N. This learner N will on a text for L_d eventually find the minimum d needed to compute the function ℓ. Once N has found this d, N will on each input σ conjecture the set

$$W_{N(\sigma)} = \{x : x \geq \max(range(\sigma)) \vee \exists y \in range(\sigma) \, [\ell(x) \leq \ell(y)]\}$$

In case $L_d = L_{d,e}$ for some e, $L_{d,e}$ is infinite, and for each text for $L_{e,d}$, almost all prefixes σ of this text satisfy $\max\{\ell(y) : y \in range(\sigma)\} = e$ and $L_{d,e} \subseteq W_{N(\sigma)}$. So almost all conjectures are correct on the infinite set L_d itself. Furthermore, $W_{N(\sigma)}$ does not contain any $x < \max(range(\sigma))$ with $\ell(x) > e$, hence N eventually becomes correct also on any $x \notin L_{d,e}$ and therefore N approximates $L_{d,e} = L_d$.

In case $L_d = \{d, d+1, \ldots\}$, all $L_{d,e}$ are finite. Then consider the infinite set $S = \{x : \forall y > x\,[\ell(y) > \ell(x)]\}$. Let $x \in S$ and consider any σ with $\min(range(\sigma)) = d$. If $x \geq \max(range(\sigma))$ then $x \in W_{N(\sigma)}$. If $x < \max(range(\sigma))$ then $\ell(\max(range(\sigma))) \geq \ell(x)$ and again $x \in W_{N(\sigma)}$. Thus $W_{N(\sigma)}$ contains S. Furthermore, for all $x \geq d$ and sufficiently long prefixes σ of the text, $\ell(\max(range(\sigma))) \geq \ell(x)$ and therefore all $x \in W_{N(\sigma)}$ for almost all prefixes σ of the text. So again N approximates L_d. Thus claim (iii) is true. ∎

One can further show that the class in the above proof is explanatorily learnable if the learner has access to an oracle for the jump of the halting set.

While these negative results suggest that approximate and weakly approximate learning imposes constraints that are too stringent for combining with partial learning, at least partly positive results can be obtained. For example, the following theorem shows that *ConsvPart* learnable classes are *ApproxPart* learnable (thus dropping only the conservativeness constraint) by BC^* learners. This considerably improves an earlier result by Gao, Stephan and Zilles [7] which states that every *ConsvPart* learnable class is also BC^* learnable.

Theorem 16. *If* C *is ConsvPart learnable then* C *is ApproxPart learnable by a* BC^* *learner.*

Proof. Let M be a *ConsvPart* learner for C. For a text T for a language $L \in C$, one considers the sequence $e_0, e_1, \ldots$ of distinct hypotheses issued by M; it contains one correct hypothesis while all others are not indices of supersets of L. For each hypothesis e_n one has two numbers tracking its quality: $b_{n,t}$ is the maximal $s \leq n + t$ such that all $T(u)$ with $u < s$ are in $W_{e_n, n+t} \cup \{\#\}$ and $a_{n,t} = 1 + \max\{b_{m,t} : m < n\}$.

Now one defines the hypothesis set $H_{e_n,\sigma}$ for any sequence σ. Let $e_{n,0}, e_{n,1}, \ldots$ be a sequence with $e_{n,0} = e_n$ and $e_{n,u}$ be the e_m for the minimum m such that $m = n$ or W_{e_m} has enumerated all members of $range(\sigma)$ within $u + t$ time steps. The set $H_{e_n,\sigma}$ contains all x for which there is a $u \geq x$ with $x \in W_{e_{n,u}}$.

An intermediate learner O now conjectures some canonical index of a set $H_{e_n,\sigma}$ at least k times iff there is a t with $\sigma = T(0)T(1)\ldots T(a_{n,t})$ and $b_{n,t} > k$. Thus O conjectures $H_{e_n,\sigma}$ infinitely often iff W_{e_n} contains $range(T)$ and $a_{n,t} = |\sigma|$ for almost all t.

If e_n is the correct index for the set to be learnt then, by conservativeness, the sets W_{e_m} with $m < n$ are not supersets of the target set. So the values $b_{m,t}$ converge which implies that $a_{n,t}$ converges to some s. It follows that for the prefix σ of T of length s, the canonical index of $H_{e_n,\sigma}$ is conjectured infinitely often while no other index is conjectured infinitely often. Thus O is a partial learner. Furthermore, for all sets $H_{e_m,\tau}$ conjectured after $a_{n,t}$ has reached its final value s, it holds that the $e_{m,u}$ in the construction of $H_{e_m,\tau}$ converge to e_n. Thus $H_{e_m,\tau}$ is the union of W_{e_n} and a finite set. Hence O is a BC^* learner. To guarantee the third condition on approximate learning, O will be translated into another learner N.

Let $d_0, d_1, \dots$ be the sequence of O output on the text T. Now N will copy this sequence but with some delay. Assume that $N(\sigma_k) = d_k$ and σ_k is a prefix of T. Then N will keep the hypothesis d_k until the current prefix σ_{k+1} considered satisfies either $range(\sigma_{k+1}) \not\subseteq range(\sigma_k)$ or $W_{d_k, |\sigma_{k+1}|} \neq range(\sigma_{k+1})$.

If $range(T)$ is infinite, the sequence of hypotheses of N will be the same as that of O, only with some additional delay. Furthermore, almost all W_{d_n} contain $range(T)$, thus the resulting learner N learns $range(T)$ and is almost always correct on the infinite set $range(T)$; in addition, N learns $range(T)$ partially and is also BC^*. If $range(T)$ is finite, there will be some correct index that equals infinitely many d_n. There is a step t by which all elements of $range(T)$ have been seen in the text and enumerated into W_{d_n}. Therefore, when the learner conjectures this correct index again, it will never withdraw it; furthermore, it will replace eventually every incorrect conjecture due to the comparison of the two sets. Thus the learner converges explanatorily to $range(T)$ and is also in this case learning $range(T)$ in a BC^* way, partially and approximately. From the proof of Theorem 10, one can see that N may be translated into a learner satisfying all the three requirements (a), (b) and (c). ∎

Case and Smith [4] published Harrington's observation that the class of recursive functions is BC^* learnable. This result does not carry over to the class of r.e. sets; for example, Gold's class consisting of the set of natural numbers and all finite sets is not BC^* learnable. In light of Theorem 5, which established that the class of recursive functions can be BC^* and $Part$ learnt simultaneously, it is interesting to know whether any BC^* learnable class of r.e. sets can be both BC^* and $Part$ learnt at the same time. While this question in its general form remains open, the next result shows that BC^n learning is indeed combinable with partial learning.

Theorem 17. Let $n \in \mathbb{N}$. If C is BC^n learnable, then C is $Part$ learnable by a BC^n learner.

Proof. Fix any n such that C is BC^n learnable. Given a recursive BC^n learner M of C, one can construct a new learner N_1 as follows. First, let $F_0, F_1, F_2, \dots$ be a one-one enumeration of all finite sets such that $|F_i| \leq n$ for all i. Fix a text T, and let $e_0, e_1, e_2, \dots$ be the sequence of M's conjectures on T.

For each set of the form $W_{e_i} \cup F_j$ (respectively $W_{e_i} - F_j$), N_1 outputs a canonical index for $W_{e_i} \cup F_j$ (respectively $W_{e_i} - F_j$) at least m times iff the following two conditions hold.

1. There is a stage $s > j$ for which the number of distinct $x < j$ such that either $x \in W_{e_i,s} \wedge x \notin range(T[s+1])$ or $x \in range(T[s+1]) \wedge x \notin W_{e_i,s}$ holds does not exceed n.
2. There is a stage $t > m$ such that for all $x < m$, $x \in W_{e_i,t} \cup F_j$ iff $x \in range(T[t+1])$ (respectively $x \in W_{e_i,t} - F_j$ iff $x \in range(T[t+1])$).

At any stage $T[s+1]$ where no set of the form $W_{e_i} \cup F_j$ or $W_{e_i} - F_j$ satisfies the conditions above, or each such set has already been output the required number of times (up to the present stage), N_1 outputs $M(T[s+1])$. The details showing that a $BC^n Part$ learner N for C can be constructed from N_1 are omitted. ∎

Theorems 18 and 19 show that partial BC^* learning is possible for classes that can be BC^* learned by learners that satisfy some additional constraints. The proofs are omitted.

Theorem 18. *Assume that C is BC^* learnable by a learner that outputs on each text for any $L \in C$ at least once a fully correct hypothesis. Then C is Part learnable by a BC^* learner.*

Theorem 19. *Suppose there is a recursive learner that BC^* learns C and outputs on every text for any $L \in C$ at least one index infinitely often. Then there is a recursive learner for C that BC^* and Part learns C.*

References

1. Angluin, D.: Inductive inference of formal languages from positive data. Information and Control **45**(2), 117–135 (1980)
2. Bārzdiņš, J.: Two theorems on the limiting synthesis of functions. In: Theory of Algorithms and Programs, vol. 1, pp. 82–88. Latvian State University (1974) (in Russian)
3. Case, J., Lynes, C.: Machine inductive inference and language identification. In: Nielsen, M., Schmidt, E.M. (eds.) Automata, Languages and Programming. LNCS, vol. 140, pp. 107–115. Springer, Heidelberg (1982)
4. Case, J., Smith, C.: Comparison of identification criteria for machine inductive inference. Theoretical Computer Science **25**, 193–220 (1983)
5. Fulk, M., Jain, S.: Approximate inference and scientific method. Information and Computation **114**, 179–191 (1994)
6. Gao, Z., Jain, S., Stephan, F.: On conservative learning of recursively enumerable languages. In: Bonizzoni, P., Brattka, V., Löwe, B. (eds.) CiE 2013. LNCS, vol. 7921, pp. 181–190. Springer, Heidelberg (2013)
7. Gao, Z., Stephan, F., Zilles, S.: Partial learning of recursively enumerable languages. In: Jain, S., Munos, R., Stephan, F., Zeugmann, T. (eds.) ALT 2013. LNCS, vol. 8139, pp. 113–127. Springer, Heidelberg (2013)
8. Gold, E.M.: Language identification in the limit. Information and Control **10**, 447–474 (1967)
9. Jain, S., Osherson, D., Royer, J.S., Sharma, A.: Systems that learn: an introduction to learning theory. MIT Press (1999)
10. Jain, S., Martin, E., Stephan, F.: Learning and classifying. Theoretical Computer Science **482**, 73–85 (2013)
11. Martin, E., Osherson, D.N.: Elements of scientific inquiry. MIT Press (1998)
12. Osherson, D.N., Stob, M., Weinstein, S.: Learning strategies. Information and Control **53**, 32–51 (1982)
13. Osherson, D.N., Stob, M., Weinstein, S.: Systems that learn: an introduction to learning theory for cognitive and computer scientists. MIT Press (1986)
14. Rogers, Jr., H.: Theory of recursive functions and effective computability. MIT Press (1987)

Learning from Queries, Teaching Complexity

Exact Learning of Multivalued Dependencies

Montserrat Hermo[1] and Ana Ozaki[2($\boxtimes$)]

[1] Languages and Information Systems,
University of the Basque Country, Leioa, Spain
`montserrat.hermo@ehu.es`
[2] Department of Computer Science, University of Liverpool, Liverpool, UK
`anaozaki@liverpool.ac.uk`

Abstract. The transformation of a relational database schema into fourth normal form, which minimizes data redundancy, relies on the correct identification of multivalued dependencies. In this work, we study the learnability of multivalued dependency formulas (MVDF), which correspond to the logical theory behind multivalued dependencies. As we explain, MVDF lies between propositional Horn and 2-Quasi-Horn. We prove that MVDF is polynomially learnable in Angluin et al.'s exact learning model with membership and equivalence queries, provided that counterexamples and membership queries are formulated as 2-Quasi-Horn clauses. As a consequence, we obtain that the subclass of 2-Quasi-Horn theories which are equivalent to MVDF is polynomially learnable.

1 Introduction

Among the models proposed to represent databases, since its presentation by Codd [6], the relational model has been the most successful one. In this model, data is represented by tuples which are grouped into relations. Different types of formalisms based on the concept of data dependencies have been used to design and analyse database schemas. Data dependencies can be classified as functional [6], or multivalued [7,8] (also called tuple generating), where the latter is a generalization of the first. Functional dependencies correspond to the Horn fragment of propositional logic in the sense that one can map each functional dependency to a Horn clause preserving the logical consequence relation [11,15]. The same correspondence can be established between multivalued dependencies and multivalued dependency formulas (MVDF) [5,11]. They have long been studied in the literature and it is well known that the transformation of a relational database schema into the fourth normal form (4NF), which minimizes data redundancy, relies on the identification of multivalued dependencies [8].

In this work, we cast the problem of identifying data dependencies as a learning problem and study the learnability of MVDF, which correspond to the logical theory behind data dependencies. Identification of the Horn fragment from interpretations in Angluin's exact learning model is stated in [4], and later an algorithm that learns Horn from entailments is presented in [9]. Furthermore, a variant that learns sets of functional dependencies appears in [10]. Regarding

© Springer International Publishing Switzerland 2015
K. Chaudhuri et al. (Eds.): ALT 2015, LNAI 9355, pp. 73–88, 2015.
DOI: 10.1007/978-3-319-24486-0_5

MVDF, it is known that this class cannot be learned either using equivalence [3] or membership queries alone [13], and that a particular subclass of them is learnable when both types of queries are allowed [12]. However, to the best of our knowledge, there is no positive result for the general class MVDF using membership and equivalence queries. One of main obstacles to find a learning algorithm for MVDF is the fact the MVDF theories are not closed under intersection in contrast to the Horn case [5]. In general, given a multivalued dependency formula, there is not a unique minimal model that satisfies both the formula and a particular set of variables, a property extensively exploited by Horn algorithms.

A major open problem in learning theory (and also within the exact learning model) is whether the class CNF (or the class DNF) can be efficiently learnable. Although it is known that this class cannot be polynomially learned using either membership or equivalence queries alone [2,3], it is open whether CNF can be learned using both types of queries. Several restrictions have been imposed on both CNF and DNF in order to make them polynomially learnable. For instance, the classes monotone DNF [2], i.e., DNF formulas with no negated variables, k-term DNF or k-clause CNF [1], that is, DNF or CNF formulas with at most k terms or k clauses, and read-twice DNF [14], which are DNF where each variable occurs at most twice, are all polynomially learnable via queries.

One of the most important results concerning a restriction of the class CNF appears in the mentioned article [4], where propositional Horn formulas are learned using both types of queries. In fact, Horn is a special case of a class called k-quasi-Horn, meaning that clauses may contain at most k unnegated literals. However, it is pointed in [4] that, even for $k = 2$, learning the class of k-quasi-Horn formulas is as hard as learning CNF. Thus, if exact learning CNF is indeed intractable, the boundary of what can be learned in polynomial time with queries lies between 1-quasi-Horn (or simply Horn) and 2-quasi-Horn formulas. Since MVDF is a natural restriction of 2-quasi-Horn and a non-trivial generalization of Horn, investigating how far this boundary can be extended constitutes one of our main motivations and guide for this work, which is theoretical in nature.

In this paper, we give a polynomial algorithm that exactly learns MVDF using membership and equivalence queries. Membership queries and counterexamples given by the oracle are formulated as 2-quasi-Horn clauses. As a consequence, an algorithm that efficiently learns the subclass of 2-quasi-Horn formulas which are equivalent to multivalued dependency formulas is obtained. The paper is organized as follows. In Section 2 we introduce some notation and give definitions for MVDF and the class of k-quasi-Horn formulas. Section 3 shows a property that is crucial to learn the class MVDF: (although not unique) the number of minimal models that satisfy a multivalued dependency formula and a set of variables is polynomial in the size of the formula. In Section 4 we present our algorithm that efficiently learns the class MVDF from 2-quasi-Horn clauses. We end in Section 5 with some concluding remarks and open problems.

2   Preliminaries

Exact Learning. Let E be a set of examples (also called *domain* or *instance space*). A *concept over* E is a subset of E and a *concept class* is a set $\mathcal{C}$ of concepts over E. Each concept c over E induces a dichotomy of *positive* and *negative* examples, meaning that $e \in c$ is a positive example and $e \in E \setminus c$ is a negative example. For computational purposes, concepts need to be specified by some representation. So we define a *learning framework* to be a triple $(E, \mathcal{L}, \mu)$, where E is a set of examples, $\mathcal{L}$ is a set of *concept representations* and μ is a surjective function from $\mathcal{L}$ to a concept class $\mathcal{C}$ of concepts over E.

Given a learning framework $(E, \mathcal{L}, \mu)$, for each $l \in \mathcal{L}$, denote by $\mathsf{MEM}_{l,E}$ the oracle that takes as input some $e \in E$ and returns 'yes' if $e \in \mu(l)$ and 'no' otherwise. A *membership query* is a call to an oracle $\mathsf{MEM}_{l,E}$ with some $e \in E$ as input, for $l \in \mathcal{L}$ and E. Similarly, for every $l \in \mathcal{L}$, we denote by $\mathsf{EQ}_{l,E}$ the oracle that takes as input a concept representation $h \in \mathcal{L}$ and returns 'yes', if $\mu(h) = \mu(l)$, or a counterexample $e \in \mu(h) \oplus \mu(l)$, otherwise. An *equivalence query* is a call to an oracle $\mathsf{EQ}_{l,E}$ with some $h \in \mathcal{L}$ as input, for $l \in \mathcal{L}$ and E.

We say that a learning framework $(E, \mathcal{L}, \mu)$ is *exact learnable* if there is an algorithm A such that for any target $l \in \mathcal{L}$ the algorithm A always halts and outputs $l' \in \mathcal{L}$ such that $\mu(l) = \mu(l')$ using membership and equivalence queries answered by the oracles $\mathsf{MEM}_{l,E}$ and $\mathsf{EQ}_{l,E}$, respectively. A learning framework $(E, \mathcal{L}, \mu)$ is *polynomially* exact learnable if it is exact learnable by an algorithm A such that at every step of computation the time used by A up to that step is bounded by a polynomial $p(|l|, |e|)$, where l is the target and $e \in E$ is the largest counterexample seen so far[1].

Multivalued Dependencies and K-quasi-Horn Formulas. Let V be a set of boolean variables. The logical constant *true* is represented by $\mathbf{T}$ and the logical constant *false* is represented by $\mathbf{F}$. An *mvd clause* is an implication $X \to Y \vee Z$, where X, Y and Z are pairwise disjoint conjunctions of variables from V and $X \cup Y \cup Z = V$. An *mvd formula* is a conjunction of mvd clauses. A *k-quasi-Horn clause* is a propositional clause containing at most k unnegated literals. A *k-quasi-Horn formula* is a conjunction of k-quasi-Horn clauses. To simplify the notation, we treat sometimes conjunctions as sets and vice versa. Also, if for example $V = \{v_1, v_2, v_3, v_4, v_5, v_6\}$ is a set of variables and $\varphi = (v_1 \to (v_2 \wedge v_3) \vee (v_4 \wedge v_5 \wedge v_6)) \wedge ((v_2 \wedge v_3) \to (v_1 \wedge v_5 \wedge v_6) \vee v_4)$ is a formula then we write φ in this shorter way: $\{1 \to 23 \vee 456, 23 \to 156 \vee 4\}$, where conjunctions between variables are omitted and each propositional variable $v_i \in V$ is mapped to $i \in \mathbb{N}$. From the definitions above it is easy to see that:

1. any Horn clause is logically equivalent to a set of 2 mvd clauses. For instance, the Horn clause $135 \to 4$, is equivalent to: $\{12356 \to 4, 135 \to 4 \vee 26\}$;
2. any mvd clause is logically equivalent to a conjunction of 2-quasi-Horn clauses with size polynomial in the number of variables. For instance, the

---

[1] We count each call to an oracle as one step of computation. Also, we assume some natural notion of length for an example e and a concept representation l, denoted by $|e|$ and $|l|$, respectively.

mvd clause $1 \to 23 \lor 456$, by distribution, is equivalent to: $\{1 \to 2 \lor 4, 1 \to 2 \lor 5, 1 \to 2 \lor 6, 1 \to 3 \lor 4, 1 \to 3 \lor 5, 1 \to 3 \lor 6\}$.

Remark 1. Point 1 above means that w.l.o.g. we can assume that any mvd clause is either $V \to \mathbf{F}$ or $V \setminus \{v\} \to v$ or of the form $X \to Y \lor Z$ with Y and Z non-empty. We call *Horn-like* clauses of the form $V \setminus \{v\} \to v$. Note that $\mathbf{T} \to V \equiv \{\mathbf{T} \to v \mid v \in V\}$ and each $\mathbf{T} \to v$ is equivalent to $\{\mathbf{T} \to V \setminus \{v\} \lor v, V \setminus \{v\} \to v\}$.

Formally, in this paper we study the learning framework $\mathfrak{F}_M = (E_Q, \mathcal{L}_M, \mu_M)$, where E_Q is the set of all 2-quasi-Horn clauses in the propositional variables V under consideration, $\mathcal{L}_M$ is MVDF, which is the set of all mvd formulas that can be expressed in V and, for every $\mathcal{T} \in \mathcal{L}_M$, $\mu_M(\mathcal{T}) = \{e \in E_Q \mid \mathcal{T} \models e\}$. Note that learning MVDF from 2-quasi-Horn examples also corresponds to learning the set of all 2-quasi-Horn formulas that can be constructed by distribution from any mvd formula.

An interpretation $\mathcal{I}$ is a mapping from $V \cup \{\mathbf{T}, \mathbf{F}\}$ to $\{true, false\}$, where $\mathcal{I}(\mathbf{T}) = true$ and $\mathcal{I}(\mathbf{F}) = false$. We denote by $\mathsf{true}(\mathcal{I})$ the set of variables assigned to *true* in $\mathcal{I}$. In the same way, let $\mathsf{false}(\mathcal{I})$ be the set of variables assigned to *false* in $\mathcal{I}$. Observe that $\mathsf{false}(\mathcal{I}) = V \setminus \mathsf{true}(\mathcal{I})$. Let $\mathcal{H}$ and $\mathcal{T}$ be sets of mvd clauses. If $\mathcal{I} \models \mathcal{H}$ and $\mathcal{I} \not\models \mathcal{T}$ then we say that $\mathcal{I}$ is a *negative countermodel* w.r.t. $\mathcal{T}$. We follow the terminology provided in [4] and say that an interpretation $\mathcal{I}$ *covers* $X \to Y \lor Z$ if $X \subseteq \mathsf{true}(\mathcal{I})$. An interpretation $\mathcal{I}$ *violates* $X \to Y \lor Z$ if $\mathcal{I}$ covers $X \to Y \lor Z$ and: (a) Y and Z are non-empty and there are $v \in Y$ and $w \in Z$ such that $v, w \in \mathsf{false}(\mathcal{I})$; or (b) there is v such that $\mathsf{false}(\mathcal{I}) = \{v\}$ and $X \to Y \lor Z$ is the Horn-like clause $V \setminus \{v\} \to v$; or (c) $\mathsf{false}(\mathcal{I}) = \emptyset$ and $X \to Y \lor Z$ is the clause $V \to \mathbf{F}$. Given two interpretations $\mathcal{I}$ and $\mathcal{J}$, we define $\mathcal{I} \cap \mathcal{J}$ to be the interpretation such that $\mathsf{true}(\mathcal{I} \cap \mathcal{J}) = \mathsf{true}(\mathcal{I}) \cap \mathsf{true}(\mathcal{J})$.

3   Computing Minimal Models

In this section, we present Algorithm 1, which computes in polynomial time all minimal models (i.e. models with minimal number of variables assigned to 'true') satisfying both a set $\mathcal{P}$ of mvd clauses that have the form $X' \to Y' \lor Z'$ with Y' and Z' non-empty and a set of variables X (Horn-like clauses are treated in Line 15). To ensure the existence of minimal models, we consider $\mathcal{P}$ such that $\mathcal{P}$ does not contain $V \to \mathbf{F}$. Algorithm 1 receives $\mathcal{P}$ and X as input and constructs a *semantic tree*, in the sense that each child node satisfies one of the two consequents of an mvd clause. In each iteration of the main loop we 'apply' an mvd clause, meaning that, given a tree leaf node, we pick a (not used) mvd clause $X' \to Y' \lor Z' \in \mathcal{P}$ and construct two child nodes, one of them containing variables in Y' and the other variables in Z'. We exhaustively apply mvd clauses in $\mathcal{P}$ so that in the end each leaf node contains a set of variables that need to be true in order to satisfy both X and $\mathcal{P}$.

The following information is stored for each node i: a set M_i of mvd clauses in $\mathcal{P}$ that have not yet been applied in the i-node path; a set S_i of variables implied by X and by mvd clauses that have already been applied (i.e. clauses in $\mathcal{P} \setminus M_i$); and, to simplify the presentation, we also use an auxiliary set N_i of variables which are 'new' in the path, that is, if node a is predecessor of node i in the tree then a does not have these variables in S_a. The following example illustrates how the algorithm works.

Algorithm 1. Semantic Tree

1: Let $\mathcal{S} = \emptyset$ be a set of interpretations
2: Let $\mathcal{P}$ be a set of mvd clauses without $V \to \mathbf{F}$ and X a set of variables
3: **function** SEMANTICTREE($\mathcal{P}$, X)
4:     Create a node $i = 0$ with $S_0 = X$, $N_0 = \emptyset$ and $M_0 = \mathcal{P}$
5:     **repeat**
6:         Choose a leaf i and $X' \to Y' \vee Z' \in M_i$ with $Y' \neq \emptyset$, $Z' \neq \emptyset$ and $X' \subseteq S_i$
7:         **if** there is $v \in Y' \cup Z'$ such that $v \notin S_i$ **then**
8:             Create a new node $2i + 1$ as a child of i
9:             $S_{2i+1} = S_i \cup Y'$, $N_{2i+1} = Y' \setminus S_i$, and $M_{2i+1} = M_i \setminus \{X' \to Y' \vee Z'\}$
10:            Create a new node $2i + 2$ as a child of i
11:            $S_{2i+2} = S_i \cup Z'$, $N_{2i+2} = Z' \setminus S_i$, and $M_{2i+2} = M_i \setminus \{X' \to Y' \vee Z'\}$
12:        **end if**
13:    **until** no more nodes can be created
14:    **for** every node j that is a leaf **do**
15:        Create $\mathcal{I}$ with $\mathsf{true}(\mathcal{I}) = S_j \cup \{v \mid S_j = V \setminus \{v\}$ and $V \setminus \{v\} \to v \in \mathcal{P}\}$
16:        $\mathcal{S} := \mathcal{S} \cup \{\mathcal{I}\}$
17:    **end for**
18:    **return** ($\mathcal{S}$)
19: **end function**

Example. Let $X = \{1, 2, 3, 4\}$ and $\mathcal{P} = \{c_1 = 13 \to 257 \vee 468, c_2 = 12 \to 34 \vee 5678, c_3 = 145 \to 26 \vee 378, c_4 = 1234567 \to 8\}$ be a set of mvd clauses. In Line 6, the choice of an mvd clause made by Algorithm 1 is non-deterministic and in this example we choose clauses in the same order they appear above. The root of the semantic tree of $\mathcal{P}$ and X has $S_0 = \{1, 2, 3, 4\}$, $N_0 = \{1, 2, 3, 4\}$, and $M_0 = \{c_1, c_2, c_3, c_4\}$. Choosing the first clause c_1 we have $S_1 = \{1, 2, 3, 4, 5, 7\}$, $N_1 = \{5, 7\}$, $S_2 = \{1, 2, 3, 4, 6, 8\}$, $N_2 = \{6, 8\}$, and $M_1 = M_2 = \{c_2, c_3, c_4\}$. Now we choose the second clause c_2 to obtain $S_3 = S_1$, $N_3 = \emptyset$, $S_4 = \{1, 2, 3, 4, 5, 6, 7, 8\}$, $N_4 = \{6, 8\}$, $S_5 = S_2$, $N_5 = \emptyset$, $S_6 = S_4$, $N_6 = \{5, 7\}$ and $M_3 = M_4 = M_5 = M_6 = \{c_3, c_4\}$.

Finally, we choose third clause. Figure 1 illustrates the sets N_i, which represent the new variables in each node. In Line 15, Algorithm 1 checks that the antecedent of c_4 is satisfied in node 7 and adds variable 8 to its corresponding interpretation. Algorithm 1, returns $\mathcal{S} = \{\mathcal{I}_1, \mathcal{I}_2, \mathcal{I}_3\}$, with $\mathsf{true}(\mathcal{I}_1) = \{1, 2, 3, 4, 5, 6, 7, 8\}$, $\mathsf{true}(\mathcal{I}_2) = \{1, 2, 3, 4, 5, 7, 8\}$ and $\mathsf{true}(\mathcal{I}_3) = \{1, 2, 3, 4, 6, 8\}$.

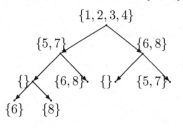

Fig. 1. Semantic Tree

The following Theorem shows that Algorithm 1 runs in polynomial time and that the returned set $\mathcal{S}$ includes all minimal models satisfying $\mathcal{P}$ and X.

Theorem 1. *Let $\mathcal{P}$ be a set of mvd clauses and X a set of variables. One can construct in polynomial time a set of interpretations $\mathcal{S}$ that verifies the following properties:*

1. *if $\mathcal{I} \in \mathcal{S}$ then $\mathcal{I} \models \mathcal{P}$;*
2. *if $\mathcal{J} \models \mathcal{P}$ and $X \subseteq \mathsf{true}(\mathcal{J})$ then there is $\mathcal{I} \in \mathcal{S}$ such that $\mathsf{true}(\mathcal{I}) \subseteq \mathsf{true}(\mathcal{J})$.*

Proof. Let $\mathcal{S}$ be the return value of Algorithm 1 with $\mathcal{P}$ and X as input. Point (1) is a corollary of a more general one: for all nodes i in a semantic tree, the interpretation $\mathcal{J}$ defined as $\mathsf{true}(\mathcal{J}) = S_i$ is a model of the set of mvd clauses $X' \rightarrow Y' \vee Z' \in \mathcal{P} \setminus M_i$ with Y' and Z' non-empty. The proof of this fact is by induction in the number of levels of the semantic tree. (Algorithm 1 treats Horn-like clauses in Line 15.) Point (2) is a corollary of a more general one: if $\mathcal{J} \models \mathcal{P}$ and $X \subseteq \mathsf{true}(\mathcal{J})$, then there exists a path from the root to a node k that is leaf, in such a way that $S_k \subseteq \mathsf{true}(\mathcal{J})$. The proof of this fact is again by induction in the number of levels of the semantic tree.

Now, it remains to show that the construction of $\mathcal{S}$ is in polynomial time. Let $n = |V|$ and $m = |\mathcal{P}|$. The fact that the number of nodes of any semantic tree for $\mathcal{P}$ and X is bounded by $2 \times n \times m$ follows from the next 3 claims.

Claim 1. For any level j whose nodes are $j_1, j_2, \ldots j_k$, the set $\{N_{j_1}, N_{j_2}, \ldots, N_{j_k}\}$ is pairwise disjoint. .

Suppose at level j there are nodes a_1 and a_2 and a variable v such that $v \in N_{a_1} \cap N_{a_2}$. This means that in the lowest common ancestor c of both nodes a_1 and a_2, we have that $v \notin S_c$. Then, by construction of the semantic tree, this also means that exactly one of the two children b_1 or b_2 of the ancestor c must have introduced v in S_{b_1} or in S_{b_2}. Therefore, it is not possible to find $v \in N_{a_1} \cap N_{a_2}$.

Claim 2. For any level j whose nodes are $j_1, j_2, \ldots j_k$, the set $\{N_{j_1}, N_{j_2}, \ldots, N_{j_k}\}$ contains at most $k/2$ empty sets. .

In the execution of Algorithm 1, whenever two children of a node are created, at least one new variable is introduced in at least one of the siblings.

Thus, the number of nodes in each level is bounded by $2 \times n$.

Claim 3. The depth of a semantic tree for $\mathcal{P}$ is bounded by m. .

This is because any mvd clause is used at most once along a branch. Note that this claim also ensures termination.   ❏

It is worth saying that Algorithm 1 allows us to decide whether a set of mvd clauses $\mathcal{P}$ is satisfiable. Note that if $V \rightarrow \mathbf{F} \notin \mathcal{P}$ then $\mathcal{P}$ is trivially satisfiable. Otherwise, we only need to set the input X as empty and check whether $\mathcal{S}$ (the return of Algorithm 1) contains only $\mathcal{I}$ such that $\mathsf{true}(\mathcal{I}) = V$. If so then $\mathcal{P} \cup \{V \rightarrow \mathbf{F}\}$ is unsatisfiable.

4   Learning MVDF from 2-Quasi-Horn

In this section we present an algorithm that learns the class MVDF from 2-quasi-Horn. More precisely, we show that the learning framework $\mathfrak{F}_M = (E_Q, \mathcal{L}_M, \mu_M)$ (defined in the Preliminaries) is polynomially exact learnable.

Algorithm 2 maintains a sequence $\mathfrak{L}$ of interpretations used to construct the learner's hypothesis $\mathcal{H}$. In each iteration of the main loop, if $\mathcal{H}$ is not equivalent to the target $\mathcal{T}$ then the oracle $\mathsf{EQ}_{\mathcal{T}, E_Q}$ provides the learner with a 2-quasi-Horn clause c that is a positive counterexample. That is, $\mathcal{T} \models c$ and $\mathcal{H} \not\models c$. The assumption that the counterexample is positive is justified by the construction of $\mathcal{H}$, which ensures at all times that $\mathcal{T} \models \mathcal{H}$. Each *positive* counterexample is used to construct a *negative* countermodel that either refines an element of $\mathfrak{L}$, or is added to the end of $\mathfrak{L}$. In order to learn all of the clauses in $\mathcal{T}$, we would like the clauses induced by the elements in $\mathfrak{L}$ to approximate distinct clauses in $\mathcal{T}$. This will happen if at most polynomially many elements in $\mathfrak{L}$ violate the same clause in $\mathcal{T}$. As explained in [4], overzealous refinement may result in several elements in $\mathfrak{L}$ violating the same clause of $\mathcal{T}$. This is avoided in Algorithm 2 by (1) refining at most one (the first) element of $\mathfrak{L}$ per iteration and (2) previously refining the constructed countermodel with the elements of $\mathfrak{L}$.

The following notion essentially describes under which conditions we say that it is 'good' to refine an interpretation (which can be either a countermodel or an element of $\mathfrak{L}$). There are two cases this can happen in our algorithm: (1) an element in $\mathfrak{L}$ is refined with a countermodel (Line 10 of Algorithm 2) or (2) the countermodel is refined with some element in $\mathfrak{L}$ (Line 5 of Algorithm 3).

Definition 1. *We say that a pair* $(\mathcal{I}, \mathcal{J})$ *of interpretations is a* goodCandidate *if: (i)* $\mathsf{true}(\mathcal{I} \cap \mathcal{J}) \subset \mathsf{true}(\mathcal{I})$; *(ii)* $\mathcal{I} \cap \mathcal{J} \models \mathcal{H}$; *and (iii)* $\mathcal{I} \cap \mathcal{J} \not\models \mathcal{T}$.

Algorithms for Horn formulas in [4,9] use a notion of 'goodCandidate' that is more relaxed than ours. They only need conditions (i) and (iii). The reason is that (ii) always holds because the intersection of two models of a set of Horn clauses $\mathcal{H}$ is also a model of $\mathcal{H}$. The lack of this property in the case of MVDF has two consequences. The first one is that there is not a unique minimal model that satisfies both an mvd formula and a particular set of variables. We solved this problem in the previous section, by constructing a semantic tree. The second consequence is that in the Horn algorithm [4] only a single interpretation of the sequence that the algorithm maintains can violate a Horn clause from the target. However, in our algorithm any mvd clause of the target $\mathcal{T}$ can be violated by polynomially many interpretations of the sequence $\mathfrak{L}$. This fact causes that, eventually, a polynomial amount of interpretations can be removed from $\mathfrak{L}$.

Remark 2. In the rest of this paper we consider interpretations $\mathcal{I}$ such that $|\mathsf{false}(\mathcal{I})| \geq 1$. This is justified by the fact that in Line 1 of Algorithm 2 we check whether $\mathcal{T} \models V \rightarrow \mathbf{F}$ and if so we add it to $\mathcal{H}_0$. Then, any negative countermodel $\mathcal{I}$ computed in Line 6 is such that $|\mathsf{false}(\mathcal{I})| \geq 1$.

Lemma 1 shows how the learner can decide Point (iii) of Definition 1 with polynomially many 2-quasi-Horn entailment queries. It also shows how Line 6 of Algorithm 2 can be implemented.

Lemma 1. *Let $\mathcal{I}$ be an interpretation and $\mathcal{T}$ the target (set of mvd clauses). One can decide in polynomial time whether $\mathcal{I}$ satisfies $\mathcal{T}$ using polynomially many 2-quasi-Horn entailment queries.*

Proof. Let $C = \{\text{true}(\mathcal{I}) \to v \vee w \mid v, w \in \text{false}(\mathcal{I}), \mathcal{T} \models \text{true}(\mathcal{I}) \to v \vee w\} \cup \{\text{true}(\mathcal{I}) \to v \mid \text{true}(\mathcal{I}) = V \setminus \{v\}, \mathcal{T} \models \text{true}(\mathcal{I}) \to v\}$. We show that $\mathcal{I}$ does not satisfy $\mathcal{T}$ if, and only if, there is $c \in C$ such that $\mathcal{T} \models c$. ($\Rightarrow$) If $\mathcal{I}$ does not satisfy $\mathcal{T}$ then there is $X \to Y \vee Z \in \mathcal{T}$ that is violated by $\mathcal{I}$. That is, $X \subseteq \text{true}(\mathcal{I})$ and: (a) Y and Z are non-empty and there are $v \in Y$ and $w \in Z$ such that $v, w \in \text{false}(\mathcal{I})$; or (b) there is v such that $\text{false}(\mathcal{I}) = \{v\}$ and $X \to Y \vee Z$ is the Horn-like clause $V \setminus \{v\} \to v$. In case (a) we have that $\mathcal{T} \models \text{true}(\mathcal{I}) \to v \vee w$. In case (b) we have $\mathcal{T} \models \text{true}(\mathcal{I}) \to v$ (meaning that $\mathcal{T} \models V \setminus \{v\} \to v$). ($\Leftarrow$) Follows from the fact that for all $c \in C$, $\mathcal{I} \not\models c$. ❏

In Line 5 Algorithm 2 calls a function that builds a semantic tree (Algorithm 1) for $\mathcal{H}$ and the antecedent of the counterexample given in Line 4. Using this tree, by Lemma 2, one can create in polynomial time a set of interpretations $\mathcal{S}$ such that (i) for all $\mathcal{I} \in \mathcal{S}$, $\mathcal{I} \models \mathcal{H}$ and (ii) there is an interpretation $\mathcal{I} \in \mathcal{S}$ such that $\mathcal{I} \not\models \mathcal{T}$. That is, there is $\mathcal{I} \in \mathcal{S}$ such that $\mathcal{I}$ is a negative countermodel.

Lemma 2. *Let c be a positive counterexample received in Line 4 of Algorithm 2. That is, $\mathcal{T} \models c$ and $\mathcal{H} \not\models c$. One can construct in polynomial time a set of interpretations $\mathcal{S}$ such that all elements of $\mathcal{S}$ satisfy $\mathcal{H}$ and, at least, one element of $\mathcal{S}$ does not satisfy $\mathcal{T}$.*

Proof. Let $\mathcal{S}$ be return of Algorithm 1 with $\mathcal{H}$ and X as the input, where X is the set of variables in the antecedent of c. $\mathcal{S}$ verifies the following properties: (1) if $\mathcal{I} \in \mathcal{S}$ then $\mathcal{I} \models \mathcal{H}$; (2) if $\mathcal{J} \models \mathcal{H}$ and $X \subseteq \text{true}(\mathcal{J})$ then there is $\mathcal{I} \in \mathcal{S}$ such that $\text{true}(\mathcal{I}) \subseteq \text{true}(\mathcal{J})$; and (3) there is an interpretation $\mathcal{I} \in \mathcal{S}$ such that $\mathcal{I} \not\models \mathcal{T}$. By Theorem 1 we have Points (1) and (2) and the fact that $\mathcal{S}$ is computed in polynomial time w.r.t. $|\mathcal{T}|$. For Point (3), we show that there is an interpretation $\mathcal{I} \in \mathcal{S}$ such that $\mathcal{I} \not\models \mathcal{T}$. As $\mathcal{H} \not\models c$, there must exist an interpretation $\mathcal{J}$ such that $\mathcal{J} \models \mathcal{H}$ and $\mathcal{J} \not\models c$. Thus, $X \subseteq \text{true}(\mathcal{J})$ and by Points (1) and (2), there must exist $\mathcal{I} \models \mathcal{H}$ such that $\text{true}(\mathcal{I}) \subseteq \text{true}(\mathcal{J})$. Since for all $\mathcal{I} \in \mathcal{S}$, we have that $X \subseteq \text{true}(\mathcal{I})$, the latter fact ensures that $\mathcal{I} \not\models c$ and therefore $\mathcal{I} \not\models \mathcal{T}$. ❏

Given a set of 2-quasi-Horn clauses, Algorithm 4 transforms each 2-quasi-Horn clause c into an mvd clause c' such that $\{c'\} \models c$ with polynomially many 2-quasi-Horn entailment queries. This property is exploited by the learner to generate mvd clauses for the hypothesis in Line 15 of Algorithm 2. Lines 5-11 of Algorithm 4 rely on Lemma 4. Lemma 4 requires the following technical lemma.

Lemma 3. *Let $\mathcal{T}$ be a set of mvd clauses. If $\mathcal{I}_1$ and $\mathcal{I}_2$ are models such that $\mathcal{I}_1 \models \mathcal{T}$ and $\mathcal{I}_2 \models \mathcal{T}$, but $\mathcal{I}_1 \cap \mathcal{I}_2 \not\models \mathcal{T}$, then $\text{true}(\mathcal{I}_1) \cup \text{true}(\mathcal{I}_2) = V$.*

Proof. If $\mathcal{I}_1 \models \mathcal{T}, \mathcal{I}_2 \models \mathcal{T}$ and $\mathcal{I}_1 \cap \mathcal{I}_2 \not\models \mathcal{T}$ then there is $X \rightarrow Y \vee Z \in \mathcal{T}$ such that $X \subseteq \mathsf{true}(\mathcal{I}_1 \cap \mathcal{I}_2)$ and $v, w \in \mathsf{false}(\mathcal{I}_1 \cap \mathcal{I}_2)$ with $v \in Y$ and $w \in Z$. Suppose $v \notin \mathsf{true}(\mathcal{I}_1) \cup \mathsf{true}(\mathcal{I}_2)$. Then, as $\mathcal{I}_1 \models \mathcal{T}$ and $\mathcal{I}_2 \models \mathcal{T}$, we have that $w \in \mathsf{true}(\mathcal{I}_1 \cap \mathcal{I}_2)$. This contradicts the fact that $w \in \mathsf{false}(\mathcal{I}_1 \cap \mathcal{I}_2)$. Thus, either $v \in \mathsf{true}(\mathcal{I}_1)$ or $v \in \mathsf{true}(\mathcal{I}_2)$. By the same argument we also have $w \in \mathsf{true}(\mathcal{I}_1) \cup \mathsf{true}(\mathcal{I}_2)$. As $v, w \in \mathsf{false}(\mathcal{I}_1 \cap \mathcal{I}_2)$ are arbitrary variables such that $v \in Y$ and $w \in Z$, this holds for any v, w with these properties. Since $X \subseteq \mathsf{true}(\mathcal{I}_1 \cap \mathcal{I}_2)$, we also have that $X \subseteq \mathsf{true}(\mathcal{I}_1) \cup \mathsf{true}(\mathcal{I}_2)$. Then $\mathsf{true}(\mathcal{I}_1) \cup \mathsf{true}(\mathcal{I}_2) = X \cup Y \cup Z = V$.   ❑

Lemma 4. *Let $\mathcal{T}$ be a set of mvd clauses. If $\mathcal{T} \models V_1 \rightarrow V_2 \vee V_3$ then either $\mathcal{T} \models V_1 \rightarrow V_2\{v\} \vee V_3$ or $\mathcal{T} \models V_1 \rightarrow V_2 \vee V_3\{v\}$, where $V_1, V_2, V_3, \{v\} \subseteq V$ and V_2, V_3 are non-empty[2].*

Proof. We can assume that $v \notin V_1$ and the proof is by reduction to the absurd: suppose $\mathcal{T} \models V_1 \rightarrow V_2 \vee V_3$ but $\mathcal{T} \not\models V_1 \rightarrow V_2\{v\} \vee V_3$ and $\mathcal{T} \not\models V_1 \rightarrow V_2 \vee V_3\{v\}$. Then, there are two models $\mathcal{I}_1$ and $\mathcal{I}_2$ such that: (a) $\mathcal{I}_1 \models \mathcal{T}$ and $\mathcal{I}_2 \models \mathcal{T}$; (b) $\mathcal{I}_1 \models V_1 \rightarrow V_2 \vee V_3$ and $\mathcal{I}_2 \models V_1 \rightarrow V_2 \vee V_3$; and (c) $\mathcal{I}_1 \not\models V_1 \rightarrow V_2\{v\} \vee V_3$ and $\mathcal{I}_2 \not\models V_1 \rightarrow V_2 \vee V_3\{v\}$. If $\mathcal{I}$ is a model such that $\mathcal{I} \not\models V_1 \rightarrow V_2 \vee V_3$, then $V_1 \subseteq \mathsf{true}(\mathcal{I})$. Then, by Point (c), we can ensure that $V_1 \subseteq \mathsf{true}(\mathcal{I}_1) \cap \mathsf{true}(\mathcal{I}_2)$ and using Point (b), also that: $V_2 \subseteq \mathsf{true}(\mathcal{I}_1)$ or $V_3 \subseteq \mathsf{true}(\mathcal{I}_1)$; and; $V_2 \subseteq \mathsf{true}(\mathcal{I}_2)$ or $V_3 \subseteq \mathsf{true}(\mathcal{I}_2)$.

We analyze all possibilities: $V_3 \subseteq \mathsf{true}(\mathcal{I}_1)$ and $V_2 \subseteq \mathsf{true}(\mathcal{I}_2)$ do not occur because it is a contradiction with Point (c). W.l.o.g. the only possibility is that: $V_2 \subseteq \mathsf{true}(\mathcal{I}_1)$, $V_3 \subseteq \mathsf{true}(\mathcal{I}_2)$ and $v \notin \mathsf{true}(\mathcal{I}_1) \cap \mathsf{true}(\mathcal{I}_2)$. As v is neither in $\mathsf{true}(\mathcal{I}_1)$ nor in $\mathsf{true}(\mathcal{I}_2)$, we have that $\mathsf{true}(\mathcal{I}_1) \cup \mathsf{true}(\mathcal{I}_2) \neq V$. By Lemma 3 and Point (a), necessarily we have $\mathcal{I}_1 \cap \mathcal{I}_2 \models \mathcal{T}$. But this is a contradiction with our assumption that $\mathcal{T} \models V_1 \rightarrow V_2 \vee V_3$, because we have found the model $\mathcal{I}_1 \cap \mathcal{I}_2$, which is a model of $\mathcal{T}$, but is not a model of $V_1 \rightarrow V_2 \vee V_3$.   ❑

If Algorithm 2 terminates, then it obviously has found a $\mathcal{H}$ that is logically equivalent to $\mathcal{T}$. It thus remains to show that the algorithm terminates in polynomial time. We can see the hypothesis $\mathcal{H}$ as a sequence of sets of entailments, where each $\mathcal{H}_i$ corresponds to the transformation of the set "BuildClauses" with $\mathcal{I}_i$ in $\mathcal{L}$ as input into mvd clauses (Line 15 of Algorithm 2). By Line 2 of Algorithm 2 the number of entailments created by "BuildClauses" is bounded by $|V|^2 + 1$. Lemmas 5 to 10 show that at all times the number of interpretations that violate a clause in $\mathcal{T}$ is bounded by $|V|$.

Lemma 5. *Let $\mathcal{L}$ be the sequence produced by Algorithm 2. Assume an interpretation $\mathcal{I}$ violates $c \in \mathcal{T}$. For all $\mathcal{I}_i \in \mathcal{L}$ such that $\mathcal{I}_i$ covers c, $\mathsf{true}(\mathcal{I}_i) \subseteq \mathsf{true}(\mathcal{I})$ if, and only if, $\mathcal{I} \not\models BuildClauses(\mathcal{I}_i)$.*

Proof. Let c be the mvd clause $X \rightarrow Y \vee Z$. The ($\Leftarrow$) direction is trivial. Now, suppose that $\mathsf{true}(\mathcal{I}_i) \subseteq \mathsf{true}(\mathcal{I})$ to prove ($\Rightarrow$). As $\mathcal{I} \not\models X \rightarrow Y \vee Z$, we have that $X \subseteq \mathsf{true}(\mathcal{I})$ and: (a) Y and Z are non-empty and there are $v \in Y$ and

---

[2] $V_i\{v\}$ is the conjunction of variables in V_i and v, where $i \in \{1, 2\}$.

Algorithm 2. Learning algorithm for MVDF by 2-quasi-Horn

1: Let $\mathcal{H}_0 = \{V \to \mathbf{F} \mid \mathcal{T} \models V \to \mathbf{F}\}$, $\mathfrak{L} = \emptyset$, $\mathcal{H} = \mathcal{H}_0$
2: Let BuildClauses($\mathcal{I}$) be the function that given an interpretation $\mathcal{I}$ with $|\mathsf{false}(\mathcal{I})| \geq$
   1 returns $\{\mathsf{true}(\mathcal{I}) \to v \vee w \mid v, w \in \mathsf{false}(\mathcal{I}), \mathcal{T} \models \mathsf{true}(\mathcal{I}) \to v \vee w\} \cup \{\mathsf{true}(\mathcal{I}) \to$
   $v \mid \mathsf{true}(\mathcal{I}) = V \setminus \{v\}, \mathcal{T} \models \mathsf{true}(\mathcal{I}) \to v\}$
3: **while** $\mathcal{H} \not\equiv \mathcal{T}$ **do**
4:     Let $X \to v \vee w$ (or $X \to v$) be a 2-quasi-Horn positive counterexample
5:     Let S be the return value of SEMANTICTREE($\mathcal{H}, X$)
6:     Find $\mathcal{I} \in S$ such that $\mathcal{I} \not\models \mathcal{T}$ (we know that for all $\mathcal{I} \in S, \mathcal{I} \models \mathcal{H}$)
7:     $\mathcal{J} := $ REFINECOUNTERMODEL($\mathcal{I}$)
8:     **if** there is $\mathcal{I}_k \in \mathfrak{L}$ such that goodCandidate($\mathcal{I}_k, \mathcal{J}$) **then**
9:         Let $\mathcal{I}_i$ be the first in $\mathfrak{L}$ such that goodCandidate($\mathcal{I}_i, \mathcal{J}$)
10:        Replace $\mathcal{I}_i$ by ($\mathcal{I}_i \cap \mathcal{J}$)
11:        Remove all $\mathcal{I}_j \in \mathfrak{L}$ such that $\mathcal{I}_j \not\models$ BuildClauses($\mathcal{I}_i \cap \mathcal{J}$)
12:    **else**
13:        Append $\mathcal{J}$ to $\mathfrak{L}$
14:    **end if**
15:    Construct $\mathcal{H} = \mathcal{H}_0 \cup$ TRANSFORMMVDF($\bigcup_{\mathcal{I} \in \mathfrak{L}}$ BuildClauses($\mathcal{I}$))
16: **end while**

Algorithm 3. Function to Refine the Countermodel

1: **function** REFINECOUNTERMODEL($\mathcal{I}$)
2:     Let $\mathcal{J} := \mathcal{I}$
3:     **if** there is $\mathcal{I}_k \in \mathfrak{L}$ such that goodCandidate($\mathcal{I}, \mathcal{I}_k$) **then**
4:         Let $\mathcal{I}_i$ be the first in $\mathfrak{L}$ such that goodCandidate($\mathcal{I}, \mathcal{I}_i$)
5:         $\mathcal{J} := $ REFINECOUNTERMODEL($\mathcal{I} \cap \mathcal{I}_i$)
6:     **end if**
7:     return($\mathcal{J}$)
8: **end function**

$w \in Z$ such that $v, w \in \mathsf{false}(\mathcal{I})$; or (b) there is v such that $\mathsf{false}(\mathcal{I}) = \{v\}$ and $X \to Y \vee Z$ is the Horn-like clause $V \setminus \{v\} \to v$. In case (a), as $\mathsf{true}(\mathcal{I}_i) \subseteq \mathsf{true}(\mathcal{I})$, we have that $v \in Y \setminus \mathsf{true}(\mathcal{I}_i)$ and $w \in Z \setminus \mathsf{true}(\mathcal{I}_i)$. Since $\mathcal{I}_i$ covers $X \to Y \vee Z$, $X \subseteq \mathsf{true}(\mathcal{I}_i)$. Then $\mathcal{T} \models \mathsf{true}(\mathcal{I}_i) \to v \vee w$. By definition of $\mathcal{H}_i$, there is $X_i \to Y_i \vee Z_i \in \mathcal{H}_i$ such that $v \in Y_i$ and $w \in Z_i$. But this means that $\mathcal{I} \not\models$ BuildClauses($\mathcal{I}_i$). Case (b) is similar, we have that $\{v\} = \mathsf{false}(\mathcal{I}_i)$ and $V \setminus \{v\} = \mathsf{true}(\mathcal{I}_i)$. Then $\mathcal{T} \models \mathsf{true}(\mathcal{I}_i) \to v$ and we also have $\mathcal{I} \not\models$ BuildClauses($\mathcal{I}_i$).   ❏

Lemma 6. *Let $\mathfrak{L}$ be the sequence produced by Algorithm 2. Assume a negative countermodel $\mathcal{I}$ violates $c \in \mathcal{T}$. For all $\mathcal{I}_i \in \mathfrak{L}$ such that $\mathcal{I}_i$ covers c the following holds: (1) $\mathsf{true}(\mathcal{I}_i \cap \mathcal{I}) \subset \mathsf{true}(\mathcal{I}_i)$; and (2) $\mathcal{I}_i \cap \mathcal{I} \not\models \mathcal{T}$.*

Proof. As $\mathcal{I}$ is a negative countermodel, $\mathcal{I} \models \mathcal{H}$, and then $\mathcal{I} \models$ BuildClauses($\mathcal{I}_i$). By Lemma 5, $\mathsf{true}(\mathcal{I}_i) \not\subseteq \mathsf{true}(\mathcal{I})$. Therefore, $\mathsf{true}(\mathcal{I}_i \cap \mathcal{I}) \subset \mathsf{true}(\mathcal{I}_i)$. For Point 2, as $\mathcal{I}_i$ covers $c \in \mathcal{T}$ and $\mathcal{I}$ violates $c \in \mathcal{T}$, we have that $\mathcal{I}_i \cap \mathcal{I}$ violates $c \in \mathcal{T}$. Then, $\mathcal{I}_i \cap \mathcal{I} \not\models \mathcal{T}$.   ❏

Algorithm 4. Transform a 2-quasi-Horn clause into an mvd clause

```
1: function TRANSFORMMVDF( H' )
2:     H := {c ∈ H' | c is of the form V \ {v} → v}
3:     for every X → v ∨ w ∈ H' do
4:         Let W = V \ (X ∪ {v, w}), Y = {v} and Z = {w}
5:         for each w' ∈ W do
6:             if T ⊨ X → Y{w'} ∨ Z then
7:                 add w' to Y
8:             else
9:                 add w' to Z
10:            end if
11:        end for
12:        add X → Y ∨ Z to H
13:    end for
14:    return (H)
15: end function
```

Lemma 7. *Let $\mathfrak{L}$ be the sequence produced by Algorithm 2. At all times, for all $\mathcal{I}_i \in \mathfrak{L}$, $\mathcal{I}_i \models \mathcal{H} \setminus \mathcal{H}_i$.*

Proof. By Lemma 2 and Algorithm 3, the interpretation $\mathcal{J}$ computed in Line 7 of Algorithm 2 is a negative countermodel. So if Algorithm 2 executes Line 13 then it holds that $\mathcal{J} \models \mathcal{H}$. If there exists $\mathcal{I}_j \in \mathfrak{L}$ such that $\mathcal{I}_j \not\models \text{BuildClauses}(\mathcal{J})$, then $\text{true}(\mathcal{J}) \subset \text{true}(\mathcal{I}_j)$ and the pair $(\mathcal{I}_j, \mathcal{J})$ is a goodCandidate. This contradicts the fact that the algorithm did not replace some interpretation in $\mathfrak{L}$. Otherwise, Algorithm 2 executes Line 10 and, then, an interpretation $\mathcal{I}_i \in \mathfrak{L}$ is replaced with $\mathcal{I}_i \cap \mathcal{J}$, where the pair $(\mathcal{I}_i, \mathcal{J})$ is a goodCandidate. In this case, by Definition 1 part (ii), $\mathcal{I}_i \cap \mathcal{J} \models \mathcal{H}$. It remains to check that for any other $\mathcal{I}_j \in \mathfrak{L}$ it holds $\mathcal{I}_j \models \text{BuildClauses}(\mathcal{I}_i \cap \mathcal{J})$, but this is always true because of Line 11. ☐

Lemma 8. *Let $\mathfrak{L}$ be the sequence produced by Algorithm 2. If Algorithm 2 replaces some $\mathcal{I}_i \in \mathfrak{L}$ with $\mathcal{I} \cap \mathcal{I}_i$ then $\text{false}(\mathcal{I}_i) \subseteq \text{false}(\mathcal{I})$ ($\mathcal{I}_i$ before the replacement).*

Proof. If $\text{true}(\mathcal{I}_i \cap \mathcal{I}) = \text{true}(\mathcal{I})$ then $\text{false}(\mathcal{I}_i) \subseteq \text{false}(\mathcal{I})$. We thus suppose to the contrary that (∗) $\text{true}(\mathcal{I} \cap \mathcal{I}_i) \subset \text{true}(\mathcal{I})$. If Algorithm 2 replaced $\mathcal{I}_i \in \mathfrak{L}$ then $(\mathcal{I}_i, \mathcal{I})$ is a goodCandidate. Then, $\mathcal{I}_i \cap \mathcal{I} \not\models T$ and $\mathcal{I}_i \cap \mathcal{I} \models \mathcal{H}$. If (i) $\text{true}(\mathcal{I} \cap \mathcal{I}_i) \subset \text{true}(\mathcal{I})$ (by (∗)), (ii) $\mathcal{I} \cap \mathcal{I}_i \models \mathcal{H}$ and (iii) $\mathcal{I}_i \cap \mathcal{I} \not\models T$; then $(\mathcal{I}, \mathcal{I}_i)$ is a goodCandidate. This contradicts the condition in Line 3 of Algorithm 3, which would not return $\mathcal{I}$ but make a recursive call with $\mathcal{I} \cap \mathcal{I}_i$ and, thus, $\text{true}(\mathcal{I}_i \cap \mathcal{I}) = \text{true}(\mathcal{I})$. ☐

Lemma 9. *Let $\mathfrak{L}$ be the sequence produced by Algorithm 2. At all times the following holds. Let $\mathcal{J}$ be a countermodel computed in Line 7 of Algorithm 2 that violates $c \in T$. Then, either:*

- *Algorithm 2 replaces some $\mathcal{I}_j \in \mathfrak{L}$ with $\mathcal{J} \cap \mathcal{I}_j$, or*
- *Algorithm 2 appends $\mathcal{J}$ in $\mathfrak{L}$ and, for all $\mathcal{I}_j \in \mathfrak{L}$ such that $\mathcal{I}_j$ covers c, we have that* $\mathsf{false}(\mathcal{I}_j) \cap \mathsf{false}(\mathcal{J}) = \emptyset$.

Proof. If Algorithm 2 replaces some $\mathcal{I}_j \in \mathfrak{L}$ with $\mathcal{J} \cap \mathcal{I}_j$ or appends $\mathcal{J}$ in $\mathfrak{L}$ and, for all $\mathcal{I}_j \in \mathfrak{L}$ such that $\mathcal{I}_j$ covers c, we have that $\mathsf{false}(\mathcal{I}_j) \cap \mathsf{false}(\mathcal{J}) = \emptyset$, then we are done. Suppose this does not happen. Then,

- Algorithm 2 appends $\mathcal{J}$ in $\mathfrak{L}$, meaning that for all $\mathcal{I}_j \in \mathfrak{L}$, $(\mathcal{I}_j, \mathcal{J})$ is not a goodCandidate; and
- there exists $\mathcal{I}_j \in \mathfrak{L}$ that covers c and $\mathsf{false}(\mathcal{I}_j) \cap \mathsf{false}(\mathcal{J}) \neq \emptyset$.

By Lemma 6 we obtain conditions (i) and (iii) of Definition 1 (goodCandidate) for the pair $(\mathcal{I}_j, \mathcal{J})$: $\mathsf{true}(\mathcal{I}_j \cap \mathcal{J}) \subset \mathsf{true}(\mathcal{I}_j)$; and $\mathcal{I}_j \cap \mathcal{J} \not\models \mathcal{T}$. By Lemma 7, we have $\mathcal{I}_j \models \mathcal{H} \setminus \mathcal{H}_j$. As $\mathcal{J} \models \mathcal{H}$ and $\mathsf{true}(\mathcal{I}_j \cap \mathcal{J}) \subset \mathsf{true}(\mathcal{I}_j)$, by Lemma 3, if $\mathsf{false}(\mathcal{I}_j) \cap \mathsf{false}(\mathcal{J}) \neq \emptyset$ then we have $\mathcal{I}_j \cap \mathcal{J} \models \mathcal{H}$. This is condition (ii) of Definition 1, and therefore the pair $(\mathcal{I}_j, \mathcal{J})$ is a goodCandidate. This contradicts the fact that the algorithm did not refine some interpretation in $\mathfrak{L}$. ❑

Lemma 10. *Let $\mathfrak{L}$ be the sequence produced by Algorithm 2. At all times, for all $i \neq j$, if $\mathcal{I}_i \in \mathfrak{L}$ and $\mathcal{I}_j \in \mathfrak{L}$ violate $c \in \mathcal{T}$ then* $\mathsf{false}(\mathcal{I}_i) \cap \mathsf{false}(\mathcal{I}_j) = \emptyset$.

Proof. Suppose Algorithm 2 modifies the sequence in response to receiving a counterexample $\mathcal{I}$. Consider the possibilities.

- If Algorithm 2 appends $\mathcal{I}$ and there is $\mathcal{I}_i \in \mathfrak{L}$ such that $\mathcal{I}_i$ violates c then, by Lemma 9, $\mathsf{false}(\mathcal{I}_i) \cap \mathsf{false}(\mathcal{I}) = \emptyset$.
- Otherwise, the Algorithm 2 replaces $\mathcal{I}_i$ by $\mathcal{I}_i \cap \mathcal{I}$. Suppose the lemma fails to hold. Then, we have that $\mathcal{I}_i \cap \mathcal{I}$ and $\mathcal{I}_j$ violate c, $\mathsf{false}(\mathcal{I}_j) \cap \mathsf{false}(\mathcal{I}_i \cap \mathcal{I}) \neq \emptyset$, and $\mathcal{I}_j$ is not removed in Line 11.

The above means that $\mathcal{I}_j \models \mathrm{BuildClauses}(\mathcal{I}_i \cap \mathcal{I})$, and by Lemma 5 $\mathsf{true}(\mathcal{I}_i \cap \mathcal{I}) \not\subseteq \mathsf{true}(\mathcal{I}_j)$. As $\mathcal{I}_i \cap \mathcal{I} = \mathcal{I}$ (by Lemma 8), we actually have that: (a) $\mathcal{I}$ violates c, (b) $\mathsf{false}(\mathcal{I}_j) \cap \mathsf{false}(\mathcal{I}) \neq \emptyset$ and (c) $\mathsf{true}(\mathcal{I}) \not\subseteq \mathsf{true}(\mathcal{I}_j)$, or equivalently, $\mathsf{false}(\mathcal{I}_j) \not\subseteq \mathsf{false}(\mathcal{I})$. As $\mathcal{I}$ violates c (Point (a)), by Lemma 6, we obtain: $\mathsf{true}(\mathcal{I}_j \cap \mathcal{I}) \subset \mathsf{true}(\mathcal{I}_j)$ (condition (i) of Definition 1); and $\mathcal{I}_j \cap \mathcal{I} \not\models \mathcal{T}$ (condition (iii) of Definition 1). By Lemma 7, we have $\mathcal{I}_j \models \mathcal{H} \setminus \mathcal{H}_j$. As $\mathcal{I} \models \mathcal{H}$ and $\mathsf{true}(\mathcal{I}_j \cap \mathcal{I}) \subset \mathsf{true}(\mathcal{I}_j)$, by Lemma 3 and Point (b) we have $\mathcal{I}_j \cap \mathcal{I} \models \mathcal{H}$. This is condition (ii) of Definition 1, and therefore the pair $(\mathcal{I}, \mathcal{I}_j)$ is a goodCandidate. So Algorithm 3 makes a recursive call with $\mathcal{I} \cap \mathcal{I}_j$, which contradicts that $\mathsf{false}(\mathcal{I}_j) \not\subseteq \mathsf{false}(\mathcal{I})$ (Point (c)). ❑

By Lemma 10 if any two interpretations $\mathcal{I}_i, \mathcal{I}_j \in \mathfrak{L}$ violate the same clause in $\mathcal{T}$ then their sets of false variables are disjoint. As for each interpretation $|\mathsf{false}(\mathcal{I})| \geq 1$, the number of mutually disjoint interpretations violating any mvd clause in $\mathcal{T}$ is bounded by $|V|$. Since every $\mathcal{I}_i \in \mathfrak{L}$ is such that $\mathcal{I}_i \not\models \mathcal{T}$, we have every $\mathcal{I}_i \in \mathfrak{L}$ violates at least one $c \in \mathcal{T}$. This bounds the number of elements in $\mathfrak{L}$ to the number of mvd clauses in $\mathcal{T}$.

Corollary 1. *Let $\mathfrak{L}$ be the sequence produced by Algorithm 2. At all times every $c \in \mathcal{T}$ is violated by at most $|V|$ interpretations $\mathcal{I}_i \in \mathfrak{L}$.*

Then, at all times the number of elements in $\mathfrak{L}$ is bounded by $|\mathcal{T}| \cdot |V|$. As in each replacement the number of variables in the antecedent is strictly smaller, we have that the number of replacements that can be done for each $\mathcal{I}_i \in \mathfrak{L}$ is bounded by the number of variables, $|V|$. To ensure the progress of the algorithm, we also need to show that the number of iterations is polynomial in the size of $\mathcal{T}$.

The rest of this section is devoted to show an upper bound polynomial in $|\mathcal{T}|$ on the total number iterations of Algorithm 2. Before showing our upper bound in Lemma 13, we need the following two lemmas. Essentially, Lemma 11 state the main property obtained by our refinement conditions (Definition 1). Lemma 12 shows that (1) if an interpretation $\mathcal{I}_i$ is replaced and an element $\mathcal{I}_j$ is removed from $\mathfrak{L}$ then they are mutually disjoint; and (2) if any two elements are removed then they are mutually disjoint.

Lemma 11. *At all times the following holds. Let $\mathfrak{L}$ be the sequence produced by Algorithm 2. Let $\mathcal{I}_i, \mathcal{I}_j \in \mathfrak{L}$. W.l.o.g. assume $i < j$. Then, the pair $(\mathcal{I}_i, \mathcal{I}_j)$ is not a* goodCandidate.

Proof. Consider the possibilities. If Algorithm 2 appends $\mathcal{J}$ to $\mathfrak{L}$, then for all $\mathcal{I}_k \in \mathfrak{L}$ the pair $(\mathcal{I}_k, \mathcal{J})$ cannot be a goodCandidate, because the algorithm would satisfy the condition in Line 8 and instead of appending $\mathcal{J}$, it should replace an interpretation $\mathcal{I}_k$ in $\mathfrak{L}$. Otherwise, either Algorithm 2 replaces (a) $\mathcal{I}_i$ by $\mathcal{I}_i \cap \mathcal{J}$ or (b) $\mathcal{I}_j$ by $\mathcal{I}_j \cap \mathcal{J}$. Suppose the lemma fails to hold in case (a). The pair $((\mathcal{I}_i \cap \mathcal{J}), \mathcal{I}_j)$ is a goodCandidate. By Lemma 8, false$(\mathcal{I}_i) \subseteq$ false$(\mathcal{J})$ and $\mathcal{I}_i \cap \mathcal{J} = \mathcal{J}$. Then, the pair $(\mathcal{J}, \mathcal{I}_j)$ is a goodCandidate too. This contradicts the condition in Line 3 of Algorithm 3, which would not return $\mathcal{J}$ but make a recursive call with $\mathcal{J} \cap \mathcal{I}_j$. Now, suppose the lemma fails to hold in case (b). The pair $(\mathcal{I}_i, (\mathcal{I}_j \cap \mathcal{J}))$ is a goodCandidate. By Lemma 8, false$(\mathcal{I}_j) \subseteq$ false$(\mathcal{J})$ and $\mathcal{I}_j \cap \mathcal{J} = \mathcal{J}$. Therefore, the pair $(\mathcal{I}_i, \mathcal{J})$ is a goodCandidate too. This contradicts the fact that in Line 9 of Algorithm 2, the first goodCandidate is replaced and since $i < j$, $\mathcal{I}_i$ should be replaced instead of $\mathcal{I}_j$. ❑

Lemma 12. *Let $\mathfrak{L}$ be the sequence produced by Algorithm 2. In Algorithm 2 Line 11, the following holds:*

1. *if $\mathcal{I}_j$ is removed during the replacement of some $\mathcal{I}_i \in \mathfrak{L}$ by $\mathcal{I}_i \cap \mathcal{J}$ (Line 10) then* false$(\mathcal{I}_i) \cap$ false$(\mathcal{I}_j) = \emptyset$ *($\mathcal{I}_i$ before the replacement);*
2. *if $\mathcal{I}_j, \mathcal{I}_k$ with $j < k$ are removed during the replacement of some $\mathcal{I}_i \in \mathfrak{L}$ by $\mathcal{I}_i \cap \mathcal{J}$ (Line 10) then* false$(\mathcal{I}_j) \cap$ false$(\mathcal{I}_k) = \emptyset$.

Proof. We argue that under the conditions stated by this lemma if false$(\mathcal{I}_i) \cap$ false$(\mathcal{I}_j) = \emptyset$ (respectively, false$(\mathcal{I}_j) \cap$ false$(\mathcal{I}_k) = \emptyset$) does not hold then the pair $(\mathcal{I}_i, \mathcal{I}_j)$ (respectively, $(\mathcal{I}_j, \mathcal{I}_k)$) is a goodCandidate (Definition 1), which contradicts Lemma 11. In our proof by contradiction, we show that conditions (i), (ii) and (iii) of Definition 1 hold for both $(\mathcal{I}_i, \mathcal{I}_j)$ and $(\mathcal{I}_j, \mathcal{I}_k)$.

- For condition (i): assume to the contrary that $\mathsf{true}(\mathcal{I}_i) \subseteq \mathsf{true}(\mathcal{I}_j)$. As $\mathcal{I}_j \not\models$ BuildClauses$(\mathcal{I}_i \cap \mathcal{J})$, there is $c \in \mathcal{T}$ such that $\mathcal{I}_j$ and $\mathcal{I}_i \cap \mathcal{J}$ violate c. Therefore, $\mathcal{I}_i$ covers c and by Lemma 5, $\mathcal{I}_j \not\models$ BuildClauses$(\mathcal{I}_i)$, which is a contradiction with Lemma 7. Similarly, assume to the contrary that $\mathsf{true}(\mathcal{I}_j) \subseteq \mathsf{true}(\mathcal{I}_k)$. As $\mathcal{I}_k \not\models$ BuildClauses$(\mathcal{I}_i \cap \mathcal{J})$, there is $c \in \mathcal{T}$ such that $\mathcal{I}_k$ and $\mathcal{I}_i \cap \mathcal{J}$ violate c. Therefore, $\mathcal{I}_i$ covers c and by Lemma 5, $\mathcal{I}_k \not\models$ BuildClauses$(\mathcal{I}_i)$, which is a contradiction with Lemma 7.
- For condition (ii): as $\mathcal{I}_j \models \mathcal{H} \setminus \mathcal{H}_j$ (Lemma 7) we have $\mathcal{I}_j \models \mathcal{H} \setminus (\mathcal{H}_i \cup \mathcal{H}_j)$. By the same argument $\mathcal{I}_i \models \mathcal{H} \setminus (\mathcal{H}_i \cup \mathcal{H}_j)$. If $\mathsf{false}(\mathcal{I}_i) \cap \mathsf{false}(\mathcal{I}_j) \neq \emptyset$ then, by Lemma 3, $\mathcal{I}_i \cap \mathcal{I}_j \models \mathcal{H}$. Similarly, as $\mathcal{I}_j \models \mathcal{H} \setminus \mathcal{H}_j$ (Lemma 7) we have $\mathcal{I}_j \models \mathcal{H} \setminus (\mathcal{H}_j \cup \mathcal{H}_k)$. By the same argument $\mathcal{I}_k \models \mathcal{H} \setminus (\mathcal{H}_j \cup \mathcal{H}_k)$. If $\mathsf{false}(\mathcal{I}_j) \cap \mathsf{false}(\mathcal{I}_k) \neq \emptyset$ then, by Lemma 3, $\mathcal{I}_j \cap \mathcal{I}_k \models \mathcal{H}$.
- For condition (iii): by Line 11 of Algorithm 2, $\mathcal{I}_j \not\models$BuildClauses$(\mathcal{I}_i \cap \mathcal{J})$. Then there is $c \in \mathcal{T}$ such that $\mathcal{I}_j$ and $\mathcal{I}_i \cap \mathcal{J}$ violate c. If $\mathcal{I}_i \cap \mathcal{J}$ violates c then $\mathcal{I}_i$ covers c. Then, $\mathcal{I}_i \cap \mathcal{I}_j \not\models \mathcal{T}$. Similarly, by Line 11 of Algorithm 2, we have that $\mathcal{I}_k \not\models$BuildClauses$(\mathcal{I}_i \cap \mathcal{J})$. Then there is $c \in \mathcal{T}$ such that $\mathcal{I}_k$ and $\mathcal{I}_i \cap \mathcal{J}$ violate c. As $\mathcal{I}_j \not\models$BuildClauses$(\mathcal{I}_i \cap \mathcal{J})$, $\mathsf{false}(\mathcal{I}_j) \subseteq \mathsf{false}(\mathcal{I}_i \cap \mathcal{J})$. So $\mathcal{I}_j$ covers c. Then, $\mathcal{I}_k \cap \mathcal{I}_j \not\models \mathcal{T}$.

So conditions (i), (ii) and (iii) of Definition 1 hold for $(\mathcal{I}_i, \mathcal{I}_j)$ and $(\mathcal{I}_j, \mathcal{I}_k)$, which contradicts Lemma 11. Then, $\mathsf{false}(\mathcal{I}_i) \cap \mathsf{false}(\mathcal{I}_j) = \emptyset$ and $\mathsf{false}(\mathcal{I}_j) \cap \mathsf{false}(\mathcal{I}_k) = \emptyset$.
❑

We present a polynomial upper bound on the number of iterations of the main loop via a bound function. That is, an expression that decreases on every iteration and is always ≥ 0 inside the loop body. Note that we obtain this upper bound even though the learner does not know the size $|\mathcal{T}|$ of the target.

Lemma 13. *Let $\mathfrak{L}$ be the sequence produced by Algorithm 2. Let N be the constant $2 \cdot |V| \cdot |V| \cdot |\mathcal{T}|$. The expression $E = |\mathfrak{L}| + (N - 2 \cdot \sum_{\mathcal{I} \in \mathfrak{L}} |\mathsf{false}(\mathcal{I})|)$ always evaluates to a natural number inside the loop body and decreases on every iteration.*

Proof. By Corollary 1, the size of $\mathfrak{L}$ is bounded at all times by $|V| \cdot |\mathcal{T}|$. Thus, by Corollary 1, N is an upper bound for $2 \cdot \sum_{\mathcal{I} \in \mathfrak{L}} |\mathsf{false}(\mathcal{I})|$. There are three possibilities: (1) an element $\mathcal{I}$ is appended. Then, $|\mathfrak{L}|$ increases by one but $|\mathsf{false}(\mathcal{I})| \geq 1$ and, therefore, E decreases; (2) an element is replaced and no element is removed. Then, E trivially decreases. Otherwise, (3) we have that an element $\mathcal{I}_i$ is replaced and p interpretations are removed from $\mathfrak{L}$ in Line 11 of Algorithm 2. By Point 2 of Lemma 12, if $\mathcal{I}_i$ is replaced by $\mathcal{I}_i \cap \mathcal{J}$ and $\mathcal{I}_j, \mathcal{I}_k$ are removed then $\mathsf{false}(\mathcal{I}_j) \cap \mathsf{false}(\mathcal{I}_k) = \emptyset$. This means that if p interpretations are removed then their sets of false variables are all mutually disjoint. By Point 1 of Lemma 12, if $\mathcal{I}_i$ is replaced by $\mathcal{I}_i \cap \mathcal{J}$ and some $\mathcal{I}_j$ is removed then $\mathsf{false}(\mathcal{I}_j) \cap \mathsf{false}(\mathcal{I}_i) = \emptyset$. Then, the p interpretations also have sets of false variables disjoint from $\mathsf{false}(\mathcal{I}_i)$. For each interpretation $\mathcal{I}_j$ removed we have $\mathsf{false}(\mathcal{I}_j) \subseteq \mathsf{false}(\mathcal{J} \cap \mathcal{I}_i)$ (because $\mathcal{I}_j \not\models$BuildClauses$(\mathcal{J} \cap \mathcal{I}_i)$). Then, the number of 'falses' is at least as large as before. However $|\mathfrak{L}|$ decreases and, thus, we can ensure that E decreases.
❑

By Lemma 13, the total number of iterations of Algorithm 2 is bounded by a polynomial in $|\mathcal{T}|$ and $|V|$. We now state our main result.

Theorem 2. *The learning MVDF from 2-quasi-Horn framework $\mathfrak{F}_M$ is polynomially exact learnable.*

5   Conclusions and Open Problems

We presented an algorithm that exactly learns the class MVDF in polynomial time from 2-quasi-Horn clauses. As this class is a generalization of Horn and a restriction of 2-quasi-Horn, we extended the boundary between 1 and 2-quasi-Horn of what can be efficiently learned in the exact model. We would like to find similar algorithms where the examples are either mvd clauses or interpretations. A more general open problem is whether the ideas presented here can be extended to handle other restrictions of 2-quasi-Horn. Another direction is to use our algorithm to develop software to support the design of database schemas in 4NF.

Acknowledgments. We thank the reviewers. Hermo was supported by the Spanish Project TIN2013-46181-C2-2-R and the Basque Project GIU12/26 and grant UFI11/45. Ozaki is supported by the Science without Borders scholarship programme.

References

1. Angluin, D.: Learning k-term dnf formulas using queries and counterexamples. Technical report, Department of Computer Science, Yale University, August 1987
2. Angluin, D.: Queries and concept learning. Mach. Learn. **2**(4), 319–342 (1988)
3. Angluin, D.: Negative results for equivalence queries. Machine Learning **5**, 121–150 (1990)
4. Angluin, D., Frazier, M., Pitt, L.: Learning conjunctions of Horn clauses. Machine Learning **9**, 147–164 (1992)
5. Balcázar, J.L., Baixeries, J.: Characterizations of multivalued dependencies and related expressions. In: Suzuki, E., Arikawa, S. (eds.) DS 2004. LNCS (LNAI), vol. 3245, pp. 306–313. Springer, Heidelberg (2004)
6. Codd, E.F.: A relational model of data for large shared data banks. Communications of the ACM **13**(6), 377–387 (1970)
7. Delobel, C.: Normalization and hierarchical dependencies in the relational data model. ACM Transactions on Database Systems **3**(3), 201–222 (1978)
8. Fagin, R.: Multivalued dependencies and a new normal form for relational databases. ACM Transactions on Database Systems **2**, 262–278 (1977)
9. Frazier, M., Pitt, L.: Learning from entailment: an application to propositional Horn sentences. In: Machine Learning, Proceedings of the Tenth International Conference, University of Massachusetts, Amherst, MA, USA, June 27–29, 1993, pp. 120–127 (1993)
10. Hermo, M., Lavín, V.: Learning minimal covers of functional dependencies with queries. In: Watanabe, O., Yokomori, T. (eds.) ALT 1999. LNCS (LNAI), vol. 1720, p. 291. Springer, Heidelberg (1999)
11. Khardon, R., Mannila, H., Roth, D.: Reasoning with Examples: Propositional Formulae and Database Dependencies. Acta Informatica **36**(4), 267–286 (1999)

12. Lavín, V.: On learning multivalued dependencies with queries. Theor. Comput. Sci. **412**(22), 2331–2339 (2011)
13. Lavín, V., Hermo, M.: Negative results on learning multivalued dependencies with queries. Information Processing Letters **111**(19), 968–972 (2011)
14. Pillaipakkamnatt, K., Raghavan, V.: Read-twice DNF formulas are properly learnable. Information and Computation **122**(2), 236–267 (1995)
15. Sagiv, Y., Delobel, C., Parker Jr, D.S., Fagin, R.: An equivalence between relational database dependencies and a fragment of propositional logic. Journal of the ACM **28**(3), 435–453 (1981)

Non-adaptive Learning of a Hidden Hypergraph

Hasan Abasi[1], Nader H. Bshouty[1]([✉]), and Hanna Mazzawi[2]

[1] Department of Computer Science Technion, 32000 Haifa, Israel
bshouty@cs.technion.ac.il
[2] Google, 76 Buckingham Palace Rd, London, USA

Abstract. We give a new deterministic algorithm that non-adaptively learns a hidden hypergraph from edge-detecting queries. All previous non-adaptive algorithms either run in exponential time or have non-optimal query complexity. We give the first polynomial time non-adaptive learning algorithm for learning hypergraph that asks an almost optimal number of queries.

1 Introduction

Let $\mathcal{G}_{s,r}$ be a set of all labeled hypergraphs of rank at most r (the maximum size of an edge $e \subseteq V$ in the hypergraph) on the set of vertices $V = \{1, 2, \ldots, n\}$ with at most s edges. Given a hidden Sperner hypergraph[1] $G \in \mathcal{G}_{s,r}$, we need to identify it by asking *edge-detecting queries*. An edge-detecting query $Q_G(S)$, for $S \subseteq V$ is: Does S contain at least one edge of G? Our objective is to *non-adaptively* learn the hypergraph G by asking as few queries as possible.

This problem has many applications in chemical reactions, molecular biology and genome sequencing, where deterministic non-adaptive algorithms are most desirable. In chemical reactions, we are given a set of chemicals, some of which react and some which do not. When multiple chemicals are combined in one test tube, a reaction is detectable if and only if at least one set of the chemicals in the tube reacts. The goal is to identify which sets react using as few experiments as possible. The time needed to compute which experiments to do is a secondary consideration, though it is polynomial for the algorithms we present. See [2,4–7,13,15,17–22,26,28,29,31,34] for more details and many other applications in molecular biology.

In all of the above applications the rank of the hypergraph and the number of edges are much smaller than the number of vertices n. Therefore, throughout the paper, we will assume that $n \geq (\max(r,s))^2$. In the full paper we show that the results in this paper are also true for all values of r, s and[2] n.

The above hypergraph $\mathcal{G}_{s,r}$ learning problem is equivalent to the problem of exact learning a monotone DNF with at most s monomials (monotone terms),

---

[1] The hypergraph is Sperner hypergraph if no edge is a subset of another. If it is not Sperner hypergraph then learning is not possible.
[2] It is easy to see from Lemma 4 that all the results in this paper are also true for any r, s and $n \geq r + s$.

© Springer International Publishing Switzerland 2015
K. Chaudhuri et al. (Eds.): ALT 2015, LNAI 9355, pp. 89–101, 2015.
DOI: 10.1007/978-3-319-24486-0_6

where each monomial contains at most r variables (s-term r-MDNF) from membership queries [1,7]. In this paper we will use the later terminology rather than the hypergraph one.

The non-adaptive learnability of s-term r-MDNF was studied in [12,20,21, 25,28,29,34]. Torney, [34], first introduced the problem and gave some applications in molecular biology. The first explicit non-adaptive learning algorithm for s-term r-MDNF was given by Gao et. al., [25]. They showed that this class can be learned using a $(n,(s,r))$-cover-free family $((n,(s,r))$-CFF). This family is a set $A \subseteq \{0,1\}^n$ of assignments such that for every distinct $i_1,\ldots,i_s,j_1,\ldots,j_r \in \{1,\ldots,n\}$ there is $a \in A$ such that $a_{i_1} = \cdots = a_{i_s} = 0$ and $a_{j_1} = \cdots = a_{j_r} = 1$. Given such a set, the "folklore algorithm" simply takes all the monomials M of size at most r that satisfy $(\forall a \in A)(M(a) = 1 \Rightarrow f(a) = 1)$. The disjunction of all such monomials is equivalent to the target function. Assuming a set of $(n,(s,r))$-CFF of size N can be constructed in time T, this algorithm learns s-term r-MDNF with N membership queries in time $O(\binom{n}{r} + T)$. Notice that, no matter what is the time complexity of constructing the $(n,(s,r))$-CFF, the folklore algorithm runs in time at least $n^{\Theta(r)}$, which is nonpolynomial for nonconstant r.

In [9,21], it is shown that any set $A \subset \{0,1\}^n$ that non-adaptively learns s-term r-MDNF is an $(n,(s-1,r))$-CFF. We also show that $A \subset \{0,1\}^n$ is an $(n,(s,r-1))$-CFF. Therefore, the minimum size of an $(n,(s-1,r))$-CFF and $(n,(s,r-1))$-CFF is also a lower bound for the number of queries (and therefore, for the time complexity) of any non-adaptively learning algorithm for s-term r-MDNF. It is known, [32], that any $(n,(s,r))$-CFF, $n \geq (\max(r,s))^2$, must have size at least $\Omega(N(s,r)\log n)$ where

$$N(s,r) = \frac{s+r}{\log \binom{s+r}{r}} \binom{s+r}{r}. \tag{1}$$

Therefore, any non-adaptive algorithm for learning s-term r-DNF must ask at least $\max(N(s-1,r),N(s,r-1))\log n = \Omega(N(s,r)\log n)$ membership queries and runs in at least $\Omega(N(s,r)n\log n)$ time.

To improve the query complexity of the folklore algorithm, many tried to construct $(n,(s,r))$-CFF with optimal size in polynomial time. That is, time $poly(n,N(s,r))$. Gao et. al. constructed an $(n,(s,r))$-CFF of size $S = (2s\log n/\log(s\log n))^{r+1}$ in time $\tilde{O}(S)$. It follows from [33] that an $(n,(s,r))$-CFF of size $O\left((sr)^{\log^* n}\log n\right)$ can be constructed in polynomial time. Polynomial time almost optimal constructions of size $N(s,r)^{1+o(1)}\log n$ for $(n,(s,r))$-CFF were given in [10–12,24]. Those constructions give almost optimal query complexity for the folklore algorithm, but still, the running time is nonpolynomial for nonconstant r.

Chin et. al. claim in [20] that they have a polynomial time algorithm that constructs an $(n,(s,r))$-CFF of optimal size. Their analysis is misleading.[3] The size is indeed optimal but the time complexity of the construction is $O(\binom{n}{r+s})$.

---

[3] Some parts of the construction can indeed be performed in polynomial time, but not the whole construction.

But, as we mentioned above, even if an $(n, (s, r))$-CFF can be constructed in polynomial time, the folklore learning algorithm still takes nonpolynomial time.

The first polynomial time randomized non-adaptive learning algorithm was given by Macula et. al., [28,29]. They gave several randomized non-adaptive algorithms that are not optimal in the number of queries but use a different learning algorithm that runs in polynomial time. They show that for every s-term r-MDNF f and for every monomial M in f there is an assignment a in a $(n, (s - 1, r))$-CFF A such that $M(a) = 1$ and all the other monomials of f are zero on a. To learn this monomial they compose every assignment in A with a set of assignments that learns one monomial.

We first use the algorithm of Macula et. al., [28,29], combined with the deterministic constructions of $(n, (r, s))$-CFF in [10–12,24] and the fact that the assignments used in any non-adaptive algorithm must be $(n, (s, r - 1))$-CFF and $(n, (s - 1, r))$-CFF to change their algorithm to a deterministic non-adaptive algorithm and show that it asks $N(s, r)^{1+o(1)} \log^2 n$ queries and runs in polynomial time. The query complexity of this algorithm is almost optimal in s and r but quadratic in $\log n$. We then use a new technique, similar to the one in [16], that changes any non-adaptive learning algorithm that asks $Q(r, s, n)$ queries and runs in polynomial time to a non-adaptive learning algorithm that asks $(rs)^2 \cdot Q(r, s, (rs)^2) \log n$ queries and runs in polynomial time. This gives a non-adaptive learning algorithm that asks $N(s, r)^{1+o(1)} \log n$ queries and runs in $n \log n \cdot poly(N(s, r))$ time. Notice that the time complexity of this algorithm is almost linear in n compared to the folklore algorithm that runs in nonpolynomial time $n^{\Theta(r)} + n \log n \cdot N(s, r)^{1+o(1)}$. Our algorithm has almost optimal query complexity and time complexity.

The following table summarizes the results mentioned above for non-adaptive learning algorithms for s-term r-MDNF. In the table we assume that $n \geq (\max(r, s))^2$ and $r = \omega(1)$. For $r = O(1)$ the folklore algorithm is almost optimal and runs in polynomial time.[4]

References	Query Complexity	Time Complexity
[25]	$N(s, r) \cdot (r \log n / \log(s \log n))^{r+1}$	$\binom{n}{r}$
[20]	$N(s, r) \log n$	$\binom{n}{r+s}$
[10–12,24]	$N(s, r)^{1+o(1)} \log n$	$\binom{n}{r}$
Ours+[28,29]+[12]	$N(s, r)^{1+o(1)} \log^2 n$	$poly(n, N(s, r))$
Ours	$N(s, r)^{1+o(1)} \log n$	$(n \log n) \cdot poly(N(s, r))$
Ours, $r = o(s)$	$N(s, r)^{1+o(1)} \log n$	$(n \log n) \cdot N(s, r)^{1+o(1)}$
Lower Bound [21]	$N(s, r) \log n$	$(n \log n) \cdot N(s, r)$

The *adaptive* learnability of s-term r-MDNF was already studied in many papers which gave almost optimal algorithms [5–7,21]. In [5], Abasi et. al. gave a polynomial time adaptive learning algorithm for s-term r-MDNF with almost optimal query complexity.

---

[4] The factor of n in the lower bound of the time complexity comes from the length n of the queries.

This paper is organized as follows. Section 2 gives some definitions and preliminary results that will be used throughout the paper. Section 3 gives the first algorithm that asks $N(s,r)^{1+o(1)} \log^2 n$ membership queries and runs in time $poly(n, N(s,r))$. Section 4 gives the reduction and shows how to use it to give the second algorithm that asks $N(s,r)^{1+o(1)} \log n$ membership queries and runs in time $(n \log n) \cdot N(s,r)^{1+o(1)}$.

All the algorithms in this paper are deterministic. One can consider a randomized construction of CFF that gives a randomized algorithm that slightly improves (in the $o(1)$ of the exponent) the query and time complexity.

2 Definitions

2.1 Monotone Boolean Functions

For a vector w, we denote by w_i the ith entry of w. Let $\{e^{(i)} \mid i = 1, \ldots, n\} \subset \{0,1\}^n$ be the standard basis. That is, $e_j^{(i)} = 1$ if $i = j$ and $e_j^{(i)} = 0$ otherwise. For a positive integer j, we denote by $[j]$ the set $\{1, 2, \ldots, j\}$. For two assignments $a, b \in \{0,1\}^n$ we denote by $(a \wedge b) \in \{0,1\}^n$ the bitwise AND assignment. That is, $(a \wedge b)_i = a_i \wedge b_i$.

Let $f(x_1, x_2, \ldots, x_n)$ be a boolean function from $\{0,1\}^n$ to $\{0,1\}$. For $1 \le i_1 < i_2 < \cdots < i_k \le n$ and $\sigma_1, \ldots, \sigma_k \in \{0,1\} \cup \{x_1, \ldots, x_n\}$ we denote by

$$f|_{x_{i_1} \leftarrow \sigma_1, x_{i_2} \leftarrow \sigma_2, \cdots, x_{i_k} \leftarrow \sigma_k}$$

the function $f(y_1, \ldots, y_n)$ where $y_{i_j} = \sigma_j$ for all $j \in [k]$ and $y_i = x_i$ for all $i \in [n] \backslash \{i_1, \ldots, i_k\}$. We say that the variable x_i is *relevant* in f if $f|_{x_i \leftarrow 0} \not\equiv f|_{x_i \leftarrow 1}$. A variable x_i is *irrelevant* in f if it is not relevant in f. We say that the class is *closed under variable projections* if for every $f \in C$ and every two variables x_i and x_j, $i, j \le n$, we have $f|_{x_i \leftarrow x_j} \in C$.

For two assignments $a, b \in \{0,1\}^n$, we write $a \le b$ if for every $i \in [n]$, $a_i \le b_i$. A Boolean function $f : \{0,1\}^n \to \{0,1\}$ is *monotone* if for every two assignments $a, b \in \{0,1\}^n$, if $a \le b$ then $f(a) \le f(b)$. Recall that every monotone boolean function f has a unique representation as a reduced monotone DNF, [1]. That is, $f = M_1 \vee M_2 \vee \cdots \vee M_s$ where each *monomial* M_i is an AND of input variables, and for every monomial M_i there is a unique assignment $a^{(i)} \in \{0,1\}^n$ such that $f(a^{(i)}) = 1$ and for every $j \in [n]$ where $a_j^{(i)} = 1$ we have $f(a^{(i)}|_{x_j \leftarrow 0}) = 0$. We call such assignment a *minterm* of the function f. Notice that every monotone DNF can be uniquely determined by its minterms [1]. That is, $a \in \{0,1\}^n$ is a minterm of f iff $M := \wedge_{i \in \{j : a_j = 1\}} x_i$ is a monomial in f.

An *s-term r-MDNF* is a monotone DNF with at most s monomials, where each monomial contains at most r variables. It is easy to see that the class s-term r-MDNF is closed under variable projections.

2.2 Learning from Membership Queries

Consider a *teacher* that has a *target function* $f : \{0,1\}^n \rightarrow \{0,1\}$ that is s-term r-MDNF. The teacher can answer *membership queries*. That is, when receiving $a \in \{0,1\}^n$ it returns $f(a)$. A *learning algorithm* is an algorithm that can ask the teacher membership queries. The goal of the learning algorithm is to *exactly learn* (exactly find) f with a minimum number of membership queries and optimal time complexity.

Let c and $H \supset C$ be classes of boolean formulas. We say that C is *learnable from* H in time $T(n)$ with $Q(n)$ membership queries if there is a learning algorithm that, for a target function $f \in C$, runs in time $T(n)$, asks at most $Q(n)$ membership queries and outputs a function h in H that is equivalent to C. When $H = C$ then we say that C is *properly learnable* in time $T(n)$ with $Q(n)$ membership queries.

In adaptive algorithms the queries can depend on the answers to the previous queries where in non-adaptive algorithms the queries are independent of the answers to the previous queries and therefore all the queries can be asked in parallel, that is, in one step.

2.3 Learning a Hypergraph

Let $\mathcal{G}_{s,r}$ be a set of all labeled hypergraphs on the set of vertices $V = \{1, 2, \ldots, n\}$ with s edges of rank (size) at most r. A hypergraph is called Sperner hypergraph if no edge is a subset of another. Given a hidden Sperner hypergraph $G \in \mathcal{G}_{s,r}$, we need to identify it by asking *edge-detecting queries*. An edge-detecting query $Q_G(S)$, for $S \subseteq V$ is: does S contain at least one edge of G? Our objective is to learn (identify) the hypergraph G by asking as few queries as possible.

This problem is equivalent to learning s-term r-MDNF f from membership queries. Each edge e in the hypergraph corresponds to the monotone term $\wedge_{i \in e} x_i$ in f and the edge-detecting query $Q_G(S)$ corresponds to asking membership queries of the assignment $a^{(S)}$ where $a_i^{(S)} = 1$ if and only if $i \in S$. Therefore, the class $\mathcal{G}_{s,r}$ can be regarded as the set of s-term r-MDNF. The class of s-term r-MDNF is denoted by $\mathcal{G}_{s,r}^*$. Now it obvious that any learning algorithm for $\mathcal{G}_{s,r}^*$ is also a learning algorithm for $\mathcal{G}_{s,r}$.

The following example shows that learning is not possible for hypergraphs that are not Sperner hypergraphs. Let G_1 be a graph where $V_1 = \{1, 2\}$ and $E_1 = \{\{1\}, \{1, 2\}\}$. This graph corresponds to the function $f = x_1 \vee x_1 x_2$ that is equivalent to x_1 which corresponds to the graph G_2 where $V_2 = \{1, 2\}$ and $E_2 = \{\{1\}\}$. Also, no edge-detecting query can distinguish between G_1 and G_2.

We say that $A \subseteq \{0, 1\}^n$ is an *identity testing set* for $\mathcal{G}_{s,r}^*$ if for every two distinct s-term r-MDNF f_1 and f_2 there is $a \in A$ such that $f_1(a) \neq f_2(a)$. Obviously, every identity testing set for $\mathcal{G}_{s,r}^*$ can be used as queries to nonadaptively learns $\mathcal{G}_{s,r}^*$. Also, if $\mathcal{A}$ is a nonadaptive algorithm that learns $\mathcal{G}_{s,r}^*$ from membership queries A then A is identity testing set for $\mathcal{G}_{s,r}^*$. We denote by $\text{OPT}(\mathcal{G}_{s,r}^*)$ the minimum size of an identity testing set for $\mathcal{G}_{s,r}^*$. We say that a non-adaptive algorithm $\mathcal{A}$ is *almost optimal* if it runs in $poly(\mathrm{OPT}(\mathcal{G}^*_{s,r}), n)$ time and asks $\mathrm{OPT}(\mathcal{G}^*_{s,r})^{1+o(1)}$ queries.

2.4   Cover Free Families

An $(n, (s, r))$-cover free family $((n, (s, r))$-CFF), [23], is a set $A \subseteq \{0, 1\}^n$ such that for every $1 \le i_1 < i_2 < \cdots < i_d \le n$ where $d = s + r$ and every $J \subseteq [d]$ of size $|J| = s$ there is $a \in A$ such that $a_{i_k} = 0$ for all $k \in J$ and $a_{i_j} = 1$ for all $j \in [d] \backslash J$. Denote by $N(n, (s, r))$ the minimum size of such set. The lower bound in [30,32] is, for $n \ge (\max(r, s))^2$,

$$N(n, (s, r)) \ge \Omega\left(N(s, r) \cdot \log n\right) \tag{2}$$

where $N(s, r)$ is as defined in (1). It is known that a set of random

$$m = O\left(r^{1.5}\left(\log\left(\frac{s}{r} + 1\right)\right)\left(N(s, r) \cdot \log n + \frac{N(s, r)}{s + r}\log\frac{1}{\delta}\right)\right)$$
$$= N(s, r)^{1+o(1)}(\log n + \log(1/\delta)) \tag{3}$$

assignments $a^{(i)} \in \{0, 1\}^n$, where each $a_j^{(i)}$ is 1 with probability $r/(s + r)$, is an $(n, (s, r))$-CFF with probability at least $1 - \delta$.

It follows from [10–12,24] that for $n \ge (rs)^c$, for some constant c, there is a polynomial time (in the size of the CFF) deterministic construction algorithm of $(n, (s, r))$-CFF of size

$$N(s, r)^{1+o(1)}\log n \tag{4}$$

where the $o(1)$ is with respect to r. When $r = o(s)$ the construction runs in linear time [10,12].

We now show

Lemma 1. *For any r, s and $n \ge r + s$, there is a polynomial time deterministic construction algorithm of $(n, (s, r))$-CFF of size*

$$N(s, r)^{1+o(1)}\log n.$$

Proof. For $n \ge (rs)^c$, for some constant c, the result follows from [10–12,24]. For $r + s \le n \le (rs)^c$ we construct the $(n, (s, r))$-CFF for $N = (rs)^c$ and then truncate the vectors to length n. The size is $N(s, r)^{1+o(1)}\log(rs)^c = N(s, r)^{1+o(1)}\log n$. □

2.5   Perfect Hash Function

Let H be a family of functions $h : [n] \rightarrow [q]$. For $d \le q$ we say that H is an (n, q, d)-*perfect hash family* $((n, q, d)$-PHF) [8] if for every subset $S \subseteq [n]$ of size $|S| = d$ there is a *hash function* $h \in H$ such that $h|_S$ is injective (one-to-one) on S, i.e., $|h(S)| = d$.

In [10] Bshouty shows

Lemma 2. *Let $q \geq 2d^2$. There is a (n, q, d)-PHF of size*

$$O\left(\frac{d^2 \log n}{\log(q/d^2)}\right)$$

that can be constructed in time $O(qd^2 n \log n / \log(q/d^2))$.

We now give the following folklore results that will be used for randomized learning algorithms

Lemma 3. *Let $q > d(d-1)/2$ be any integer. Fix any set $S \subset [n]$ of d integers. Consider*

$$N := \frac{\log(1/\delta)}{\log\left(\frac{1}{1-g(q,d)}\right)} \leq \frac{\log(1/\delta)}{\log \frac{2q}{d(d-1)}}$$

uniform random hash functions $h_i : [n] \to [q]$, $i = 1, \ldots, N$ where

$$g(q, d) := \left(1 - \frac{1}{q}\right)\left(1 - \frac{2}{q}\right) \cdots \left(1 - \frac{d-1}{q}\right)$$

With probability at least $1 - \delta$ one of the hash functions is one-to-one on S.

2.6 A Lower Bound For Learning

In this subsection we prove the following lower bound

Lemma 4. *Let $n \geq r + s$. Any identity testing set $A \subseteq \{0,1\}^n$ for s-term r-MDNF is $(n, (s, r-1))$-CFF and $(n, (s-1, r))$-CFF.*
In particular, for $w = \max(r, s)$ and $d = \min(r, s)$,

1. *if $n > w^2$ then $|A| = \Omega(N(s, r) \log n)$,*
2. *if $n = w^{1+\epsilon}$ for some $1/d < \epsilon < 1$, then $|A| = \Omega(N(s, r) / \log^{O(1/\epsilon)} w)$,*
3. *if $r + s \leq n \leq w^{1+1/d}$ then $|A| = \Omega\left(\binom{n}{d}\right)$, and*
4. *for all $n \geq r + s$ we have $|A| = \Omega(\binom{s+r}{r})$.*

Proof. Consider any distinct $1 \leq i_1, \cdots, i_{r+s-1} \leq n$. To be able to distinguish between the two functions $f_1 = (x_{i_1} \cdots x_{i_r}) \vee x_{i_{r+1}} \vee \cdots \vee x_{i_{r+s-1}}$ and $f_2 = (x_{i_1} \cdots x_{i_{r-1}}) \vee x_{i_{r+1}} \vee \cdots \vee x_{i_{r+s-1}}$ we must have an assignment a that satisfies $a_{i_1} = \cdots = a_{i_{r-1}} = 1$ and $a_{i_r} = \cdots = a_{i_{r+s-1}} = 0$. Therefore A is $(n, (s, r-1))$-CFF. To be able to distinguish between the two functions $g_1 = (x_{i_1} \cdots x_{i_r}) \vee x_{i_{r+1}} \vee \cdots \vee x_{i_{r+s-1}}$ and $g_2 = x_{i_{r+1}} \vee \cdots \vee x_{i_{r+s-1}}$ we must have an assignment a that satisfies $a_{i_1} = \cdots = a_{i_r} = 1$ and $a_{i_{r+1}} = \cdots = a_{i_{r+s-1}} = 0$. Therefore A is $(n, (s-1, r))$-CFF.

The bounds *1-4* follows from the lower bounds of $(n, (s-1, r))$-CFF and $(n, (s, r-1))$-CFF in [3, 32]. □

2.7   The Folklore Algorithm

The "folklore algorithm" simply construct a $(n, (s, r))$-CFF, A, and takes all the monomials M of size at most r that satisfy $(\forall a \in A)(M(a) = 1 \Rightarrow f(a) = 1)$. The disjunction of all such monomials is equivalent to the target function. This follows from the following two facts: (1) For any monomial M' where $M' \not\Rightarrow f$, there is an assignment $a \in A$ such that $f(a) = 0$ and $M'(a) = 1$. (2) $f = \wedge_{M \Rightarrow f} M$. Assuming a set of $(n, (s, r))$-CFF of size N can be constructed in time T, this algorithm learns s-term r-MDNF with N membership queries in time $O(\binom{n}{r} + T)$. In particular we have

Lemma 5. *Let $n \geq r + s$. A $(n, (s, r))$-CFF is an identity testing set for s-term r-MDNF.*

We now show

Theorem 1. *Let $n \geq r + s$. For constant r or constant s there is a non-adaptive proper learning algorithm for s-term r-MDNF that asks*

$$N(s, r)^{1 + o(1)} \log n$$

queries and runs in time $poly(n, N(s, r))$.
For $n \geq (\max(r, s))^2$ the algorithm is almost optimal.

Proof. For constant r the folklore algorithm runs in polynomial time and, by Lemma 1, asks $N(s, r)^{1 + o(1)} \log n$ queries. By Lemma 4 it is almost optimal for $n \geq (\max(r, s))^2$.

For constant s, every s-term r-MDNF is r^s-clause s-MCNF. We first learn f as r^s-clause s-MCNF. This takes polynomial time and $N(s, r)^{1 + o(1)} \log n$ queries.[5] We then change the target to s-term r-MDNF. This can be done in polynomial time $(rs)^s$. By Lemma 4, it is almost optimal for $n \geq (\max(r, s))^2$.   □

3   The First Algorithm

In this section we give the first algorithm that asks $N(s, r)^{1 + o(1)} \log^2 n$ queries and runs in time $poly(n, N(s, r))$

The first algorithm is similar to the algorithm in [28, 29] that were used to give a Monte Carlo randomized algorithm. Here we use the deterministic construction of CFF to change it to a deterministic algorithm and give a full analysis for its complexity.

We first prove

Lemma 6. *Let A be an $(n, (1, r))$-CFF and B be an $(n, (s - 1, r))$-CFF. There is a non-adaptive proper learning algorithm for s-term r-MDNF that asks all the queries in $A \wedge B := \{a \wedge b \mid a \in A, b \in B\}$ and finds the target function in time $|A \wedge B| \cdot n$.*
In particular, $A \wedge B$ is an identity testing set for s-term r-MDNF.

---

[5] To show that a clause $C = x_{i_1} \vee \cdots \vee x_{i_s}$ is not in f and $f \not\Rightarrow C$ we need an assignment a such that $C(a) = 0$ and $f(a) = 1$. Therefore, f can be learned with an $(n, (s, r))$-CFF.

Proof. Let f be the target function. For every $b \in B$, let $A_b = A \wedge b := \{a \wedge b \mid a \in A\}$. Let I_b be the set of all $i \in [n]$ such that $(a \wedge b)_i \geq f(a \wedge b)$ for all $a \in A$. Let $T_b := \wedge_{i \in I_b} x_i$. We will show that

1. If T is a term in f then there is $b \in B$ such that $T_b \equiv T$.
2. Either $T_b = \wedge_{i \in [n]} x_i$ or T_b is a subterm of one of terms of f.

To prove 1, let T be a term in f and let $b \in B$ be an assignment that satisfies T and does not satisfy the other terms. Such assignment exists because B is $(n, (s-1, r))$-CFF. Notice that $f(x \wedge b) = T(x) = T(x \wedge b)$. If x_i is in T and $f(a \wedge b) = 1$ then $T(a \wedge b) = T(a) = f(a \wedge b) = 1$ and $(a \wedge b)_i = 1$. Therefore $i \in I_b$ and x_i in T_b. If x_i not in T then since A is $(n, (1, r))$-CFF there is $a' \in A$ such that $T(a') = 1$ and $a'_i = 0$. Then $(a' \wedge b)_i = 0$ where $f(a' \wedge b) = 1$. Therefore i is not in I_b and x_i is not in T_b. Thus, $T_b \equiv T$.

We now prove 2. We have shown in 1 that if b satisfies one term T then $T_b \equiv T$. If b does not satisfy any one of the terms in f then $f(a \wedge b) = 0$ for all $a \in A$ and then $T_b = \wedge_{i \in [n]} x_i$. Now suppose b satisfies at least two terms T_1 and T_2. Consider any variable x_i. If x_i is not in T_1 then as before x_i will not be in T_b. This shows that T_b is a subterm of T_1. $\qquad \square$

This gives the following algorithm

Learn$(\mathcal{G}^*_{s,r})$
1) Construct an $(n, (1, r))$-CFF A and an $(n, (s-1, r))$-CFF B.
2) Ask membership queries for all $a \wedge b$, $a \in A$ and $b \in B$.
3) For every $b \in B$.
4) $T_b \leftarrow 1$.
5)   For every $i \in [n]$.
6)   If for all $a \in A$, $(a \wedge b)_i \geq f(a \wedge b)$
7)   then $T_b \leftarrow T_b \wedge x_i$.
8) $\mathcal{T} \leftarrow \mathcal{T} \cup \{T_b\}$.
9) Remove from $\mathcal{T}$ the term $\wedge_{i \in [n]} x_i$
       and all subterms of a larger term.

Fig. 1. An algorithm for learning $\mathcal{G}^*_{s,r}$.

We now have

Theorem 2. *Let $n \geq r + s$. There is a non-adaptive proper learning algorithm for s-term r-MDNF that asks*

$$N(s, r)^{1+o(1)} \log^2 n$$

queries and runs in time $poly(n, N(s, r))$.
For $n \geq (\max(r, s))^2$ the algorithm is almost optimal.

Proof. For constant r or constant s the result follows from Theorem 1. Let $r, s = \omega(1)$. By Lemma 1, constructing a $(n, (1, r))$-CFF of size $|A| = r^2 \log n$ and a $(n, (s-1, r))$-CFF of size $|B| = N(s-1, r)^{1+o(1)} \log n = N(s, r)^{1+o(1)} \log n$ takes $poly(n, N(s, r))$ time. By Lemma 6, the learning takes time $|A \wedge B| \cdot n = poly$ $(n, N(s, r))$ time. The number of queries of the algorithm is $|A \wedge B| \leq |A| \cdot |B| = N(s, r)^{1+o(1)} \log^2 n$ for $r, s = \omega(1)$.

By Lemma 4 the algorithm is almost optimal for $n \geq (\max(r, s))^2$.   □

A randomized algorithm with a better query complexity can be obtained using a randomized $(n, (1, r))$-CFF, B, and a randomized $(n, (s-1, r))$-CFF, A.

4   The Second Algorithm

In this section we give the second algorithm
    We first prove the following result

Lemma 7. *Let C be a class of boolean functions that is closed under variable projection. Let H be a class of boolean functions and suppose there is an algorithm that, given $f \in H$ as an input, finds the relevant variables of f in time $R(n)$.*

If C is non-adaptively learnable from H in time $T(n)$ with $Q(n)$ membership queries then C is non-adaptively learnable from H in time

$$O\left(qd^2n \log n + \frac{d^2 \log n}{\log(q/(d+1)^2)}(T(q) + Q(q)n + R(q))\right)$$

with

$$O\left(\frac{d^2Q(q)}{\log(q/(d+1)^2)} \log n\right)$$

membership queries where d is an upper bound on the number of relevant variables in $f \in C$ and q is any integer such that $q \geq 2(d+1)^2$.

Proof. Consider the algorithm in Figure 2. Let $\mathcal{A}(n)$ be a non-adaptive algorithm that learns C from H in time $T(n)$ with $Q(n)$ membership queries. Let $f \in C_n$ be the target function. Consider the $(n, q, d+1)$-PHF P that is constructed in Lemma 2 (Step 1 in the algorithm). Since C is closed under variable projection, for every $h \in P$ the function $f_h := f(x_{h(1)}, \ldots, x_{h(n)})$ is in C_q. Since the membership queries to f_h can be simulated by membership queries to f there is a set of $|P| \cdot Q(q)$ assignments from $\{0, 1\}^n$ that can be generated from $\mathcal{A}(q)$ that non-adaptively learn f_h for all $h \in P$ (Step 2 in the algorithm). The algorithm $\mathcal{A}(q)$ learns $f'_h \in H$ that is equivalent to f_h.

Then the algorithm finds the relevant variables of each $f'_h \in H$ (Step 3 in the algorithm). Let V_h be the set of relevant variables of f'_h and let $d_{max} = \max_h |V_h|$. Suppose $x_{i_1}, \ldots, x_{i_{d'}}$, $d' \leq d$ are the relevant variables in the target function f. There is a map $h' \in P$ such that $h'(i_1), \ldots, h'(i_{d'})$ are distinct and therefore $f'_{h'}$ depends on d' variables. In particular, $d' = d_{max}$ (Step 4 in the algorithm).

After finding $d' = d_{max}$ we have: Every h for which f'_h depends on d' variables necessarily satisfies $h(i_1), \ldots, h(i_{d'})$ are distinct. Consider any other non-relevant variable $x_j \notin \{x_{i_1}, \ldots, x_{i_{d'}}\}$. Since P is $(n, q, d+1)$-PHF, there is $h'' \in P$ such that $h''(j), h''(i_1), \ldots, h''(i_{d'})$ are distinct. Then $f'_{h''}$ depends on $x_{h''(i_1)}, \ldots, x_{h''(i_{d'})}$ and not on $x_{h''(j)}$. This way the non-relevant variables can be eliminated. This is Step 6 in the algorithm. Since the above is true for every non-relevant variable, after Step 6 in the algorithm, the set X contains only the relevant variables of f. Then in Steps 7 and 8, the target function f can be recovered from any f'_{h_0} that satisfies $|V(h_0)| = d'$.   □

Algorithm Reduction I

$\mathcal{A}(n)$ is a non-adaptive learning algorithm for C from H.

1) Construct an $(n, q, d+1)$-PHF P.
2) For each $h \in P$
       Run $\mathcal{A}(q)$ to learn $f_h := f(x_{h(1)}, \ldots, x_{h(n)})$.
       Let $f'_h \in H$ be the output of $\mathcal{A}(q)$.
3) For each $h \in P$
       $V_h \leftarrow$ the relevant variables in f'_h
4) $d_{max} \leftarrow \max_h |V_h|$.
5) $X \leftarrow \{x_1, x_2, \ldots, x_n\}$.
6) For each $h \in P$
       If $|V_h| = d_{max}$ then $X \leftarrow X \backslash \{x_i \mid x_{h(i)} \notin V_h\}$
7) Take any h_0 with $|V_{h_0}| = d_{max}$
8) Replace each relevant variable x_i in f'_{h_0} by $x_j \in X$ where $h_0(j) = i$.
9) Output the function constructed in step (8).

Fig. 2. Algorithm Reduction.

We now prove

Theorem 3. *Let $n \geq r + s$. There is a non-adaptive proper learning algorithm for s-term r-MDNF that asks*

$$N(s, r)^{1+o(1)} \log n$$

queries and runs in time $(n \log n) \cdot poly(N(s, r))$ time.
    For $n \geq (\max(r, s))^2$ the algorithm is almost optimal.

Proof. For constant r or constant s the result follows from Theorem 1. Let $r, s = \omega(1)$. We use Lemma 7. $C = H$ is the class of s-term r-MDNF. This class is closed under variable projection. Given f that is s-term r-MDNF, one can find all the relevant variables in $R(n) = O(sr)$ time. The algorithm in the previous section runs in time $T(n) = poly(n, N(s, r))$ and asks $Q(n) = N(s, r)^{1+o(1)} \log^2 n$ queries. The number of variables in the target is bounded by

$d = rs$. Let $q = 3r^2s^2 \geq 2d^2$. By Lemma 7, and since $r, s = \omega(1)$, there is a non-adaptive algorithm that runs in time

$$O\left(qd^2 n \log n + \frac{d^2 \log n}{\log(q/d^2)}(T(q)n + R(q))\right) = (n \log n)poly(N(r,s))$$

and asks

$$O\left(\frac{d^2 Q(q)}{\log(q/d^2)} \log n\right) = N(s,r)^{1+o(1)} \log n$$

membership queries.

By Lemma 4 the algorithm is almost optimal for $n \geq (\max(r,s))^2$.   □

A randomized algorithm with a better query complexity can be obtained using a randomized $(n, q, d+1)$-PHF, a randomized $(n, (1, r))$-CFF, B, and a randomized $(n, (s-1, r))$-CFF, A.

Acknowledgments. We would like to thank the reviewers for their helpful comments.

References

1. Angluin, D.: Queries and Concept Learning. Machine Learning **2**(4), 319–342 (1987)
2. Alon, N., Asodi, V.: Learning a Hidden Subgraph. SIAM J. Discrete Math. **18**(4), 697–712 (2005)
3. Abdi, A.Z., Bshouty, N.H.: Lower Bounds for Cover-Free Families. CoRR abs/1502.03578 (2015)
4. Alon, N., Beigel, R., Kasif, S., Rudich, S., Sudakov, B.: Learning a Hidden Matching. SIAM J. Comput. **33**(2), 487–501 (2004)
5. Abasi, H., Bshouty, N.H., Mazzawi, H.: On exact learning monotone DNF from membership queries. In: Auer, P., Clark, A., Zeugmann, T., Zilles, S. (eds.) ALT 2014. LNCS, vol. 8776, pp. 111–124. Springer, Heidelberg (2014)
6. Angluin, D., Chen, J.: Learning a Hidden Hypergraph. Journal of Machine Learning Research **7**, 2215–2236 (2006)
7. Angluin, D., Chen, J.: Learning a Hidden Graph using $O(\log n)$ Queries per Edge. J. Comput. Syst. Sci. **74**(4), 546–556 (2008)
8. Alon, N., Moshkovitz, D., Safra, S.: Algorithmic construction of sets for k-restrictions. ACM Transactions on Algorithms **2**(2), 153–177 (2006)
9. Bshouty, N.H.: Exact learning from membership queries: some techniques, results and new directions. In: Jain, S., Munos, R., Stephan, F., Zeugmann, T. (eds.) ALT 2013. LNCS, vol. 8139, pp. 33–52. Springer, Heidelberg (2013)
10. Bshouty, N.H.: Linear time constructions of some d-Restriction problems. In: Paschos, V.T., Widmayer, P. (eds.) CIAC 2015. LNCS, vol. 9079, pp. 74–88. Springer, Heidelberg (2015)
11. Bshouty, N.H.: Testers and their Applications. Electronic Collouium on Computational Complexity (ECCC) 19:11 (2012). ITCS 2014, pp. 327–352 (2014)
12. Bshouty, N.H., Gabizon, A.: Almost Optimal Cover-Free Family (In preperation)

13. Beigel, R., Alon, N., Kasif, S., Serkan Apaydin, M., Fortnow, L.: An optimal procedure for gap closing in whole genome shotgun sequencing. In: RECOMB 2001, pp. 22–30 (2001)
14. Bshouty, N.H., Goldman, S.A., Hancock, T.R., Matar, S.: Asking Questions to Minimize Errors. J. Comput. Syst. Sci. **52**(2), 268–286 (1996)
15. Bouvel, M., Grebinski, V., Kucherov, G.: Combinatorial search on graphs motivated by bioinformatics applications: a brief survey. In: Kratsch, D. (ed.) WG 2005. LNCS, vol. 3787, pp. 16–27. Springer, Heidelberg (2005)
16. Bshouty, N.H., Hellerstein, L.: Attribute-Efficient Learning in Query and Mistakebound Models. COLT **1996**, 235–243 (1996)
17. Chang, H., Chen, H.-B., Fu, H.-L., Shi, C.-H.: Reconstruction of hidden graphs and threshold group testing. J. Comb. Optim. **22**(2), 270–281 (2011)
18. Chang, H., Fu, H.-L., Shih, C.-H.: Learning a hidden graph. Optim, Lett. (2014)
19. Chen, H.-B., Hwang, F.K.: A survey on nonadaptive group testing algorithms through the angle of decoding. J. Comb. Optim. **15**(1), 49–59 (2008)
20. Chin, F.Y.L., Leung, H.C.M., Yiu, S.-M.: Non-adaptive complex group testing with multiple positive sets. Theor. Comput. Sci. **505**, 11–18 (2013)
21. Du, D.Z., Hwang, F.: Pooling Design and Nonadaptive Group Testing: Important Tools for DNA Sequencing. World Scientific, Singapore (2006)
22. D'yachkov, A., Vilenkin, P., Macula, A., Torney, D.: Families of finite sets in which no intersection of ℓ sets is covered by the union of s others. J. Comb Theory Ser A. **99**, 195–218 (2002)
23. Kautz, W.H., Singleton, R.C.: Nonrandom binary superimposed codes. IEEE Trans. Inform. Theory **10**(4), 363–377 (1964)
24. Fomin, F.V., Lokshtanov, D., Saurabh, S.: Efficient Computation of Representative Sets with Applications in Parameterized and Exact Algorithms. SODA **2014**, 142–151 (2014)
25. Gao, H., Hwang, F.K., Thai, M.T., Wu, W., Znati, T.: Construction of d(H)-disjunct matrix for group testing in hypergraphs. J. Comb. Optim. **12**(3), 297–301 (2006)
26. Grebinski, V., Kucherov, G.: Reconstructing a Hamiltonian Cycle by Querying the Graph: Application to DNA Physical Mapping. Discrete Applied Mathematics **88**(1–3), 147–165 (1998)
27. Kleitman, D.J., Spencer, J.: Families of k-independent sets. Discrete Mathematics **6**(3), 255–262 (1972)
28. Macula, A.J., Popyack, L.J.: A group testing method for finding patterns in data. Discret Appl Math. **144**, 149–157 (2004)
29. Macula, A.J., Rykov, V.V., Yekhanin, S.: Trivial two-stage group testing for complexes using almost disjunct matrices. Discrete Applied Mathematics **137**(1), 97–107 (2004)
30. Ma, X., Wei, R.: On Bounds of Cover-Free Families. Designs, Codes and Cryptography **32**, 303–321 (2004)
31. Reyzin, L., Srivastava, N.: Learning and verifying graphs using queries with a focus on edge counting. In: Hutter, M., Servedio, R.A., Takimoto, E. (eds.) ALT 2007. LNCS (LNAI), vol. 4754, pp. 285–297. Springer, Heidelberg (2007)
32. Stinson, D.R., Wei, R., Zhu, L.: Some New Bounds for Cover free Families. Journal of Combinatorial Theory, Series A **90**(1), 224–234 (2000)
33. Stinson, D.R., Wei, R., Zhu, L.: New constructions for perfect hash families and related structures using combintorial designs and codes. J. Combin. Designs **8**(3), 189–200 (2000)
34. Torney, D.C.: Sets pooling designs. Ann. Comb. **3**, 95–101 (1999)

On the Teaching Complexity of Linear Sets

Ziyuan Gao[1]([⊠]), Hans Ulrich Simon[2], and Sandra Zilles[1]

[1] Department of Computer Science, University of Regina,
Regina, SK S4S 0A2, Canada
{gao257,zilles}@cs.uregina.ca

[2] Horst Görtz Institute for IT Security and Faculty of Mathematics,
Ruhr-Universität Bochum, 44780 Bochum, Germany
hans.simon@rub.de

Abstract. Linear sets are the building blocks of semilinear sets, which are in turn closely connected to automata theory and formal languages. Prior work has investigated the learnability of linear sets and semilinear sets in three models – Valiant's *PAC-learning* model, Gold's *learning in the limit* model, and Angluin's *query learning* model. This paper considers a *teacher-learner* model of learning families of linear sets, whereby the learner is assumed to know all the smallest sets $T_1, T_2, \ldots$ of labelled examples that are consistent with exactly one language in the class $\mathcal{L}$ to be learnt, and is always presented with a sample S of labelled examples such that S is contained in at least one of $T_1, T_2, \ldots$; the learner then interprets S according to some fixed protocol. In particular, we will apply a generalisation of a recently introduced model – the *recursive teaching model* of teaching and learning – to several infinite classes of linear sets, and show that the maximum sample complexity of teaching these classes can be drastically reduced if each of them is taught according to a carefully chosen sequence. A major focus of the paper will be on determining two relevant teaching parameters, the *teaching dimension* and *recursive teaching dimension*, for various families of linear sets.

1 Introduction

A *linear set* L is defined by a nonnegative lattice point (called a *constant*) and a finite set of nonnegative lattice sets (called *periods*); the members of L are generated by adding to the constant an arbitrary finite sequence of the periods (allowing repetitions of the same period in the sequence). A *semilinear set* is a finite union of linear sets. Semilinear sets are not only objects of mathematical interest, but have also been linked to finite-state machines and formal languages. One of the earliest and most important results on the connection between semilinear sets and context-free languages is *Parikh's theorem* [9], which states that any context-free language is mapped to a semilinear set via a function known as the Parikh vector of a string. Another interesting result, due to Ibarra [6], characterises semilinear sets in terms of reversal-bounded multicounter machines. Moving beyond abstract theory, semilinear sets have also recently been applied in the fields of DNA self-assembly [3] and membrane computing [7].

© Springer International Publishing Switzerland 2015
K. Chaudhuri et al. (Eds.): ALT 2015, LNAI 9355, pp. 102–116, 2015.
DOI: 10.1007/978-3-319-24486-0_7

The learnabilities of linear sets and semilinear sets have been investigated in Valiant's PAC-learning model [1], Gold's learning in the limit model [12], and Angluin's query learning model [12]. Abe [1] showed that when the integers are encoded in unary, the class of semilinear sets of dimension 1 or 2 is polynomially PAC-learnable; on the other hand, the question as to whether classes of semilinear sets of higher dimensions are PAC-learnable is open. Takada [12] established that for any fixed dimension, the family of linear sets is learnable from positive examples but the family of semilinear sets is not learnable from only positive examples. Takada also showed the existence of a learning procedure via restricted subset and restricted superset queries that identifies any semilinear set and halts; however, he proved at the same time that any such algorithm must necessarily be time consuming.

This paper is primarily concerned with the *sample complexity of teaching* classes of linear sets with a fixed dimension, which we determine mainly with two combinatorial parameters (and some variants), the *teaching dimension* (TD) and the *recursive teaching dimension* (RTD). These teaching complexity measures are based on a variant of the online learning model in which a cooperative teacher selects the instances presented to the learner [5,11,13]. In the teacher-learner model, the teacher must present a finite set of instances so that the learner achieves *exact identification* of the target concept via some consistent teaching-learning protocol. To preclude any unnatural collusion between the teacher and learner that could arise from, say, encoding concepts in examples, the teaching-learning protocol must, in some definite sense, be "collusion-free." To this end, Zilles, Lange, Holte and Zinkevich [13] proposed a rigorous definition of a "collusion-free" teaching-learning protocol. They designed a protocol – the *recursive teaching protocol* – that only exploits an inherent hierarchical structure of any concept class, and showed that this protocol is collusion-free. The RTD of a concept class is the maximum sample complexity derived by applying the recursive teaching protocol to the class. The RTD possesses several regularity properties and has been fruitfully applied to the analysis of pattern languages [4,8]. A somewhat simpler protocol, the teaching set protocol [5,11], only requires that the teacher present, for each target concept C, a sample S from C of smallest possible size so that C is the only concept in the class consistent with S. The teaching set protocol is also collusion-free, although the maximum sample complexity in this case – the TD – is generally larger than the RTD.

Our results may be of interest from a formal language perspective as well as from a computational learning theory perspective. First, they uncover a number of structural properties of linear sets, especially in the one-dimensional case, which could be applied to study formal languages via the Parikh vector function. Consider, for example, the set $L(\pi)$ of all words obtained by substituting nonempty strings over $\{a\}$ for variables in some nonempty string π of symbols chosen from $\{a\} \cup X$, where X is an infinite set of variables.[1] As will be seen later, the Parikh vector maps $L(\pi)$ to a linear subset L of the natural numbers such that the sum of L's periods does not exceed the constant associated

---

[1] $L(\pi)$ is known as a *non-erasing pattern language*.

to L. Thus one could determine various teaching complexity measures of any non-erasing pattern language from the teaching complexity measures of a certain linear subset of the natural numbers. Second, the class of linear sets affords quite a natural setting to study models of teaching and learning over infinite concept classes. Besides showing that the RTD can be significantly lower than the TD for many infinite classes of linear sets, we will consider a more stringent variant of the RTD, the RTD^+, which considers sequential teaching of classes using only positive examples. It will be shown that there are natural classes of linear sets that cannot even be taught sequentially using only positive examples while the RTD is finite; these examples illustrate how supplying negative information may sometimes be indispensable to successful teaching and learning.

2   Preliminaries

$\mathbb{N}_0$ denotes the set of all nonnegative integers and $\mathbb{N}$ denotes the set of all positive integers. For each integer $m \geq 1$, let $\mathbb{N}_0^m = \mathbb{N}_0 \times \ldots \times \mathbb{N}_0$ (m times). $\mathbb{N}_0^m$ is regarded as a subset of the vector space of all m-tuples of rational numbers over the rational numbers. For each $r \in \mathbb{N}$, $[r]$ denotes $\{1, \ldots, r\}$. For any $v = (a_1, \ldots, a_m) \in \mathbb{N}_0^m$, define $\|v\|_1 = \sum_{i=1}^m a_i$. 0 will denote the zero vector in $\mathbb{N}_0^m$ when there is no possibility of confusion.

2.1   Linear Sets

A subset L of $\mathbb{N}_0^m$ is said to be *linear* iff there exist an element c and a finite subset P of $\mathbb{N}_0^m$ such that $L = c + \langle P \rangle := \{q : q = c + n_1 p_1 + \ldots + n_k p_k, n_i \in \mathbb{N}_0, p_i \in P\}$. c is called the *constant* and each p_i is called a *period* of $c + \langle P \rangle$. Denote $0 + \langle P \rangle$ by $\langle P \rangle$. For any linear set L, if $L = c + \langle P \rangle$, then (c, P) is called a *representation of* L. Any finite $P \subset \mathbb{N}_0^m$ is *independent* iff for all $P' \subsetneq P$, it holds that $\langle P' \rangle \neq \langle P \rangle$. A representation (c, P) of a linear set L is *canonical* iff P is independent. A linear subset of $\mathbb{N}_0^m$ will also be called a *linear set of dimension* m. $\langle \{p_1, \ldots, p_k\} \rangle$ will often be written as $\langle p_1, \ldots, p_k \rangle$ and $d\langle p_1, \ldots, p_k \rangle$ will denote $\langle dp_1, \ldots, dp_k \rangle$.

Our paper will focus on the linear subsets of $\mathbb{N}_0$. The main classes of linear sets investigated are denoted as follows. In these definitions, $k \in \mathbb{N}$.

(I) $\mathrm{LINSET}_k := \{\langle P \rangle : P \subset \mathbb{N}_0 \wedge \exists p \in P[p \neq 0] \wedge |P| \leq k\}$.
(II) $\mathrm{LINSET} := \bigcup_{k \in \mathbb{N}} \mathrm{LINSET}_k$.
(III) $\mathrm{CF\text{-}LINSET}_k{}^2 := \{\langle P \rangle : \emptyset \neq P \subset \mathbb{N} \wedge gcd(P) = 1 \wedge |P| \leq k\}$.
(IV) $\mathrm{CF\text{-}LINSET} := \bigcup_{k \in \mathbb{N}} \mathrm{CF\text{-}LINSET}_k$.
(V) $\mathrm{NE\text{-}LINSET}_k{}^3 := \{c + \langle P \rangle : c \in \mathbb{N}_0 \wedge P \subset \mathbb{N}_0 \wedge |P| \leq k \wedge \sum_{p \in P} p \leq c\}$.
(VI) $\mathrm{NE\text{-}LINSET} := \bigcup_{k \in \mathbb{N}} \mathrm{NE\text{-}LINSET}_k$.

Note that the classes in items (I) to (IV) exclude singleton linear sets; the reason for this omission will be explained later. The motivation for studying each subfamily in items (III) to (VI) will be explained as it is introduced in the forthcoming sections.

---

[2] CF stands for "cofinite."
[3] NE stands for "non-erasing."

2.2   Teaching Dimension and Recursive Teaching Dimension

The two main teaching parameters studied in this paper are the *teaching dimension* and the *recursive teaching dimension*.

Let $\mathcal{L}$ be a family of subsets of $\mathbb{N}_0^m$. Let $L \in \mathcal{L}$ and T be a subset of $\mathbb{N}_0^m \times \{+, -\}$. Furthermore, let T^+ (resp. T^-) be the set of vectors in T that are labelled "+" (resp. "−"). $X(T)$ is defined to be $T^+ \cup T^-$. A subset L of $\mathbb{N}_0^m$ is said to be *consistent* with T iff $T^+ \subseteq L$ and $T^- \cap L = \emptyset$. T is a *teaching set* for L w.r.t. $\mathcal{L}$ iff T is consistent with L and for all $L' \in \mathcal{L} \setminus \{L\}$, T is not consistent with L'. Every element of $\mathbb{N}_0^m \times \{+, -\}$ is known as a *labelled example*.

Definition 1. [5,11] Let $\mathcal{L}$ be any family of subsets of $\mathbb{N}_0^m$. Let $L \in \mathcal{L}$. The size of a smallest teaching set for L w.r.t. $\mathcal{L}$ is called the *teaching dimension of L w.r.t. $\mathcal{L}$*, denoted by $\mathrm{TD}(L, \mathcal{L})$. The *teaching dimension of $\mathcal{L}$* is defined as $\sup\{\mathrm{TD}(L, \mathcal{L}) : L \in \mathcal{L}\}$ and is denoted by $\mathrm{TD}(\mathcal{L})$.

Another complexity parameter recently studied in computational learning theory is the recursive teaching dimension. It refers to the maximum size of teaching sets in a series of nested subfamilies of the family.

Definition 2. (Based on [8,13]) Let $\mathcal{L}$ be any family of subsets of $\mathbb{N}_0^m$. A *teaching sequence for $\mathcal{L}$* is any sequence $R = ((\mathcal{F}_0, d_0), (\mathcal{F}_1, d_1), \ldots)$ where (i) the families $\mathcal{F}_i$ form a partition of $\mathcal{L}$ and each $\mathcal{F}_i$ is nonempty, and (ii) $d_i = \mathrm{TD}(L, \mathcal{L} \setminus \bigcup_{0 \leq j < i} \mathcal{F}_j)$ for all i and all $L \in \mathcal{F}_i$. $\sup\{d_i : i \in \mathbb{N}_0\}$ is called the *order of R*, and is denoted by $ord(R)$. The *recursive teaching dimension of $\mathcal{L}$* is defined as $\inf\{ord(R) : R \text{ is a teaching sequence for } \mathcal{L}\}$ and is denoted by $\mathrm{RTD}(\mathcal{L})$.

One can also restrict the instances of the teaching sets in a teaching sequence to positive examples; the best possible order of such a teaching sequence will be denoted by RTD^+.

Definition 3. Let $\mathcal{L}$ be any family of subsets of $\mathbb{N}_0^m$. A *teaching sequence with positive examples for $\mathcal{L}$* (or *positive teaching sequence for $\mathcal{L}$*) is any sequence $P = ((\mathcal{F}_0, d_0), (\mathcal{F}_1, d_1), \ldots)$ such that (i) the families $\mathcal{F}_i$ form a partition of $\mathcal{L}$ and each $\mathcal{F}_i$ is nonempty, and (ii) for all i and all $L \in \mathcal{F}_i$, there is a subset $S_L \subseteq L$ with $|S_L| = d_i < \infty$ such that for all $L' \in \bigcup_{j \geq i} \mathcal{F}_j$, it holds that $S_L \subseteq L' \Rightarrow L = L'$. $\sup\{d_i : i \in \mathbb{N}_0\}$ is called the *order of P*, and is denoted by $ord(P)$. If $\mathcal{L}$ has at least one teaching sequence with positive examples, then the *positive recursive teaching dimension of $\mathcal{L}$* is defined as $\inf\{ord(P) : P \text{ is a teaching sequence with positive examples for } \mathcal{L}\}$ and is denoted by $\mathrm{RTD}^+(\mathcal{L})$. If $\mathcal{L}$ does not have any teaching sequence with positive examples, define $\mathrm{RTD}^+(\mathcal{L}) = \infty$.

A *teaching plan for $\mathcal{L}$* is a teaching sequence $((\mathcal{F}_0, d_0), (\mathcal{F}_1, d_1), \ldots)$ for $\mathcal{L}$ such that $|\mathcal{F}_i| = 1$ for each i. A teaching plan $(({L_0}, d_0), ({L_1}, d_1), ({L_2}, d_2), \ldots)$ for $\mathcal{L}$ will often be written as $((L_0, S_0), (L_1, S_1), (L_2, S_2), \ldots)$, where S_i is a teaching set for L_i w.r.t. $\mathcal{L} \setminus \{L_j : 0 \leq j < i\}$. A *teaching plan with positive examples* *for* $\mathcal{L}$ is defined analogously. Note that for any family $\mathcal{L}$, if $\mathrm{TD}(L, \mathcal{L}) = \infty$ for some $L \in \mathcal{L}$, then any teaching plan for $\mathcal{L}$ must have an infinite order. Moreover, TD, RTD and RTD^+ are monotonic, that is, for all $\mathcal{L}' \subseteq \mathcal{L}$ and $K \in \{\mathrm{TD}, \mathrm{RTD}, \mathrm{RTD}^+\}$, $K(\mathcal{L}') \leq K(\mathcal{L})$. Another useful fact is that for any family $\mathcal{L}$, $\inf\{\mathrm{TD}(L, \mathcal{L}) : L \in \mathcal{L}\} \leq \mathrm{RTD}(\mathcal{L})$.

For any $\mathcal{L}' \subseteq \mathcal{L}$, call $R = ((\mathcal{F}_0, d_0), (\mathcal{F}_1, d_1), \ldots)$ a *teaching subsequence* for $\mathcal{L}$ covering $\mathcal{L}'$ iff $\mathcal{F}_0, \mathcal{F}_1, \ldots$ are nonempty, pairwise disjoint subsets of $\mathcal{L}$ such that $\mathcal{L}' \subseteq \bigcup_{i \in \mathbb{N}_0} \mathcal{F}_i$ and $d_i = \mathrm{TD}(L, \mathcal{L} \setminus \bigcup_{0 \leq j < i} \mathcal{F}_j)$ for all i and all $L \in \mathcal{F}_i$. Define $ord(R) = \sup\{d_i : i \in \mathbb{N}_0\}$ and $\mathrm{RTD}(\mathcal{L}', \mathcal{L}) = \inf\{ord(R) : R$ is a teaching subsequence for $\mathcal{L}$ covering $\mathcal{L}'\}$.

A family $\mathcal{L}$ of subsets of $\mathbb{N}_0^m$ is said to have *finite thickness* [2] iff for every $v \in \mathbb{N}_0^m$, the class of linear sets in $\mathcal{L}$ that contain v is finite. Note that finite thickness is a sufficient condition for families that do not contain the empty set to have a teaching plan with positive examples. The proof is omitted.

Proposition 4. *Let $\mathcal{L}$ be a family of subsets of $\mathbb{N}_0^m$ such that $\mathcal{L}$ has finite thickness and $\emptyset \notin \mathcal{L}$. Then there exists a teaching plan with positive examples for $\mathcal{L}$, Q, such that $ord(Q) = RTD^+(\mathcal{L})$.*

The next proposition provides a necessary condition for any family to have a teaching sequence with positive examples. This condition will be used later to establish the non-existence of positive teaching sequences for some families of linear sets.

Proposition 5. *Let $\mathcal{L}$ be a family of subsets of $\mathbb{N}_0^m$ that has at least one positive teaching sequence. Then for every $L \in \mathcal{L}$, there does not exist any infinite descending chain $H_0 \supsetneq H_1 \supsetneq H_2 \supsetneq \ldots$ such that $\{H_0, H_1, H_2, \ldots\} \subseteq \mathcal{L}$ and $L \subsetneq H_i$ for each i.*

Proof. Suppose there is some $L \in \mathcal{L}$ for which there exists an infinite descending chain $H_0 \supsetneq H_1 \supsetneq H_2 \supsetneq \ldots$ with $\{H_0, H_1, H_2, \ldots\} \subseteq \mathcal{L}$ and $L \subsetneq H_i$ for each i. Assume by way of a contradiction that $((\mathcal{L}_0, d_0), (\mathcal{L}_1, d_1), \ldots)$ were a positive teaching sequence for $\mathcal{L}$. Suppose $L \in \mathcal{L}_i$ for some i. Note that for all $j \in \{0, \ldots, i\}$, $L \subsetneq H_j$ implies that $H_j \in \mathcal{L}_{k_j}$ for some $k_j < i$. Further, for all $j \in \{0, \ldots, i-1\}$, since $H_{j+1} \subsetneq H_j$, it must hold that $k_j < k_{j+1}$. This contradicts the fact that $0 \leq k_j < i$ for all $j \in \{0, \ldots, i\}$. ∎

3   Linear Subsets of $\mathbb{N}_0$ With Constant 0

This section will analyse the class LINSET of linear subsets of $\mathbb{N}_0$ with constant 0. Even in the apparently simple one-dimensional case, the teaching complexity measures can vary quite widely across families of linear sets. Many proofs will exploit the fact that linear sets of dimension 1 are ultimately periodic, a property that has no exact analogue for linear sets of higher dimensions.

Proposition 6. *[10] Let $P \subset \mathbb{N}$ be a finite set such that $\gcd(P) = 1$. Then $\mathbb{N} \setminus \langle P \rangle$ is finite. The largest number in $\mathbb{N} \setminus \langle P \rangle$ is known as the* Frobenius *number of $\langle P \rangle$.*

For any $P = \{p_1, \ldots, p_k\}$ with $gcd(P) = 1$, $F(P)$ and $F(p_1, \ldots, p_k)$ will denote the Frobenius number of $\langle P \rangle$. We will characterise the teaching sets of all linear sets $\langle P \rangle$ such that $gcd(P) = 1$ with respect to LINSET in terms of P and a certain (finite) subset of $\mathbb{N} \setminus \langle P \rangle$. The following variation of a notion from the theory of numerical semigroups will help to formulate the characterisation.

The *partial ordering induced by* P (modified from [10]) is defined as follows: $x \leq_P y \iff \exists a \in \mathbb{N} : y - ax \in \langle P \rangle$. We write $x <_P y$ as an abbreviation of $x \leq_P y \wedge x \neq y$. One has $x \in \langle P \rangle \wedge x \leq_P y \Rightarrow y \in \langle P \rangle$, or equivalently, $y \notin \langle P \rangle \wedge x \leq_P y \Rightarrow x \notin \langle P \rangle$.

For the rest of this section, "maximal" (resp. "minimal") always means "maximal w.r.t. $\leq_P$" (resp. "minimal w.r.t. $\leq_P$") unless specified otherwise. Let MAX_P be the set of maximal elements in $\mathbb{N} \setminus \langle P \rangle$ and let MIN_P denote the set of minimal elements in $\langle P \rangle \setminus \{0\}$. The following lemma collects some useful known facts. Many of these facts are proven in [10], or may be directly deduced from related results proven in [10].

Lemma 7.   (I) $\mathbb{N} \setminus \langle P \rangle$ *contains an infinite ascending chain* $x_0 <_P x_1 <_P x_2 <_P \ldots$ *(e.g.* $x_i = 1 + ip$ *with an arbitrary choice of* $p \in P$*) iff* $gcd(P) > 1$.

(II) *If* $H \subseteq \langle P \rangle$, *then the following hold:*
   (a) $\langle H \rangle \subseteq \langle P \rangle$.
   (b) *The partial ordering* $\leq_P$ *is a refinement of the partial ordering* $\leq_H$; *that is, for any* x, y, $x \leq_H y$ *implies* $x \leq_P y$.
   (c) $\text{MIN}_P \cap \langle H \rangle \subseteq \text{MIN}_H$.

(III) *Let* $s = a_1 p_1 + \ldots + a_r p_r \in \langle P \rangle$ *and let* $I = \{i \in [r] | \ a_i \neq 0\}$. *Then* $p_i \leq_P s$ *for each* $i \in I$. *This implies that* $\text{MIN}_P \subseteq P$.

(IV) *If* $p, p' \in P$ *and* $p <_P p'$, *then* p' *is* superfluous, *i.e.,* $\langle P \rangle = \langle P \setminus \{p'\} \rangle$.

(V) *If* P *is independent, then* $P = \text{MIN}_P$.

For the rest of this section, it will always be assumed that P is independent. We now study teaching sets.

Lemma 8.   1. *Let* T *be a teaching set for* $\langle P \rangle$ *w.r.t. LINSET. Then* $P \subseteq T^+$.
   2. *Let* T *be a teaching set for* $\langle P \rangle$ *w.r.t.* LINSET_k *for* $k = |P| + 1$. *Then, for each* $x \in \mathbb{N} \setminus \langle P \rangle$, *there exists* $y \in T^-$ *such that* $x \leq_P y$.

Proof.

1. Since the labelling in T is consistent with $\langle P \rangle$, it follows that $T^+ \subseteq \langle P \rangle$ and $T^- \cap \langle P \rangle = \emptyset$. Therefore, $\langle T^+ \rangle \subseteq \langle P \rangle$ so that $\langle T^+ \rangle$ is consistent with T. Since T is a teaching set for $\langle P \rangle$, we may conclude that $\langle T^+ \rangle = \langle P \rangle$, which implies that $\leq_{T^+}$ is the same partial ordering as $\leq_P$. Since P (by the general convention made above) is independent, it follows that $P = \text{MIN}_P = \text{MIN}_{T^+} \subseteq T^+$. Thus $P \subseteq T^+$, as desired.

2. Pick an arbitrary but fixed $x \in \mathbb{N} \setminus \langle P \rangle$. We have to show that T^- contains some element y such that $x \leq_P y$. Let $P' = P \cup \{x\}$. Clearly, $\langle P \rangle$ is a proper subset of $\langle P' \rangle$ and $\langle P' \rangle$ is consistent with T^+. Since T is a teaching

set for $\langle P \rangle$ w.r.t. LINSET_k and $\langle P \rangle, \langle P' \rangle \in \text{LINSET}_k$, it follows that T^- contains an element $y \in \langle P' \rangle \setminus \langle P \rangle$. Thus y can be written in the form $y = ax + \sum_{p \in P} a(p)p$ for some properly chosen $a \in \mathbb{N}$ and $a(p) \in \mathbb{N}_0$. It follows that $x \leq_P y$, as desired.   ∎

Corollary 9. *If* $\gcd(P) > 1$ *and* $k = 1 + |P|$, *then* $TD(\langle P \rangle, \text{LINSET}_k) = \infty$.

Proof. According to Lemma 7, $\mathbb{N} \setminus \langle P \rangle$ contains an infinite ascending chain $x_1 <_P x_2 <_P x_3 <_P \ldots$. According to the second assertion in Lemma 8, a teaching set for $\langle P \rangle$ must contain infinitely many elements of this chain.   ∎

Corollary 10. *If* $\gcd(P) = 1$, *then the set* $T(P)$ *given by* $T(P)^+ = P$ *and* $T(P)^- = MAX_P$ *is the unique smallest teaching set for* $\langle P \rangle$ *w.r.t. LINSET.*

Proof. Suppose that $\langle H \rangle$ is consistent with $T(P)$. We show that $\langle H \rangle = \langle P \rangle$ (implying that $T(P)$ is a teaching set for $\langle P \rangle$). Since $P \subseteq \langle H \rangle$ (by consistency), it follows that $\langle P \rangle \subseteq \langle H \rangle$. Pick an arbitrary but fixed element x from $\mathbb{N} \setminus \langle P \rangle$. Recall that $\mathbb{N} \setminus \langle P \rangle$ is finite. Thus, by the definition of MAX_P, there must exist an element $y \in MAX_P$ such that $x \leq_P y$. From $x \leq_P y$ and $\langle P \rangle \subseteq \langle H \rangle$, we may conclude that $x \leq_H y$. The number $y \in MAX_P$ cannot belong to $\langle H \rangle$ (because $\langle H \rangle$ is consistent with $T(P)$). Now, $x \leq_H y$ implies that $x \notin \langle H \rangle$. Thus, $\langle H \rangle$ does not contain any element outside $\langle P \rangle$. It follows that $\langle P \rangle = \langle H \rangle$. Thus, $T(P)$ is a teaching set for $\langle P \rangle$ w.r.t. LINSET, indeed. According to Lemma 8, any other teaching set must contain $T(P)$ as a subset.   ∎

Remark 11. If T is a teaching set for $L' \subseteq \mathbb{N}_0^m$ w.r.t. $\mathcal{L}$, then for any $c \in \mathbb{N}_0^m$, $c + T = \{(c+x, +) : x \in T^+\} \cup \{(c+y, -) : y \in T^-\}$ is a teaching set for $c + L'$ w.r.t. $\mathcal{L}[n] = \{c + L : L \in \mathcal{L}\}$. Thus Lemma 8 and Corollaries 9 and 10 may be readily generalised, *mutatis mutandis*, to the class $\text{LINSET}[c] = \{c + L : L \in \text{LINSET}\}$ for any $c \in \mathbb{N}_0$.

The proof of [8, Theorem 6] provides a construction that may be slightly modified to show that $TD(\text{LINSET}_1) = \infty$ even though $TD(\langle q \rangle, \text{LINSET}_1) < \infty$ for any $q > 0$. By the monotonicity of TD, $TD(\text{LINSET}) = TD(\text{LINSET}_k) = \infty$ for all $k > 0$.

By Corollary 9, LINSET contains infinitely many members that have an infinite TD w.r.t. LINSET. Thus for any $\mathcal{L} \subseteq \text{LINSET}$, it may be difficult to interpret a value of ∞ for $TD(\mathcal{L})$: (1) on the one hand, all cofinite subclasses of $\mathcal{L}$ may have an infinite TD w.r.t. $\mathcal{L}$; (2) on the other hand, there may be a cofinite subclass of $\mathcal{L}$ that has a finite TD w.r.t. $\mathcal{L}$. Intuitively, it seems that $\mathcal{L}$ in Case (2) is unteachable in a weaker sense than in Case (1), but the TD makes no such distinction. It shall be shown, however, that the RTD is a bit more well-behaved when applied to LINSET. In particular, for all $\mathcal{L} \subset \text{LINSET}$, $RTD(\mathcal{L}, \text{LINSET})$ grows only linearly with $\sup\{\min(P) : \langle P \rangle \in \mathcal{L} \wedge \min(P) > 0\}$. We will also give a finer analysis of LINSET_k for $k \in \{1, 2, 3\}$, showing that while LINSET_2 does not have any positive teaching sequence, $RTD(\text{LINSET}_k) < \infty$ for $k \in \{1, 2, 3\}$. In addition, $RTD(\text{LINSET}_k)$ grows at least linearly in k, implying that

RTD(LINSET) $= \infty$. The question of whether RTD(LINSET$_k$) $< \infty$ for any $k > 3$ remains open.

First, the following proposition explains why the singleton $\{0\}$ was excluded from the definition of LINSET. The proof is quite similar to that of Proposition 21, which will be proven later.

Proposition 12. *RTD*($\{\{0\}\}, \{\{0\}\} \cup LINSET_1) = \infty$.

Proposition 12 should be contrasted with the observation that RTD(LINSE-T$_1$) $= 1$: $((\langle 1 \rangle, 1), (\langle 2 \rangle, 1), \ldots)$ is a teaching plan with positive examples for LINSET$_1$, where the ith linear set in the plan is $\langle i \rangle$ and $\{(i, +)\}$ is a teaching set for $\langle i \rangle$ w.r.t. $\{\langle j \rangle : j \geq i\}$. The next theorem shows on the other hand that for any finite $\mathcal{L} \subset$ LINSET, RTD($\mathcal{L}$, LINSET) $< \infty$; in fact, for *any* $\mathcal{L} \subset$ LINSET, RTD($\mathcal{L}$, LINSET) is at most linear in $\sup\{\min(P) : \langle P \rangle \in \mathcal{L} \wedge \min(P) > 0\}$.

Theorem 13. *Let* $\mathcal{F}_n = \{\langle P \rangle : P \text{ is independent } \wedge \min(P) \leq n\}$. *Then* *RTD*($\mathcal{F}_n$, *LINSET*) $\leq 2n - 1$.

Proof. (Sketch.) To streamline the proof, we will adopt some graph terminology. Let $\mathcal{L} =$ LINSET. Let $L \mapsto T(L)$ be a mapping that assigns a set of labelled examples to every $L \in \mathcal{L}$. Define the *digraph induced by* T as the graph $G = (V_G, A_G)$, where the nodes of G are identified with the members L of $\mathcal{L}$, i.e., $V_G = \mathcal{L}$, and a pair $(L', L) \in \mathcal{L} \times \mathcal{L}$ is included in A_G iff L' is consistent with $T(L)$. Define the *depth* of a node v in a digraph as the length of the longest path ending in v (or as ∞ if the paths ending in v can become arbitrarily long). Say that the mapping $L \mapsto T(L)$ with L ranging over all members of $\mathcal{L}$ is *RTD-admissible for* $\mathcal{L}$ if the digraph G induced by T is acyclic and every node in G has a finite depth.

We shall use the following two facts (proofs omitted due to space constraints): (1) there exists a partition of $\mathcal{L}$ into $\mathcal{L}_0, \mathcal{L}_1, \ldots$ such that, for all i and all $L \in \mathcal{L}_i$, it holds that $T(L)$ is a teaching set for L w.r.t. $\bigcup_{j \geq i} \mathcal{L}_j$ iff T is RTD-admissible; (2) if $P = \{p_1, \ldots, p_k\} \subseteq \mathbb{N}_0$ is independent, $p_1 = \min(P)$ and $d = \gcd(P)$, then $k = |P| \leq p_1$ and $|\text{MAX}_P| \leq p_1 - 1$.

Let P range over finite independent subsets of $\mathbb{N}_0$. We shall show that the mapping $\langle P \rangle \mapsto T(\langle P \rangle)$ given by $T(\langle P \rangle)^+ = P$ and $T(\langle P \rangle)^- = \text{MAX}_P$ is RTD-admissible for $\mathcal{L}$. It suffices to show that the digraph $G = (\mathcal{L}, A)$ induced by the mapping $\langle P \rangle \mapsto T(\langle P \rangle)$ is acyclic and every node $\langle P \rangle \in \mathcal{L}$ has a finite depth. Suppose that $(\langle P' \rangle, \langle P \rangle) \in A$. It follows from the construction of G that $\langle P' \rangle$ is consistent with $T(\langle P \rangle)$. The consistency with $T(\langle P \rangle)^+ = P$ implies that $d' = \gcd(P')$ is a divisor of $d = \gcd(P)$. It suffices to show that d' is a proper divisor of d since this implies that G is acyclic and that the depth of $\langle P \rangle$ is bounded by the number of prime power divisors of $d = \gcd(P)$. Suppose for sake of contradiction that $d' = d$ so that $\langle P' \rangle, \langle P \rangle \subseteq d \cdot \mathbb{N}_0$. Now we may argue as follows. Since $(\langle P' \rangle, \langle P \rangle) \in A$, the two linear sets do not coincide so that we may pick a point u from their symmetric difference. Since both linear sets are subsets of $d \cdot \mathbb{N}_0$, there exists $u' \in \mathbb{N}$ such that $u = du'$. But then u' is in the symmetric difference of $(1/d)\langle P' \rangle$ and $(1/d)\langle P \rangle$. On the other hand, since $\langle P' \rangle$ is consistent with $T(\langle P \rangle)$, it follows that $(1/d)\langle P' \rangle$ is consistent with $(1/d)P$ labelled "+" and $(1/d)\text{MAX}_P = \text{MAX}_{(1/d)P}$ labelled "−". According to Corollary 10, $(1/d) \cdot P$ labelled "+" and $\text{MAX}_{(1/d)P}$ labelled "−" is a teaching set for $(1/d)\langle P \rangle$ w.r.t. $\mathcal{L}$, a contradiction.

Finally, observe that from Facts (1), (2), the RTD-admissibility of T for $\mathcal{L}$, and the condition that $\min(P) \leq n$ for all $\langle P \rangle \in \mathcal{F}_n$ with P independent, one can find a teaching subsequence for $\mathcal{L}$ covering $\mathcal{F}_n$ such that the order of this teaching subsequence is at most $n + (n-1) = 2n - 1$. ∎

Theorem 15 (to be shown later) will imply that the order of any teaching sequence for LINSET must necessarily be infinite. Nonetheless, the preceding theorem shows roughly that the growth of $\text{RTD}(\mathcal{L}, \text{LINSET})$ with $\mathcal{L}$ is relatively modest if the minimum positive periods of all $L \in \mathcal{L}$ vary only linearly.

The next series of results will present a detailed study of CF–LINSET$_k$ for each k and CF–LINSET, which comprises all linear sets $\langle P \rangle$ such that P $(\neq \emptyset)$ is a finite subset of $\mathbb{N}$ and and $gcd(P) = 1$. By Proposition 6, this is precisely the class of cofinite linear subsets of $\mathbb{N}_0$ with constant 0. Since LINSET$_k$ is a union of classes of linear sets, each of which is isomorphic to CF–LINSET$_k$, it is hoped that investigating the teaching complexity of CF–LINSET$_k$ may lead to some insights into the question of whether $\text{RTD}(\text{LINSET}_k)$ is finite for each k. CF–LINSET is also perhaps interesting in its own right: on the one hand, the teaching dimension of CF–LINSET$_k$ is finite for $k \leq 3$ but infinite for $k \geq 5$; on the other hand, for all k, CF–LINSET$_k$ has a relatively simple teaching sequence that gives it an RTD$^+$ of k. The first result gives an almost complete analysis of $\text{TD}(\text{CF–LINSET}_k)$ for all k; the case $k = 4$ is left open.

Theorem 14.   (I) $TD(\text{CF–LINSET}_1) = 0$;
(II) $TD(\text{CF–LINSET}_2) = 3$;
(III) $TD(\text{CF–LINSET}_3) = 5$;
(IV) *for each* $k \geq 5$, $TD(\text{CF–LINSET}_k) = \infty$.

Proof. The proofs of Assertions (III) and (IV) are quite long and will be omitted. *Assertion (I).* Note that CF–LINSET$_1 = \{\langle 1 \rangle\}$. The empty set is a teaching set for $\langle 1 \rangle$ w.r.t. CF–LINSET$_1$.

Assertion (II). We first prove the upper bound. Note that $\mathbb{N}_0$ is the only member of CF–LINSET$_2$ that is generated by one number. $\{(1, +)\}$ is a teaching set for $\mathbb{N}_0$ w.r.t. CF–LINSET$_2$, and so $\text{TD}(\mathbb{N}_0) = 1$. Now consider $L = \langle p_1, p_2 \rangle$, where $gcd(p_1, p_2) = 1$. We claim that $T = \{(p_1, +), (p_2, +), (p_1 p_2 - p_1 - p_2, -)\}$ is a teaching set for L w.r.t. CF–LINSET$_2$. Note that $F(p_1, p_2) = p_1 p_2 - p_1 - p_2$ (see, for example, [10]), and so the labelling of T is consistent with L. For any $L' \in$ CF–LINSET$_2$ such that $\{p_1, p_2\} \subseteq L'$, it must hold that $L' \subseteq L$. Suppose further that $L' \neq L$, and take any $p_3 \in L' - L$. Then there exists some k with $0 \leq k \leq p_1 - 1$ such that $p_3 \equiv kp_2 \pmod{p_1}$. Since $p_3 \notin L$, this means that for some $m \geq 1$, $p_3 = kp_2 - mp_1$. As $p_1 p_2 - p_1 - p_2 = kp_2 - mp_1 + (m-1)p_1 + (p_1 - k - 1)p_2 = p_3 + (m-1)p_1 + (p_1 - k - 1)p_2 \in \langle p_1, p_2, p_3 \rangle \subseteq L'$, it follows that $p_1 p_2 - p_1 - p_2 \in L'$, and so T cannot be consistent with L'. Hence $\text{TD}(\text{CF–LINSET}_2) \leq 3$.

For the lower bound, choose primes p_1, p_2, p_3 such that $2 < p_1 < p_2 < p_3$. Note that $\langle 2, p_1 p_2 p_3 \rangle \subsetneq \langle 2, p_1 p_2 \rangle \subsetneq \langle 2, p_1 \rangle$ is a chain in CF–LINSET$_2$. Thus if T is any teaching set for $\langle 2, p_1 p_2 \rangle$ w.r.t. CF–LINSET$_2$ such that $|T| \leq 2$, then T must contain exactly one positive example $(x_1, +)$ and exactly one negative example $(x_2, -)$. Choose any prime $p > max(\{x_1, x_2, 2, p_1 p_2\})$. Then $\langle x_1, p \rangle$ is consistent with T but $\langle 2, p_1 p_2 \rangle \neq \langle x_1, p \rangle$. Hence $|T| \geq 3$. ∎

Theorem 15. *For all $k \geq 1$, $RTD(CF$–$LINSET_k) \in \{k-1, k\}$ and $RTD^+(CF$–$LINSET_k) = k$. Moreover, $RTD(CF$–$LINSET_2) = 2$.*

Proof. (Sketch.) We prove $RTD^+(CF$–$LINSET_k) = k$. First, a teaching plan with positive examples for CF–LINSET$_k$ is constructed. Let $\langle P_0 \rangle, \langle P_1 \rangle, \dots$ be a one-one enumeration of CF–LINSET$_k$ such that for all i, j with $i < j$, P_i is independent and $\langle P_i \rangle \not\subseteq \langle P_j \rangle$. Such an enumeration exists because for each $\langle P \rangle \in$ CF–LINSET$_k$, there are only finitely many $\langle P' \rangle \in$ CF–LINSET$_k$ such that $\langle P \rangle \subseteq \langle P' \rangle$, which implies that there are only finitely many chains $C_1, \dots$ such that $\langle P \rangle$ is the least member (with respect to set inclusion) of each C_l, and each of these chains has finite length. Let Q be the teaching plan $((\langle P_0 \rangle, S_0), (\langle P_1 \rangle, S_1), (\langle P_2 \rangle, S_2), \dots)$ where, for each P_i, $S_i = \{(p, +) : p \in P_i\}$. Since $\langle P_i \rangle \not\subseteq \langle P_j \rangle$ for all $j > i$, S_i is a teaching set for $\langle P_i \rangle$ w.r.t. $\{\langle P_j \rangle : j \geq i\}$. Further, as $|S_i| \leq k$ for all i, Q is a teaching plan for CF–LINSET$_k$ of order at most k.

Now it is shown that $RTD^+(CF$–$LINSET_k) \geq k$. Let $P = \{k, k+1, \dots, 2k-1\}$, and consider the class $\mathcal{C}_k = \{H \in$ CF–LINSET$_k : H \subseteq \langle P \rangle\}$. For any positive teaching sequence Q' of $\mathcal{C}_k$, $\langle P \rangle$ must be contained in the first nonempty subclass of $\mathcal{C}_k$ removed. If $\langle P \rangle$ has a teaching set with positive examples S (w.r.t. the subclass of linear sets in $\mathcal{C}_k$ that do not occur before $\langle P \rangle$ in Q') such that $|S| \leq k - 1$, then $\langle X(S) \rangle$ is a proper subset of $\langle P \rangle$ that is consistent with S; further, there exists some prime $p \notin X(S)$ such that $\langle X(S) \cup \{p\} \rangle \in$ CF–LINSET$_k$ and $\langle X(S) \cup \{p\} \rangle \subsetneq \langle P \rangle$. Hence $|S| \geq k$. This proves that $RTD^+(CF$–$LINSET_k) \geq RTD^+(\mathcal{C}_k) \geq k$. We skip the proof that $RTD(CF$–$LINSET_2) = 2$.

The construction that witnesses $RTD(CF$–$LINSET_k) \geq k - 1$ is based on a hitherto unpublished proof [4]. For each k, let $\mathcal{L}_k = \{\langle k, p_1, \dots, p_{k-1} \rangle : \forall i \in \{1, \dots, k-1\}[p_i \in \{k + i, 2k + i\}]\}$. One can show that $RTD(CF$–$LINSET_k) \geq RTD(\mathcal{L}_k) \geq min(\{TD(L, \mathcal{L}_k) : L \in \mathcal{L}_k\}) \geq k - 1$. ∎

Corollary 16. *$TD(CF$–$LINSET) = RTD(CF$–$LINSET) = RTD^+(CF$–$LINSET) = RTD(LINSET) = \infty$.*

For each $k \in \{1, 2, 3\}$, the result on $TD(CF$–$LINSET_k)$ may be directly applied to construct a teaching sequence of finite order for LINSET$_k$.

Theorem 17.   (I) *LINSET$_2$ does not have any positive teaching sequence.*
(II) *$RTD^+(LINSET_1) = RTD(LINSET_1) = 1$, $RTD(LINSET_2) = 3$ and $3 \leq RTD(LINSET_3) \leq 5$.*

Proof. *Assertion (I).* Let $p_1, p_2, p_3, \dots$ be a strictly increasing infinite sequence of primes with $p_1 > 2$. Note that for all j, $\langle 2 \rangle \subsetneq \langle 2, p_1 \dots p_j \rangle$. Further, $\langle 2, p_1 \rangle \supsetneq$

$\langle 2, p_1 p_2 \rangle \supsetneq \langle 2, p_1 p_2 p_3 \rangle \supsetneq \ldots \supsetneq \langle 2, p_1 \ldots p_j \rangle \supsetneq \langle 2, p_1 \ldots p_j p_{j+1} \rangle \supsetneq \ldots$ is an infinite descending chain in LINSET_2. Thus by Proposition 5, LINSET_2 does not have a positive teaching sequence.

Assertion (II). (Sketch.) We had described earlier (after Proposition 12) a teaching plan with positive examples of order 1 for LINSET_1. Here, a teaching sequence for LINSET_2 of order no more than 3 is given; a teaching sequence for LINSET_3 can be constructed analogously using teaching sets of size at most 5 for linear sets in CF-LINSET_3. Define the sequence $((\mathcal{L}_0, d_0), (\mathcal{L}_1, d_1), \ldots)$ where, for all $i \in \mathbb{N}_0$, $\mathcal{L}_i = \{\langle p_1, p_2 \rangle : p_1, p_2 \in \mathbb{N} \wedge gcd(p_1, p_2) = i + 1\}$. Consider any $\langle p_1, p_2 \rangle \in \mathcal{L}_i$. The proof of Assertion (II) in Theorem 14 gives a teaching set T of size no more than 3 for $\left\langle \frac{p_1}{i+1}, \frac{p_2}{i+1} \right\rangle$ w.r.t. CF-LINSET_2. Now let $T' = \{((i+1)x, +) : x \in T^+\} \cup \{((i+1)y, -) : y \in T^-\}$. Note that since T' contains the two positive examples $(p_1, +), (p_2, +)$ and $gcd(p_1, p_2) = i + 1$, no linear set $\langle P \rangle \in \text{LINSET}_2$ with $gcd(P) > i + 1$ can be consistent with T'. Moreover, T' is a teaching set for $\langle p_1, p_2 \rangle$ w.r.t. all $\langle P \rangle \in \text{LINSET}_2$ with $gcd(P) = i + 1$. Hence $d_i \leq 3$ for all $i \in \mathbb{N}_0$. We omit the proof that $\text{RTD}(\text{LINSET}_2) \geq 3$. ∎

4   Linear Subsets of $\mathbb{N}_0$ with Bounded Period Sums

The present section will examine a special family of linear subsets of $\mathbb{N}_0$ that arises from studying an invariant property of the class of non-erasing pattern languages over varying unary alphabets. Recall that the *commutative* image, or *Parikh* image, of $w \in \{a\}^*$ is the number of times that a appears in w, or the length of w. Thus the commutative image of the language generated by a non-erasing pattern $a^{k_0} x_1^{k_1} \ldots x_n^{k_n}$ is the linear subset $(k_0 + k_1 + \ldots + k_n) + \langle k_1, \ldots, k_n \rangle$ of $\mathbb{N}_0$, which is in NE–LINSET. Conversely, any $L \in$ NE–LINSET is the commutative image of a non-erasing pattern language. This gives a one-to-one correspondence between the class of non-erasing pattern languages and the class NE–LINSET, so that the two classes have equivalent teachability properties. The following theorem gives the exact value of $\text{RTD}^+(\text{NE–LINSET}_k)$.

Theorem 18. $RTD^+(NE\text{--}LINSET_k) = k + 1$.

Proof. (Sketch.) Note that as NE–LINSET$_k$ has finite thickness and $\emptyset \notin$ NE–LINSET$_k$, Proposition 4 implies that NE–LINSET$_k$ has a teaching plan with positive examples. $\text{RTD}^+(\text{NE–LINSET}_k) \leq k + 1$ is shown. A teaching plan Q for NE–LINSET$_k$ is built in stages as follows. Let Q_c denote the segment of Q that has been defined up to stage c and let A_c denote the class of all $L \in$ NE–LINSET$_k$ such that Q_c does not contain L. It is assumed inductively that A_c does not contain any linear subset of $\mathbb{N}_0$ with constant less than c. The idea of the construction is to design a teaching plan for the finite subclass $P_c^0 = \{c + \langle p_1, \ldots, p_k \rangle : p_1 + \ldots + p_k \leq c\}$ at stage c.

Inductively, assume that P_c^g has been defined for some $g \geq 0$. If $P_c^g = \emptyset$, then the teaching plan for P_c^0 is complete. Otherwise, suppose that P_c^g is nonempty. Choose $c + \langle p_1, \ldots, p_k \rangle \in P_c^g$, the next linear set to be taught in the teaching plan for P_c^0, so that $c + \langle p_1, \ldots, p_k \rangle$ is maximal in P_c^g with respect to the subset inclusion relation. Set $P_c^{g+1} = P_c^g \setminus \{c + \langle p_1, \ldots, p_k \rangle\}$. We claim that $S = \{(c, +)\} \cup \{(c + p_i, +) : 1 \leq i \leq k\}$ is a teaching set for $c + \langle p_1, \ldots, p_k \rangle$ w.r.t. $P_c^g \cup (A_c \setminus P_c^0)$. Assume that $\{c\} \cup \{c + a_i : 1 \leq i \leq k\} \subseteq c' + \langle b_1, \ldots, b_l \rangle \in P_c^g \cup (A_c \setminus P_c^0)$. Since A_c does not contain any linear subset of $\mathbb{N}_0$ with constant less than c, $c' = c$ and so $c' + \langle b_1, \ldots, b_l \rangle \in P_c^g$. Then for all $i \in \{1, \ldots, k\}$, $a_i = \sum_{i=1}^{l} q_i b_i$ for some non-negative integers $q_1, \ldots, q_l$, and therefore $c + \langle a_1, \ldots, a_k \rangle \subseteq c + \langle b_1, \ldots, b_l \rangle$. By the maximality of $c + \langle a_1, \ldots, a_k \rangle$, one has $c + \langle a_1, \ldots, a_k \rangle = c + \langle b_1, \ldots, b_l \rangle$. The construction continues until a stage g' is reached where $P_c^{g'} = \emptyset$. Q_{c+1} is then defined as the concatenation of Q_c and the teaching plan for P_c^0 (with Q_c as the prefix). We omit the somewhat long proof that $\mathrm{RTD}^+(\mathrm{NE\text{--}LINSET}_k) \geq k + 1$. ∎

The lower bound on $\mathrm{RTD}(\mathrm{NE\text{--}LINSET}_k)$ in the following theorem may be obtained by adapting the proof of the corresponding result for $\mathrm{CF\text{--}LINSET}_k$; the proof of [8, Theorem 6] immediately implies that $\mathrm{TD}(\mathrm{NE\text{--}LINSET}_k) = \infty$.

Theorem 19. *For all $k \geq 1$, $k - 1 \leq \mathrm{RTD}(\mathrm{NE\text{--}LINSET}_k) \leq k + 1$ and $\mathrm{TD}(\mathrm{NE\text{--}LINSET}_k) = \infty$.*

Remark 20. Our results on NE–LINSET may be generalised to classes of linear subsets of $\mathbb{N}_0^m$ for any $m > 1$ in the following way. For each m, define

(I) $\mathrm{NE\text{--}LINSET}_k^m := \{c + \langle P \rangle : c \in \mathbb{N}_0^m \wedge P \subset \mathbb{N}_0^m \wedge |P| \leq k \wedge \| \sum_{p \in P} p \|_1 \leq \|c\|_1\}$.
(II) $\mathrm{NE\text{--}LINSET}^m := \bigcup_{k \in \mathbb{N}} \mathrm{NE\text{--}LINSET}_k^m$.

Note that $\mathrm{NE\text{--}LINSET}_k^1 = \mathrm{NE\text{--}LINSET}_k$ and $\mathrm{NE\text{--}LINSET}^1 = \mathrm{NE\text{--}LINSET}$. Then one has $\mathrm{RTD}^+(\mathrm{NE\text{--}LINSET}_k^m) = \mathrm{RTD}^+(\mathrm{NE\text{--}LINSET}_k) = k + 1$, $k - 1 \leq \mathrm{RTD}(\mathrm{NE\text{--}LINSET}_k^1) \leq \mathrm{RTD}(\mathrm{NE\text{--}LINSET}_k^m)$ and $\mathrm{RTD}(\mathrm{NE\text{--}LINSET}^m) = \mathrm{RTD}(\mathrm{NE\text{--}LINSET}) = \mathrm{RTD}^+(\mathrm{NE\text{--}LINSET}^m) = \mathrm{RTD}^+(\mathrm{NE\text{--}LINSET}) = \infty$.

5   Linear Subsets of $\mathbb{N}_0^2$ with Constant 0

Finally, we consider how our preceding results may be extended to general classes of linear subsets of higher dimensions. Finding teaching sequences for families of linear sets with dimension $m > 1$ seems to present a new set of challenges, as many of the proof methods for the case $m = 1$ do not carry over directly to the higher dimensional cases. The classes of linear subsets of $\mathbb{N}_0^2$ briefly studied in this section are denoted as follows. In the first definition, $k \in \mathbb{N}$.

(I) $\mathrm{LINSET}_k^2 := \{\langle P \rangle : P \subset \mathbb{N}_0^2 \wedge \exists p \in P[p \neq 0] \wedge |P| \leq k\}$.
(II) $\mathrm{LINSET}_{=2}^2 := \mathrm{LINSET}_2^2 \setminus \mathrm{LINSET}_1^2$.

The following result suggests that to identify interesting classes of linear subsets of $\mathbb{N}_0^m$ for $m > 1$ that have finite teaching complexity measures, it might be a good idea to first exclude certain linear sets.

Proposition 21. $\mathrm{RTD}(\{\langle (0, 1) \rangle\}, \mathrm{LINSET}_2^2) = \infty$.

Proof. Assume that $R = ((\mathcal{L}_0, d_0), (\mathcal{L}_1, d_1), \ldots)$ were a teaching subsequence for LINSET_2^2 covering $\{\langle (0,1) \rangle\}$. Suppose that $\langle (0,1) \rangle \in \mathcal{L}_i$ and T were a teaching set for $\langle (0,1) \rangle$ w.r.t. $\text{LINSET}_2^2 \setminus \bigcup_{j<i} \mathcal{L}_j$. Choose any $N > \max(\{d_j : j < i\})$ such that N is larger than every component of any instance $(a, b) \in \mathbb{N}_0^2$ in T. Further, let $p_0, \ldots, p_{N+i}$ be a strictly increasing sequence of primes. Observe that by the choice of N, $\langle (0,1), p_0 \ldots p_{N+i}(1,1) \rangle$ is consistent with T. Hence this linear set occurs in some $\mathcal{L}_{j_0}$ with $j_0 < i$. In addition, since, for any two distinct $(N+i)$-subsets S, S' of $\{p_0, \ldots, p_{N+i}\}$, $\langle (0,1), p_0 \ldots p_{N+i}(1,1) \rangle \subsetneqq \langle (0,1), \prod_{x \in S} x(1,1) \rangle$ and $\langle (0,1), \prod_{x \in S} x(1,1) \rangle \cap \langle (0,1), \prod_{x \in S'} x(1,1) \rangle \subseteq \langle (0,1), p_0 \ldots p_{N+i}(1,1) \rangle$, the choice of N again gives that for some $(N+i)$-subset S_1 of $\{p_0, \ldots, p_{N+i}\}$, $\langle (0,1), \prod_{x \in S_1} x(1,1) \rangle \in \mathcal{L}_{j_1}$ for some $j_1 < j_0$. The preceding line of argument can be applied again to show that for some $(N+i-1)$-subset S_2 of S_1, $\langle (0,1), \prod_{x \in S_2} x(1,1) \rangle \in \mathcal{L}_{j_2}$ for some $j_2 < j_1$. Repeating the argument successively thus yields a chain $S_1 \supsetneqq S_2 \supsetneqq \ldots \supsetneqq S_i$ of subsets of $\{p_0, \ldots, p_{N+i}\}$ such that $\langle (0,1), \prod_{x \in S_l} x(1,1) \rangle \in \mathcal{L}_{j_l}$ for all $l \in \{1, \ldots, i\}$, where $j_i < \ldots < j_1 < j_0 < i$, which is impossible as $j_i \geq 0$. Hence there is no teaching subsequence of LINSET_2^2 covering $\{\langle (0,1) \rangle\}$.   ∎

One can define quite a meaningful subclass of LINSET_2^2 that does have a finite RTD. $\text{LINSET}_{=2}^2$ consists of all linear sets in LINSET_2^2 that are *strictly 2-generated*. Examples of strictly 2-generated linear subsets include $\langle (1,0), (0,1) \rangle$ and $\langle (4,6), (6,9) \rangle$. $\langle (0,1) \rangle$ is not a strictly 2-generated linear subset.

Theorem 22.   (I) $TD(LINSET_{=2}^2) = \infty$.
  (II) $LINSET_{=2}^2$ does not have any positive teaching sequence.
  (III) $RTD(LINSET_{=2}^2) \in \{3, 4\}$.

Proof. *Assertion (I)*. Observe from the proof of Proposition 21 that for any N distinct primes $p_0, p_1, \ldots, p_{N-1}$, $\text{TD}(\langle (0,1), p_0 p_1 \ldots p_{N-1}(1,0) \rangle, \text{LINSET}_{=2}^2) \geq N$. Hence $\text{TD}(\mathcal{L}, \text{LINSET}_{=2}^2) = \infty$ for any cofinite subclass $\mathcal{L}$ of $\text{LINSET}_{=2}^2$.

Assertion (II). Let $p_1, p_2, p_3, \ldots$ be a strictly increasing infinite sequence of primes. Note that for all j, $\langle (2,0), (3,0) \rangle \subsetneqq \langle (1,0), p_1 \ldots p_j(0,1) \rangle$. Further, $\langle (1,0), p_1(0,1) \rangle \supsetneqq \langle (1,0), p_1 p_2(0,1) \rangle \supsetneqq \langle (1,0), p_1 p_2 p_3(0,1) \rangle \supsetneqq \ldots \supsetneqq \langle (1,0), p_1 \ldots p_j(0,1) \rangle \supsetneqq \langle (1,0), p_1 \ldots p_j p_{j+1}(0,1) \rangle \supsetneqq \ldots$ is an infinite descending chain in $\text{LINSET}_{=2}^2$. Thus by Proposition 5, $\text{LINSET}_{=2}^2$ does not have a positive teaching sequence.

Assertion (III). (Sketch.) The main idea is that for each strictly 2-generated linear subset S of $\mathbb{N}_0^2$ with canonical representation $(0, P)$, if M denotes the class of all $S' \in \text{LINSET}_{=2}^2$ for which each $S' \in M$ with canonical representation $(0, P')$ satisfies $\|\sum_{p' \in P'} p'\|_1 \geq \|\sum_{p \in P} p\|_1$, then $\text{TD}(S, M) \leq 4$. The sequence $((\mathcal{L}_0, d_0), (\mathcal{L}_1, d_1), \ldots)$ defined by $\mathcal{L}_i = \{\langle u_1, u_2 \rangle : \|u_1 + u_2\|_1 = i + 2\}$ would then be a teaching sequence for $\text{LINSET}_{=2}^2$ of order at most 4. To prove this assertion, it suffices to find a teaching set of size at most 4 for any $\langle u_1, u_2 \rangle$ w.r.t. the class of all $S' \in \text{LINSET}_{=2}^2$ such that if S' has the canonical representation $(0, P')$, then $\|\sum_{p' \in P'} p'\|_1 \geq \|u_1 + u_2\|_1$.

Owing to space constraints, we will only give a proof for the case when $\{u_1, u_2\}$ is linearly independent. For a given linear set L with canonical representation

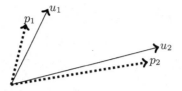

Fig. 1. p_1 and p_2 (not drawn to scale)

(c, P), call each $p \in P$ a *minimal period of L*. Assume that u_1 lies to the left of u_2. Consider the set A of linear sets L in M such that $\langle u_1, u_2 \rangle \subsetneq L$. Since no single vector in $\mathbb{N}_0^2$ can generate two linearly independent vectors in $\mathbb{N}_0^2$, each $L \in A$ must have two linearly independent periods p_1 and p_2, neither of which lies strictly between u_1 and u_2; in addition, $\max(\{\|p_1\|_1, \|p_2\|_1\}) \leq \max(\{\|u_1\|_1, \|u_2\|_1\})$. Thus A is finite. Furthermore, for each $L \in A$ with canonical representation $(0, P')$, at least one of the periods in P' is not parallel to u_1 and also not parallel to u_2, for otherwise $\|\sum_{p' \in P'} p'\|_1 < \|u_1 + u_2\|_1$. If $A = \emptyset$, then $\{(u_1, +), (u_2, +)\}$ is a teaching set for $\langle u_1, u_2 \rangle$ w.r.t. M. Assume that $A \neq \emptyset$. Consider the set $Q = \bigcup_{L \in A} \{w \ : \ w$ is a minimal period of L not parallel to u_1 and not parallel to $u_2\}$. Choose some p_1 among the periods in Q that are closest to u_1 to the left of u_1, and choose p_2 so that p_2 is among the periods in Q that are closest to u_2 to the right of u_2 (see Figure 1); note that at least one of p_1, p_2 exists. For every $L \in A$ with canonical representation $(0, \{v_1, v_2\})$, at least one of p_1 and p_2 lies between (not necessarily strictly) v_1 and v_2, and $\{kp_1, k'p_2\} \cap \langle u_1, u_2 \rangle = \emptyset$ for all $k, k' \in \mathbb{N}$. Thus there is a sufficiently large $K \in \mathbb{N}$ such that for all $L \in A$, either $Kp_1 \in L \setminus \langle u_1, u_2 \rangle$ or $Kp_2 \in L \setminus \langle u_1, u_2 \rangle$. Therefore a teaching set for $\langle u_1, u_2 \rangle$ w.r.t. M is $\{(u_1, +), (u_2, +), (Kp_1, -), (Kp_2, -)\}$. If p_i does not exist for exactly one i, then remove $(Kp_i, -)$ from this teaching set. ∎

6   Conclusion

We have studied two main teaching parameters, the TD and RTD (and its variant RTD$^+$), of classes of linear sets with a fixed dimension. Notice that in Table 1, even though all the classes have an infinite TD, there are finer notions of teachability that occasionally yield different finite sample complexity measures. In particular, there are families of linear sets that have an infinite TD and RTD$^+$ and yet have a finite RTD. We broadly interpret a class that has an infinite RTD as being "unteachable" in a stronger sense than merely having an infinite TD. Quite interestingly, the fact that some classes in Table 1 have an infinite RTD contrasts with Takada's [12] theorem that the family of linear subsets of $\mathbb{N}_0^m$ is learnable in the limit from just positive examples. One possible interpretation of this contrast is that classes of linear sets may be generally harder to teach than to learn. Further, a number of quantitative problems remain open. For example, we did not solve the question of whether RTD(LINSET$_k$) is finite for each $k > 3$. A more precise analysis of the values of RTD for various families of linear sets studied in the present paper (see Table 1) would also be desirable.

Table 1. Partial summary of results.

Class	TD	RTD	RTD$^+$
CF–LINSET$_k$, $k \geq 5$	∞ (Thm 14(IV))	RTD $\in \{k-1, k\}$ (Thm 15)	k (Thm 15)
LINSET$_1$	∞ (Rem 11)	1 (Thm 17(II))	1 (Thm 17(II))
LINSET$_2$	∞ (Rem 11)	3 (Thm 17(II))	∞ (Thm 17(I))
LINSET$_3$	∞ (Rem 11)	RTD $\in \{3, 4, 5\}$ (Thm 17(II))	∞ (Thm 17(I))
LINSET	∞ (Rem 11)	∞ (Cor 16)	∞ (Thm 17(I))
NE–LINSET$_k^m$, $m, k \geq 1$	∞ (Thm 19)	RTD $\in \{k-1, k, k+1\}$ (Rem 20)	$k+1$ (Rem 20)
NE–LINSET^m, $m \geq 1$	∞ (Thm 19)	∞ (Rem 20)	∞ (Rem 20)
LINSET$_{=2}^2$	∞ (Thm 22)	RTD $\in \{3, 4\}$ (Thm 22)	∞ (Thm 22)

Acknowledgments. We thank the referees of ALT 2015 for their critical reading of the manuscript; special thanks go to one referee for pointing out an error in the original version of Proposition 12 and for helping to simplify our definition of a teaching set. Sandra Zilles was partially supported by the Natural Sciences and Engineering Research Council of Canada (NSERC).

References

1. Abe, N.: Polynomial learnability of semilinear sets. In: Computational Learning Theory (COLT), pp. 25–40 (1989)
2. Angluin, D.: Inductive inference of formal languages from positive data. Information and Control **45**(2), 117–135 (1980)
3. Doty, D., Patitz, M.J., Summers, S.M.: Limitations of self-assembly at temperature 1. Theoretical Computer Science **412**(1–2), 145–158 (2011)
4. Gao, Z., Mazadi, Z., Meloche, R., Simon, H.U., Zilles, S.: Distinguishing pattern languages with membership examples, Manuscript (2014)
5. Goldman, S.A., Kearns, M.J.: On the complexity of teaching. Journal of Computer and System Sciences **50**(1), 20–31 (1995)
6. Ibarra, O.H.: Reversal-bounded multicounter machines and their decision problems. Journal of the Association for Computing Machinery **25**(1), 116–133 (1978)
7. Ibarra, O.H., Dang, Z., Egecioglu, O.: Catalytic P systems, semilinear sets, and vector addition systems. Theoretical Computer Science **312**(2–3), 379–399 (2004)
8. Mazadi, Z., Gao, Z., Zilles, S.: Distinguishing pattern languages with membership examples. In: Dediu, A.-H., Martín-Vide, C., Sierra-Rodríguez, J.-L., Truthe, B. (eds.) LATA 2014. LNCS, vol. 8370, pp. 528–540. Springer, Heidelberg (2014)
9. Parikh, R.J.: On context-free languages. Journal of the Association for Computing Machinery **13**(4), 570–581 (1966)
10. Rosales, J.C., García-Sánchez, P.A.: Numerical semigroups. Springer, New York (2009)
11. Shinohara, A., Miyano, S.: Teachability in computational learning. New Generation Computing **8**(4), 337–347 (1991)
12. Takada, Y.: Learning semilinear sets from examples and via queries. Theoretical Computer Science **104**(2), 207–233 (1992)
13. Zilles, S., Lange, S., Holte, R., Zinkevich, M.: Models of cooperative teaching and learning. Journal of Machine Learning Research **12**, 349–384 (2011)

Computational Learning
Theory and Algorithms

Learning a Random DFA from Uniform Strings and State Information

Dana Angluin and Dongqu Chen$^{(\boxtimes)}$

Department of Computer Science, Yale University, New Haven, CT 06520, USA
dongqu.chen@yale.edu

Abstract. Deterministic finite automata (DFA) have long served as a fundamental computational model in the study of theoretical computer science, and the problem of learning a DFA from given input data is a classic topic in computational learning theory. In this paper we study the learnability of a random DFA and propose a computationally efficient algorithm for learning and recovering a random DFA from uniform input strings and state information in the statistical query model. A random DFA is uniformly generated: for each state-symbol pair $(q \in Q, \sigma \in \Sigma)$, we choose a state $q' \in Q$ with replacement uniformly and independently at random and let $\varphi(q, \sigma) = q'$, where Q is the state space, Σ is the alphabet and φ is the transition function. The given data are string-state pairs (x, q) where x is a string drawn uniformly at random and q is the state of the DFA reached on input x starting from the start state q_0. A theoretical guarantee on the maximum absolute error of the algorithm in the statistical query model is presented. Extensive experiments demonstrate the efficiency and accuracy of the algorithm.

Keywords: Deterministic finite automaton · Random DFA · Statistical queries · Regular languages · PAC learning

1 Introduction

Deterministic finite automata are one of the most elementary computational models in the study of theoretical computer science. The important role of DFA leads to the classic problem in computational learning theory, the learnability of DFA. The applications of this learning problem include formal verification, natural language processing, robotics and control systems, computational biology, data mining and music. Exploring the learnability of DFA is significant to both theoretical and applied realms. In the classic PAC learning model defined by Valiant [21], unfortunately, the concept class of DFAs is known to be inherently unpredictable [14,15]. In a modified version of Valiant's model which allows the learner to make membership queries, Angluin [1] has shown that the concept class of DFAs is efficiently PAC learnable. Subsequent efforts have searched for nontrivial properly PAC learnable subfamilies of regular languages [2,6,16].

Since learning all DFAs is computationally intractable, it is natural to ask whether we can pursue positive results for "almost all" DFAs. This is addressed

© Springer International Publishing Switzerland 2015
K. Chaudhuri et al. (Eds.): ALT 2015, LNAI 9355, pp. 119–133, 2015.
DOI: 10.1007/978-3-319-24486-0_8

by studying high-probability properties of uniformly generated random DFAs. The same approach has been used for learning random decision trees and random DNFs from uniform strings [11,12,17,18]. However, the learnability of random DFAs has long been an open problem. Few formal results about random walks on random DFAs are known. Grusho [9] was the first work establishing an interesting fact about this problem. Since then, very little progress was made until a recent subsequent work by Balle [4]. Our work connects these two problems and contributes an algorithm for efficiently learning random DFAs, in addition to positive theoretical results on random walks on random DFAs.

Trakhtenbrot and Barzdin [20] first introduced two random DFA models with different sources of randomness: one with a random automaton graph, one with random output labeling. In this paper we study the former model. A random DFA is uniformly generated: for each state-symbol pair $(q \in Q, \sigma \in \Sigma)$, we choose a state $q' \in Q$ with replacement uniformly and independently at random and let $\varphi(q, \sigma) = q'$, where Q is the state space, Σ is the alphabet and φ is the transition function. Given data are of form (x, q) where x is a string drawn uniformly at random and q is the state of the DFA reached on input x starting from the start state q_0.

Previous work by Freund et al. [8] has studied a different model under different settings. First, the DFAs are generated with arbitrary transition graphs and random output labeling, which is the latter model in [20]. Second, in their work, the learner predicts and observes the exact label sequence of the states along each walk. Such sequential data are crucial to the learner walking on the graph. In our paper, the learner is given noisy statistical data on the ending state, with no information about any intermediate states along the walk.

Like most spectral methods, the theoretical error bound of our algorithm contains a spectral parameter ($\||P_A^\dagger\||_\infty$ in Section 4.1), which reflects the asymmetry of the underlying graph. This leads to a potential future work of eliminating this parameter using random matrix theory techniques. Another direction of subsequent works is to consider the more general case where the learner only observes the accept/reject bits of the final states reached, which under arbitrary distributions has been proved to be hard in the statistical query model by Angluin et al. [3] but remains open under the uniform distribution [4]. Our contribution narrows this gap and pushes forward the study of the learnability of random DFAs.

2   Preliminaries

Deterministic Finite Automaton (DFA) is a powerful and widely studied computational model in computer science. Formally, a DFA is a quintuple $A = (Q, \varphi, \Sigma, q_0, F)$ where Q is a finite set of states, Σ is the finite alphabet, $q_0 \in Q$ is the start state, $F \subseteq Q$ is the set of accepting states, and φ is the transition function: $Q \times \Sigma \to Q$. Let λ be the empty string. Define the extended transition function $\varphi^* : Q \times \Sigma^* \to Q$ by $\varphi^*(q, \lambda) = q$ and inductively $\varphi^*(q, x\sigma) = \varphi(\varphi^*(q, x), \sigma)$ where $\sigma \in \Sigma$ and $x \in \Sigma^*$. Denote by $s = |\Sigma|$ the size of the alphabet and by

$n = |Q|$ the number of states. In this paper we assume $s \geq 2$. Let $G = (V, E)$ be the underlying directed multi-graph of DFA A (also called an *automaton graph*). We say a vertex set $V_0 \subseteq V$ is *closed* if for any $u \in V_0$ and any v such that $(u, v) \in E$, we must have $v \in V_0$.

A *walk* on an automaton graph G is a sequence of states $(v_0, v_1, \ldots, v_\ell)$ such that $(v_{i-1}, v_i) \in E$ for all $1 \leq i \leq \ell$, where v_0 is the corresponding vertex in G of the start state q_0. A *random walk* on graph G is defined by a transition probability matrix P with $P(u, v) = \#\{(u, v) \in E\} \cdot s^{-1}$ denoting the probability of moving from vertex u to vertex v, where $\#\{(u, v) \in E\}$ is the number of edges from u to v. For an automaton graph, a random walk always starts from the start state q_0. In this paper random walks on a DFA refer to the random walks on the underlying automaton graph. A vertex u is *aperiodic* if $\gcd\{t \geq 1 \mid P^t(u, u) > 0\} = 1$. Graph G (or a random walk on G) is *irreducible* if for every pair of vertices u and v in V there exists a directed cycle in G containing both u and v, and is *aperiodic* if every vertex is aperiodic. A distribution vector ϕ satisfying $\phi P = \phi$ is called a *Perron vector* of the walk. An irreducible and aperiodic random walk has a unique Perron vector ϕ and $\lim_{t \to +\infty} P^t(u, \cdot) = \phi$ (called the *stationary distribution*) for any $u \in V$. In the study of rapidly mixing walks, the *convergence rate* in L_2 distance $\Delta_{L_2}(t) = \max_{u \in V} \|P^t(u, \cdot) - \phi\|_2$ is often used. A stronger notion in L_1 distance is measured by the *total variation distance*, given by $\Delta_{TV}(t) = \frac{1}{2} \max_{u \in V} \sum_{v \in V} |P^t(u, v) - \phi(v)|$. Another notion of distance for measuring convergence rate is the *χ-square distance*:

$$\Delta_{\chi^2}(t) = \max_{u \in V} \left(\sum_{v \in V} \frac{(P^t(u, v) - \phi(v))^2}{\phi(v)} \right)^{\frac{1}{2}}$$

As the Cauchy-Schwarz inequality gives $\Delta_{L_2}(t) \leq 2\Delta_{TV}(t) \leq \Delta_{\chi^2}(t)$, a convergence upper bound for $\Delta_{\chi^2}(t)$ implies ones for $\Delta_{L_2}(t)$ and $\Delta_{TV}(t)$.

Trakhtenbrot and Barzdin [20] first introduced the model of random DFA by employing a uniformly generated automaton graph as the underlying graph and labeling the edges uniformly at random. In words, for each state-symbol pair $(q \in Q, \sigma \in \Sigma)$, we choose a state $q' \in Q$ with replacement uniformly and independently at random and let $\varphi(q, \sigma) = q'$.

In a computational learning model, an algorithm is usually given access to an oracle providing information about the target concept. Kearns [13] modified Valiant's model and introduced the *statistical query oracle STAT*. Kearns' oracle takes as input a statistical query of the form (χ, τ). Here χ is any mapping of a labeled example to $\{0, 1\}$ and $\tau \in [0, 1]$ is called the noise *tolerance*. Let c be the target concept and $\mathcal{D}$ be the distribution over the instance space. Oracle $STAT(c, \mathcal{D})$ returns to the learner an estimate for the expectation $\mathbf{E}\chi$, that is, the probability that $\chi = 1$ when the labeled example is drawn according to $\mathcal{D}$. A statistical query can have a condition, in which case $\mathbf{E}\chi$ is a conditional probability. This estimate is accurate within additive error τ. A statistical query χ is *legitimate* and *feasible* if and only if:

1. Query χ maps a labeled example $\langle x, c(x) \rangle$ to $\{0, 1\}$;
2. Query χ can be efficiently evaluated in polynomial time;

3. The condition of χ, if any, can be efficiently evaluated in polynomial time;
4. The probability of the condition of χ, if any, should be at least inverse polynomially large.

Kearns [13] proved that the statistical query model is weaker than the classic PAC model. That is, PAC learnability from oracle $STAT$ implies PAC learnability from the classic example oracle, but not vice versa.

3   Random Walks on a Random DFA

Random walks have proven to be a simple, yet powerful mathematical tool for extracting information from well connected graphs. Since automaton graphs are long known to be of strong connectivity with high probability [9], it's interesting to explore the possibilities of applying random walks to DFA learning. In this section we will show that with high probability, a random walk on a random DFA converges to the stationary distribution ϕ polynomially fast in χ-square distance as in Theorem 1.

Theorem 1. *With probability $1 - o(1)$, a random walk on a random DFA has $\Delta_{\chi^2}(t) \leq e^{-k}$ after $t \geq 2C(C+1)sn^{1+C}(\log n + k) \cdot \log_s n$, where constant $C > 0$ depends on s and approaches unity with increasing s.*

A standard proof of fast convergence consists of three parts: irreducibility, aperiodicity and convergence rate. Grusho [9] first proved the irreducibility of a random automaton graph.

Lemma 1. *With probability $1-o(1)$, a random automaton graph G has a unique strongly connected component, denoted by $\tilde{G} = (\tilde{V}, \tilde{E})$, of size $\tilde{n}$, and a) $\lim_{n \to +\infty} \frac{\tilde{n}}{n}$ $= C$ for some constant $C > 0.7968$ when $s \geq 2$ or some $C > 0.999$ when $s > 6$; b) $\tilde{V}$ is closed.*

A subsequent work by Balle [4] proved the aperiodicity.

Lemma 2. *With probability $1 - o(1)$, the strongly connected component $\tilde{G}$ in Lemma 1 is aperiodic.*

However, the order of the convergence rate of random walks on a random DFA was left as an open question. One canonical technique for bounding the convergence rate of a random walk is to bound the smallest nonzero eigenvalue of the *Laplacian matrix* $\mathcal{L}$ of the graph G, defined by

$$\mathcal{L} = I - \frac{\Phi^{\frac{1}{2}} P \Phi^{-\frac{1}{2}} + \Phi^{-\frac{1}{2}} P^* \Phi^{\frac{1}{2}}}{2}$$

where Φ is an $n \times n$ diagonal matrix with entries $\Phi(u, u) = \phi(u)$ and P^* denotes the transpose of matrix P. For a random walk P, define the *Rayleigh quotient* for any function $f : V \to \mathbb{R}$ as follows.

$$R(f) = \frac{\sum_{u \to v} |f(u) - f(v)|^2 \phi(u) P(u, v)}{\sum_v |f(v)|^2 \phi(v)}$$

Chung [7] proved the connection between the Rayleigh quotient and the Laplacian matrix of a random walk.

Lemma 3

$$R(f) = 2\frac{\langle g\mathcal{L}, g\rangle}{\|g\|_2^2}$$

where $g = f\Phi^{\frac{1}{2}}$ and $\langle \cdot, \cdot \rangle$ means the inner product of two vectors.

On top of this lemma we can further infer the relation between the Rayleigh quotient and the Laplacian eigenvalues. Suppose the Laplacian matrix $\mathcal{L}$ has eigenvalues $0 = \lambda_0 \leq \lambda_1 \leq \ldots \leq \lambda_{n-1}$.

Lemma 4. *For all $1 \leq i \leq n-1$, let vector η_i be the unit eigenvector of λ_i and vector $f_i = \eta_i \Phi^{-\frac{1}{2}}$. Then $\lambda_i = \frac{1}{2}R(f_i)$ and f_i satisfies $\langle f_i, \phi \rangle = 0$.*

Proof. By Lemma 3 we know $\frac{1}{2}R(f) = \frac{\langle g\mathcal{L}, g\rangle}{\|g\|^2}$. From the symmetry of Laplacian matrix $\mathcal{L}$, there exists a set of eigenvectors of $\mathcal{L}$ that forms an orthogonal basis. We denote this set of eigenvectors by $\eta_0, \eta_1, \ldots, \eta_{n-1}$ where η_i is the corresponding eigenvector of λ_i. Notice that for all $0 \leq i \leq n-1$ we have

$$\frac{1}{2}R(\eta_i\Phi^{-\frac{1}{2}}) = \frac{\langle \eta_i\mathcal{L}, \eta_i\rangle}{\|\eta_i\|_2^2} = \frac{\lambda_i\|\eta_i\|_2^2}{\|\eta_i\|_2^2} = \lambda_i$$

We let $f_i = \eta_i\Phi^{-\frac{1}{2}}$. According to the definition of $R(f)$, we have $R(f) \geq 0$. We know $\lambda_0 = R(f_0) = 0$. Thus f_0 is the all-one vector and $\eta_0 = \phi^{\frac{1}{2}}$ is the unit eigenvector of eigenvalue 0. For all $1 \leq i \leq n-1$ we have $\langle \eta_i, \eta_0 \rangle = 0$, i.e., $(f_i\Phi^{\frac{1}{2}}) \cdot \phi^{\frac{1}{2}} = \langle f_i, \phi \rangle = 0$. Hence, for all $1 \leq i \leq n-1$, we have $\lambda_i = \frac{1}{2}R(f_i)$ where f_i satisfies $\langle f_i, \phi \rangle = 0$. ∎

From this we can see that the Rayleigh quotient serves as an important tool for bounding the Laplacian eigenvalues. A lower bound on $R(f_1)$ is equivalent to one on λ_1. We present a lower bound of λ_1 in terms of the diameter and the maximum out-degree of the vertices in the graph.

Lemma 5. *For a random walk on a strongly connected graph G, let λ_1 be the smallest nonzero eigenvalue of its Laplacian matrix $\mathcal{L}$. Denote by $Diam$ the diameter of graph G and by s_0 the maximum out-degree of the vertices in the graph. Then*

$$\lambda_1 \geq \frac{1}{2n \cdot Diam \cdot s_0^{1+Diam}}$$

Proof. Denote $u_0 = \arg\max_{x \in V} \phi(x)$ and $v_0 = \arg\min_{x \in V} \phi(x)$. Let ℓ_0 be the distance from u_0 to v_0. As $\phi P^{\ell_0} = \phi$, we have $\phi(v_0) \geq P^{\ell_0}(u_0, v_0)\phi(u_0) \geq s_0^{-\ell_0}\phi(u_0) \geq s_0^{-Diam}\phi(u_0)$. We then have $1 = \sum_{x \in V} \phi(x) \leq n\phi(u_0) \leq ns_0^{Diam}\phi(v_0)$ and $\phi(v_0) \geq n^{-1}s_0^{-Diam}$.

From Lemma 4 we have $\lambda_1 = \frac{1}{2}R(f_1)$ and $\langle f_1, \phi \rangle = 0$. As $\phi(x) > 0$ for any vertex $x \in V$, there must exist some vertex u with $f_1(u) > 0$ and some vertex v whose $f_1(v) < 0$. Let $y = \arg\max_{x \in V} |f_1(x)|$. Then there must exist some vertex z such that $f_1(y)f_1(z) < 0$. Let $\boldsymbol{r} = (y, x_1, x_2 \ldots, x_{\ell-1}, z)$ be the shortest directed path from y to z, which must exist due to the strong connectivity. Then the length of path $\boldsymbol{r}$ is ℓ. Therefore,

$$\lambda_1 = \frac{1}{2}R(f_1) = \frac{1}{2}\frac{\sum_{u \to v} |f_1(u) - f_1(v)|^2 \phi(u) P(u,v)}{\sum_v |f_1(v)|^2 \phi(v)}$$

$$\left(\text{due to } \min_{x \in V} \phi(x) \geq n^{-1} s_0^{-Diam} \text{ and } \min_{(u,v) \in E} P(u,v) \geq \frac{1}{s_0} \right)$$

$$\geq \frac{1}{2ns_0^{1+Diam}} \frac{\sum_{u \to v} |f_1(u) - f_1(v)|^2}{\sum_v |f_1(v)|^2 \phi(v)}$$

$$\geq \frac{1}{2ns_0^{1+Diam}} \frac{\sum_{u \to v \in r} |f_1(u) - f_1(v)|^2}{\sum_v |f_1(v)|^2 \phi(v)}$$

(by letting $x_0 = y$ and $x_\ell = z$)

$$= \frac{1}{2ns_0^{1+Diam}} \frac{\sum_{i=0}^{\ell-1} |f_1(x_i) - f_1(x_{i+1})|^2}{\sum_v |f_1(v)|^2 \phi(v)}$$

$$\geq \frac{1}{2ns_0^{1+Diam}} \frac{\left[\sum_{i=0}^{\ell-1} (f_1(x_i) - f_1(x_{i+1})) \right]^2}{\ell \cdot \sum_v |f_1(v)|^2 \phi(v)}$$

$$= \frac{1}{2ns_0^{1+Diam}} \frac{[f_1(y) - f_1(z)]^2}{\ell \cdot \sum_v |f_1(v)|^2 \phi(v)}$$

(for $f_1(y)f_1(z) < 0$)

$$\geq \frac{1}{2n \cdot Diam \cdot s_0^{1+Diam}} \frac{|f_1(y)|^2}{\sum_v |f_1(v)|^2 \phi(v)}$$

$$\geq \frac{1}{2n \cdot Diam \cdot s_0^{1+Diam}} \frac{|f_1(y)|^2}{|f_1(y)|^2 \sum_v \phi(v)}$$

$$= \frac{1}{2n \cdot Diam \cdot s_0^{1+Diam}}$$

which completes the proof. ∎

As a canonical technique, a lower bound of the smallest nonzero eigenvalue of the Laplacian matrix implies a lower bound of the convergence rate. Chung [7] proved

Theorem 2. *A lazy random walk on a strongly connected graph G has convergence rate of order $2\lambda_1^{-1}(-\log\min_u \phi(u))$. Namely, after at most $t \geq 2\lambda_1^{-1} ((-\log\min_u \phi(u)) + 2k)$ steps, we have $\Delta_{\chi^2}(t) \leq e^{-k}$.*

In the paper Chung used lazy walks to avoid periodicity. If the graph is irreducible and aperiodic, we let $\widehat{P} = \frac{1}{2}(I + P)$ be the transition probability matrix of the lazy random walk and vector $\widehat{\phi}$ be its Perron vector, matrix $\widehat{\varPhi}$ be the diagonal matrix of $\widehat{\phi}$, matrix $\widehat{\mathcal{L}}$ be its Laplacian matrix.

We know ϕ is the solution of $\phi P = \phi$ or equivalently $\phi(I - P) = 0$ and $\sum_i \phi(i) = 1$. Similarly, $\widehat{\phi}$ is the solution of $\widehat{\phi}(I - \widehat{P}) = 0$ and $\sum_i \widehat{\phi}(i) = 1$. Observe that $I - \widehat{P} = I - \frac{1}{2}(I + P) = \frac{1}{2}(I - P)$ and $\widehat{\phi}(I - \widehat{P}) = \frac{1}{2}\widehat{\phi}(I_P) = 0$, which is equivalently $\widehat{\phi}(I - P) = 0$. Thus $\widehat{\phi} = \phi$ and $\widehat{\varPhi} = \varPhi$. Then

$$
\begin{aligned}
\widehat{\mathcal{L}} &= I - \frac{1}{2}\left(\widehat{\varPhi}^{\frac{1}{2}}\widehat{P}\widehat{\varPhi}^{-\frac{1}{2}} + \widehat{\varPhi}^{-\frac{1}{2}}\widehat{P}^*\widehat{\varPhi}^{\frac{1}{2}}\right) \\
&= I - \frac{1}{2}\left(\varPhi^{\frac{1}{2}} \cdot \frac{1}{2}(I + P) \cdot \varPhi^{-\frac{1}{2}} + \varPhi^{-\frac{1}{2}} \cdot \frac{1}{2}(I + P^*) \cdot \varPhi^{\frac{1}{2}}\right) \\
&= I - \frac{1}{2}\left(\frac{1}{2}I + \frac{1}{2}\varPhi^{\frac{1}{2}}P\varPhi^{-\frac{1}{2}} + \frac{1}{2}I + \frac{1}{2}\varPhi^{-\frac{1}{2}}P^*\varPhi^{\frac{1}{2}}\right) \\
&= I - \frac{1}{2}\left(I + \frac{1}{2}\varPhi^{\frac{1}{2}}P\varPhi^{-\frac{1}{2}} + \frac{1}{2}\varPhi^{-\frac{1}{2}}P^*\varPhi^{\frac{1}{2}}\right) \\
&= \frac{1}{2}I - \frac{1}{4}\left(\varPhi^{\frac{1}{2}}P\varPhi^{-\frac{1}{2}} + \varPhi^{-\frac{1}{2}}P^*\varPhi^{\frac{1}{2}}\right) \\
&= \frac{1}{2}\mathcal{L}
\end{aligned}
$$

Let $\widehat{\lambda}_1$ be the smallest positive eigenvalue of $\widehat{\mathcal{L}}$. Then $\lambda_1 = 2\widehat{\lambda}_1$. Therefore, combining this with Lemma 5, we have

Theorem 3. *A random walk on a strongly connected and aperiodic directed graph G has convergence rate of order $2n \cdot Diam \cdot s_0^{1+Diam}(\log(ns_0^{Diam}))$, where $s_0 = \arg\max_{u \in V} d_u$ is the maximum out-degree of a vertex in G. Namely, after at most $t \geq 2n \cdot Diam \cdot s_0^{1+Diam}((\log(ns_0^{Diam}) + 2k))$ steps, we have $\Delta_{\chi^2}(t) \leq e^{-k}$.*

However, the convergence rate is still exponential in s_0 and $Diam$. Fortunately, in our case $s_0 = s$ and Trakhtenbrot and Barzdin [20] proved the diameter of a random DFA is logarithmic.

Theorem 4. *With probability $1 - o(1)$, the diameter of a random automaton graph is $O(\log_s n)$.*

With the logarithmic diameter we complete the poof of Theorem 1. The constant C in Theorem 1 is the constant used in the proof of Theorem 4 by Trakhtenbrot and Barzdin [20]. It depends on s and approaches unity with increasing s.

Notice that the diameter of an automaton graph won't increase after state-merging operations, thus with high probability, a random DFA has at most logarithmic diameter after DFA minimization. It is also easy to see an irreducible DFA still maintains irreducibility after minimization. Besides, Balle [4] proved DFA minimization preserves aperiodicity. Now we also have Corollary 1.

Corollary 1. *With probability $1 - o(1)$, a random walk on a random DFA after minimization has $\Delta_{\chi^2}(t) \leq e^{-k}$ after $t \geq 2C(C + 1)sn^{1+C}(\log n + k) \cdot \log_s n$, where constant $C > 0$ depends on s and approaches unity with increasing s.*

4 Reconstructing a Random DFA

In this section we present a computationally efficient algorithm for recovering random DFAs from uniform input strings in the statistical query model with a theoretical guarantee on the maximum absolute error and supporting experimental results.

4.1 The Learning Algorithm

In our learning model, the given data are string-state pairs (x, q) where x is a string drawn uniformly at random from Σ^t and q is the state of the DFA reached on input x starting from the start state q_0. Here $t = poly(n, s)$ is the length of the example strings. Our goal is to recover the unique irreducible and closed component of the target DFA from the given data in the statistical query model. The primary constraint on our learning model is the need to estimate the distribution of the ending state, while the advantage is that our algorithm reconstructs the underlying graph structure of the automaton. Let quintuple $A = (Q, \varphi, \Sigma, q_0, F)$ be the target DFA we are interested in. We represent the transition function φ as a collection of $n \times n$ binary matrices M_σ indexed by symbols $\sigma \in \Sigma$ as follows. For each pair of states (i, j), the element $M_\sigma(i, j)$ is 1 if $\varphi(i, \sigma) = j$ and 0 otherwise. For a string of m symbols $y = y_1 y_2 \ldots y_m$, define M_y to be the matrix product $M_y = M_{y_1} \cdot M_{y_2} \ldots M_{y_m}$. Then $M_y(i, j)$ is 1 if $\varphi^*(i, y) = j$ and 0 otherwise.

A uniform input string $x \in \Sigma^t$ corresponds to a random walk of length t on the states of the DFA A starting from the start state q_0. By Lemma 1 and 2, we can assume the irreducibility and aperiodicity of the random walk. Due to the uniqueness of the strongly connected component, the walk will finally converge to the stationary distribution ϕ with any start state q_0. For any string $y = y_1 y_2 \ldots y_m$, we define the distribution vector p_y over the state space Q obtained by starting from the stationary distribution ϕ and inputting string y to the automaton. That is, $p_y = \phi M_y$ and $p_\lambda = \phi$. Consequently, each string $y \in \Sigma^*$ and symbol $\sigma \in \Sigma$ contribute a linear equation $p_y M_\sigma = p_{y\sigma}$ where $y\sigma$ is the concatenation of y and σ. Due to Theorem 4, the diameter of a random DFA is $O(\log_s n)$ with high probability. The complete set of $\Theta(\log_s n)$-step walks should have already traversed the whole graph and no new information can be retrieved after $\Theta(\log_s n)$ steps. Hence, we can only consider the equation set $\{p_y M_\sigma = p_{y\sigma} \mid y \in \Sigma^{O(\log_s n)}\}$ for each $\sigma \in \Sigma$. We further observe that the equation system $\{p_y M_\sigma = p_{y\sigma} \mid y \in \Sigma^{\Theta(\log_s n)}\}$ shares the same solution with $\{p_y M_\sigma = p_{y\sigma} \mid y \in \Sigma^{O(\log_s n)}\}$. Let vector z be the i-th column of matrix M_σ, matrix P_A be the $s^{\Theta(\log_s n)} \times n$ coefficient matrix whose rows are $\{p_y \mid y \in \Sigma^{\Theta(\log_s n)}\}$ and vector b be the vector consisting of $\{p_{y\sigma}(i) \mid y \in \Sigma^{\Theta(\log_s n)}\}$. The task reduces to solving the linear equation system $P_A z = b$ for z. Let ϕ_t be the distribution vector over Q after t steps of random walk. As the random walk always starts from the start state q_0, the initial distribution ϕ_0 is a coordinate vector whose entry of q_0 is 1 and the rest are 0, for which

$$2\|\phi_t - \phi\|_{TV} \leq \left(\sum_{v \in V} \frac{(\phi_t(v) - \phi(v))^2}{\phi(v)}\right)^{\frac{1}{2}} \leq \max_{u \in V} \left(\sum_{v \in V} \frac{(P^t(u,v) - \phi(v))^2}{\phi(v)}\right)^{\frac{1}{2}}$$

Theorem 1 claims that a polynomially large $t_0 = 2C(C+1)sn^{1+C}(\log n + \log \frac{2}{\tau}) \cdot \log_s n$ is enough to have the random walk converge to $p_\lambda = \phi$ within any polynomially small χ-square distance $\frac{\tau}{2}$ with high probability where $C > 0$ is the constant in the theorem. Let $t = t_0 + C\log_s n$, which is still polynomially large. We can estimate the stationary distribution for a state i by the fraction of examples (x,q) such that $q = i$. In general, for any string y, we can estimate the value of p_y for a state i as the ratio between the number of pairs (x,q) such that y is a suffix of x and $q = i$ and the number of examples (x,q) where y is a suffix of x.

In the statistical query model we are unable to directly observe the data; instead we are given access to the oracle $STAT$. Define a conditional statistical query $\chi_{y,i}(x,q) = \mathbb{1}\{q = i \mid y \text{ is a suffix of } x\}$ where $\mathbb{1}$ is the boolean indicator function. It's easy to see the legitimacy and feasibility of query $\chi_{y,i}(x,q)$ for any $y \in \Sigma^{\Theta(\log_s n)}$ because: (1) it is a boolean function mapping an example (x,q) to $\{0,1\}$; (2) the proposition $\mathbb{1}\{q = i\}$ can be tested in $O(1)$ time; (3) the condition $\mathbb{1}\{y \text{ is a suffix of } x\}$ can be tested within $\Theta(\log_s n)$ time; (4) the probability of the condition that y is a suffix of x is inverse polynomially large $s^{-|y|} = s^{-\Theta(\log_s n)} = \Theta(n^{-C})$ for some constant $C > 0$.

Let $\tilde{p}_\lambda$ be the distribution vector over the states after t steps and $\tilde{p}_y = \tilde{p}_\lambda M_y$. Also denote by vector $\widehat{p}_y$ the query result returned by oracle $STAT$ where $\widehat{p}_y(i)$ is the estimate $\mathbf{E}\chi_{y,i}$, and by $\widehat{P}_A$ and $\widehat{b}$ the estimates for P_A and b respectively from oracle $STAT$. We infer the solution z by solving the perturbed linear least squares problem: $\min_z \|\widehat{P}_A z - \widehat{b}\|_2$. Let $\widehat{z}$ be the solution we obtain from this perturbed problem. According to the main theorem, the distance $\|p_\lambda - \tilde{p}_\lambda\|_1 = 2\|\phi_t - \phi\|_{TV} \leq \Delta_{\chi^2}(t) \leq \frac{\tau}{2}$. Then for any string y, $\|p_y - \tilde{p}_y\|_\infty = \|(p_\lambda - \tilde{p}_\lambda)M_y\|_\infty \leq \|p_\lambda - \tilde{p}_\lambda\|_1 \leq \frac{\tau}{2}$. If we do the statistical queries with tolerance $\frac{\tau}{2}$, the maximum additive error will be $\|\tilde{p}_y - \widehat{p}_y\|_\infty \leq \frac{\tau}{2}$ for any string y. Thus we have $\|p_y - \widehat{p}_y\|_\infty \leq \tau$. To conclude a theoretical upper bound on the error, we use the following theorem by Björck [5], which was later refined by Higham [10].

Theorem 5. *Let z be the optimal solution of least squares problem $\min_z \|Mz - b\|_2$ and $\widehat{z}$ be the optimal solution of $\min_z \|\widehat{M}z - \widehat{b}\|_2$. If $|M - \widehat{M}| \precsim \omega E$ and $|b - \widehat{b}| \precsim \omega f$ for some element-wise non-negative matrix E and vector f, where $|\cdot|$ refers to element-wise absolute value and $\precsim$ means element-wise $\leq$ comparison, then*

$$\|z - \widehat{z}\|_\infty \leq \omega(\||M^\dagger|(E|z| + f)\|_\infty + \||(M^\top M)^{-1}|E^\top|Mz - b|\|_\infty) + O(\omega^2)$$

when M has full column rank, or

$$\|z - \widehat{z}\|_\infty \leq \omega(\||\widehat{M}^\dagger|(E|\widehat{z}| + f)\|_\infty + \||(\widehat{M}^\top \widehat{M})^{-1}|E^\top|\widehat{M}\widehat{z} - \widehat{b}|\|_\infty) + O(\omega^2)$$

when $\widehat{M}$ has full column rank, where $M^\dagger$ is the MoorePenrose pseudoinverse of matrix M.

Applying Theorem 5 to our case gives an upper bound on the maximum absolute error.

Corollary 2. *If P_A has full rank with high probability,*

$$\|z - \hat{z}\|_\infty \leq \frac{(1+\varepsilon)\log ns}{\log\log ns}\||P_A^\dagger\||_\infty \tau + O(\tau^2)$$

with probability $1 - o(1)$ for any constant $\varepsilon > 0$.

Proof. First in our case the offset $|P_A z - b| = 0$ and $\omega = \tau$. Matrix E is the all-one matrix and vector f is the all-one vector. As a consequence, $\|f\|_\infty = 1$ and $\||E|z|\|_\infty = \|z\|_1$. Now it remains to prove with high probability $\|z\|_1 \leq \frac{(1+\varepsilon)\log ns}{\log\log ns}$ for all columns in all $M_\sigma, \sigma \in \Sigma$.

Let θ be the largest 1-norm of the columns in M_σ. According to the properties of a random DFA, the probability of $\theta > n$ is 0 and $\Pr[\theta = n] \leq n \cdot n^{-n}$ is exponentially small. For any $k < n$,

$$\Pr[\theta \geq k] \leq n \cdot \Pr[\text{a particular column has 1-norm at least } k]$$

$$\leq n \cdot \binom{n}{k}\left(\frac{1}{n}\right)^k$$

$$\leq \frac{\sqrt{2\pi n}\left(\frac{n}{e}\right)^n e^{\frac{1}{12n}}}{\sqrt{2\pi k}\left(\frac{k}{e}\right)^k e^{\frac{1}{12k+1}} \cdot \sqrt{2\pi(n-k)}\left(\frac{n-k}{e}\right)^{n-k} e^{\frac{1}{12(n-k)+1}}} \cdot n\left(\frac{1}{n}\right)^k$$

$$\leq \sqrt{\frac{n^3 s^2}{2\pi k(n-k)s^2}} \cdot \frac{e^{\frac{1}{12n}}(n)^n}{(nk)^k(n-k)^{n-k}}$$

$$\leq \frac{1}{s} \cdot e^{\log ns + n\log n - k\log k - (n-k)\log(n-k) - k\log n + \frac{1}{12n}}$$

We only need to choose a k such that the exponent goes to $-\infty$, which is equal to

$$\log ns + k\left(1 - \frac{n}{k}\right)\log\left(1 - \frac{k}{n}\right) - k\log k + \frac{1}{12n}$$

If $k \geq n$ then $\Pr[\theta \geq k]$ is exponentially small as discussed above. Otherwise we have $\left(1 - \frac{n}{k}\right)\log\left(1 - \frac{k}{n}\right) \leq 1$ in our case. Also notice that $\frac{1}{12n} \leq 1$. Let $k = \frac{(1+\varepsilon)\log ns}{\log\log ns}$. The expression is upper bounded by

$$\log ns + \frac{(1+\varepsilon)\log ns}{\log\log ns} - \frac{(1+\varepsilon)\log ns}{\log\log ns}\log\frac{(1+\varepsilon)\log ns}{\log\log ns} + 1$$

$$= \log ns + \frac{(1+\varepsilon)\log ns}{\log\log ns}(1 - \log(1+\varepsilon) - \log\log ns + \log\log\log ns) + 1$$

$$= -\varepsilon \log ns + \left(\frac{1 - \log(1+\varepsilon)}{\log\log ns} + \frac{\log\log\log ns}{\log\log ns}\right)(1+\varepsilon)\log ns + 1$$

With respect to n and s, the expression goes to $-\infty$. There are in total s matrices $\{M_\sigma \mid \sigma \in \Sigma\}$. Using a union bound we have $\|z\|_1 \leq \frac{(1+\varepsilon)\log ns}{\log\log ns}$ for all columns in all M_σ with probability $1 - o(1)$, and plugging this upper bound into the conclusion of Theorem 5 completes the proof. ∎

This further implies that if we set the tolerance $\tau = \frac{\log \log ns}{3 |||P_A^\dagger|||_\infty \log ns}$, the solution error $\|z - \widehat{z}\|_\infty < \frac{1}{2}$ with high probability. Based on the prior knowledge we have on z, we could refine $\widehat{z}$ by rounding up $\widehat{z}$ to a binary vector $\tilde{z}$, i.e., for each $1 \le i \le n$, $\tilde{z}(i) = 1$ if $\widehat{z}(i) > \frac{1}{2}$ and 0 otherwise, whereby we will have $\tilde{z}(q) = z(q)$ for any state q in the strongly connected component. A toy example is provided in the appendices to demonstrate how the algorithm works.

Our algorithm only recovers the strongly connected component $\tilde{A}$ of a random DFA A because it relies on the convergence of the random walk and any state $q \notin \tilde{A}$ will have zero probability after the convergence. We have no information for reconstructing the disconnected part. In the positive direction, due to Lemma 1, with high probability we are able to recover at least 79.68% of the DFA for any $s \ge 2$ and at least 99.9% of the whole automaton if $s > 6$. Because $\tilde{A}$ is unique and closed, it is also a well defined DFA. In Section 3 we have proved $\min_{q \in Q} \{p_\lambda(q) \mid p_\lambda(q) > 0\} \ge n^{-1} s^{-Diam} = n^{-C}$ for some constant $C > 0$ with high probability. This means we have a polynomially large gap so that we are able to distinguish the recurrent states from the transient ones by making a query to estimate $\tilde{p}_\lambda(q)$ for each state $q \in Q$. In our result $|||P_A^\dagger|||_\infty$ is regarded as a parameter. It might be possible to improve the result by polynomially bounding $|||P_A^\dagger|||_\infty$ with other given parameters n and s using random matrix theory techniques. The full-rank assumption is reasonable because a random matrix is usually well conditioned and full-rank. From the empirical results in Section 4.2, the coefficient matrix P_A is almost surely full-rank and $|||P_A^\dagger|||_\infty$ is conjecturally $\le ns \log s$. Furthermore, according to Corollary 1, our algorithm is also applicable to learning a random DFA after minimization.

4.2   Experiments and Empirical Results

In this section we present a series of experimental results to study the empirical performance of the learning algorithm, which was run in MATLAB on a workstation built with Intel i5-2500 3.30GHz CPU and 8GB memory. To be more robust against fluctuation from randomness, each test was run for 20 times and the medians were taken. The automata are generated uniformly at random as defined and the algorithm solves the equation system $\{p_y M_\sigma = p_{y\sigma} \mid y \in \Sigma^{\le \lceil \log_s n \rceil}\}$ using the built-in linear least squares function in MATLAB. We simulate the statistical query oracle with uniform additive noise. The experiments start with an empirical estimate for the norm $|||P_A^\dagger|||_\infty$. We first vary the automaton size n from 32 to 4300 with fixed alphabet size $s = 2$. Figure 3 (in the appendices) shows the curve of $|||P_A^\dagger|||_\infty$ versus n with fixed s. Notice that the threshold phenomenon in the plot comes from the ceiling operation in the algorithm configuration. When n is much smaller than the threshold $s^{\lceil \log_s n \rceil}$, the system is overdetermined with many extra equations. Thus it is robust to perturbation and well-conditioned. When n grows up and approaches the threshold $s^{\lceil \log_s n \rceil}$, the system has fewer extra equa-

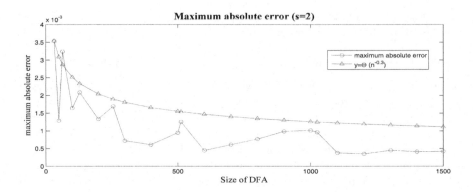

Fig. 1. Maximum absolute error versus n with fixed $s = 2$

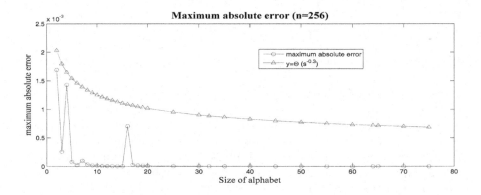

Fig. 2. Maximum absolute error versus s with fixed $n = 256$

tions and becomes relatively more sensitive to perturbations, for which the condition number increases until the automaton size reaches $n = s^i$ of the next integer i. One can avoid this threshold phenomenon by making the size of the equation system grow smoothly as n increases. We then fix n to be 256 and vary s from 2 to 75, as shown in Figure 4 (in the appendices). Similarly there is the threshold phenomenon resulting from the ceiling strategy. All peaks where $n = s^i$ are included and plotted. Meanwhile the rank of P_A is measured to support the full-rank assumption. Matrix P_A is almost surely full-rank for large n or s and both figures suggest an upper bound $ns \log s$ for $\||P_A^\dagger|\|_\infty$. We set the query tolerance τ as $\frac{\log \log ns}{ns \log ns \log_2 s}$ in the algorithm and measure the maximum absolute error $\|z - \widehat{z}\|_\infty$ at each run. Figures 1 and 2 demonstrate the experimental results. Along with the error curve in each figure a function is plotted to approximate the asymptotic order of the decline rate of the error. An empirical error bound is $O(n^{-0.3})$ with fixed s and $O(s^{-0.3})$ with fixed n.

5   Discussion

In this paper we prove fast convergence of random walks on a random DFA and apply this theoretical result to learning a random DFA in the statistical query model. One potential future work is to validate the full-rank assumption or to polynomially bound $\||P_A^\dagger\||_\infty$ using the power of random matrix theory. Note that $\||P_A^\dagger\||_\infty$ reflects the asymmetry of the automaton graph. The class of permutation automata [19] is one example that has symmetric graph structure and degenerate P_A. Another technical question on the fast convergence result is whether it can be generalized to weighted random walks on random DFAs. An immediate benefit from this generalization is the release from the requirement of uniform input strings in the DFA learning algorithm. However, we conjecture such generalization requires a polynomial lower bound on the edge weights in the graph, to avoid exponentially small nonzero elements in the walk matrix P. A further generalization is applying this algorithm to learning random probabilistic finite automata. In this case we will have a similar linear equation system, but the solution vector z can be continuous, not necessarily being a binary vector.

Acknowledgments. We thank Borja Balle Pigem for helpful discussions and the anonymous reviewers of *ALT* 2015 for their valuable comments.

Appendix: A Toy Example

Suppose we consider the alphabet $\{0,1\}$ and a 3-state DFA with the following transition matrices.

$$M_0 = \begin{pmatrix} 0 & 1 & 0 \\ 1 & 0 & 0 \\ 1 & 0 & 0 \end{pmatrix} \text{ and } M_1 = \begin{pmatrix} 0 & 1 & 0 \\ 0 & 0 & 1 \\ 0 & 1 & 0 \end{pmatrix}$$

For this automaton, the stationary distribution p_λ is $(1/3, 4/9, 2/9)$. Since $\lceil \log_s n \rceil = \lceil \log_2 3 \rceil = 2$, the algorithm recovers the first column of matrix M_0, denoted by $z = (M_0(1,1), M_0(2,1), M_0(3,1))^\top$, by solving the overdetermined equation system

$$\begin{cases} p_{00} \cdot z = p_{000}(1) \\ p_{01} \cdot z = p_{010}(1) \\ p_{10} \cdot z = p_{100}(1) \\ p_{11} \cdot z = p_{110}(1) \end{cases} \text{, i.e., } \begin{cases} \frac{1}{3}M_0(1,1) + \frac{2}{3}M_0(2,1) + 0M_0(3,1) = \frac{2}{3} \\ 0M_0(1,1) + \frac{2}{3}M_0(2,1) + \frac{1}{3}M_0(3,1) = 1 \\ 1M_0(1,1) + 0M_0(2,1) + 0M_0(3,1) = 0 \\ 0M_0(1,1) + \frac{4}{9}M_0(2,1) + \frac{5}{9}M_0(3,1) = 1 \end{cases}$$

Similarly the algorithm recovers all columns in M_0 and M_1 and reconstructs the target automaton. Note that in the statistical query model the above equation system is perturbed but we showed the algorithm is robust to statistical query noise.

Appendix: Estimate of $|||P_A^\dagger|||_\infty$

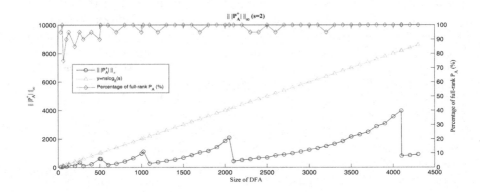

Fig. 3. $|||P_A^\dagger|||_\infty$ versus n with fixed $s = 2$

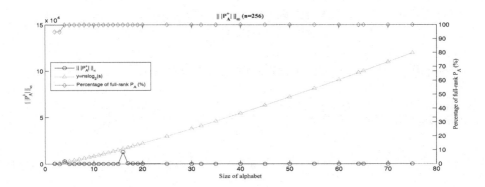

Fig. 4. $|||P_A^\dagger|||_\infty$ versus s with fixed $n = 256$

References

1. Angluin, D.: Learning regular sets from queries and counterexamples. Inf. Comput. **75**(2), 87–106 (1987)
2. Angluin, D., Aspnes, J., Eisenstat, S., Kontorovich, A.: On the learnability of shuffle ideals. Journal of Machine Learning Research **14**, 1513–1531 (2013)
3. Angluin, D., Eisenstat, D., Kontorovich, L.A., Reyzin, L.: Lower bounds on learning random structures with statistical queries. In: ALT (2010)
4. Balle, B.: Ergodicity of random walks on random DFA. CoRR, abs/1311.6830 (2013)
5. Björck, A.: Component-wise perturbation analysis and error bounds for linear least squares solutions. BIT Numerical Mathematics **31**(2), 237–244 (1991)

6. Chen, D.: Learning shuffle ideals under restricted distributions. In: NIPS (2014)
7. Chung, F.: Laplacians and the Cheeger inequality for directed graphs. Annals of Combinatorics **9**, 1–19 (2005)
8. Freund, Y., Kearns, M., Ron, D., Rubinfeld, R., Schapire, R.E., Sellie, L.: Efficient learning of typical finite automata from random walks. In: STOC (1993)
9. Grusho, A.A.: Limit distributions of certain characteristics of random automaton graphs. Mathematical notes of the Academy of Sciences of the USSR (1973)
10. Higham, N.J.: A survey of componentwise perturbation theory in numerical linear algebra. In: Proceedings of symposia in applied mathematics (1994)
11. Jackson, J.C., Lee, H.K., Servedio, R.A., Wan, A.: Learning random monotone DNF. In: Goel, A., Jansen, K., Rolim, J.D.P., Rubinfeld, R. (eds.) APPROX and RANDOM 2008. LNCS, vol. 5171, pp. 483–497. Springer, Heidelberg (2008)
12. Jackson, J.C., Servedio, R.A.: Learning random log-depth decision trees under uniform distribution. SIAM Journal on Computing **34**(5), 1107–1128 (2005)
13. Kearns, M.: Efficient noise-tolerant learning from statistical queries. J. ACM **45**(6), 983–1006 (1998)
14. Kearns, M., Valiant, L.: Cryptographic limitations on learning boolean formulae and finite automata. J. ACM **41**(1), 67–95 (1994)
15. Pitt, L., Warmuth, M.K.: The minimum consistent DFA problem cannot be approximated within any polynomial. J. ACM **40**(1), 95–142 (1993)
16. Ruiz, J., Garcia, P.: Learning k-piecewise testable languages from positive data. In: Grammatical Interference Learning Syntax from Sentences (1996)
17. Sellie, L.: Learning random monotone DNF under the uniform distribution. In: COLT, pp. 181–192 (2008)
18. Sellie, L.: Exact learning of random DNF over the uniform distribution. In: STOC, pp. 45–54. ACM (2009)
19. Thierrin, G.: Permutation automata. Theory of Computing Systems (1968)
20. Trakhtenbrot, B.A., Barzdin, I.M.: Finite automata; behavior and synthesis. Fundamental Studies in Computer Science **1** (1973)
21. Valiant, L.G.: A theory of the learnable. Commun. ACM (November 1984)

Hierarchical Design of Fast Minimum Disagreement Algorithms

Malte Darnstädt, Christoph Ries$^{(\boxtimes)}$, and Hans Ulrich Simon

Fakultät für Mathematik, Ruhr-Universität Bochum, 44780 Bochum, Germany
{malte.darnstaedt,christoph.ries,hans.simon}@rub.de

Abstract. We compose a toolbox for the design of Minimum Disagreement algorithms. This box contains general procedures which transform (without much loss of efficiency) algorithms that are successful for some d-dimensional (geometric) concept class $\mathcal{C}$ into algorithms which are successful for a $(d+1)$-dimensional extension of $\mathcal{C}$. An iterative application of these transformations has the potential of starting with a base algorithm for a trivial problem and ending up at a smart algorithm for a non-trivial problem. In order to make this working, it is essential that the algorithms are not proper, i.e., they return a hypothesis that is not necessarily a member of $\mathcal{C}$. However, the "price" for using a super-class $\mathcal{H}$ of $\mathcal{C}$ is so low that the resulting time bound for achieving accuracy ε in the model of agnostic learning is significantly smaller than the time bounds achieved by the up to date best (proper) algorithms.

We evaluate the transformation technique for $d = 2$ on both artificial and real-life data sets and demonstrate that it provides a fast algorithm, which can successfully solve practical problems on large data sets.

1 Introduction

In this paper, we are concerned with the Minimum Disagreement problem (sometimes also called Maximum Weight problem) associated with a family $\mathcal{C}$ of sets over some domain $\mathcal{X}$: given a sequence $S = [(x_1, w_1), \ldots, (x_n, w_n)] \in (\mathcal{X} \times \mathbb{R})^n$ of points in $\mathcal{X}$ along with their weights, find a set $C \in \mathcal{C}$ whose total weight $W_S(C) := \sum_{i:x_i \in C} w_i$ is as large as possible. Note that $W_S(C)$ is maximized iff

$$E_S(C) := \sum_{i:w_i>0,x_i \notin C} w_i \;-\; \sum_{i:w_i<0,x_i \in C} w_i$$

is minimized. In learning theory, $E_S(C)$ is called the *empirical error of C on S*, and this term plays a central role in statistical learning theory, especially in the model of agnostic learning [10].

Although the Minimum Disagreement problem is intractable for a wide variety of classes [10, 12], it has been noticed by several researchers in an early stage of learning theory already that relatively simple and low-dimensional classification rules (e.g. axis-parallel rectangles [16–18], unions of intervals [9], or 2-level decision trees [1]) can be quite successful on benchmark data provided that these

© Springer International Publishing Switzerland 2015
K. Chaudhuri et al. (Eds.): ALT 2015, LNAI 9355, pp. 134–148, 2015.
DOI: 10.1007/978-3-319-24486-0_9

rules are given in terms of the (few) most relevant attributes. For this reason a couple of algorithms have been developed which solve the Minimum Disagreement problem w.r.t. some simple classes and run in polynomial time [1,2,5,11].

It seems that efficient algorithms for the Minimum Disagreement problem have been found in the past mainly for geometric classes of a relatively low dimension d. The run-time of these algorithms usually exhibits an exponential dependence on d. Moreover, improving on the currently best time bounds does not appear to be an easy job. For instance, the algorithm from [11] solves the Minimum Disagreement problem for axis-parallel rectangles in time[1] $O(n^2 \log(n))$. It was not until recently [2] that a faster algorithm has been found (time $O(n^2)$ in case of axis-parallel rectangles or, more generally, time $O(n^d)$ in case of d-dimensional axis-parallel hyper-rectangles). Thus, one may easily get the impression that the early attempts of designing efficient Minimum Disagreement algorithms got stuck, and even modest improvements on the existing time bounds are not easy to obtain.

One means of escape from the marshy grounds of intractability is opened up by the usage of convex surrogate loss functions at the place of the discrete loss function underlying the Minimum Disagreement problem. This option is taken, for instance, by the Support Vector Machine [13,14]. In this paper, we investigate another relaxation of the original problem: instead of searching for a set $C \in \mathcal{C}$ with the smallest possible value of $E_S(C)$, we bring suitably chosen classes $\mathcal{H}$ into play and search for a set $H \in \mathcal{H}$ such that $E_S(H) \leq \min_{C \in \mathcal{C}} E_S(C)$. While this approach is well known in the context of Boolean classes [12] and standard in the theory of agnostic learning [10], it is apparently not exploited to full extent in the context of geometric classes. Here is a short summary of our approach:

– We make use of the clever data structures that have been invented in the past in order to solve the Minimum Disagreement problem for low-dimensional geometric classes. We observe that these data structures naturally lead to the concept of "flexible" algorithms. Here, "flexibility" means that the underlying data structure can easily be updated in reaction to a modified weight parameter.
– We show that a flexible algorithm which solves the Minimum Disagreement problem for two d-dimensional classes, say $\mathcal{C}$ and $\mathcal{H}$, can be transformed (without much loss of efficiency) into a new flexible algorithm which solves the Minimum Disagreement problem for two (more expressive) $(d + 1)$-dimensional classes. An iterative application of these transformations has the potential of starting with a base algorithm for a trivial problem and ending up at a smart algorithm for a non-trivial problem.
– By a suitable choice of the class $\mathcal{H}$, we obtain algorithms which achieve an accuracy of ε in the model of agnostic learning considerably faster than the best currently known algorithms do. For instance, we obtain a (non-proper) algorithm that agnostically learns axis-parallel rectangles in time $\tilde{O}(1/\varepsilon^2)$ while the learning procedure based on the up to date fastest proper algorithm

---

[1] The machine model used throughout the paper is a random-access machine with unit costs (even on real arithmetic).

from [2] needs time $\tilde{O}(1/\varepsilon^4)$. In this paper, $\tilde{O}$ is defined as Landau's O but additionally hides factors logarithmic in its argument and the dependency on confidence parameter δ.

It should be mentioned that fragments of our approach heavily builds on existing work [5,11]; in particular, the employed data structures are a variant of Segment Trees[2] [3]. But it seems to be the combination of three factors—data structures that provide flexibility, iteratively applicable transformations, clever choice of the class $\mathcal{H}$—which generates a surprising amount of additional horse power.

2   Definitions, Notations and Facts

Let $\mathcal{X}$ be a set. In the parlance of learning theory, any subset of $\mathcal{X}$ is called a *concept* over the domain $\mathcal{X}$ or, alternatively, a *hypothesis* over $\mathcal{X}$. A family of concepts (resp. hypotheses) over $\mathcal{X}$ is called a *concept class* (resp. *hypothesis class*) over $\mathcal{X}$. A sequence of the form $S = [(x_1, w_1), \ldots, (x_n, w_n)] \in (\mathcal{X} \times \mathbb{R})^n$ is called a *weighted sample* over $\mathcal{X}$. We will assume throughout the paper that the domain $\mathcal{X}$ is equipped with a linear ordering and that S is ordered so that $x_1 \leq \ldots \leq x_n$.

Intuitively, a concept C "performs well" on a weighted sample S if it includes the points of S with a positive weight and excludes the points in S with a negative weight. The empirical error of C on S, denoted $E_S(C)$ and already defined in Section 1, measures to which extent the concept C does not perform well.

Let $\mathcal{C}$ and $\mathcal{H}$ be two classes over the same domain $\mathcal{X}$. The *Minimum Disagreement problem* for $\mathcal{C}$ and $\mathcal{H}$ is denoted by $\mathrm{MinDis}(\mathcal{C}, \mathcal{H})$ in the sequel. Recall from Section 1 that it is the following problem: given a sorted weighted sample S, find a hypothesis $H \in \mathcal{H}$ such that H does not perform worse on S than the best concept in $\mathcal{C}$ does, i.e., $E_S(H) \leq \min_{C \in \mathcal{C}} E_S(C)$.

Let $\mathcal{P}(k)$ denote the family of all ordered partitions of the reals into k non-empty intervals, i.e., $\mathcal{P}(k)$ consists of all k-tuples $(I_1, \ldots, I_k)$ such that $I_1, \ldots, I_k \subseteq \mathbb{R}$ are pairwise disjoint non-empty intervals whose union equals $\mathbb{R}$, and the right endpoint of I_j coincides with the left endpoint of I_{j+1} for $j = 1, \ldots, k-1$. For instance $((-\infty, 0), [0, 10), [10, \infty))$ is a member of the family $\mathcal{P}(3)$.

Analogously, let $\mathcal{P}'(k)$ denote the family of all ordered partitions of some bounded non-empty interval of the reals into k consecutive non-empty subintervals. For instance, $([-10, 0), [0, 10), [10, 20))$ is a member of the family $\mathcal{P}'(3)$.

A sub-interval $[c, d]$ of $[a, b]$ is said to be *left-aligned* (resp. *right-aligned*) in $[a, b]$ if $c = a$ (resp. $d = b$). It is called a *proper* sub-interval of $[a, b]$ if it does not coincide with $[a, b]$. If $[c, d] \subseteq (a, b)$ it is said to be *located in the interior*

---

[2] A Segment Tree is a binary tree storing a set of intervals with endpoints from a finite set of (sorted) real valued points. Each leaf of the tree corresponds to an elementary interval (either a point itself or an open interval between two points) and each internal node corresponds to the union of the intervals given by the children. Any interval over the points can easily be represented by an antichain of vertices.

of $[a, b]$. Clearly, a proper sub-interval of $[a, b]$ is either left-aligned, right-aligned, or located in the interior of $[a, b]$. In the first (resp. second or third) case, we say that it is *of type "L"* (resp. *of type "R"* or *of type "I"*). For $a, b \in \mathbb{R}$, we define $[a : b] := [a, b] \cap \mathbb{Z}$.

Let B be a complete binary tree with root r_B and with n leaves that are numbered $1, \ldots, n$ from left to right. For a node $u \in B$, let $B(u)$ be the sub-tree of B rooted at u, and let $l(u)$ (resp. $r(u)$) be the smallest (resp. largest) number of a leaf in $B(u)$. Then $[l(u) : r(u)]$ is called the *interval represented by u*. Every maximal antichain V of nodes in B represents a partition of $\{1, \ldots, n\}$ in the obvious manner. For instance $V = \{r_B\}$ represents the trivial partition with the single equivalence class $\{1, \ldots, n\}$. The set of leaves in B represents the partition of $\{1, \ldots, n\}$ into n singletons $\{1\}, \ldots, \{n\}$. The other maximal antichains induce partitions which are in between these two extremes. The following result is not hard to show:

Lemma 1. *1. For all $1 \le a \le b \le n$, there exists an antichain V of size at most $2\lfloor \log n \rfloor$ in B such that $[a : b] = \cup_{v \in V}[l(v) : r(v)]$. Moreover, given B and a, b, the smallest antichain with this property can be found in time $O(\log(n))$.*

2. Let $k \ge 2$, $\ell_2(n) = \lceil \log n \rceil + 1$ and $\ell_k(n) = (k - 1)\log(n)$ for $k \ge 3$. Then, for every partition $(I_1, \ldots, I_k) \in \mathcal{P}(k)$, there exists a maximal antichain V of size at most $\ell_k(n)$ in B such that the partition represented by V is a refinement of the partition induced by $(I_1, \ldots, I_k)$ on $\{1, \ldots, n\}$.

3 From Simple to More Complex Concept Classes

With each concept class $\mathcal{C}$ over domain $\mathcal{X}$ and with each $k \ge 1$, we associate the following concept classes over $\mathbb{R} \times \mathcal{X}$:

$$\mathcal{C}[k] = \left\{ \bigcup_{j=1}^{k'} (I_j \times C_j) : 0 \le k' \le k \wedge (I_1, \ldots, I_{k'}) \in \mathcal{P}(k') \wedge C_1, \ldots, C_{k'} \in \mathcal{C} \right\}$$

Analogously, let $\mathcal{C}'[k]$ be defined as $\mathcal{C}[k]$ with $\mathcal{P}$ replaced by $\mathcal{P}'$. Note that the empty set is a member of $\mathcal{C}[k]$ and $\mathcal{C}'[k]$.

In the sequel, $\mathcal{I}$ denotes the class of bounded intervals over the domain $\mathbb{R}$, $\mathcal{R}$ denotes the class of bounded axis-parallel rectangles over the domain $\mathbb{R}^2$, $\mathcal{I}_k$ denotes the class of unions of at most k bounded intervals, and $\mathcal{R}_k$ denotes the class of unions of at most k bounded axis-parallel rectangles.

Example 1. Let $\mathcal{X} = \{x\}$ and $\mathcal{C}_1 = \{\mathcal{X}\}$ and $\mathcal{C}_2 = \{\emptyset, \mathcal{X}\}$. We identify the domain $\mathbb{R} \times \{x\}$ with $\mathbb{R}$ in the obvious manner. Then, for each $k \ge 1$, $\mathcal{C}_1'[k]$ coincides with $\mathcal{I}$ and $\mathcal{C}_2'[2k - 1]$ coincides with $\mathcal{I}_k$. Moreover, $\mathcal{I}_k$ is a subclass of $\mathcal{C}_2[2k + 1]$.

Example 2. Obviously, $\mathcal{I}'[1] = \mathcal{R}$. The class $\mathcal{I}'[k]$ with $k \ge 2$ contains horizontally connected sequences of at most k bounded axis-parallel rectangles, i.e., it contains concepts of the form $\cup_{l=1}^{k'}(I_l \times J_l)$ with $k' \le k$, $(I_1, \ldots, I_{k'}) \in \mathcal{P}'(k')$ and $J_1, \ldots, J_{k'} \in \mathcal{I}$.

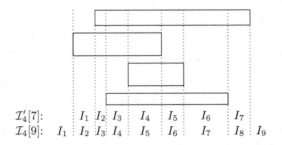

$$\mathcal{I}'_4[7]: \quad I_1 \; I_2 \; I_3 \quad I_4 \quad I_5 \quad I_6 \quad I_7$$
$$\mathcal{I}_4[9]: \quad I_1 \; I_2 \; I_3 \; I_4 \quad I_5 \quad I_6 \quad I_7 \quad I_8 \; I_9$$

Fig. 1. An example showing that a union of 4 rectangles can be viewed as a concept from $\mathcal{I}'_4[7]$ or as a concept from $\mathcal{I}_4[9]$, respectively.

Example 3. Obviously, $\mathcal{I}'_s[k]$ is the class over $\mathbb{R}^2$ whose concepts are of the form $\cup_{l=1}^{k'}(I_l \times U_l)$ with $k' \le k$, $(I_1, \ldots, I_{k'}) \in \mathcal{P}'(k')$ and $U_1, \ldots, U_{k'} \in \mathcal{I}_s$. It is easy to see that $\mathcal{R}_k$ is a subclass of $\mathcal{I}'_k[2k-1]$ and $\mathcal{I}_k[2k+1]$, respectively. See Fig. 1 for an illustration.

The *VC-dimension* of class $\mathcal{H}$ over domain $\mathcal{X}$, denoted VCdim($\mathcal{H}$), is defined as the cardinality of the largest subset $M \subseteq \mathcal{X}$ such that every subset of M can be written in the form $M \cap H$ for some $H \in \mathcal{H}$. If there is no bound on the size of such sets M, then VCdim($\mathcal{H}$) $= \infty$. Let $d =$ VCdim($\mathcal{H}$). It is well known that the number of random examples, required for achieving ε-accuracy in the model of agnostic learning, is of order $\tilde{O}(d/\varepsilon^2)$ [15], and a suitable hypothesis is found by running a Minimum Disagreement algorithm A with hypothesis class $\mathcal{H}$ on a random sample of this size. Thus, a time bound $T(n)$ for A reads as $\tilde{O}(T(d/\varepsilon^2))$ when written in terms of d and ε. We now analyze how much we have to pay in terms of the VC-dimension for moving to more complex concept classes.

Theorem 1. *Let $\mathcal{C}$ be a concept class of VC-dimension d over domain $\mathcal{X}$ such that $\emptyset \in \mathcal{C}$. Then, for all $k \ge 1$, we have that:*

- *$VCdim(\mathcal{C}[k]) \le VCdim(\mathcal{C}'[k]) \le VCdim(\mathcal{C}[k]) + 2$*
- *$dk \le VCdim(\mathcal{C}[k]) \le (d+1)k - 1$*
- *$dk \le VCdim(\mathcal{C}'[k]) \le (d+1)k + 1$*

(All these bounds can be shown to be tight.)

Proof. The inequalities $dk \le$ VCdim($\mathcal{C}[k]$) $\le$ VCdim($\mathcal{C}'[k]$) are rather obvious. Let $S = [(z_1, x_1), \ldots, (z_s, x_s)]$ be a sequence of instances from $\mathbb{R} \times \mathcal{X}$ ordered according to non-decreasing z-coordinates. Suppose that S is shattered by $\mathcal{C}'[k]$. Thus each label combination $(b_1, \ldots, b_s) \in \{0,1\}^s$ can be realized by some concept in $\mathcal{C}'[k]$. Let $\cup_{j=1}^{k'}(I'_j \times C_j) \in \mathcal{C}'[k]$ be a concept realizing the bit pattern $(1, b_2, \ldots, b_{s-1}, 1)$ on S. Then the same bit pattern can be realized by $\cup_{j=1}^{k'}(I_j \times C_j) \in \mathcal{C}[k]$ where I_1 is the interval from $-\infty$ to the right endpoint of I'_1, I_s is the interval from the left endpoint of I'_s to ∞, and $I_j = I'_j$ for $j \notin \{1, s\}$. It follows that VCdim($\mathcal{C}'[k]$) $\le$ VCdim($\mathcal{C}[k]$) $+ 2$. We still have to show that the above sequence S of length s cannot be shattered by $C[k]$ if $s = (d + 1)k$. This can be seen as follows. Split S into k segments of length $d + 1$. For each segment, choose a label combination that cannot be realized by any concept from C. It is then easy to see that the resulting label combination for the full sequence S cannot be realized by any concept from $C[k]$.[3] From this discussion, it follows that $\text{VCdim}(C[k]) \leq (d + 1)k - 1$ and therefore $\text{VCdim}(C'[k]) \leq \text{VCdim}(C[k]) + 2 \leq (d + 1)k + 1$. $\qquad\square$

4 From Trivial to Smart Algorithms

An algorithm that solves $\text{MinDis}(C, C)$ is called a *proper* Minimum Disagreement algorithm for C. An algorithm A that solves $\text{MinDis}(C, \mathcal{H})$ is called a *flexible* Minimum Disagreement algorithm with time bounds T_1, T_2, T_3 if the following holds:

1. Given a sorted weighted sample $S = [(x_1, w_1), \ldots, (x_n, w_n)] \in (\mathcal{X} \times \mathbb{R})^n$, A builds a data structure $\text{DS}(S)$ in time $T_1(n)$.
2. After a modification of one of the weights in S, the data structure $\text{DS}(S)$ can be updated accordingly in time $T_2(n)$.
3. $\text{DS}(S)$ implicitly represents a hypothesis $H(S) \in \mathcal{H}$ which satisfies

$$E_S(H(S)) \leq \min_{C \in C} E_S(C) \ . \tag{1}$$

   Given $\text{DS}(S)$ and $x \in \mathcal{X}$, it can be decided in time $T_3(n)$ whether $x \in \mathcal{H}(S)$.
4. Given $\text{DS}(S)$, the quantity $E_S(H(S))$ can be computed in constant time.

Moreover we say that the data structure DS *can be merged efficiently* if, for every pair S_1, S_2 of samples, the data structure for the composition of S_1 and S_2 can be built in constant time from $\text{DS}(S_1)$ and $\text{DS}(S_2)$.

Here is a trivial example for a proper and flexible Minimum Disagreement algorithm, that we will use as a building block for the design of clever and highly non-trivial algorithms:

Example 4. Let $C_1 = \{\mathcal{X}\}$ for $\mathcal{X} = \{x\}$ be the trivial class that we had considered in Example 1 already. We claim that the (trivial) problem $\text{MinDis}(C_1, C_1)$ can be solved by a flexible algorithm with time bounds $T_1(n) = O(n)$, $T_2(n) = O(1)$ and $T_3(n) = O(1)$:

- For $S = [(x, w_1), \ldots, (x, w_n)]$, set $\text{DS}(S) := W_S^- := \sum_{i:w_i<0} w_i$. Thus, $\text{DS}(S)$ is simply a real number that can be determined in time $O(n)$.
- If a weight w_k is replaced by a new weight w_k', then $\text{DS}(S)$ is updated in constant time by setting $W_S^- := W_S^- + \min\{w_k', 0\} - \min\{w_k, 0\}$.
- $\text{DS}(S)$ represents $H(S) := \{x\}$, the only hypothesis in $\mathcal{H}$. The evaluation problem for $H(S)$ is trivial.

---

[3] The same argument was used in [10] in connection with a class of piecewise defined functions over the real domain.

– Note that $E_S(\{x\}) = |W_S^-|$. Thus, given $DS(S) = W_S^-$, $E_S(H(S))$ is computed in constant time.

If the sample S is the composition of the samples S_1 and S_2, then $W_S^- = W_{S_1}^- + W_{S_2}^-$. Thus, the data structure DS can be merged efficiently.

Let $\mathcal{C}_2$ be the other trivial class that we had considered in Example 1. We briefly note that there is a flexible algorithm for $MinDis(\mathcal{C}_2, \mathcal{C}_2)$ which has the same time bounds as the algorithm for $MinDis(\mathcal{C}_1, \mathcal{C}_1)$.

In the sequel, we assume that $T_i(n) = o(n)$ for $i = 2, 3$ and $T_1(n)$ is of the form $nh(n)$ for some monotonically non-decreasing function $h(n) \geq 1$. From the latter assumption, it follows that

$$\sum_{i=1}^{s} n_i = n \implies \left(\sum_{i=1}^{s} T_1(n_i) \leq \sum_{i=1}^{s} (n_i h(n)) = nh(n) = T_1(n) \right) . \quad (2)$$

Here comes the first main result of this section:

Theorem 2. *A flexible algorithm A solving $MinDis(\mathcal{C}, \mathcal{H})$ with time bounds T_1, T_2, T_3 can be transformed into a flexible algorithm A' that solves $MinDis(\mathcal{C}'[1], \mathcal{H}'[2\lfloor \log n \rfloor])$ with time bounds $T_i'(n) = O(\log(n) T_i(n))$ for $i = 1, 2$ and $T_3'(n) = O(\log(n) + T_3(n))$. Moreover, if the data structure used by A can be merged efficiently, then the first two time bounds for A' are even better, namely $T_1'(n) = O(T_1(n))$ and $T_2'(n) = O(\log(n) + T_2(n))$.*

Proof. We write vectors from $\mathbb{R} \times \mathcal{X}$ in the form $x' = (z, x)$ with $z \in \mathbb{R}$ and $x \in \mathcal{X}$, and we equip $\mathbb{R} \times \mathcal{X}$ with the lexicographic ordering. Let

$$S' = [(x_1', w_1), \dots, (x_n', w_n)] \in (\mathbb{R} \times \mathcal{X} \times \mathbb{R})^n$$

be a lexicographically sorted weighted sample. Let $S = [(x_1, w_1), \dots, (x_n, w_n)] \in (\mathcal{X} \times \mathbb{R})^n$ be the sequence obtained by stripping off the z-coordinates of the items in S'. Note that segments of S with the same z-coordinate are sorted according to the linear ordering of $\mathcal{X}$. Let $z_1' < \dots < z_{n'}'$ with $n' \leq n$ be the sorted sequence of distinct z-coordinates of the items in S'. For each interval $[l : r] \subseteq [1 : n']$, we define $S'[l : r]$ as the coherent sub-sequence of S' consisting of all items in S' whose z-coordinate lies in the interval $[z_l' : z_r']$, i.e.,

$$S'[l : r] = \{(x_k', w_k) : z_l' \leq z_k \leq z_r'\} .$$

Let $S[l : r]$ be the corresponding list with z-coordinates omitted. Let B be a complete binary tree with n' leaves which are numbered $1, \dots, n'$ from left to right. With each node u in B, we associate the following pieces of information:

1. $l(u) \in [1 : n']$ (resp. $r(u) \in [1 : n']$) is the number of the leftmost (resp. rightmost) leaf in the sub-tree of B induced by u.
2. $S(u)$ is defined as the "sorted version" of $S[l(u) : r(u)]$, i.e., it contains the same items as $S[l(u) : r(u)]$ but in $S(u)$ they are ordered according to non-decreasing x-values.

3. $DS(u) = DS(S(u))$, i.e., $DS(u)$ is the data structure returned by the algorithm A on input $S(u)$.

4. $d_0(u) = E_{S(u)}(\emptyset)$. Note that $d_0(u)$ equals the sum of all positive weights that are found in $S(u)$.

5. Let $H(u) = H(S(u))$, i.e., $H(u)$ is the hypothesis which is represented by the data structure $DS(u)$. Let $d_1(u) = E_{S(u)}(H(u))$. We may conclude from (1) that, for all nodes u in B,

$$d_1(u) = E_{S(u)}(H(u)) \leq \min_{C \in \mathcal{C}} E_{S(u)}(C) \ . \tag{3}$$

6. Let J be a sub-interval of $[l(u) : r(u)]$. Let $V[J]$ be the smallest antichain V in B that satisfies $J = \cup_{v \in V}[l(v) : r(v)]$. We say that

$$H_J^u = \bigcup_{v \in V[J]} [z'_{l(v)}, z'_{r(v)}] \times H(v) \tag{4}$$

is the *hypothesis induced by J at node u*. Note that $|V[J]| \leq 2\lfloor \log n \rfloor$ according to Lemma 1 so that $H_J^u \in \mathcal{H}'[2\lfloor \log n \rfloor]$. Given the convention $\min \emptyset = \infty$, we set

$$d_I(u) = \min_{(a,b):l(u)<a\leq b<r(u)} E_{S(u)}(H_{[a:b]}^u) \ ,$$

$$d_L(u) = \min_{b:l(u)\leq b<r(u)} E_{S(u)}(H_{[l(u):b]}^u) \ ,$$

$$d_R(u) = \min_{a:l(u)<a\leq r(u)} E_{S(u)}(H_{[a:r(u)]}^u) \ ,$$

i.e., $d_I(u)$ is the error on $S(u)$ of the best hypothesis among the ones which are induced by some sub-interval of $[l(u) : r(u)]$ of type "I". The analogous remark applies to $d_L(u)$ and $d_R(u)$, respectively. The sub-interval J of $[l(u) : r(u)]$ of type "I" that satisfies $d_I(u) = E_{S(u)}(H_J^u)$ is denoted $J_I(u)$ in what follows. The notations $J_L(u)$ and $J_R(u)$ are understood analogously.

The tree B augmented by $K = (K(u))_{u \in B}$ for

$$K(u) = [\, l(u), r(u), DS(u),$$
$$d_0(u), d_1(u), d_L(u), d_R(u), d_I(u),$$
$$J_L(u), J_R(u), J_I(u)\,]$$

constitutes the data structure $DS(S')$. The remaining part of the proof is sketched only. Leaving out of account the computation of K, B can be built in time $O(n)$. The additional pieces of information, K, can be computed as follows:

1. The quantities $(l(u), r(u), d_0(u))_{u \in B}$ are easy to compute within $O(n)$ steps in a bottom-up fashion. The sorted sequences $(S(u))_{u \in B}$ can be computed bottom-up in time $O(n \log(n))$ (in the same way as it is done by "Mergesort").

2. Making use of (2), it is easy to show that, within $T_1(n)$ steps, we can compute $DS(u)$ for all nodes at a fixed level. Thus, it takes time $O(T_1(n)\log(n))$ to compute $(DS(u))_{u \in B}$. Moreover, if DS can be merged efficiently, then it is easy to see that the sequences $(S(u))_{u \in B}$ are not needed because $(DS(u))_{u \in B}$ can be computed directly in time $O(T_1(n) + n) = O(T_1(n))$.

3. Given $(DS(u))_{u \in B}$, it is easy to compute $(d_1(u))_{u \in B}$ in time $O(n)$.

4. Given $(d_1(u))_{u \in B}$, we can compute the quantities $(d_L(u), d_R(u), d_I(u),$ $J_L(u), J_R(u), J_I(u))_{u \in B}$ in a bottom-up fashion in time $O(n)$. For instance, if u is a node with left child u_0 and right child u_1, then $d_R(u)$ is computed according to $d_R(u) = \min\{d_0(u_0) + d_1(u_1), d_R(u_0) + d_1(u_1), d_0(u_0) + d_R(u_1)\}$. Similar equations can be set up for $d_L(u)$ and $d_I(u)$. Moreover, for each $X \in \{L, R, I\}$, the computation of $J_X(u)$ is just as easy as the computation of $d_X(u)$.

It follows from the previous discussion that $(K(u))_{u \in B}$ can be computed in time $O(\log(n)T_1(n))$. Moreover, if DS can be merged efficiently, then time $O(T_1(n))$ is sufficient.

Suppose that for one item in S', say the item (z_k, x_k, w_k), the weight parameter is modified. Let j be the unique index with $z'_j = z_k$ and let v be the leaf in B numbered j. Since $K(u)$ need not be changed for all nodes in B but only for those which are located on the path P from v to the root r_B of B, it easily follows that $(K(u))_{u \in B}$ can be updated in time $O(\log(n)T_2(n))$, or even in time $O(\log(n) + T_2(n))$ if DS can be merged efficiently.

Let $d_{min} = \min\{d_0(r_B), d_1(r_B), d_L(r_B), d_R(r_B), d_I(r_B)\}$. We now claim that $DS(S')$ represents an easy-to-evaluate hypothesis $H(S') \in \mathcal{H}[2\lfloor\log n\rfloor]$ that satisfies $d_{min} = E_{S'}(H(S'))$. This can be seen as follows. If $d_{min} = d_0(r_B)$, we set $H(S') = \emptyset$. If $d_{min} = d_1(r_B)$, we set $H(S') = H^{r_B}_{[1:n']}$. Finally, if $d_{min} = d_X(r_B)$ for $X \in \{I, L, R\}$, then we set $H(S') = H^{r_B}_{J_X(r_B)}$. The evaluation problem for $H(S') = \emptyset$ is trivial. As for the remaining cases, note first that $(z, x) \in H^{r_B}_{[1:n']}$ iff $z \in [z'_1, z'_{n'}]$ and $x \in H(r_B)$. Suppose now that $d_{min} = d_X(r_B)$. Let $J_X = [a : b]$. If $z \notin [z'_a, z'_b]$, then clearly $(z, x) \notin H^{r_B}_{a,b}$. Otherwise, we use B as a search tree and follow the search path for z until we reach a node v satisfying $d_1(v) = \min\{d_0(v), d_1(v), d_L(v), d_R(v), d_I(v)\}$. This must be a node from the antichain $V[a : b]$. An inspection of (4) shows that $(z, x) \in H^{r_B}_{a,b}$ iff $x \in H(v)$. It follows from this discussion that, in any case, the evaluation problem for $H(S')$ can be solved in time $O(\log(n) + T_3(n))$.

Clearly, d_{min} is retrieved from $DS(S')$ in constant time. Making use of (3), it is not hard to show that $d_{min} \leq \min_{C \in \mathcal{C}'[1]} E_{S'}(C)$, which concludes the proof. $\qquad \square$

Note that the proof of Theorem 2 is completely constructive. The minimum disagreement and learning algorithms given in the following arise from an iterative application of Theorem 2 and the trivial Example 4.

Recall that $\mathcal{I}$ denotes the class of bounded real intervals. As discussed in Example 1, $\mathcal{I} = \mathcal{C}'_1[1] = \mathcal{C}'_1[2\lfloor\log n\rfloor]$. We immediately obtain the following result:

Theorem 3. *The transformation from Theorem 2 applied to the flexible algorithm for* $\mathrm{MinDis}(\mathcal{C}_1, \mathcal{C}_1)$ *from Example 4 yields a flexible algorithm that solves* $\mathrm{MinDis}(\mathcal{I}, \mathcal{I})$ *with time bounds* $T_1(n) = O(n)$ *and* $T_i(n) = O(\log(n))$ *for* $i = 2, 3$.

The flexible algorithm for $\mathrm{MinDis}(\mathcal{I}, \mathcal{I})$ resulting from Theorem 3 basically coincides with the algorithm for $\mathrm{MinDis}(\mathcal{I}, \mathcal{I})$ from [11]. However, since our transformation is general and can be iterated, we can now go one step further and obtain the following result:

Theorem 4. *The problem* $\mathrm{MinDis}(\mathcal{R}, \mathcal{I}'[2\lfloor\log n\rfloor])$ *can be solved by a flexible algorithm with time bounds* $T_1(n) = O(n \log(n))$, *and* $T_2(n) = O(\log^2(n))$ *and* $T_3(n) = O(\log(n))$.

Proof. Recall from Example 2 that $\mathcal{R} = \mathcal{I}'[1]$. The theorem now follows immediately from Theorems 2 and 3.   □

By following the construction in the paragraph before Theorem 1 and applying the bounds on the VC-dimension from Theorem 1 and on the running time T_1 from Theorem 4, one immediately obtains the fast (non-proper) agnostic learner for the class of axis-parallel rectangles promised in the introduction:

Theorem 5. *Our algorithm for the problem* $\mathrm{MinDis}(\mathcal{R}, \mathcal{I}'[2\lfloor\log n\rfloor])$ *agnostically learns concepts from class* $\mathcal{R}$ *with accuracy* ε *in time* $\tilde{O}(1/\varepsilon^2)$ *if a random sample of size* $n = \tilde{O}(1/\varepsilon^2)$ *is provided.*

The proof of the following result (omitted here because of space constraints) bears some similarity to the proof of Theorem 2:

Theorem 6. *Let the function* $\ell_k(n)$ *be defined as in Lemma 1. Suppose that there is a flexible algorithm A for* $\mathrm{MinDis}(\mathcal{C}, \mathcal{H})$ *with time bounds* T_1, T_2, T_3. *Then the problem* $\mathrm{MinDis}(\mathcal{C}[k], \mathcal{H}[\ell_k(n)])$ *can be solved by a flexible algorithm with time bounds* $T_1'(n) = O(\log(n)T_1(n) + k^2 n \log^2(n))$, $T_2'(n) = O(\log(n)T_2(n) + k^2 \log^3(n))$ *and* $T_3'(n) = O(k + \log(n) + T_3(n))$. *Moreover, if the data structure used by A can be merged efficiently, then the first two time bounds are even better, namely* $T_1'(n) = O(T_1(n) + k^2 n \log^2(n))$ *and* $T_2'(n) = O(T_2(n) + k^2 \log^3(n))$.

Recall that $\mathcal{I}_k$ denotes the class of unions of at most k bounded intervals. As mentioned in Example 1, $\mathcal{I}_k$ is a subclass of $\mathcal{C}_2[2k + 1]$. A flexible algorithm that successfully competes with the best concept from $\mathcal{I}_k$ is obtained when we apply the transformation from Theorem 6 to the (trivial) flexible algorithm for $\mathrm{MinDis}(\mathcal{C}_2, \mathcal{C}_2)$. The resulting time bounds are $T_1(n) = O(k^2 n \log^2(n))$, $T_2(n) = O(k^2 \log^3(n))$ and $T_3(n) = O(k + \log(n))$. However, the algorithm resulting from this general transformation is inferior to the algorithm from [11] (which is specialized to the class $\mathcal{I}_k$):[4]

---

[4] In [11], flexibility of algorithms is not an issue. An inspection of the algorithm for $\mathrm{MinDis}(\mathcal{I}_k, \mathcal{I}_k)$ reveals, however, that the underlying data structure provides flexibility.

Theorem 7 ([11]). *The problem* $\mathrm{MinDis}(\mathcal{I}_k, \mathcal{I}_k)$ *can be solved by a flexible algorithm with* $T_1(n, k) = O(k^2 n)$, $T_2(n, k) = O(k^2 \log(n))$ *and* $T_3(n, k) = O(k)$.

As a final application, we consider unions of axis-parallel rectangles:

Theorem 8. *Let the function* $\ell_k(n)$ *be defined as in Lemma 1. Then the problem* $\mathrm{MinDis}(\mathcal{R}_k, \mathcal{I}_k[\ell_{2k+1}(n)])$ *can be solved by a flexible algorithm with* $T_1(n, k) = O(k^2 n \log^2(n))$, $T_2(n, k) = O(k^2 \log^3(n))$ *and* $T_3(n, k) = O(k + \log(n))$.

Proof. Recall from Example 3 that $\mathcal{R}_k$ is a subclass of $\mathcal{I}_k[2k + 1]$. Combining Theorems 7 and 6, we may conclude that the problem $\mathrm{MinDis}(\mathcal{I}_k[2k + 1], \mathcal{I}_k[\ell_{2k+1}(n)])$ can be solved by a flexible algorithm with time bounds as given in the assertion of the theorem. $\qquad\square$

The transformations described in Theorems 2 and 6 preserve flexibility but destroy properness. As for the transformation described in the following theorem, we have the reverse situation:

Theorem 9. *A flexible algorithm A for* $\mathrm{MinDis}(\mathcal{C}, \mathcal{H})$ *with time bounds* T_1, T_2, T_3 *can be transformed into an algorithm* A' *that solves the problem* $\mathrm{MinDis}(\mathcal{C}[2], \mathcal{H}[2])$ *in time* $O(n \log(n) + T_1(n) + n T_2(n))$.

Proof. Let $S' = [(x'_1, w_1), \ldots, (x'_n, w_n)] = [(z_1, x_1, w_1), \ldots, (z_n, x_n, w_n)] \in (\mathbb{R} \times \mathcal{X} \times \mathbb{R})^n$ be a given instance of $\mathrm{MinDis}(\mathcal{C}[2], \mathcal{H}[2])$. Let n' be the number of distinct z-coordinates in S', and let $z'_1 < z'_2 < \ldots < z'_{n'}$ be the corresponding sorted sequence. For sake of convenience, let $z'_{n'+1} = z'_{n'} + 1$. For $k = 1, \ldots, n' + 1$, let $S'_1(k) = \{(x'_i, w_i) : z_i < z'_k\}$ and $S'_2(k) = \{(x'_i, w_i) : z_i \geq z'_k\}$. Similarly, let $S_1(k) = \{(x_i, w_i) : z_i < z'_k\}$ and $S_2(k) = \{(x_i, w_i) : z_i \geq z'_k\}$.

Without loss of generality let $C^* = ((-\infty, z'_{k'}) \times C_1) \cup ([z'_{k'}, +\infty) \times C_2) \in \mathcal{C}[2]$ be the concept with the smallest empirical error on S' among all concepts from $\mathcal{C}[2]$. For each $k \in \{1, \ldots, n' + 1\}$, let H^k_1 (resp. H^k_2) be the hypothesis represented by $\mathrm{DS}(S_1(k))$ (resp. by $\mathrm{DS}(S_2(k))$). Let

$$H^k = ((-\infty, z'_k) \times H^k_1) \cup ([z'_k, \infty) \times H^k_2) \ . \tag{5}$$

Furthermore, let k'' be a minimizer of

$$W(k) := E_{S_1(k)}(H^k_1) + E_{S_2(k)}(H^k_2) \tag{6}$$

and let $H^* = H^{k''}$. With these notations, we get

$$E_{S'}(H^*) = E_{S_1(k'')}(H^{k''}_1) + E_{S_2(k'')}(H^{k''}_2) \leq E_{S_1(k')}(H^{k'}_1) + E_{S_2(k')}(H^{k'}_2)$$
$$\leq E_{S_1(k')}(C_1) + E_{S_2(k')}(C_2) = E_{S'}(C^*) \ .$$

Thus the empirical error of $H^* \in \mathcal{H}[2]$ on S' is not larger than the empirical error of $C^* \in \mathcal{C}[2]$ on S'. Suppose that we know the values $W(k)$ for all $k = 1, \ldots, n' + 1$. Then we can determine (a representation of) H^* as follows:

1. Set $k'' := \mathrm{argmin}\{W(k) : k \in \{1, \ldots, n' + 1\}\}$. This takes $O(n)$ steps.
2. Extract $S_1(k'')$ and $S_2(k'')$ from S' and sort each of these two sequences according to the x-coordinates of its items. This takes $O(n \log(n))$ steps.

3. Feed $S_1(k'')$ (resp. $S_2(k'')$) into A and obtain the data structure $DS(S_1(k''))$ (resp. $DS(S_2(k''))$). This takes $O(T_1(n))$ steps.
4. Recall that $DS(S_i(k''))$ represents $H_i^{k''}$ for $i = 1, 2$. These data structures augmented by $z'_{k''}$ form a suitable and easy-to-evaluate representation of H^*.

It remains to answer the question how the values $W(k)$ for $k = 1, \ldots, n' + 1$ can be computed efficiently. We observe first that the operation of deleting an item (x_k, w_k) from a set S of (at most n) items can be simulated by setting $w_k = 0$. According to (6), $W(k)$ is easy to compute from $E_{S_1(k)}(H_1^k)$ and $E_{S_2(k)}(H_2^k)$. For reasons of symmetry, it suffices to describe how the values $E_{S_1(k)}(H_1^k)$ for $k = 1, \ldots, n' + 1$ can be computed efficiently. This is done (similarly to a procedure used in [1] for learning 2-level decision trees) as follows:

1. Given S', let $k := n' + 1$, $S := S_1(k)$ and sort this sequence according to non-decreasing x-coordinates. This takes $O(n \log(n))$ steps.
2. Feed S into A and obtain $DS(S)$. This takes $O(T_1(n))$ steps.
3. Given $DS(S)$, compute $E_S(H_1^k)$ and store it in $W(k)$. This takes $O(1)$ steps.
4. If $k = 1$, then stop. Otherwise, set $w_k := 0$, update the data structure $DS(S)$ accordingly, set $k := k - 1$ and go back to Step 3. This takes $T_2(n)$ steps.

The time complexity of the whole procedure for computing H^* is dominated by the amount of time needed for computing the quantities $W(k)$ for $k = 1, \ldots, n'+1$, and this takes $O(n \log(n) + T_1(n) + nT_2(n))$ steps.   □

5   Experimental Results

We chose to investigate Minimum Disagreement algorithms for the class of axis-aligned rectangles $\mathcal{R}$ in the following experiments, because we expect to observe a considerable improvement in light of Theorem 4: the algorithm from Theorem 4 has an asymptotic worst-case time bound of only $O(n \log(n))$, where n is the number of examples, compared to the proper algorithms from [2,11] with a running time of $O(n^2 \log(n))$ and $O(n^2)$, respectively. While $O(n^2)$ is clearly better than $O(n^2 \log(n))$, we noticed that in the range of sample sizes used in the following experiments (and for our implementation) the learner from [11] is actually slightly faster than the one from [2]. Therefore, we compare our method from Theorem 4, which we denote by TRANS, with the algorithm from [11], which we denote by RECT. Note that—as in the theoretical analysis—we measured all running times without taking the time for pre-sorting the training data into account. This is justified as all considered algorithms rely on pre-sorted data. The experiments were performed on a AMD Opteron 6234 processor, running at 2400 MHz, with Oracle Java 1.8.0_31 under CentOS 6.6.

Data Sets. We are solving binary classification problems where the weight of any instance is either -1 or $+1$. We use three different data sets: one artificially generated data set, which is given by a mixture of two two-dimensional Gaussian distributions—one distribution for each weight—with identical covariance matrices. The "Glass" data set from [7], which consists of nine-dimensional instances

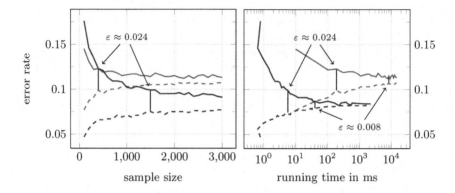

Fig. 2. Error rates of RECT (in red) and TRANS (in blue) as a function of the sample size (left-hand side) and of the running time (right-hand side) on the artificial Gaussian distribution. The x-axis on the right-hand side is logarithmic to accommodate the vast range of running times. The solid lines depict the error rates on the test set, while the dashed lines show the error rate on the training set. Note that the accuracy ε from Theorem 5 is proportional to the difference between these two error rates. All values were measured on an independent test set of size 1000 and averaged over 50 runs.

from forensic examinations of glass samples. While the original data set contains seven classes and 214 instances, we obtained a binary classification problem with 163 instances by following [9] (merging classes one and three and removing all instances from class four to seven). The "MAGIC" data set from [4,6], which consists of 19020 ten-dimensional instances of (simulated) observations of a "Cherenkov gamma telescope". The task is to discern gamma ray events from background noise. The latter two data sets are available on the UC Irvine Machine Learning Repository.

Experimental Results and Discussion. The results for the Gaussian mixture are given in Fig. 2. The error rates as a function of the sample size, shown in the left-hand side of Fig. 2, behave as expected: RECT, whose hypothesis class has the smaller VC-dimension, achieves a smaller error on the test set for small sample sizes and its error rates converge faster (as a function of the sample size). Note that TRANS' error rate outperforms RECT already at $n \approx 500$ because its higher estimation error is getting more than compensated by its lower approximation error. When computation time is the resource of consideration, as depicted in the right-hand side of Fig. 2, the differences become more drastic: TRANS consistently outperforms the much slower RECT-algorithm. Furthermore, as predicted by Theorem 5, TRANS' error rates indeed converge much faster. We would like to add that the measured running times nicely match the theoretical analysis.

The dimensions of the instances in both the Glass and MAGIC data sets are larger than two, so we cannot directly apply RECT and TRANS. We followed [9] and trained one hypothesis on every pair of coordinates choosing the hypothesis with the smallest error on the training set. (Another approach would be to

Table 1. Experimental results for the best instance pair using TRANS and RECT on the Glass and MAGIC data sets (upper table), and for AdaBoost using TRANS as the base learner on the MAGIC data set with different numbers of iterations T (lower table). Data sets were randomly split 2:1 into a training and test set. All values are averages over 50 runs, except for RECT on the MAGIC data set, which was run only 10 times.

data set	algorithm	error rate on training set	error rate on test set	time
Glass	TRANS	0.081	0.233	14 ms
	RECT	0.160	0.252	393 ms
MAGIC	TRANS	0.178	0.189	6 s
	RECT	0.214	0.219	17256 s

data set	algorithm	T	error rate on training set	error rate on test set	time
MAGIC	AdaBoost	1	0.178	0.189	6 s
	(on TRANS)	5	0.153	0.167	29 s
		10	0.138	0.156	59 s
		20	0.124	0.151	117 s
		40	0.107	0.149	235 s

iteratively transform the MinDis algorithm until we arrive at the desired dimension. However, this method introduces too much overhead using our implementation for 9 resp. 10 dimensions.) The results are given in the upper part of Table 1 and show that—for both the smaller Glass data set with 108 training instances and the larger MAGIC data set with 12680 training instances—TRANS' error rates are smaller than the ones from RECT. As expected, TRANS is considerably faster than RECT: notice the giant gap between six seconds and almost six hours on the MAGIC data set. The mean error rate of 0.233 on the Glass data set is in fact smaller than the rates reported in [9], which were 0.271 for a simple one-dimensional hypothesis and 0.257 for a (more complex) decision tree. Our mean error rate of 0.189 on the MAGIC data set is considerably larger than the ones reported in [6], which were in the range of 0.16 to 0.138 (on the test set) for different variants of decision trees. We try to close this gap in the following by using the well-known AdaBoost [8] scheme with TRANS as the base learner. The results for AdaBoost on the MAGIC data set are given in the lower part of Table 1. Obviously, one round of boosting is equivalent to the previous approach of choosing the hypothesis with the smallest empirical error. Note that already twenty iterations provide an error rate on the test set that is comparable with the rates from [6], and that TRANS is obviously fast enough for boosting to be practical on 12680 instances. Furthermore, our boosted classifier surpasses all methods considered in [4] in all but one measure of merit (we omit details due to space constrains). Admittedly, we suspect that this is mostly due to boosting and independent from the choice of base learners, as experiments using decision stumps yield similar error rates with a larger number of iterations but a smaller over-all running time.

References

1. Auer, P., Holte, R.C., Maass, W.: Theory and applications of agnostic PAC-learning with small decision trees. In: ICML 1995, pp. 21–29 (1995)
2. Barbay, J., Chan, T.M., Navarro, G., Pérez-Lantero, P.: Maximum-weight planar boxes in $O(n^2)$ time (and better). Information Processing Letters **114**(8), 437–445 (2014)
3. de Berg, M., Cheong, O., van Kreveld, M., Overmars, M.: Computational Geometry: Algorithms and Applications. Springer-Verlag, Santa Clara (2008)
4. Bock, R., Chilingarian, A., Gaug, M., Hakl, F., Hengstebeck, T., Jiřina, M., Klaschka, J., Kotrč, E., Savický, P., Towers, S., Vaiciulis, A., Wittek, W.: Methods for multidimensional event classification: a case study using images from a cherenkov gamma-ray telescope. Nuclear Instruments and Methods in Physics Research A **516**(2–3), 511–528 (2004)
5. Cortés, C., Díaz-Báñez, J.M., Pérez-Lantero, P., Seara, C., Urrutia, J., Ventura, I.: Bichromatic separability with two boxes: A general approach. Journal of Algorithms **64**(2–3), 79–88 (2009)
6. Dvořák, J., Savický, P.: Softening splits in decision trees using simulated annealing. In: Beliczynski, B., Dzielinski, A., Iwanowski, M., Ribeiro, B. (eds.) ICANNGA 2007. LNCS, vol. 4431, pp. 721–729. Springer, Heidelberg (2007)
7. Evett, I.W., Spiehler, E.J.: Rule induction in forensic science. Tech. rep, Central Research Establishment, Home Office Forensic Science Service (1987)
8. Freund, Y., Schapire, R.E.: A decision-theoretic generalization of on-line learning and an application to boosting. Journal of Computer and System Sciences **55**(1), 119–139 (1997)
9. Holte, R.C.: Very simple classification rules perform well on most commonly used datasets. Machine Learning **11**(1), 63–91 (1993)
10. Kearns, M.J., Schapire, R.E., Sellie, L.M.: Toward efficient agnostic learning. Machine Learning **17**(2), 115–141 (1994)
11. Maass, W.: Efficient agnostic PAC-learning with simple hypothesis. In: COLT 1994, pp. 67–75 (1994)
12. Pitt, L., Valiant, L.G.: Computational limitations on learning from examples. Journal of the Association on Computing Machinery **35**(4), 965–984 (1988)
13. Shalev-Shwartz, S., Ben-David, S.: Understanding Machine Learning: From Theory to Algorithms. Cambridge University Press (2014)
14. Vapnik, V.: Statistical learning theory. Wiley & Sons (1998)
15. Vapnik, V.N., Chervonenkis, A.Y.: On the uniform convergence of relative frequencies of events to their probabilities. Theory of Probability and its Applications **XVI**(2), 264–280 (1971)
16. Weiss, S.M., Galen, R.S., Tadepalli, P.: Maximizing the predictive value of production rules. Artificial Intelligence **45**(1–2), 47–71 (1990)
17. Weiss, S.M., Kapouleas, I.: An empirical comparison of pattern recognition, neural nets, and machine learning classification methods. In: IJCAI 1989, pp. 781–787 (1989)
18. Weiss, S.M., Kulikowski, C.A.: Computer Systems That Learn: Classification and Prediction Methods from Statistics, Neural Nets, Machine Learning and Expert Systems. Morgan Kaufmann (1990)

Learning with a Drifting Target Concept

Steve Hanneke[1]([⊠]), Varun Kanade[2], and Liu Yang [3]

[1] Princeton, NJ, USA
steve.hanneke@gmail.com
[2] Département d'informatique, École normale supérieure, Paris, France
varun.kanade@ens.fr
[3] IBM T.J. Watson Research Center, Yorktown Heights, NY, USA
yangli@us.ibm.com

Abstract. We study the problem of learning in the presence of a drifting target concept. Specifically, we provide bounds on the error rate at a given time, given a learner with access to a history of independent samples labeled according to a target concept that can change on each round. One of our main contributions is a refinement of the best previous results for polynomial-time algorithms for the space of linear separators under a uniform distribution. We also provide general results for an algorithm capable of adapting to a variable rate of drift of the target concept. Some of the results also describe an active learning variant of this setting, and provide bounds on the number of queries for the labels of points in the sequence sufficient to obtain the stated bounds on the error rates.

1 Introduction

Much of the work on statistical learning has focused on learning settings in which the concept to be learned is static over time. However, there are many application areas where this is not the case. For instance, in the problem of face recognition, the concept to be learned actually changes over time as each individual's facial features evolve over time. In this work, we study the problem of learning with a drifting target concept. Specifically, we consider a statistical learning setting, in which data arrive i.i.d. in a stream, and for each data point, the learner is required to predict a label for the data point at that time. We are then interested in obtaining low error rates for these predictions. The target labels are generated from a function known to reside in a given concept space, and at each time t the target function is allowed to change by at most some distance Δ_t: that is, the probability the new target function disagrees with the previous target function on a random sample is at most Δ_t.

This framework has previously been studied in a number of articles. The classic works of [5,6,15,16] and [7] together provide a general analysis of a very-much related setting. Though the objectives in these works are specified slightly differently, the results established there are easily translated into our present framework, and we summarize many of the relevant results in Section 3.

© Springer International Publishing Switzerland 2015
K. Chaudhuri et al. (Eds.): ALT 2015, LNAI 9355, pp. 149–164, 2015.
DOI: 10.1007/978-3-319-24486-0_10

While the results in these classic works are general, the best guarantees on the error rates are only known for methods having no guarantees of computational efficiency. In a more recent effort, the work of [8] studies this problem in the specific context of learning a homogeneous linear separator, when all the Δ_t values are identical. They propose a polynomial-time algorithm (based on the modified Perceptron algorithm of [9]), and prove a bound on the number of mistakes it makes as a function of the number of samples, when the data distribution satisfies a certain condition called "λ-good" (which generalizes a useful property of the uniform distribution on a sphere). However, their result is again worse than that obtainable by the known computationally-inefficient methods.

Thus, the natural question is whether there exists a polynomial-time algorithm achieving roughly the same guarantees on the error rates known for the inefficient methods. In the present work, we resolve this question in the case of learning homogeneous linear separators under the uniform distribution, by proposing a polynomial-time algorithm that indeed achieves roughly the same bounds on the error rates known for the inefficient methods in the literature. This represents the main technical contribution of this work.

We also study the interesting problem of *adaptivity* of an algorithm to the sequence of Δ_t values, in the setting where Δ_t may itself vary over time. Since the values Δ_t might typically not be accessible in practice, it seems important to have learning methods having no explicit dependence on the sequence Δ_t. We propose such a method below, and prove that it achieves roughly the same bounds on the error rates known for methods in the literature which require direct access to the Δ_t values. Also in the context of variable Δ_t sequences, we discuss conditions on the sequence Δ_t necessary and sufficient for there to exist a learning method guaranteeing *sublinear* growth of the number of mistakes.

We additionally study an *active learning* extension to this framework, in which, at each time, after making its prediction, the algorithm may decide whether or not to request access to the label assigned to the data point at that time. In addition to guarantees on the error rates (for *all* times, including those for which the label was not observed), we are also interested in bounding the number of labels we expect the algorithm to request, as a function of the number of samples encountered thus far.

2 Definitions and Notation

Formally, in this setting, there is a fixed distribution $\mathcal{P}$ over the instance space $\mathcal{X}$, and there is a sequence of independent $\mathcal{P}$-distributed unlabeled data $X_1, X_2, \ldots$. There is also a concept space $\mathbb{C}$, and a sequence of target functions $\mathbf{h}^* = \{h_1^*, h_2^*, \ldots\}$ in $\mathbb{C}$. Each t has an associated target label $Y_t = h_t^*(X_t)$. In this context, a (passive) learning algorithm is required, on each round t, to produce a classifier $\hat{h}_t$ based on the observations $(X_1, Y_1), \ldots, (X_{t-1}, Y_{t-1})$, and we denote by $\hat{Y}_t = \hat{h}_t(X_t)$ the corresponding prediction by the algorithm for the label of X_t. For any classifier h, we define $\mathrm{er}_t(h) = \mathcal{P}(x : h(x) \neq h_t^*(x))$.

We also say the algorithm makes a "mistake" on instance X_t if $\hat{Y}_t \neq Y_t$; thus, $\mathrm{er}_t(\hat{h}_t) = \mathbb{P}(\hat{Y}_t \neq Y_t | (X_1, Y_1), \ldots, (X_{t-1}, Y_{t-1}))$.

For notational convenience, we will suppose the h_t^* sequence is chosen independently from the X_t sequence (i.e., h_t^* is chosen prior to the "draw" of $X_1, X_2, \ldots \sim \mathcal{P}$), and is not random. In each results, we will suppose $\mathbf{h}^*$ is chosen from some set S of sequences in $\mathbb{C}$. In particular, we are interested in describing the sequence $\mathbf{h}^*$ in terms of the magnitudes of *changes* in h_t^* from one time to the next. Specifically, for any sequence $\boldsymbol{\Delta} = \{\Delta_t\}_{t=2}^{\infty}$ in $[0, 1]$, we denote by $S_{\boldsymbol{\Delta}}$ the set of all sequences $\mathbf{h}^*$ in $\mathbb{C}$ such that, $\forall t \in \mathbb{N}$, $\mathcal{P}(x : h_t(x) \neq h_{t+1}(x)) \leq \Delta_{t+1}$. Throughout this article, we denote by d the VC dimension of $\mathbb{C}$ [18], and we suppose $1 \leq d < \infty$. Also, $\forall x \in \mathbb{R}$, define $\mathrm{Log}(x) = \ln(\max\{x, e\})$.

3   Background: (ϵ, S)-Tracking Algorithms

As mentioned, the classic literature on learning with a drifting target concept is expressed in terms of a slightly different model. In order to relate those results to our present setting, we first introduce the classic setting. Specifically, we consider a model introduced by [15], presented here in a more-general form inspired by [5]. For a set S of sequences $\{h_t\}_{t=1}^{\infty}$ in $\mathbb{C}$, and a value $\epsilon > 0$, an algorithm $\mathcal{A}$ is said to be (ϵ, S)-*tracking* if $\exists t_\epsilon \in \mathbb{N}$ such that, for any choice of $\mathbf{h}^* \in S$, $\forall T \geq t_\epsilon$, the prediction $\hat{Y}_T$ produced by $\mathcal{A}$ at time T satisfies $\mathbb{P}\left(\hat{Y}_T \neq Y_T\right) \leq \epsilon$. Note that the value of this probability may be influenced by $\{X_t\}_{t=1}^{T}$, $\{h_t^*\}_{t=1}^{T}$, and any internal randomness of the algorithm $\mathcal{A}$.

The focus of the results expressed in this classical model is determining sufficient conditions on the set S for there to exist an (ϵ, S)-tracking algorithm, along with bounds on the sufficient size of t_ϵ. These conditions on S typically take the form of an assumption on the drift rate, expressed in terms of ϵ. Below, we summarize several of the strongest known results for this setting.

Bounded Drift Rate: The simplest, and perhaps most elegant, results for (ϵ, S)-tracking algorithms is for the set S of sequences with a bounded drift rate. Specifically, for any $\Delta \in [0, 1]$, define $S_\Delta = S_{\boldsymbol{\Delta}}$, where $\boldsymbol{\Delta}$ is such that $\Delta_{t+1} = \Delta$ for every $t \in \mathbb{N}$. The study of this problem was initiated in the original work of [15]. The best known general results are due to [16]: namely, that for some $\Delta_\epsilon = \Theta(\epsilon^2/d)$, for every $\epsilon \in (0, 1]$, there exists an (ϵ, S_Δ)-tracking algorithm for all values of $\Delta \leq \Delta_\epsilon$.[1] This refined an earlier result of [15] by a logarithmic factor. [16] further argued that this result can be achieved with $t_\epsilon = \Theta(d/\epsilon)$. The algorithm itself involves a beautiful modification of the one-inclusion graph prediction strategy of [14]; since its specification is somewhat involved, we refer the interested reader to the original work of [16] for the details.

---

[1] In fact, [16] also allowed the distribution $\mathcal{P}$ to vary gradually over time. For simplicity, we will only discuss the case of fixed $\mathcal{P}$.

Varying Drift Rates (Nonadaptive Algorithm): In addition to the concrete bounds for the case $\mathbf{h}^* \in S_\Delta$, [15] additionally present an elegant general result. Specifically, they argue that, for any $\epsilon > 0$, and any $m = \Omega\left(\frac{d}{\epsilon}\mathrm{Log}\frac{1}{\epsilon}\right)$, if $\sum_{i=1}^{m} \mathcal{P}(x : h_i^*(x) \neq h_{m+1}^*(x)) \leq m\epsilon/24$, then for $\hat{h} = \mathrm{argmin}_{h \in \mathbb{C}} \sum_{i=1}^{m} \mathbb{1}[h(X_i) \neq Y_i]$, $\mathbb{P}(\hat{h}(X_{m+1}) \neq h_{m+1}^*(X_{m+1})) \leq \epsilon$. This result immediately inspires an algorithm $\mathcal{A}$ which, at every time t, chooses a value $m_t \leq t-1$, and predicts $\hat{Y}_t = \hat{h}_t(X_t)$, for $\hat{h}_t = \mathrm{argmin}_{h \in \mathbb{C}} \sum_{i=t-m_t}^{t-1} \mathbb{1}[h(X_i) \neq Y_i]$. We are then interested in choosing m_t to minimize the value of ϵ obtainable via the result of [15]. However, that method is based on the values $\mathcal{P}(x : h_i^*(x) \neq h_t^*(x))$, which would typically not be accessible to the algorithm. However, suppose instead we have access to a sequence $\boldsymbol{\Delta}$ such that $\mathbf{h}^* \in S_{\boldsymbol{\Delta}}$. In this case, we could approximate $\mathcal{P}(x : h_i^*(x) \neq h_t^*(x))$ by its *upper bound* $\sum_{j=i+1}^{t} \Delta_j$. In this case, we are interested choosing m_t to minimize the smallest value of ϵ such that $\sum_{i=t-m_t}^{t-1} \sum_{j=i+1}^{t} \Delta_j \leq m_t\epsilon/24$ and $m_t = \Omega\left(\frac{d}{\epsilon}\mathrm{Log}\frac{1}{\epsilon}\right)$. One can easily verify that this minimum is obtained at a value

$$m_t = \Theta\left(\operatorname*{argmin}_{1 \leq m \leq t-1} \frac{1}{m} \sum_{i=t-m}^{t-1} \sum_{j=i+1}^{t} \Delta_j + \frac{d\mathrm{Log}(m/d)}{m}\right),$$

and via the result of [15] (applied to $X_{t-m_t}, \ldots, X_t$) the resulting algorithm has

$$\mathbb{P}\left(\hat{Y}_t \neq Y_t\right) \leq O\left(\min_{1 \leq m \leq t-1} \frac{1}{m} \sum_{i=t-m}^{t-1} \sum_{j=i+1}^{t} \Delta_j + \frac{d\mathrm{Log}(m/d)}{m}\right). \tag{1}$$

As a special case, if every t has $\Delta_t = \Delta$ for a fixed value $\Delta \in [0,1]$, this result recovers the bound $\sqrt{d\Delta\mathrm{Log}(1/\Delta)}$, which is only slightly larger than the best bound of [16]. It also applies to far more general and more intersting sequences $\boldsymbol{\Delta}$, including some that allow periodic large jumps (i.e., $\Delta_t = 1$ for some indices t), others where the sequence Δ_t converges to 0, and so on. Note, however, that the algorithm obtaining this bound directly depends on the sequence $\boldsymbol{\Delta}$. One of the contributions of the present work is to remove this requirement, while maintaining essentially the same bound, though in a slightly different form.

Computational Efficiency: [15] also proposed a reduction-based approach, which sometimes yields computationally efficient methods, though the tolerable Δ value is smaller. Specifically, given any (randomized) polynomial-time algorithm $\mathcal{A}$ that produces a classifier $h \in \mathbb{C}$ with $\sum_{t=1}^{m} \mathbb{1}[h(x_t) \neq y_t] = 0$ for any sequence $(x_1, y_1), \ldots, (x_m, y_m)$ for which such a classifier h exists (called the *consistency problem*), they propose a polynomial-time algorithm that is (ϵ, S_Δ)-tracking for all values of $\Delta \leq \Delta_\epsilon'$, where $\Delta_\epsilon' = \Theta\left(\frac{\epsilon^2}{d^2\mathrm{Log}(1/\epsilon)}\right)$. This is slightly worse (by a factor of $d\mathrm{Log}(1/\epsilon)$) than the drift rate tolerable by the (typically inefficient) algorithm mentioned above. However, it does sometimes yield computationally-efficient methods. For instance, there are known polynomial-time algorithms for the consistency problem for the classes of linear separators, conjunctions, and axis-aligned rectangles.

Lower Bounds: [15] additionally prove *lower bounds* for specific concept spaces: namely, linear separators and axis-aligned rectangles. They specifically argue that, for $\mathbb{C}$ a concept space $\mathrm{BASIC}_n = \{\cup_{i=1}^{n}[i/n, (i + a_i)/n) : \mathbf{a} \in [0, 1]^n\}$ on $[0, 1]$, under $\mathcal{P}$ the uniform distribution on $[0, 1]$, for any $\epsilon \in [0, 1/e^2]$ and $\Delta_\epsilon \geq e^4\epsilon^2/n$, for any algorithm $\mathcal{A}$, and any $T \in \mathbb{N}$, there exists a choice of $\mathbf{h}^* \in S_{\Delta_\epsilon}$ such that the prediction $\hat{Y}_T$ produced by $\mathcal{A}$ at time T satisfies $\mathbb{P}\left(\hat{Y}_T \neq Y_T\right) > \epsilon$. Based on this, they conclude that no $(\epsilon, S_{\Delta_\epsilon})$-tracking algorithm exists. They further observe that BASIC_n is embeddable in many common concept spaces, including halfspaces and axis-aligned rectangles in $\mathbb{R}^n$, so that for $\mathbb{C}$ equal to either of these, there also is no $(\epsilon, S_{\Delta_\epsilon})$-tracking algorithm.

4   Adapting to Arbitrarily Varying Drift Rates

This section presents a general bound on the error rate at each time, expressed as a function of the rates of drift, which are allowed to be *arbitrary*. Most-importantly, in contrast to the methods from the literature discussed above, the method achieving this general result is *adaptive* to the drift rates, so that it requires no information about the drift rates in advance. This is an appealing property, as it essentially allows the algorithm to learn under an *arbitrary* sequence $\mathbf{h}^*$ of target concepts; the difficulty of the task is then simply reflected in the resulting bounds on the error rates: that is, faster-changing sequences of target functions result in larger bounds on the error rates, but do not require a change in the algorithm itself.

4.1   Adapting to a Changing Drift Rate

Recall that the method yielding (1) (based on the work of [15]) required access to the sequence $\boldsymbol{\Delta}$ of changes to achieve the stated guarantee on the expected number of mistakes. That method is based on choosing a classifier to predict $\hat{Y}_t$ by minimizing the number of mistakes among the previous m_t samples, where m_t is a value chosen based on the $\boldsymbol{\Delta}$ sequence. Thus, the key to modifying this algorithm to make it adaptive to the $\boldsymbol{\Delta}$ sequence is to determine a suitable choice of m_t without reference to the $\boldsymbol{\Delta}$ sequence. The strategy we adopt here is to use the *data* to determine an appropriate value $\hat{m}_t$ to use. Roughly (ignoring logarithmic factors for now), the insight that enables us to achieve this feat is that, for the m_t used in the above strategy, one can show that $\sum_{i=t-m_t}^{t-1} \mathbb{1}[h_t^*(X_i) \neq Y_i]$ is roughly $\tilde{O}(d)$, and that making the prediction $\hat{Y}_t$ with *any* $h \in \mathbb{C}$ with roughly $\tilde{O}(d)$ mistakes on these samples will suffice to obtain the stated bound on the error rate (up to logarithmic factors). Thus, if we replace m_t with the largest value m for which $\min_{h \in \mathbb{C}} \sum_{i=t-m}^{t-1} \mathbb{1}[h(X_i) \neq Y_i]$ is roughly $\tilde{O}(d)$, then the above observation implies $m \geq m_t$. This then implies that, for $\hat{h} = \operatorname{argmin}_{h \in \mathbb{C}} \sum_{i=t-m}^{t-1} \mathbb{1}[h(X_i) \neq Y_i]$, we have that $\sum_{i=t-m_t}^{t-1} \mathbb{1}[\hat{h}(X_i) \neq Y_i]$ is also roughly $\tilde{O}(d)$, so that the stated bound on the error rate will be achieved (aside from logarithmic factors) by choosing $\hat{h}_t$ as this classifier $\hat{h}$. There are a few technical modifications to this argument needed to get the logarithmic factors to work out properly, and for this reason the actual algorithm below (and proof) is somewhat more involved. Specifically, consider the following algorithm (the value of the universal constant $K \geq 1$ will be specified below).

---

0. For $T = 1, 2, \ldots$

1.   Let $\hat{m}_T = \max\left\{ m \in \{1, \ldots, T-1\} : \min_{h \in \mathbb{C}} \max_{m' \leq m} \frac{\sum_{t=T-m'}^{T-1} \mathbb{1}[h(X_t) \neq Y_t]}{d\mathrm{Log}(m'/d) + \mathrm{Log}(1/\delta)} < K \right\}$

2.   Let $\hat{h}_T = \operatorname*{argmin}_{h \in \mathbb{C}} \max_{m' \leq \hat{m}_T} \frac{\sum_{t=T-m'}^{T-1} \mathbb{1}[h(X_t) \neq Y_t]}{d\mathrm{Log}(m'/d) + \mathrm{Log}(1/\delta)}$

Note that the classifiers $\hat{h}_t$ chosen by this algorithm have no dependence on $\boldsymbol{\Delta}$, or anything other than the data $\{(X_i, Y_i) : i < t\}$, and the concept space $\mathbb{C}$. For space, the proof is deferred to the full version of this paper [13].

Theorem 1. *Fix any $\delta \in (0,1)$, and let $\mathcal{A}$ be the above algorithm. For any sequence $\boldsymbol{\Delta}$ in $[0,1]$, for any $\mathcal{P}$ and any choice of $\mathbf{h}^* \in S_{\boldsymbol{\Delta}}$, for every $T \in \mathbb{N} \setminus \{1\}$, with probability at least $1 - \delta$,*

$$\mathrm{er}_T\left(\hat{h}_T\right) \leq O\left(\min_{1 \leq m \leq T-1} \frac{1}{m} \sum_{i=T-m}^{T-1} \sum_{j=i+1}^{T} \Delta_j + \frac{d\mathrm{Log}(m/d) + \mathrm{Log}(1/\delta)}{m} \right).$$

One immediate implication of Theorem 1 is that, if the sum of Δ_t values grows sublinearly, then there exists an algorithm achieving an expected number of mistakes growing sublinearly in the number of predictions. Formally, we have the following corollary. The proof is deferred to the full version [13].

Corollary 1. *If $\sum_{t=1}^{T} \Delta_t = o(T)$, then there exists an algorithm $\mathcal{A}$ such that, for every $\mathcal{P}$ and every choice of $\mathbf{h}^* \in S_{\boldsymbol{\Delta}}$, $\mathbb{E}\left[\sum_{t=1}^{T} \mathbb{1}\left[\hat{Y}_t \neq Y_t\right]\right] = o(T)$.*

For many concept spaces of interest, the condition $\sum_{t=1}^{T} \Delta_t = o(T)$ in Corollary 1 is also a *necessary* condition for *any* algorithm to guarantee a sublinear number of mistakes. In the full version of this paper [13], we establish that for the class of *homogeneous linear separators* on $\mathbb{R}^2$ with $\mathcal{P}$ the uniform distribution on the unit circle, there exists an algorithm with $\mathbb{E}\left[\sum_{t=1}^{T} \mathbb{1}\left[\hat{Y}_t \neq Y_t\right]\right] = o(T)$ for *every* choice of $\mathbf{h}^* \in S_{\boldsymbol{\Delta}}$ if and only if $\sum_{t=1}^{T} \Delta_t = o(T)$.

5   Polynomial-Time Algorithms for Linear Separators

In this section, we suppose $\Delta_t = \Delta$ for every $t \in \mathbb{N}$, for a fixed constant $\Delta > 0$, and we consider the special case of learning homogeneous linear separators in $\mathbb{R}^k$ under a uniform distribution on the origin-centered unit sphere. In this case, the analysis of [15] mentioned in Section 3 implies that it is possible to achieve a bound on the error rate that is $\tilde{O}(d\sqrt{\Delta})$, using an algorithm that runs in time $\mathrm{poly}(d, 1/\Delta, \log(1/\delta))$ (and independent of t) for each prediction. This also implies that it is possible to achieve expected number of mistakes among T predictions that is $\tilde{O}(d\sqrt{\Delta})T$. [8][2] have since proven that a variant of the Perceptron algorithm achieves an expected number of mistakes $\tilde{O}((d\Delta)^{1/4})T$.

Below, we improve on this result by showing that there exists an efficient algorithm that achieves a bound on the error rate that is $\tilde{O}(\sqrt{d\Delta})$, as was possible with the inefficient algorithm of [15,16]. This leads to a bound $\tilde{O}(\sqrt{d\Delta})T$ on the expected number of mistakes. Furthermore, our approach also allows us to present the method as an *active learning* algorithm, and to bound the expected number of queries, as a function of the number of samples T, by $\tilde{O}(\sqrt{d\Delta})T$. The technique is based on modifying the algorithm of [15], replacing an ERM step with (a modification of) the computationally-efficient algorithm of [1].

Formally, define the class of homogeneous linear separators as the set of classifiers $h_w : \mathbb{R}^d \to \{-1, +1\}$, for $w \in \mathbb{R}^d$ with $\|w\| = 1$, such that $h_w(x) = \mathrm{sign}(w \cdot x)$ for every $x \in \mathbb{R}^d$. We have the following result.

Theorem 2. *When $\mathbb{C}$ is the space of homogeneous linear separators (with $d \geq 4$) and $\mathcal{P}$ is the uniform distribution on the surface of the origin-centered unit sphere in $\mathbb{R}^d$, for any fixed $\Delta > 0$, for any $\delta \in (0, 1/e)$, there is an algorithm that runs in time $\mathrm{poly}(d, 1/\Delta, \log(1/\delta))$ for each time t, such that for any $\mathbf{h}^* \in S_\Delta$, for every sufficiently large $t \in \mathbb{N}$, with probability at least $1 - \delta$,*

$$\mathrm{er}_t(\hat{h}_t) = O\left(\sqrt{\Delta d \log\left(\tfrac{1}{\delta}\right)}\right).$$

Also, choosing $\delta = \sqrt{\Delta d} \wedge 1/e$, the expected number of mistakes among the first T predictions is $O\left(\sqrt{\Delta d \log\left(\tfrac{1}{\Delta d}\right)}T\right)$. Furthermore, the algorithm can be run as an active learning algorithm, in which case, for this δ, the expected number of labels requested by the algorithm among the first T instances is $O\left(\sqrt{\Delta d}\log^{3/2}\left(\tfrac{1}{\Delta d}\right)T\right)$.

We first state the algorithm used to obtain this result. It is primarily based on a margin-based learning strategy of [1], combined with an initialization step based on a modified Perceptron rule from [8,9]. For $\tau > 0$ and $x \in \mathbb{R}$, define $\ell_\tau(x) = \max\left\{0, 1 - \frac{x}{\tau}\right\}$. Consider the following algorithm and subroutine; parameters δ_k, m_k, τ_k, r_k, b_k, α, and κ will all be specified in the context of the proof (see Lemmas 2 and 6); we suppose $M = \sum_{k=0}^{\lceil \log_2(1/\alpha) \rceil} m_k$.

Algorithm: DriftingHalfspaces
0. Let $\tilde{h}_0$ be an arbitrary classifier in $\mathbb{C}$
1. For $i = 1, 2, \ldots$
2. $\tilde{h}_i \leftarrow \mathrm{ABL}(M(i-1), \tilde{h}_{i-1})$

---

[2] This work in fact studies a much broader model of drift, which allows the distribution $\mathcal{P}$ to vary with time as well. However, this $\tilde{O}((d\Delta)^{1/4})T$ result can be obtained from their theorem by calculating the various parameters for this particular setting.

Subroutine: ModPerceptron$(t, \tilde{h})$

0. Let w_t be any element of $\mathbb{R}^d$ with $\|w_t\| = 1$
1. For $m = t+1, t+2, \ldots, t+m_0$
2.   Choose $\hat{h}_m = \tilde{h}$ (i.e., predict $\hat{Y}_m = \tilde{h}(X_m)$ as the prediction for Y_m)
3.   Request the label Y_m
4.   If $h_{w_{m-1}}(X_m) \neq Y_m$
5. $w_m \leftarrow w_{m-1} - 2(w_{m-1} \cdot X_m)X_m$
6.   Else $w_m \leftarrow w_{m-1}$
7. Return w_{t+m_0}

Subroutine: ABL$(t, \tilde{h})$

0. Let w_0 be the return value of ModPerceptron$(t, \tilde{h})$
1. For $k = 1, 2, \ldots, \lceil \log_2(1/\alpha) \rceil$
2. $W_k \leftarrow \{\}$
3.   For $s = t + \sum_{j=0}^{k-1} m_j + 1, \ldots, t + \sum_{j=0}^{k} m_j$
4.   Choose $\hat{h}_s = \tilde{h}$ (i.e., predict $\hat{Y}_s = \tilde{h}(X_s)$ as the prediction for Y_s)
5.   If $|w_{k-1} \cdot X_s| \leq b_{k-1}$, Request label Y_s and let $W_k \leftarrow W_k \cup \{(X_s, Y_s)\}$
6.   Find $v_k \in \mathbb{R}^d$ with $\|v_k - w_{k-1}\| \leq r_k$, $0 < \|v_k\| \leq 1$, and
$$\sum_{(x,y) \in W_k} \ell_{\tau_k}(y(v_k \cdot x)) \leq \inf_{v: \|v-w_{k-1}\| \leq r_k} \sum_{(x,y) \in W_k} \ell_{\tau_k}(y(v \cdot x)) + \kappa |W_k|$$
7.   Let $w_k = \frac{1}{\|v_k\|} v_k$
8. Return $h_{w_{\lceil \log_2(1/\alpha) \rceil - 1}}$

The general idea here is to replace empirical risk minimization in the method of [15] discussed above with a computationally efficient method of [1]: namely, the subroutine ABL above. For technical reasons, we apply this method to batches of M samples at a time, and simply use the classifier learned from the previous batch to make the predictions. The method of [1] was originally proposed for the problem of *agnostic* learning, to error rate within a constant factor of the optimal. To use this for our purposes, we set up an analogy between the best achievable error rate in agnostic learning and a value $O(\Delta M)$ here (which bounds the best achievable *average* error rate in a given batch).

The analysis of [1] required this method to be initialized with a reasonably accurate classifier (constant bound on its error rate). For this, we find (in Lemma 1) that the modified Perceptron algorithm (of [8,9]) suffices. The ABL algorithm then iteratively refines a hypothesis w_k by taking a number of samples within a slab of width $b_{k-1} \propto 2^{-k}/\sqrt{d}$ around the previous hypothesis separator w_{k-1}, and optimizing a weighted hinge loss (subject to a constraint that the new hypothesis not be too far from the previous). The analysis (Lemma 6) then reveals that the hypothesis w_k approaches a classifier w^* with error rate $O(\Delta M)$ with respect to all of the target concepts in the batch.

We note that, even with the above-described analogy between $O(\Delta M)$ and the noise rate in agnostic learning, the analysis below does not follow immediately from that of [1]. This is because the sample size M that would be required by the analysis of [1] to achieve error rate within a constant factor of the noise rate would be too large (by a factor of d) for our purposes. In particular, noting that ΔM is increasing in M, converting that original analysis to our present setting would result in a bound on $\mathrm{er}_t(\hat{h}_t)$ larger than that stated in Theorem 2 by roughly a factor of $\sqrt{d}$. The analysis below refines several aspects of the analysis, using stronger concentration arguments for the weighted hinge loss, and being more careful in bounding the error rate in terms of the weighted hinge loss performance. We thereby reduce the bound to the result stated above.

We have a few lemmas that will be needed for the proof. With some effort, the following result can be derived from the analysis of ModPerceptron by [8]. The details are included in the full version of this article [13].

Lemma 1. *Suppose* $\Delta \leq \frac{\pi^2}{400 \cdot 2^{27}(d + \ln(4/\delta))}$. *For* $m_0 = \max\{\lceil 128(1/c_1) \ln(32)\rceil,$ $\lceil 512 \ln(\frac{4}{\delta})\rceil\}$, *with probability at least* $1 - \delta/4$, *ModPerceptron*$(t, \tilde{h})$ *returns a vector* w *with* $\mathcal{P}(x : h_w(x) \neq h^*_{t+m_0+1}(x)) \leq 1/16$.

Next, we consider the execution of ABL$(t, \tilde{h})$, and let the sets W_k be as in that execution. We will denote by w^* the weight vector with $\|w^*\| = 1$ such that $h^*_{t+m_0+1} = h_{w^*}$. Also denote by $M_1 = M - m_0$.

The proof relies on a few results proven in the work of [1], which we summarize in the following lemmas. Although the results were proven in a slightly different setting in that work (agnostic learning under a fixed joint distribution), one can easily verify that their proofs remain valid in our present context as well.

Lemma 2. *[1] Fix any* $k \in \{1, \ldots, \lceil \log_2(1/\alpha)\rceil\}$. *For a universal constant* $c_7 > 0$, *suppose* $b_{k-1} = c_7 2^{1-k}/\sqrt{d}$, *and let* $z_k = \sqrt{r_k^2/(d-1) + b_{k-1}^2}$. *For a universal constant* $c_1 > 0$, *if* $\|w^* - w_{k-1}\| \leq r_k$,

$$\left| \mathbb{E}\left[\sum_{(x,y) \in W_k} \ell_{\tau_k}(|w^* \cdot x|) \Big| w_{k-1}, |W_k| \right] - \mathbb{E}\left[\sum_{(x,y) \in W_k} \ell_{\tau_k}(y(w^* \cdot x)) \Big| w_{k-1}, |W_k| \right] \right|$$
$$\leq c_1 |W_k| \sqrt{2^k \Delta M_1} \frac{z_k}{\tau_k}.$$

Lemma 3. *[4]* $\forall c > 0$, *there exists* $c' > 0$ *depending only on* c *(i.e., not depending on* d) *such that, for any* $u, v \in \mathbb{R}^d$ *with* $\|u\| = \|v\| = 1$, *letting* $\sigma = \mathcal{P}(x : h_u(x) \neq h_v(x))$, *if* $\sigma < 1/2$, *then* $\mathcal{P}\left(x : h_u(x) \neq h_v(x) \text{ and } |v \cdot x| \geq c' \frac{\sigma}{\sqrt{d}}\right) \leq c\sigma$.

The following is a well-known lemma concerning concentration around the equator for the uniform distribution (see e.g., [1,3,9]).

Lemma 4. *For any* $C > 0$, *there are constants* $c_2, c_3 > 0$ *depending only on* C *(i.e., independent of* d) *such that, for any* $w \in \mathbb{R}^d$ *with* $\|w\| = 1$, $\forall \gamma \in [0, C/\sqrt{d}]$,

$$c_2 \gamma \sqrt{d} \leq \mathcal{P}(x : |w \cdot x| \leq \gamma) \leq c_3 \gamma \sqrt{d}.$$

Based on this lemma, [1] prove the following.

Lemma 5. *[1] For* $X \sim \mathcal{P}$, $\forall w \in \mathbb{R}^d$ *with* $\|w\| = 1$, $\forall C > 0$ *and* $\tau, b \in [0, C/\sqrt{d}]$, *for* c_2, c_3 *as in Lemma 4*, $\mathbb{E}\left[\ell_\tau(|w^* \cdot X|) \Big| |w \cdot X| \leq b\right] \leq \frac{c_3 \tau}{c_2 b}$.

The following is a stronger version of a result of [1]; specifically, the size of m_k, and the bound on $|W_k|$, are smaller by a factor of d compared to the original.

Lemma 6. *Fix any $\delta \in (0, 1/e)$. For universal constants $c_4, c_5, c_6, c_7, c_8, c_9, c_{10} \in (0, \infty)$, for an appropriate choice of $\kappa \in (0, 1)$ (a universal constant), if $\alpha = c_9 \sqrt{\Delta d \log\left(\frac{1}{\kappa\delta}\right)}$, for every $k \in \{1, \ldots, \lceil \log_2(1/\alpha) \rceil\}$, if $b_{k-1} = c_7 2^{1-k}/\sqrt{d}$, $\tau_k = c_8 2^{-k}/\sqrt{d}$, $r_k = c_{10} 2^{-k}$, $\delta_k = \delta/(\lceil \log_2(4/\alpha) \rceil - k)^2$, and $m_k = \left\lceil c_5 \frac{2^k}{\kappa^2} d \log\left(\frac{1}{\kappa\delta_k}\right) \right\rceil$, and if $\mathcal{P}(x : h_{w_{k-1}}(x) \neq h_{w^*}(x)) \leq 2^{-k-3}$, then with probability at least $1 - (4/3)\delta_k$, $|W_k| \leq c_6 \frac{1}{\kappa^2} d \log\left(\frac{1}{\kappa\delta_k}\right)$ and $\mathcal{P}(x : h_{w_k}(x) \neq h_{w^*}(x)) \leq 2^{-k-4}$.*

Proof. By Lemma 4, and a Chernoff and union bound, for an appropriately large choice of c_5 and any $c_7 > 0$, letting c_2, c_3 be as in Lemma 4 (with $C = c_7 \vee (c_8/2)$), with probability at least $1 - \delta_k/3$,

$$c_2 c_7 2^{-k} m_k \leq |W_k| \leq 4 c_3 c_7 2^{-k} m_k. \tag{2}$$

The claimed upper bound on $|W_k|$ follows from this second inequality.

Next note that, if $\mathcal{P}(x : h_{w_{k-1}}(x) \neq h_{w^*}(x)) \leq 2^{-k-3}$, then

$$\max\{\ell_{\tau_k}(y(w^* \cdot x)) : x \in \mathbb{R}^d, |w_{k-1} \cdot x| \leq b_{k-1}, y \in \{-1, +1\}\} \leq c_{11} \sqrt{d}$$

for some universal constant $c_{11} > 0$. Furthermore, since $\mathcal{P}(x : h_{w_{k-1}}(x) \neq h_{w^*}(x)) \leq 2^{-k-3}$, we know that the angle between w_{k-1} and w^* is at most $2^{-k-3}\pi$, so that $\|w_{k-1} - w^*\| = \sqrt{2 - 2w_{k-1} \cdot w^*} \leq \sqrt{2 - 2\cos(2^{-k-3}\pi)} \leq \sqrt{2 - 2\cos^2(2^{-k-3}\pi)} = \sqrt{2}\sin(2^{-k-3}\pi) \leq 2^{-k-3}\pi\sqrt{2}$. For $c_{10} = \pi\sqrt{2}2^{-3}$, this is r_k. By Hoeffding's inequality (under the conditional distribution given $|W_k|$), the law of total probability, Lemma 2, and linearity of conditional expectations, with probability at least $1 - \delta_k/3$, for $X \sim \mathcal{P}$,

$$\sum_{(x,y) \in W_k} \ell_{\tau_k}(y(w^* \cdot x)) \leq |W_k| \mathbb{E}\left[\ell_{\tau_k}(|w^* \cdot X|) \big| w_{k-1}, |w_{k-1} \cdot X| \leq b_{k-1}\right]$$
$$+ c_1 |W_k| \sqrt{2^k \Delta M_1 \frac{z_k}{\tau_k}} + \sqrt{|W_k|(1/2)c_{11}^2 d \ln(3/\delta_k)}. \tag{3}$$

We bound each term on the right separately. By Lemma 5, the first term is at most $|W_k| \frac{c_3 \tau_k}{c_2 b_{k-1}} = |W_k| \frac{c_3 c_8}{2 c_2 c_7}$. Next, $\frac{z_k}{\tau_k} = \frac{\sqrt{c_{10}^2 2^{-2k}/(d-1) + 4c_7^2 2^{-2k}/d}}{c_8 2^{-k}/\sqrt{d}} \leq \frac{\sqrt{2c_{10}^2 + 4c_7^2}}{c_8}$, while $2^k \leq \frac{2}{\alpha}$, so the second term is at most $\sqrt{2}c_1 \frac{\sqrt{2c_{10}^2 + 4c_7^2}}{c_8} |W_k| \sqrt{\frac{\Delta m}{\alpha}}$. Noting

$$M_1 = \sum_{k'=1}^{\lceil \log_2(1/\alpha) \rceil} m_{k'} \leq \frac{32 c_5}{\kappa^2} \frac{1}{\alpha} d \log\left(\frac{1}{\kappa\delta}\right), \tag{4}$$

the second term on the right of (3) is at most $\sqrt{\frac{c_5}{c_9}} \frac{8c_1}{\kappa} \frac{\sqrt{2c_{10}^2 + 4c_7^2}}{c_8} |W_k| \sqrt{\frac{\Delta d \log\left(\frac{1}{\kappa\delta}\right)}{\alpha^2}}$

$$= \frac{8c_1 \sqrt{c_5}}{\kappa} \frac{\sqrt{2c_{10}^2 + 4c_7^2}}{c_8 c_9} |W_k|. \text{ Since } d \ln(3/\delta_k) \leq 2d \ln(1/\delta_k) \leq \frac{2\kappa^2}{c_5} 2^{-k} m_k,$$

and (2) implies $2^{-k}m_k \leq \frac{1}{c_2c_7}|W_k|$, the third term on the right of (3) is at most $|W_k|\frac{c_{11}\kappa}{\sqrt{c_2c_5c_7}}$. Altogether, $\sum_{(x,y)\in W_k} \ell_{\tau_k}(y(w^* \cdot x)) \leq$

$$|W_k|\left(\frac{c_3c_8}{2c_2c_7} + \frac{8c_1\sqrt{c_5}}{\kappa}\frac{\sqrt{2c_{10}^2+4c_7^2}}{c_8c_9} + \frac{c_{11}\kappa}{\sqrt{c_2c_5c_7}}\right).$$ For $c_9 = 1/\kappa^3$, $c_8 = \kappa$, this is at most

$$\kappa|W_k|\left(\frac{c_3}{2c_2c_7} + 8c_1\sqrt{c_5}\sqrt{2c_{10}^2+4c_7^2} + \frac{c_{11}}{\sqrt{c_2c_5c_7}}\right).$$

Next, note that because $h_{w_k}(x) \neq y \Rightarrow \ell_{\tau_k}(y(v_k \cdot x)) \geq 1$, and because (as proven above) $\|w^* - w_{k-1}\| \leq r_k$, $|W_k|\mathrm{er}_{W_k}(h_{w_k}) \leq \sum_{(x,y)\in W_k} \ell_{\tau_k}(y(v_k \cdot x)) \leq \sum_{(x,y)\in W_k} \ell_{\tau_k}(y(w^* \cdot x)) + \kappa|W_k|$. Combined with the above, we have

$$|W_k|\mathrm{er}_{W_k}(h_{w_k}) \leq \kappa|W_k|\left(1 + \frac{c_3}{2c_2c_7} + 8c_1\sqrt{c_5}\sqrt{2c_{10}^2+4c_7^2} + \frac{c_{11}}{\sqrt{c_2c_5c_7}}\right).$$

Let $c_{12} = 1 + \frac{c_3}{2c_2c_7} + 8c_1\sqrt{c_5}\sqrt{2c_{10}^2+4c_7^2} + \frac{c_{11}}{\sqrt{c_2c_5c_7}}$. Furthermore, $|W_k|\mathrm{er}_{W_k}(h_{w_k}) = \sum_{(x,y)\in W_k}\mathbb{1}[h_{w_k}(x) \neq y] \geq \sum_{(x,y)\in W_k}\mathbb{1}[h_{w_k}(x) \neq h_{w^*}(x)] - \sum_{(x,y)\in W_k}\mathbb{1}[h_{w^*}(x) \neq y]$. For an appropriately large value of c_5, by a Chernoff bound, with probability at least $1 - \delta_k/3$, $\sum_{s=t+\sum_{j=0}^{k-1} m_j+1}^{t+\sum_{j=0}^{k} m_j} \mathbb{1}[h_{w^*}(X_s) \neq Y_s] \leq 2e\Delta M_1 m_k + \log_2(3/\delta_k)$. In particular, this implies $\sum_{(x,y)\in W_k} \mathbb{1}[h_{w^*}(x) \neq y] \leq 2e\Delta M_1 m_k + \log_2(3/\delta_k)$, so that $\sum_{(x,y)\in W_k} \mathbb{1}[h_{w_k}(x) \neq h_{w^*}(x)] \leq |W_k|\mathrm{er}_{W_k}(h_{w_k}) + 2e\Delta M_1 m_k + \log_2(3/\delta_k)$. Noting that (4) and (2) imply

$$\Delta M_1 m_k \leq \Delta \frac{32c_5}{\kappa^2}\frac{d\log\left(\frac{1}{\kappa\delta}\right)}{c_9\sqrt{\Delta d\log\left(\frac{1}{\kappa\delta}\right)}}\frac{2^k}{c_2c_7}|W_k| \leq \frac{32c_5}{c_2c_7c_9\kappa^2}\sqrt{\Delta d\log\left(\frac{1}{\kappa\delta}\right)}2^k|W_k|$$

$$= \frac{32c_5}{c_2c_7c_9^2\kappa^2}\alpha 2^k|W_k| = \frac{32c_5\kappa^4}{c_2c_7}\alpha 2^k|W_k| \leq \frac{32c_5\kappa^4}{c_2c_7}|W_k|,$$

and (2) implies $\log_2(3/\delta_k) \leq \frac{2\kappa^2}{c_2c_5c_7}|W_k|$, altogether we have

$$\sum_{(x,y)\in W_k} \mathbb{1}[h_{w_k}(x) \neq h_{w^*}(x)] \leq |W_k|\mathrm{er}_{W_k}(h_{w_k}) + \frac{64ec_5\kappa^4}{c_2c_7}|W_k| + \frac{2\kappa^2}{c_2c_5c_7}|W_k|$$

$$\leq \kappa|W_k|\left(c_{12} + \frac{64ec_5\kappa^3}{c_2c_7} + \frac{2\kappa}{c_2c_5c_7}\right).$$

Letting $c_{13} = c_{12} + \frac{64ec_5}{c_2c_7} + \frac{2}{c_2c_5c_7}$, and noting $\kappa \leq 1$, we have $\sum_{(x,y)\in W_k}\mathbb{1}[h_{w_k}(x) \neq h_{w^*}(x)] \lesssim c_{13}\kappa|W_k|$.

Applying a classic ratio-type VC bound (see [17], Section 4.9.2) under the conditional distribution given $|W_k|$, combined with the law of total probability, we have that with probability at least $1 - \delta_k/3$,

$$|W_k|\mathcal{P}\left(x : h_{w_k}(x) \neq h_{w^*}(x)\big||w_{k-1} \cdot x| \leq b_{k-1}\right)$$
$$\leq \sum_{(x,y)\in W_k}\mathbb{1}[h_{w_k}(x) \neq h_{w^*}(x)] + c_{14}\sqrt{|W_k|(d\log(|W_k|/d) + \log(1/\delta_k))},$$

for a universal constant $c_{14} > 0$. Combined with the above, and the fact that (2) implies $\log(1/\delta_k) \leq \frac{\kappa^2}{c_2c_5c_7}|W_k|$ and $d\log(|W_k|/d) \leq d\log\left(\frac{8c_3c_5c_7\log\left(\frac{1}{\kappa\delta_k}\right)}{\kappa^2}\right) \leq$

$d \log(\frac{8c_3c_5c_7}{\kappa^3\delta_k}) \le 3\log(8\max\{c_3,1\}c_5)c_5 d\log(\frac{1}{\kappa\delta_k}) \le 3\log(8\max\{c_3,1\})\kappa^2 2^{-k}m_k$
$\le \frac{3\log(8\max\{c_3,1\})}{c_2c_7}\kappa^2|W_k|$, we have $|W_k|\mathcal{P}\big(x : h_{w_k}(x) \ne h_{w^*}(x)\big||w_{k-1}\cdot x| \le b_{k-1}\big)$
$\le c_{13}\kappa|W_k| + c_{14}\sqrt{|W_k|\Big(\frac{3\log(8\max\{c_3,1\})}{c_2c_7}\kappa^2|W_k| + \frac{\kappa^2}{c_2c_5c_7}|W_k|\Big)}$. Letting $c_{15} =$
$\Big(c_{13} + c_{14}\sqrt{\frac{3\log(8\max\{c_3,1\})}{c_2c_7} + \frac{1}{c_2c_5c_7}}\Big)$, this is $c_{15}\kappa|W_k|$, which implies

$$\mathcal{P}\big(x : h_{w_k}(x) \ne h_{w^*}(x)\big||w_{k-1}\cdot x| \le b_{k-1}\big) \le c_{15}\kappa. \tag{5}$$

Next, note that $\|v_k - w_{k-1}\|^2 = \|v_k\|^2 + 1 - 2\|v_k\|\cos(\pi\mathcal{P}(x : h_{w_k}(x) \ne h_{w_{k-1}}(x)))$. Thus, one implication of the fact that $\|v_k - w_{k-1}\| \le r_k$ is that $\frac{\|v_k\|}{2} + \frac{1-r_k^2}{2\|v_k\|} \le \cos(\pi\mathcal{P}(x : h_{w_k}(x) \ne h_{w_{k-1}}(x)))$; since the left hand side is positive, we have $\mathcal{P}(x : h_{w_k}(x) \ne h_{w_{k-1}}(x)) < 1/2$. Additionally, by differentiating, one can easily verify that for $\phi \in [0,\pi]$, $x \mapsto \sqrt{x^2 + 1 - 2x\cos(\phi)}$ is minimized at $x = \cos(\phi)$, in which case $\sqrt{x^2 + 1 - 2x\cos(\phi)} = \sin(\phi)$. Thus, $\|v_k - w_{k-1}\| \ge \sin(\pi\mathcal{P}(x : h_{w_k}(x) \ne h_{w_{k-1}}(x)))$. Since $\|v_k - w_{k-1}\| \le r_k$, we have $\sin(\pi\mathcal{P}(x : h_{w_k}(x) \ne h_{w_{k-1}}(x))) \le r_k$. Since $\sin(\pi x) \ge x$ for all $x \in [0, 1/2]$, combining this with the fact (proven above) that $\mathcal{P}(x : h_{w_k}(x) \ne h_{w_{k-1}}(x)) < 1/2$ implies $\mathcal{P}(x : h_{w_k}(x) \ne h_{w_{k-1}}(x)) \le r_k$.

In particular, we have that both $\mathcal{P}(x : h_{w_k}(x) \ne h_{w_{k-1}}(x)) \le r_k$ and $\mathcal{P}(x : h_{w^*}(x) \ne h_{w_{k-1}}(x)) \le 2^{-k-3} \le r_k$. Now Lemma 3 implies that, for any universal constant $c > 0$, there exists a corresponding universal constant $c' > 0$ such that $\mathcal{P}(x : h_{w_k}(x) \ne h_{w_{k-1}}(x)$ and $|w_{k-1}\cdot x| \ge c'\frac{r_k}{\sqrt{d}}) \le cr_k$ and $\mathcal{P}(x : h_{w^*}(x) \ne h_{w_{k-1}}(x)$ and $|w_{k-1}\cdot x| \ge c'\frac{r_k}{\sqrt{d}}) \le cr_k$, so that $\mathcal{P}(x : h_{w_k}(x) \ne h_{w^*}(x)$ and $|w_{k-1}\cdot x| \ge c'\frac{r_k}{\sqrt{d}}) \le \mathcal{P}(x : h_{w_k}(x) \ne h_{w_{k-1}}(x)$ and $|w_{k-1}\cdot x| \ge c'\frac{r_k}{\sqrt{d}}) + \mathcal{P}(x : h_{w^*}(x) \ne h_{w_{k-1}}(x)$ and $|w_{k-1}\cdot x| \ge c'\frac{r_k}{\sqrt{d}}) \le 2cr_k$. In particular, letting $c_7 = c'c_{10}/2$, we have $c'\frac{r_k}{\sqrt{d}} = b_{k-1}$. Combining this with (5), Lemma 4, and a union bound, we that have that $\mathcal{P}(x : h_{w_k}(x) \ne h_{w^*}(x)) \le \mathcal{P}(x : h_{w_k}(x) \ne h_{w^*}(x)$ and $|w_{k-1}\cdot x| \ge b_{k-1}) + \mathcal{P}(x : h_{w_k}(x) \ne h_{w^*}(x)$ and $|w_{k-1}\cdot x| \le b_{k-1}) \le 2cr_k + \mathcal{P}(x : h_{w_k}(x) \ne h_{w^*}(x)||w_{k-1}\cdot x| \le b_{k-1})\mathcal{P}(x : |w_{k-1}\cdot x| \le b_{k-1}) \le 2cr_k + c_{15}\kappa c_3 b_{k-1}\sqrt{d} = (2^5 cc_{10} + c_{15}\kappa c_3 c_7 2^5)2^{-k-4}$. Taking $c = \frac{1}{2^6 c_{10}}$ and $\kappa = \frac{1}{2^6 c_3 c_7 c_{15}}$, we have $\mathcal{P}(x : h_{w_k}(x) \ne h_{w^*}(x)) \le 2^{-k-4}$, as required.

By a union bound, this occurs with probability at least $1 - (4/3)\delta_k$. □

Proof (Proof of Theorem 2). We begin with the bound on the error rate. If $\Delta > \frac{\pi^2}{400\cdot 2^{27}(d+\ln(4/\delta))}$, the result trivially holds, since then $1 \le \frac{400\cdot 2^{27}}{\pi^2}\sqrt{\Delta(d+\ln(4/\delta))}$. Otherwise, suppose $\Delta \le \frac{\pi^2}{400\cdot 2^{27}(d+\ln(4/\delta))}$. Fix any $i \in \mathbb{N}$. Lemma 1 implies that, with probability at least $1 - \delta/4$, the w_0 returned in Step 0 of ABL$(M(i-1), \tilde{h}_{i-1})$ satisfies $\mathcal{P}(x : h_{w_0}(x) \ne h^*_{M(i-1)+m_0+1}(x)) \le 1/16$. Taking this as a base case, Lemma 6 then inductively implies that, with probability at least $1 - \frac{\delta}{4} - \sum_{k=1}^{\lceil\log_2(1/\alpha)\rceil}(4/3)\frac{\delta}{2(\lceil\log_2(4/\alpha)\rceil - k)^2} \ge 1 - \delta, \forall k \in \{0, 1, \ldots, \lceil\log_2(1/\alpha)\rceil\}$,

$$\mathcal{P}(x : h_{w_k}(x) \ne h^*_{M(i-1)+m_0+1}(x)) \le 2^{-k-4}, \tag{6}$$

and furthermore the number of labels requested during $ABL(M(i-1), \tilde{h}_{i-1})$ total to at most (for appropriate universal constants $\hat{c}_1, \hat{c}_2$) $m_0 + \sum_{k=1}^{\lceil \log_2(1/\alpha) \rceil} |W_k| \leq$
$\hat{c}_1 \left(d + \ln\left(\frac{1}{\delta}\right) + \sum_{k=1}^{\lceil \log_2(1/\alpha) \rceil} d \log\left(\frac{(\lceil \log_2(4/\alpha) \rceil - k)^2}{\delta} \right) \right) \leq \hat{c}_2 d \log\left(\frac{1}{\Delta d}\right) \log\left(\frac{1}{\delta}\right)$. In particular, by a union bound, (6) implies that $\forall k \in \{1, \ldots, \lceil \log_2(1/\alpha) \rceil\}$, $\forall m \in \left\{ M(i-1) + \sum_{j=0}^{k-1} m_j + 1, \ldots, M(i-1) + \sum_{j=0}^{k} m_j \right\}$, $\mathcal{P}(x : h_{w_{k-1}}(x) \neq h_m^*(x)) \leq$
$\mathcal{P}(x : h_{w_{k-1}}(x) \neq h_{M(i-1)+m_0+1}^*(x)) + \mathcal{P}(x : h_{M(i-1)+m_0+1}^*(x) \neq h_m^*(x)) \leq 2^{-k-3} + \Delta M$.
Since $M = \sum_{k=0}^{\lceil \log_2(1/\alpha) \rceil} m_k = \Theta\left(d + \log(\frac{1}{\delta}) + \sum_{k=1}^{\lceil \log_2(1/\alpha) \rceil} 2^k d \log\left(\frac{\lceil \log_2(1/\alpha) \rceil - k}{\delta} \right) \right)$
$= \Theta\left(\frac{1}{\alpha} d \log(\frac{1}{\delta})\right) = \Theta(\sqrt{(d/\Delta)} \log(1/\delta))$, with probability at least $1 - \delta$, $\mathcal{P}(x : h_{w_{\lceil \log_2(1/\alpha) \rceil - 1}}(x) \neq h_{Mi}^*(x)) \leq O(\alpha + \Delta M) = O(\sqrt{\Delta d \log(1/\delta)})$. This implies that, with probability at least $1 - \delta$, $\forall t \in \{Mi+1, \ldots, M(i+1) - 1\}$, $\mathrm{er}_t(\hat{h}_t) \leq \mathcal{P}(x : h_{w_{\lceil \log_2(1/\alpha) \rceil - 1}}(x) \neq h_{Mi}^*(x)) + \mathcal{P}(x : h_{Mi}^*(x) \neq h_t^*(x)) \leq O(\sqrt{\Delta d \log(1/\delta)}) +$
$\Delta M = O\left(\sqrt{\Delta d \log\left(\frac{1}{\delta}\right)}\right)$, which completes the proof of the error rate bound.

Setting $\delta = \sqrt{\Delta d}$, and noting that $\mathbb{1}[\hat{Y}_t \neq Y_t] \leq 1$, we have that for any $t > M$, $\mathbb{P}\left(\hat{Y}_t \neq Y_t\right) = \mathbb{E}\left[\mathrm{er}_t(\hat{h}_t)\right] \leq O\left(\sqrt{\Delta d \log(\frac{1}{\delta})}\right) + \delta = O\left(\sqrt{\Delta d \log(\frac{1}{\Delta d})}\right)$. The bound on $\mathbb{E}\left[\sum_{t=1}^{T} \mathbb{1}[\hat{Y}_t \neq Y_t]\right]$ follows by linearity of the expectation. Furthermore, as mentioned, with probability at least $1 - \delta$, an execution of $ABL(M(i-1), \tilde{h}_{i-1})$ requests at most $O\left(d \log\left(\frac{1}{\Delta d}\right) \log\left(\frac{1}{\delta}\right)\right)$ labels. Thus, since the number of queries during the execution of $ABL(M(i-1), \tilde{h}_{i-1})$ cannot exceed M, letting $\delta = \sqrt{\Delta d}$, the expected number of queries during an execution is at most $O\left(d \log^2\left(\frac{1}{\Delta d}\right)\right) + \sqrt{\Delta d} M \leq O\left(d \log^2\left(\frac{1}{\Delta d}\right)\right)$. The bound on the expected number of queries among T samples follows by linearity of the expectation. $\quad\square$

Remark: The original work of [8] additionally allowed some number K of *jumps*: times t with $\Delta_t = 1$. In the above algorithm, since the influence of each sample is localized to the predictors trained within that batch of M instances, the effect of allowing such jumps would only change the bound on the number of mistakes to $\tilde{O}(\sqrt{d\Delta}T + \sqrt{d/\Delta}K)$. This compares favorably to the result of [8], which is roughly $O((d\Delta)^{1/4}T + \frac{d^{1/4}}{\Delta^{3/4}}K)$. However, that result was proven for a more general setting, allowing certain nonuniform distributions $\mathcal{P}$ (though they do require a relation between the angle between separators and the probability mass they disagree on, similar to that holding for the uniform distribution). It is not clear whether Theorem 2 generalizes to this larger family of distributions.

6   General Results for Active Learning

As mentioned, the above results on linear separators also provide results for the number of queries in *active learning*. One can also state quite general results on the expected number of queries and mistakes achievable by an active learning algorithm. This section provides such results, for an algorithm based on the the well-known strategy of *disagreement-based* active learning. Throughout this section, we suppose $\mathbf{h}^* \in S_\Delta$, for a given $\Delta \in (0, 1]$.

First, a few definitions. For any set $\mathcal{H} \subseteq \mathbb{C}$, define the *region of disagreement*:

$$\text{DIS}(\mathcal{H}) = \{x \in \mathcal{X} : \exists h, g \in \mathcal{H} \text{ s.t. } h(x) \neq g(x)\}.$$

This section focuses on the following algorithm. The Active subroutine is from the work of [12] (slightly modified here), and is a variant of the A^2 (Agnostic Acive) algorithm of [2]; the values of M and $\hat{T}_k(\cdot)$ are discussed below.

Algorithm: DriftingActive
0. For $i = 1, 2, \ldots$
1.   Active($M(i-1)$)

Subroutine: Active(t)
0. Let $\hat{h}_0$ be an arbitrary element of $\mathbb{C}$, and let $V_0 \leftarrow \mathbb{C}$
1. Predict $\hat{Y}_{t+1} = \hat{h}_0(X_{t+1})$ as the prediction for the value of Y_{t+1}
2. For $k = 0, 1, \ldots, \log_2(M/2)$
3. $Q_k \leftarrow \{\}$
4.   For $s = 2^k + 1, \ldots, 2^{k+1}$
5.   Predict $\hat{Y}_s = \hat{h}_k(X_s)$ as the prediction for the value of Y_s
6.   If $X_s \in \text{DIS}(V_k)$
7.   Request the label Y_s and let $Q_k \leftarrow Q_k \cup \{(X_s, Y_s)\}$
8.   Let $\hat{h}_{k+1} = \text{argmin}_{h \in V_k} \sum_{(x,y) \in Q_k} \mathbb{1}[h(x) \neq y]$
9.   Let $V_{k+1} \leftarrow \{h \in V_k : \sum_{(x,y) \in Q_k} \mathbb{1}[h(x) \neq y] - \mathbb{1}[\hat{h}_{k+1}(x) \neq y] \leq \hat{T}_k\}$

As in the DriftingHalfspaces algorithm above, this DriftingActive algorithm proceeds in batches, and in each batch runs an active learning algorithm designed to be robust to classification noise. This robustness to classification noise translates into our setting as tolerance for the fact that there is no classifier in $\mathbb{C}$ that perfectly classifies all of the data. The specific algorithm employed here maintains a set $V_k \subseteq \mathbb{C}$ of candidate classifiers, and requests the labels of samples X_s for which there is some disagreement on the classification among classifiers in V_k. We maintain the invariant that there is a low-error classifier contained in V_k at all times, and thus the points we query provide some information to help us determine which among these remaining candidates has low error rate. Based on these queries, we periodically (in Step 9) remove from V_k those classifiers making a relatively excessive number of mistakes on the queried samples, relative to the minimum among classifiers in V_k. Predictions are made with an element of V_k.

We establish an abstract bound on the number of labels requested by this algorithm, expressed in terms of the *disagreement coefficient* [11]. Specifically, for any $r \geq 0$ and any classifier h, define $B(h, r) = \{g \in \mathbb{C} : \mathcal{P}(x : g(x) \neq h(x)) \leq r\}$. Then for $r_0 \geq 0$ and any classifier h, define the disagreement coefficient of h with respect to $\mathbb{C}$ under $\mathcal{P}$: $\theta_h(r_0) = \sup_{r > r_0} \frac{\mathcal{P}(\text{DIS}(B(h,r)))}{r}$. Usually, the disagreement coefficient would be used with h equal the target concept; however, since the target concept is not fixed in our setting, we will use the worst-case value: $\theta_{\mathbb{C}}(r_0) = \sup_{h \in \mathbb{C}} \theta_h(r_0)$. This quantity has been bounded for a variety of $\mathbb{C}$ and $\mathcal{P}$ (see e.g., [4,10,11]). It is useful in bounding how quickly the region

$\mathrm{DIS}(V_k)$ collapses in the algorithm. Thus, since the probability the algorithm requests the label of the next instance is $\mathcal{P}(\mathrm{DIS}(V_k))$, the value $\theta_{\mathbb{C}}(r_0)$ naturally arises in bounding the number of labels the algorithm requests. Specifically, we have the following result. For space, the proof is deferred to the full version [13].

Theorem 3. *For an appropriate universal constant* $c_1 \in [1, \infty)$, *if* $\mathbf{h}^* \in S_\Delta$ *for a* $\Delta \in (0, 1]$, *then with*[3] $M = \left\lceil c_1 \sqrt{\frac{d}{\Delta}} \right\rceil_2$ *and* $\hat{T}_k = \log_2(1/\sqrt{d\Delta}) + 2^{2k+2}e\Delta$, *defining* $\epsilon_\Delta = \sqrt{d\Delta}\mathrm{Log}(1/(d\Delta))$, *among the first* T *instances, the expected number of mistakes by* DriftingActive *is* $O\left(\epsilon_\Delta \mathrm{Log}(d/\Delta)T\right) = \tilde{O}\left(\sqrt{d\Delta}\right)T$, *and the expected number of label requests is* $O\left(\theta_{\mathbb{C}}(\epsilon_\Delta)\epsilon_\Delta\mathrm{Log}(d/\Delta)T\right) = \tilde{O}\left(\theta_{\mathbb{C}}(\sqrt{d\Delta})\sqrt{d\Delta}\right)T$.

References

1. Awasthi, P., Balcan, M.F., Long, P.M.: The power of localization for efficiently learning linear separators with noise. arXiv:1307.8371v2 (2013)
2. Balcan, M.F., Beygelzimer, A., Langford, J.: Agnostic active learning. In: Proceedings of the 23rd International Conference on Machine Learning (2006)
3. Balcan, M.-F., Broder, A., Zhang, T.: Margin based active learning. In: Bshouty, N.H., Gentile, C. (eds.) COLT. LNCS (LNAI), vol. 4539, pp. 35–50. Springer, Heidelberg (2007)
4. Balcan, M.F., Long, P.M.: Active and passive learning of linear separators under log-concave distributions. In: Proceedings of the 26th Conference on Learning Theory (2013)
5. Bartlett, P.L., Ben-David, S., Kulkarni, S.R.: Learning changing concepts by exploiting the structure of change. Machine Learning **41**, 153–174 (2000)
6. Bartlett, P.L., Helmbold, D.P.: Learning changing problems (1996) (unpublished)
7. Barve, R.D., Long, P.M.: On the complexity of learning from drifting distributions. Information and Computation **138**(2), 170–193 (1997)
8. Crammer, K., Mansour, Y., Even-Dar, E., Vaughan, J.W.: Regret minimization with concept drift. In: Proceedings of the 23rd Conference on Learning Theory, pp. 168–180 (2010)
9. Dasgupta, S., Kalai, A., Monteleoni, C.: Analysis of perceptron-based active learning. Journal of Machine Learning Research **10**, 281–299 (2009)
10. El-Yaniv, R., Wiener, Y.: Active learning via perfect selective classification. Journal of Machine Learning Research **13**, 255–279 (2012)
11. Hanneke, S.: A bound on the label complexity of agnostic active learning. In: Proceedings of the 24th International Conference on Machine Learning (2007)
12. Hanneke, S.: Activized learning: Transforming passive to active with improved label complexity. Journal of Machine Learning Research **13**(5), 1469–1587 (2012)
13. Hanneke, S., Kanade, V., Yang, L.: Learning with a drifting target concept. arXiv:1505.05215 (2015)
14. Haussler, D., Littlestone, N., Warmuth, M.: Predicting $\{0, 1\}$-functions on randomly drawn points. Information and Computation **115**, 248–292 (1994)

---

[3] Here, we define $\lceil x \rceil_2 = 2^{\lceil \log_2(x) \rceil}$, for $x \geq 1$.

15. Helmbold, D.P., Long, P.M.: Tracking drifting concepts by minimizing disagreements. Machine Learning **14**(1), 27–45 (1994)
16. Long, P.M.: The complexity of learning according to two models of a drifting environment. Machine Learning **37**(3), 337–354 (1999)
17. Vapnik, V.: Statistical Learning Theory. John Wiley & Sons Inc., New York (1998)
18. Vapnik, V., Chervonenkis, A.: On the uniform convergence of relative frequencies of events to their probabilities. Theory of Probability and its Applications **16**, 264–280 (1971)

Interactive Clustering of Linear Classes and Cryptographic Lower Bounds

Ádám D. Lelkes[(✉)] and Lev Reyzin

Department of Mathematics, Statistics, and Computer Science,
University of Illinois at Chicago, Chicago, IL 60607, USA
{alelke2,lreyzin}@uic.edu

Abstract. We study an interactive model of supervised clustering introduced by Balcan and Blum [7], where the clustering algorithm has query access to a teacher. We give an efficient algorithm clustering linear functionals over finite fields, which implies the learnability of parity functions in this model. We also present an efficient clustering algorithm for hyperplanes which are a natural generalization of the problem of clustering linear functionals over $\mathbb{R}^d$. We also give cryptographic hardness results for interactive clustering. In particular, we show that, under plausible cryptographic assumptions, the interactive clustering problem is intractable for the concept classes of polynomial-size constant-depth threshold circuits, Boolean formulas, and finite automata.

Keywords: Interactive clustering · Query learning · Parity function · Cryptographic lower bounds

1 Introduction

In this paper we consider the interactive clustering model proposed by Balcan and Blum [7]. This clustering (and learning) model allows the algorithm to issue proposed explicit clusterings to an oracle, which replies by requesting two of the proposed clusters "merge" or that an impure cluster be "split". This model captures an interactive learning scenario, where one party has a target clustering in mind and communicates this information via these simple requests.

Balcan and Blum [7] give the example of a human helping a computer cluster news articles by topic by indicating to the computer which proposed different clusters are really about the same topic and which need to be split. Another motivating example is computer-aided image segmentation, where an algorithm can propose image segmentations to a human, who can show the computer which clusters need to be "fixed up" – this is likely to be much simpler than having the human segment the image manually.

Many interesting results are already known for this model [5,7], including the learnability of various concept classes and some generic, though inefficient, algorithms (for an overview, see Sect. 3).

In this paper we extend the theory of interactive clustering. Among our main results:

© Springer International Publishing Switzerland 2015
K. Chaudhuri et al. (Eds.): ALT 2015, LNAI 9355, pp. 165–176, 2015.
DOI: 10.1007/978-3-319-24486-0_11

- We show efficient algorithms for clustering parities and, more generally, linear functionals over finite fields – parities are a concept class of central importance in most models of learning. (Section 4)
- We also give an efficient algorithm for clustering hyperplanes, a generalization of linear functionals over $\mathbb{R}^d$. These capture a large and important set of concept classes whose efficient clusterability was not known in this model. (Section 5)
- We prove lower bounds for the interactive clustering model under plausible cryptographic assumptions, further illustrating the richness of this model. (Section 6)

2 The Model

In this section we describe the interactive clustering model of Balcan and Blum [7]. In this model of clustering, no distributional assumptions are made about the data; instead, it is assumed that the teacher knows the target clustering, but it is infeasible for him to label each data point by hand. Thus the goal of the learner is to learn the target clustering by making a small number of queries to the teacher. In this respect, the model is similar to the foundational query learning models introduced by Angluin [2]. (As a consequence, the classes we consider in this paper might be more familiar from the theory of query learning than from the usual models of clustering.)

More specifically, the learner is given a sample S of m points, and knows the number of target clusters which is denoted as k. The target clustering is an element of a concept class C. In each round, the learner presents a hypothesis clustering to the teacher. The answer of the teacher to this query is one of the following: either that the hypothesis clustering is correct, or a split or merge request. If this hypothesis is incorrect, that means that at least one of the following two cases has to hold: either there are *impure* hypothesis clusters, i.e. hypothesis clusters which contain points from more than one target cluster, or there are more than one distinct hypothesis clusters that are subsets of the same cluster. In the first case, the teacher can issue a split request to an impure cluster, in the second case the teacher can issue a merge request to two clusters that are both subsets of the same target cluster. If there are several valid possibilities for split or merge requests, the teacher can arbitrarily choose one of them.

Definition 1. *An interactive clustering algorithm is called **efficient** if it runs in $O(\mathrm{poly}(k, m, \log|C|))$ time and makes $O(\mathrm{poly}(k, \log m, \log|C|))$ queries.*

Observe that allowing the learner to make m queries would make the clustering task trivial: by starting from the all singleton hypothesis clustering and merging clusters according to the teacher's requests, the target clustering can be found in at most m rounds.

3   Previous Work

Extensive research on clustering has yielded a plethora of important theoretical results, including traditional hardness results [17,18], approximation algorithms [3,4,9,13,16,20], and generative models [12,14]. More recently researchers have examined properties of data that imply various notions of "clusterability" [1]. An ongoing research direction has been to find models that capture real-world behavior and success of clustering algorithms, in which many foundational open problems remain [11].

Inspired by models where clusterings satisfy certain natural relations with the data, e.g. [6], Balcan and Blum [7] introduced the notion of interactive clustering we consider in this paper – the data assumption here, of course, is that a "teacher" has a clustering in mind that the data satisfies, while the algorithm is aware of the space of possible clusterings.

In addition to defining the interactive clustering model, Balcan and Blum [7] gave some of the first results for it. In particular, they showed how to efficiently cluster intervals, disjunctions, and conjunctions (the latter only for constant k). Moreover, they gave a general, but inefficient, version space algorithm for clustering any finite concept class using $O(k^3 \log |C|)$ queries. They also gave a lower bound that showed efficient clustering was not possible if if the algorithm is required to be proper, i.e. produce k-clusterings to the teacher. These results contrast with our cryptographic lower bounds, which hold for arbitrary hypothesis clusterings.

Awasthi and Zadeh [5] later improved the generic bound of $O(k^3 \log |C|)$ to $O(k \log |C|)$ queries using a simpler version space algorithm. They presented an algorithm for clustering axis-aligned rectangles.

Awasthi and Zadeh [5] also introduced a noisy variant of this model. In the noisy version, split requests are still only issued for impure clusters, but merge requests might have "noise": a merge request might be issued if at least an η fraction of the points from both hypothesis clusters belong to the same target cluster. Alternatively, a stricter version of the noisy model allows arbitrary noise: the teacher might issue a merge request for two clusters even if they both have only one point from some target cluster. Awasthi and Zadeh [5] gave an example of a concept class that cannot be learned with arbitrary noise, and presented an algorithm for clustering intervals in the η noise model. To the best of our knowledge, our algorithm for clustering linear functionals over finite fields, presented in Sect. 4, is the first algorithm for clustering a nontrivial concept class under arbitrary noise.

Other interactive models of clustering have, of course, also been considered [10,15]. In this paper, however, we keep our analysis to the Balcan and Blum [7] interactive model.

4   Clustering Linear Functionals

In this section we present an algorithm for clustering linear functionals over finite fields. That is, the instance space is $X = GF(q)^n$ for some prime power

q and positive integer n, where $GF(q)$ denotes the finite field of order q. The concept class is the dual space $(GF(q)^n)^*$ of linear operations mapping from $GF(q)^n$ to $GF(q)$. Thus the number of clusters is $k = q$. Recall that every linear functional in $(GF(q)^n)^*$ is of the form $v \mapsto x \cdot v$, thus clustering linear functionals is equivalent to learning this unknown vector x. For the special case of $q = 2$, we get the concept class of parity functions over $\{0,1\}^n$, where there are two classes/clusters (for the positively and negatively labeled points).

The idea of the algorithm is the following: in each round we output the largest sets of elements that are already known to be pure, thus forcing the teacher to make a **merge** request. A **merge** request for two clusters will yield a linear equation for the target vector which is independent from all previously learned equations. We use a graph on the data points to keep track of the learned linear equations. Since the algorithm learns a independent equation in each round, it finds the target vector in at most n rounds. The description of the algorithm follows.

Algorithm 1. Cluster-Functional

---

initialize $G = (V, \emptyset)$, with $|V| = m$, each vertex corresponding an element from the sample.
initialize $Q = \emptyset$.
repeat
    find the connected components of G and output them as clusters.
    on a **merge** request to two clusters:
    for each pair a, b of points in the union **do**
        if $(a - b) \cdot x = 0$ is independent from all equations in Q **then**
            add $(a - b) \cdot x = 0$ to Q.
        end if
    end for
    for each non-edge (a, b), add (a, b) to G if $(a - b) \cdot x = 0$ follows from the equations in Q.
until the target clustering is found

Theorem 1. *Algorithm 1 finds the target clustering using at most n queries and $O(m^2 n^4)$ time. Moreover, the query complexity of the algorithm is optimal: every clustering algorithm needs to make at least n queries to find the target clustering.*

Proof. We claim that in each round we learn a linear equation that is independent from all previously learned equations, thus in n rounds we learn the target parity.

Assume for contradiction that there is a round where no independent equations are added. All hypothesis clusters are pure by construction so they can never be split. If two clusters are merged, then let us pick an element a from one of them and b from the other. Then $(a - b) \cdot x = 0$ has to be independent from Q

since otherwise the edge (a, b) would have been added in a previous round and the two elements would thus belong to the same cluster.

Thus after at most n rounds G will consist of two marked cliques which will give the correct clustering. Finding the connected components and outputting the hypothesis clusters takes linear time. To update the graph, $O(m^2)$ Gaussian elimination steps are needed. Hence the total running time is $O(m^2 n^4)$.

To show that at least n queries are necessary, notice that merge and split requests are equivalent to linear equations and inequalities, respectively. Since the dimension of the dual space is n, after less than n queries there are at least two linearly independent linear functionals, and therefore at least two different clusterings, that are consistent with all the queries. □

Observe that for $q > 2$ this is in fact an efficient implementation of the generic halving algorithm of Awasthi and Zadeh [5]. Every subset of element is either known to be pure, in which case it is consistent with the entire version space, or is possible impure, in which case a split request would imply that the target vector satisfies a disjunction of linear equations. Thus in the latter case the set is consistent with at most a $\frac{1}{q} < \frac{1}{2}$ fraction of the version space.

There are two other notable properties of the algorithm. One is that it works without modification in the noisy setting of Awasthi and Zadeh [5]: if any pair of elements from two pure sets belong to the same target cluster, then it follows immediately by linearity that both sets are subsets of this target cluster.

The other notable property is that the algorithm never outputs impure hypothesis clusters. This is because it is always the case that every subset of the sample is either known to be pure, or otherwise it is consistent with at most half of the version space. Any concept class that has a similar gap property can be clustered using only pure clusters in the hypotheses. The following remark formalizes this idea.

Remark 1. Consider the following generic algorithm: in each round, output the maximal subsets of S that are known to be pure, i.e. are consistent with the entire version space. The teacher cannot issue a split request since every hypothesis cluster is pure. If there is an $\varepsilon > 0$ such that in each round every subset $h \subseteq S$ of the sample is consistent with either the entire version space or at most a $(1 - \varepsilon)$ fraction of the version space, then on a merge request, by the maximality of the hypothesis clusters, we can eliminate an ε fraction of the version space. Therefore this algorithm finds the target clustering after $k \log_{\frac{1}{1-\varepsilon}} |C|$ queries using only pure clusters in the hypotheses.

5   Efficient Clustering of Hyperplanes

Now we turn to a natural generalization of linear functions over $\mathbb{R}^d$, k hyperplanes. Clustering geometric concept classes was one of the proposed open problems by Awasthi and Zadeh [5]; hyperplanes are an example of a very natural geometric concept class. The data are points in $\mathbb{R}^d$ and they are clustered $(d-1)$-dimensional affine subspaces. Every point is assumed to lie on exactly one of k hyperplanes.

First, observe that this is a nontrivial interactive clustering problem: even for $d = 2$ the cardinality of the concept class can be exponentially large as a function of k. For example, let k be an odd integer, and consider $m - 2(k-1)$ points on a line and $2(k-1)$ points arranged as vertices of n squares such that no two edges are on the same line. Then it is easy to see that the number of different possible clusterings is at least 3^k. Hence, if $k = \omega(\text{polylog}(m))$, the target clustering cannot be efficiently found by the halving algorithm of Awasthi and Zadeh [5]: since the cardinality of the initial version space is superpolynomial in m, the algorithm cannot keep track of the version space in polynomial time.

Nevertheless, the case of $d = 2$ can be solved by the following trivial algorithm: start with the all-singleton hypothesis, and on a `merge` request, merge all the points that are on the line going through the two points. This algorithm will find the target clustering after k queries. However, this idea does not even generalize to $d = 3$: the teacher might repeatedly tell the learner to merge pairs of points that define parallel lines. In this case, it is not immediately clear which pairs of lines span the planes of the target clustering, and there can be a linear number of such parallel lines.

On the other hand, in the case of $d = 3$, coplanar lines either have to be in the same target cluster, or they all have to be in different clusters. Therefore if we have $k + 1$ coplanar lines, by the pigeonhole principle we know that the plane containing them has to be one of the target planes. Moreover, since all points are clustered by the k planes, it follows by the pigeonhole principle that after $k^2 + 1$ `merge` requests for singleton pairs we will get $k + 1$ coplanar lines. This observation gives an algorithm of query complexity $O(k^3)$, although it is not immediately clear how the coplanar lines can be found efficiently.

Algorithm 2, described below, is an efficient clustering algorithm based on a similar idea which works for arbitrary dimension.

Algorithm 2. Cluster-Hyperplanes

let $H = S$.
  for $i = 1$ to $d - 1$ **do**
    for each affine subspace F of dimension i **do**
      if at least $k^i + 1$ elements of H are subsets of F **then**
        replace these elements in H by F.
      end if
    end for
  end for
  repeat
    output elements of H as hypothesis clusters.
    on a `merge` request, merge the two clusters in H.
  until the target clustering is found

Theorem 2. *Algorithm 2 finds the target clustering using at most $O(k^{d+1})$ queries and $O(d \cdot m^{d+1})$ time.*

Proof. We claim that in each iteration of the for loop, it holds for every F that every subset of $k^{i-1} + 1$ elements of H that lie on F spans F. The proof is by induction. For $i = 1$ this is clear: all pairs of points on a line span the line. Assume that the claim hold for $i - 1$. Consider $k^{i-1} + 1$ elements of H on an affine subspace F of dimension i. If they spanned an affine subspace of dimension less than i, then they would have been merged in a previous iteration. Hence they have to span F.

Now if $k^i + 1$ elements of H line on an i-dimensional affine subspace F for $i < d$, then they have to be in the same target cluster. If they were not, no hyperplane could contain more than k^{i-1} of the elements, and therefore the k target hyperplanes could cover at most k^i elements contained by F, which contradicts the assumption that all points belong to a target cluster.

Hence, at the start of the repeat loop there can be at most k^{d+1} elements in H: if there were more than $k^{d+1} + 1$ elements in H, by the pigeonhole principle there would be a target cluster containing $k^d + 1$ of them. However, this is not possible since those $k^d + 1$ elements would have been merged previously.

Therefore in the repeat loop we only need k^{d+1} queries to find the target clustering. In each iteration of the outer for loop, we have to consider every affine subspace of a certain dimension. Since every at most $(d - 1)$-dimensional subspace is defined by d points, there are at most $\binom{m}{d}$ subspaces. For each of them, we have to count the elements that are contained by them, this takes m time. Thus the total running time is $O\left(d \cdot \binom{m}{d} \cdot m\right) = O(d \cdot m^{d+1})$. □

Hence, for constant d, this is an efficient clustering algorithm.

6 Cryptographic Lower Bounds for Interactive Clustering

In this section, we show cryptographic lower bounds for interactive clustering. In particular, we prove that, under plausible cryptographic assumptions, the class of constant-depth polynomial-size threshold circuits and polynomial-size Boolean formulas are not learnable in the interactive clustering model. These lower bounds further go to show the richness of this model, which allows for both positive and negative clusterability results.

It was first observed by Valiant [24] that the existence of certain cryptographic primitives implies unlearnability results. Later, Kearns and Valiant [19] showed that, assuming the intractability of specific problems such as inverting the RSA function, some natural concept classes, for example the class of constant-depth threshold circuits, are not efficiently PAC learnable.

The hardness results for PAC learning are based on the following observation: if f is a trapdoor one-way function, and there is an efficient learning algorithm which, after seeing polynomially many labeled examples of the form $(f(x), x)$, can predict the correct label $f^{-1}(y)$ of a new unlabeled data point y, then that learning algorithm by definition breaks the one-way function f.

This observation doesn't apply to interactive clustering since here the learner doesn't have to make predictions about new examples and the teacher can give information about any of the elements in the sample. Indeed, if the learner were allowed to make a linear number of queries to the teacher, the clustering task would be computationally trivial. Instead, our proofs are based on the following counting argument: if the concept class is exponentially large in the size of the sample, then there is an immediate information-theoretic exponential lower bound on the required number of queries; therefore on average a learner would have to make an exponential number of queries to learn a randomly chosen clustering. If there exist certain pseudorandom objects, then one can construct concept classes of subexponential size such that a randomly chosen concept from the smaller class is computationally indistinguishable from a randomly chosen concept from the exponential-size class. However, on the smaller concept class the learner is only allowed to make a subexponential number of queries; consequently, this smaller class is not efficiently learnable.

First let us recall some relevant definitions.

A *one-way function* is a function $f : \{0,1\}^* \to \{0,1\}^*$ such that for every polynomial time algorithm A, every positive integer α, and every sufficiently large integer n it holds that

$$\Pr(f(A(f(x))) = f(x)) < n^{-\alpha}$$

where x is chosen uniformly at random from $\{0,1\}^n$.

A *pseudorandom function family* of hardness $h(n)$ is a sequence F_n of sets of efficiently computable functions $A_n \to B_n$ such that for every $h(n)$ time-bounded probabilistic algorithm A with oracle access to a function $A_n \to B_n$ and every $\alpha > 0$ it holds that

$$|\Pr(A^{f_n}(1^n) = 1) - \Pr(A^{r_n}(1^n) = 1)| < n^{-\alpha}$$

where f_n and r_n are chosen uniformly randomly from F_n and the set of all functions $A_n \to B_n$, respectively; and there is a probabilistic poly(n)-time algorithm that on input 1^n returns a uniformly randomly chosen element of F_n.

In this paper, we will pick $A_n = \{0,1\}^n$ and $B_n = \{0,1\}$. Sometimes it is useful to consider *keyed* pseudorandom functions, i.e. pseudorandom function families where $F_n = \{f_K : K \in \{0,1\}^n\}$ (and the sampling algorithm simply chooses a uniform $K \in \{0,1\}^n$ and returns f_K). The existence of one-way functions implies the existence of pseudorandom function families of polynomial hardness.

We will use the following information-theoretic lower bound to prove our hardness result.

Lemma 1. *For $k = 2$, every clustering algorithm has to make at least $\Omega\left(\frac{\log|C|}{\log m}\right)$ queries to find the target clustering.*

Proof. There are $\log|C|$ bits are needed to describe the clustering. To each query, the answer is `split` or `merge` and the identifier of at most two clusters. Since there are at most m clusters in any hypothesis, this means that the teacher gives at most $2 \log m + 1$ bits of information per query. Thus the required number of queries is $\Omega \left(\frac{\log |C|}{\log m} \right)$. $\qquad\qquad\qquad\qquad\qquad\qquad\qquad\qquad\qquad\qquad\qquad\Box$

We remark that Theorem 9 of Balcan and Blum [7] implies a worst-case lower bound of $\Omega(\log |C|)$. However, this weaker bound of $\Omega \left(\frac{\log |C|}{\log m} \right)$ holds for teachers that are not necessarily adversarial.

As we noted above, the existence of pseudorandom function families that can fool any polynomial time-bounded distinguishers is implied by the existence of one-way functions. Unfortunately, this hardness does not seem enough to imply a lower bound for interactive clustering for the following reason. If we take a sample of size m from $\{0,1\}^n$, then if $m = O(\text{poly}(n))$, the learner is allowed to make m queries which makes the clustering problem trivial. On the other hand, if m is superpolynomial in n, the learner is allowed to take superpolynomial time, therefore it might break pseudorandom functions that can only fool polynomial-time adversaries.

However, if there exist pseudorandom functions that can fool distinguishers that have slightly superpolynomial time, a hardness result for interactive clustering follows. Candidates for pseudorandom functions or permutations used in cryptographic practice are usually conjectured to have this property.

Theorem 3. *If there exist strongly pseudorandom permutations that can fool distinguishers which have $n^{\omega(1)}$ time, then there exists a concept class C which is not learnable in the interactive clustering model with* $\text{poly}(\log m, \log |C|)$ *queries and* $\text{poly}(m, \log C)$ *time.*

Proof. Let $f_K : \{0,1\}^n \to \{0,1\}$ be a keyed pseudorandom function that can fool distinguishers which have $t(n)$ time for some time-constructible $t(n) = n^{\omega(1)}$. Without loss of generality, assume that $t(n) = o(2^n)$. Let us fix a time-constructible function $m(n)$ such that $m(n) = n^{\omega(1)}$ and $\text{poly}(m(n)) = o(t(n))$. Let S be a subset of $\{0,1\}^n$ of cardinality $m = m(n)$ and let $k = 2$. Let U_n be a set of all functions $\{0,1\}^n \to \{0,1\}$, $F_n = \{f_K : K \in \{0,1\}^n\}$.

Let us assume for contradiction that there is an efficient interactive clustering algorithm A for the concept class $C = F_n$. Since $|C| = 2^n$, this learner has to make at most $\text{poly}(n, \log m(n)) = \text{poly}(n)$ queries and has $\text{poly}(n, m(n)) = \text{poly}(m(n))$ time. Let us assume that the learner finds the target clustering after $O(n^\alpha)$ queries.

Let B be the following algorithm: given oracle access to a function $f : \{0,1\}^n \to \{0,1\}$, pick a sample S of size $m = m(n)$ from $\{0,1\}^n$, label the sample vectors according to the value of f, and simulate the learner A for at most $n^{\alpha+1}$ queries. Accept if the learner finds the target clustering and reject otherwise.

Since $\text{poly}(m(n)) = o(t(n))$, B runs in time $t(n)$. If f is chosen from F_n, B will accept with probability 1. On the other hand, if f is chosen from U_n, then since $|U_n| = 2^{2^n}$, by Lemma 1, we have a query lower bound of $\frac{\log |U_n|}{\log m} = \frac{2^n}{\log m(n)} = \omega(n^{\alpha+1})$. Therefore after $n^{\alpha+1}$ queries there are at least two different clusterings in the version space, therefore B will reject with probability at least $\frac{1}{2}$. This contradicts the $t(n)$-hardness of f_K. □

Naor and Reingold [22] constructed pseudorandom functions with one-bit output that are not only as secure as factoring Blum integers, but also computable by TC^0 circuits. Since $\log|TC^0| = \mathrm{poly}(n)$, this, together with Theorem 3, implies the following corollary:

Corollary 1. *If factoring Blum integers is hard for $h(n)$-time bounded algorithms for some $h(n) = n^{\omega(1)}$ then the class TC^0 of constant-depth polynomial-size threshold circuits and the class of polynomial-size Boolean formulas are not learnable in the interactive clustering model.*

Proof. By Theorem 3, learning a pseudorandom function family of superpolynomial hardness is hard in the interactive clustering model. If factoring Blum integers is superpolynomially hard, then by the construction of Naor and Reingold [22], TC^0 contains such a pseudorandom function family. Furthermore, $\log|TC^0| = \mathrm{poly}(n)$, the learner is still only allowed to have $\mathrm{poly}(n, \log m)$ queries and $\mathrm{poly}(n, m)$ time, therefore the Theorem 3 also applies to TC^0. In fact, this holds for TC^0 circuits of size at most n^α for some constant α (determined by the size of the circuits implementing the pseudorandom function). The set of languages computable TC^0 circuits of size n^α is in turn a subset of the languages computable by Boolean formulas of size at most n^β for some other constant β. Thus our cryptographic lower bound also holds for polynomial-sized Boolean formulas. □

Remark 2. After Naor and Reingold's first construction of pseudorandom functions in TC^0, several others showed that it is possible to construct even more efficient PRFs, or PRFs based on different, possibly weaker cryptographic assumptions. For example, we refer the reader to the work of Lewko and Waters [21] for a construction under the so-called "decisional k-linear assumption" which is weaker than the assumption of Naor and Reingold [22], or to Banerjee et al. [8] for a construction based on the "learning with errors" problem, against which there is no known attack by efficient quantum algorithms.

Kearns and Valiant [19] used the results of Pitt and Warmuth [23] about prediction-preserving reductions to show that in the PAC model, their cryptographic hardness result for NC^1 circuits also implies the intractability of learning DFAs. Despite the fact the problem of interactive clustering is fundamentally different from prediction problems, we show that the ideas of Pitt and Warmuth [23] can be applied to show that DFAs are hard to learn in this model as well. We use the following theorem:

Theorem 4 (Pitt and Warmuth [23]). *Let k be a fixed positive constant. If T is a single-tape Turing machine of size at most s that runs in space at most $k \log n$ on inputs of length n, then there exist polynomials p and q such that for all positive integers n there exists a DFA M of size $q(s, n)$ such that M accepts $g(w) = 1^{|w|}0w^{p(|w|, s, n)}$ if and only if T accepts w.*

This theorem implies a hardness result for interactive clustering.

Corollary 2. *If there are $n^{\omega(1)}$-hard pseudorandom function families computable in logarithmic space, then polynomial-size DFAs are not efficiently learnable in the interactive clustering model.*

Proof. Let $f_K : \{0,1\}^n \to \{0,1\}$ be an $n^{\omega(1)}$-hard keyed pseudorandom function. If $S \subset \{0,1\}^n$ has cardinality $m(n)$ as defined in Theorem 3 and the concept class is $\{f_K : K \in \{0,1\}\}^n$, the interactive clustering task is hard.

For all $K \in \{0,1\}^n$, let T_K be a Turing machine of size at most s that runs in space $k \log n$ and, given w as an imput, computes $f_K(w)$. It is easy to see that there exists functions g, p and q defined as in Theorem 4 that work for T_K for all K. Consider the sample $S' = g(S)$ and the concept class C of DFAs of size $q(s,n)$. Since $|S'| = m(n)$ and $\log |C| = \mathrm{poly}(n)$, the hardness result of Theorem 3 holds here as well.   □

7   Conclusion

In this paper we studied a model of clustering with interactive feedback. We presented efficient clustering algorithms for linear functionals over finite fields, of which parity functions are a special case, and hyperplanes in $\mathbb{R}^d$, thereby showing that these two natural problems are learnable in the model. On the other hand, we also demonstrated that under a standard cryptographic assumptions, constant-depth polynomial-size threshold circuits, polynomial-size Boolean formulas, and polynomial-size deterministic finite automata are not learnable.

We propose the following open problems.

1. It would be interesting to see if the exponential dependence on d in the complexity of Algorithm 2 for clustering hyperplanes can be reduced.
2. Clustering half-spaces remains a natural and important open problem.

References

1. Ackerman, M., Ben-David, S.: Clusterability: A theoretical study. In: Proceedings of the Twelfth International Conference on Artificial Intelligence and Statistics, AISTATS 2009, Clearwater Beach, Florida, USA, April 16–18, pp. 1–8 (2009)
2. Angluin, D.: Queries and concept learning. Machine Learning **2**(4), 319–342 (1988)
3. Arora, S., Raghavan, P., Rao, S.: Approximation schemes for euclidean-medians and related problems. In: STOC, pp. 106–113 (1998)
4. Arya, V., Garg, N., Khandekar, R., Meyerson, A., Munagala, K., Pandit, V.: Local search heuristics for k-median and facility location problems. SIAM J. Comput. **33**(3), 544–562 (2004)
5. Awasthi, P., Zadeh, R.B.: Supervised clustering. In: Advances in Neural Information Processing Systems, pp. 91–99 (2010)
6. Balcan, M., Blum, A., Vempala, S.: A discriminative framework for clustering via similarity functions. In: Proceedings of the 40th Annual ACM Symposium on Theory of Computing, Victoria, British Columbia, Canada, May 17–20, pp. 671–680 (2008)

7. Balcan, M.-F., Blum, A.: Clustering with interactive feedback. In: Freund, Y., Györfi, L., Turán, G., Zeugmann, T. (eds.) ALT 2008. LNCS (LNAI), vol. 5254, pp. 316–328. Springer, Heidelberg (2008)

8. Banerjee, A., Peikert, C., Rosen, A.: Pseudorandom functions and lattices. In: Pointcheval, D., Johansson, T. (eds.) EUROCRYPT 2012. LNCS, vol. 7237, pp. 719–737. Springer, Heidelberg (2012)

9. Bartal, Y., Charikar, M., Raz, D.: Approximating min-sum-clustering in metric spaces. In: STOC, pp. pp. 11–20 (2001)

10. Basu, S., Banerjee, A., Mooney, R.J.: Active semi-supervision for pairwise constrained clustering. In: Proceedings of the Fourth SIAM International Conference on Data Mining, Lake Buena Vista, Florida, USA, April 22–24, pp. 333–344 (2004)

11. Ben-David, S.: Computational feasibility of clustering under clusterability assumptions. CoRR abs/1501.00437 (2015)

12. Brubaker, S.C., Vempala, S.I.: PCA and affine-invariant clustering. In: 49th Annual IEEE Symposium on Foundations of Computer Science, FOCS 2008, pp. 25–28, Philadelphia, PA, USA, pp. 551–560 (October 2008)

13. Charikar, M., Guha, S., Tardos, É., Shmoys, D.B.: A constant-factor approximation algorithm for the k-median problem. J. Comput. Syst. Sci. **65**(1), 129–149 (2002)

14. Dasgupta, A., Hopcroft, J., Kannan, R., Mitra, P.: Spectral clustering by recursive partitioning. In: Azar, Y., Erlebach, T. (eds.) ESA 2006. LNCS, vol. 4168, pp. 256–267. Springer, Heidelberg (2006)

15. Dasgupta, S., Ng, V.: Which clustering do you want? inducing your ideal clustering with minimal feedback. J. Artif. Intell. Res. (JAIR) **39**, 581–632 (2010)

16. de la Vega, W.F., Karpinski, M., Kenyon, C., Rabani, Y.: Approximation schemes for clustering problems. In: STOC, pp. 50–58 (2003)

17. Guha, S., Khuller, S.: Greedy strikes back: Improved facility location algorithms. J. Algorithms **31**(1), 228–248 (1999)

18. Jain, K., Mahdian, M., Saberi, A.: A new greedy approach for facility location problems. In: STOC, pp. 731–740. ACM (2002)

19. Kearns, M., Valiant, L.: Cryptographic limitations on learning boolean formulae and finite automata. Journal of the ACM (JACM) **41**(1), 67–95 (1994)

20. Kumar, A., Sabharwal, Y., Sen, S.: Linear-time approximation schemes for clustering problems in any dimensions. J. ACM **57**, 2 (2010)

21. Lewko, A.B., Waters, B.: Efficient pseudorandom functions from the decisional linear assumption and weaker variants. In: Proceedings of the 16th ACM Conference on Computer and Communications Security, pp. 112–120. ACM (2009)

22. Naor, M., Reingold, O.: Number-theoretic constructions of efficient pseudo-random functions. Journal of the ACM (JACM) **51**(2), 231–262 (2004)

23. Pitt, L., Warmuth, M.K.: Prediction-preserving reducibility. Journal of Computer and System Sciences **41**(3), 430–467 (1990)

24. Valiant, L.G.: A theory of the learnable. Communications of the ACM **27**(11), 1134–1142 (1984)

Statistical Learning Theory
and Sample Complexity

On the Rademacher Complexity
of Weighted Automata

Borja Balle[1]([⊠]) and Mehryar Mohri[2,3]

[1] School of Computer Science, McGill University, Montréal, Canada
bballe@cs.mcgill.ca
[2] Courant Institute of Mathematical Sciences, New York, NY, USA
[3] Google Research, New York, NY, USA

Abstract. Weighted automata (WFAs) provide a general framework for the representation of functions mapping strings to real numbers. They include as special instances deterministic finite automata (DFAs), hidden Markov models (HMMs), and predictive states representations (PSRs). In recent years, there has been a renewed interest in weighted automata in machine learning due to the development of efficient and provably correct spectral algorithms for learning weighted automata. Despite the effectiveness reported for spectral techniques in real-world problems, almost all existing statistical guarantees for spectral learning of weighted automata rely on a strong realizability assumption. In this paper, we initiate a systematic study of the learning guarantees for broad classes of weighted automata in an agnostic setting. Our results include bounds on the Rademacher complexity of three general classes of weighted automata, each described in terms of different natural quantities. Interestingly, these bounds underline the key role of different data-dependent parameters in the convergence rates.

1 Introduction

Weighted finite automata (WFAs) provide a general and highly expressive framework for representing functions mapping strings to real numbers. The properties of WFAs or their mathematical counterparts, rational power series, have been extensively studied in the past [12,17,25,30,33]. WFAs have also been used in a variety of applications, including speech recognition [31], image compression [2], natural language processing [23], model checking [3], and machine translation [19]. See also [9] for a recent survey of algorithms for learning WFAs.

The recent developments in spectral learning [4,21] have triggered a renewed interest in the use of WFAs in machine learning, with several recent successes in natural language processing [6,7] and reinforcement learning [13,20]. The interest in spectral learning algorithms for WFAs is driven by the many appealing theoretical properties of such algorithms, which include their polynomial-time complexity, the absence of local minima, statistical consistency, and finite sample bounds *à la* PAC [21]. However, the typical statistical guarantees given for the hypotheses used in spectral learning only hold in the realizable case. That is,

© Springer International Publishing Switzerland 2015
K. Chaudhuri et al. (Eds.): ALT 2015, LNAI 9355, pp. 179–193, 2015.
DOI: 10.1007/978-3-319-24486-0_12

these analyses assume that the labeled data received by the algorithm is sampled from some unknown WFA. While this assumption is a reasonable starting point for theoretical analyses, the results obtained in this setting fail to explain the good performance of spectral algorithms in many practical applications where the data is typically not generated by a WFA.

There exists of course a vast literature in statistical learning theory providing tools to analyze generalization guarantees for different hypothesis classes in classification, regression, and other learning tasks. These guarantees typically hold in an agnostic setting where the data is drawn i.i.d. from an arbitrary distribution. For spectral learning of WFAs, an algorithm-dependent agnostic generalization bound was proven in [8] using a stability argument. This seems to have been the first analysis to provide statistical guarantees for learning WFAs in an agnostic setting. However, while [8] proposed a broad family of algorithms for learning WFAs parametrized by several choices of loss functions and regularizations, their bounds hold only for one particular algorithm within this family.

In this paper, we start the systematic development of algorithm-independent generalization bounds for learning with WFAs, which apply to all the algorithms proposed in [8], as well as to others using WFAs as their hypothesis class. Our approach consists of providing upper bounds on the Rademacher complexity of general classes of WFAs. The use of Rademacher complexity to derive generalization bounds is standard [24] (see also [11] and [32]). It has been successfully used to derive statistical guarantees for classification, regression, kernel learning, ranking, and many other machine learning tasks (e.g. see [32] and references therein). A key benefit of Rademacher complexity analyses is that the resulting generalization bounds are data-dependent.

Our main results consist of upper bounds on the Rademacher complexity of three broad classes of WFAs. The main difference between these classes is the quantities used for their definition: the norm of the transition weight matrix or initial and final weight vectors of a WFA; the norm of the function computed by a WFA; and, the norm of the Hankel matrix associated to the function computed by a WFA. The formal definitions of these classes is given in Section 3. Let us point out that our analysis of the Rademacher complexity of the class of WFAs described in terms of Hankel matrices directly yields theoretical guarantees for a variety of spectral learning algorithms. We will return to this point when discussing the application of our results.

Related Work. To the best of our knowledge, this paper is the first to provide general tools for deriving learning guarantees for broad classes of WFAs. However, there exists some related work providing complexity bounds for some sub-classes of WFAs in agnostic settings. The VC-dimension of deterministic finite automata (DFAs) with n states over an alphabet of size k was shown by [22] to be in $O(kn \log n)$. For probabilistic finite automata (PFAs), it was shown by [1] that, in an agnostic setting, a sample of size $\widetilde{O}(kn^2/\varepsilon^2)$ is sufficient to learn a PFA with n states and k symbols whose log-loss error is at most ε away from the optimal one in the class. Learning bounds on the Rademacher complexity of DFAs and PFAs follow as straightforward corollaries of the general results we present in this paper.

Another recent line of work, which aims to provide guarantees for spectral learning of WFAs in the non-realizable setting, is the so-called low-rank spectral learning approach [27]. This has led to interesting upper bounds on the approximation error between minimal WFAs of different sizes [26]. See [10] for a polynomial-time algorithm for computing these approximations. This approach, however, is more limited than ours for two reasons. First, because it is algorithm-dependent. And second, because it assumes that the data is actually drawn from some (probabilistic) WFA, albeit one that is larger than any of the WFAs in the hypothesis class considered by the algorithm.

The following sections of this paper are organized as follows. Section 2 introduces the notation and technical concepts used throughout. Section 3 describes the three classes of WFAs for which we provide Rademacher complexity bounds, and gives a brief overview of our results. Our learning bounds are formally stated and proven in Sections 4, 5, and 6.

2 Preliminaries and Notation

2.1 Weighted Automata, Rational Functions, and Hankel Matrices

Let Σ be a finite alphabet of size k. Let ϵ denote the empty string and Σ^* the set of all finite strings over the alphabet Σ. The length of $u \in \Sigma^*$ is denoted by $|u|$. Given an integer $L \geq 0$, we denote by $\Sigma^{\leq L}$ the set of all strings with length at most L: $\Sigma^{\leq L} = \{x \in \Sigma^* : |x| \leq L\}$.

A WFA over the alphabet Σ with $n \geq 1$ states is a tuple $A = \langle \boldsymbol{\alpha}, \boldsymbol{\beta}, \{\mathbf{A}_a\}_{a \in \Sigma} \rangle$ where $\boldsymbol{\alpha}, \boldsymbol{\beta} \in \mathbb{R}^n$ are the initial and final weights, and $\mathbf{A}_a \in \mathbb{R}^{n \times n}$ the transition matrix whose entries give the weights of the transitions labeled with a. Every WFA A defines a function $f_A \colon \Sigma^* \to \mathbb{R}$ defined for all $x = a_1 \cdots a_t \in \Sigma^*$ by

$$f_A(x) = f_A(a_1 \cdots a_t) = \boldsymbol{\alpha}^\top \mathbf{A}_{a_1} \cdots \mathbf{A}_{a_t} \boldsymbol{\beta} = \boldsymbol{\alpha}^\top \mathbf{A}_x \boldsymbol{\beta} \ , \tag{1}$$

where $\mathbf{A}_x = \mathbf{A}_{a_1} \cdots \mathbf{A}_{a_t}$. A function $f \colon \Sigma^* \to \mathbb{R}$ is said to be *rational* if there exists a WFA A such that $f = f_A$. The rank of f is denoted by $\mathrm{rank}(f)$ and defined as the minimal number of states of a WFA A such that $f = f_A$. Note that minimal WFAs are not unique. In fact, it is not hard to see that, for any minimal WFA $A = \langle \boldsymbol{\alpha}, \boldsymbol{\beta}, \{\mathbf{A}_a\} \rangle$ with $f = f_A$ and any invertible matrix $\mathbf{Q} \in \mathbb{R}^{n \times n}$, $A^{\mathbf{Q}} = \langle \mathbf{Q}^\top \boldsymbol{\alpha}, \mathbf{Q}^{-1} \boldsymbol{\beta}, \{\mathbf{Q}^{-1} \mathbf{A}_a \mathbf{Q}\} \rangle$ is also a minimal WFA computing f. We will sometimes write $A(x)$ instead of $f_A(x)$ to emphasize the fact that we are considering a specific parametrization of f_A. Note that for the purpose of this paper we only consider weighted automata over the familiar field of real numbers with standard addition and multiplication (see [12,17,25,30,33] for more general definitions of WFAs over arbitrary semirings). Functions mapping strings to real numbers can also be viewed as non-commutative formal power series, which often helps deriving rigorous proofs in formal language theory [12,25,33]. We will not favor that point of view here, however, since we will not make use of the algebraic properties offered by that perspective.

An alternative method to represent rational functions independently of any WFA parametrization is via their *Hankel matrices*. The Hankel matrix $\mathbf{H}_f \in \mathbb{R}^{\Sigma^* \times \Sigma^*}$ of a function $f : \Sigma^* \to \mathbb{R}$ is the infinite matrix with rows and columns indexed by all strings with $\mathbf{H}_f(u, v) = f(uv)$ for all $u, v \in \Sigma^*$. By the theorem of Fliess [18] (see also [14] and [12]), $\mathbf{H}_f$ has finite rank n if and only if f is rational and there exists a WFA A with n states computing f, that is, $\text{rank}(f) = \text{rank}(\mathbf{H}_f)$.

2.2   Rademacher Complexity

Our objective is to derive learning guarantees for broad families of weighted automata or rational functions used as hypothesis sets in learning algorithms. To do so, we will derive upper bounds on the Rademacher complexity of different classes $\mathcal{F}$ of rational functions $f : \Sigma^* \to \mathbb{R}$. Thus, we first briefly introduce the definition of the Rademacher complexity of an arbitrary class of functions $\mathcal{F}$. Let D be a probability distribution over Σ^*. Suppose $S = (x_1, \ldots, x_m) \overset{\text{iid}}{\sim} D^m$ is a sample of m i.i.d. strings drawn from D. The *empirical Rademacher complexity* of $\mathcal{F}$ on S is defined as follows:

$$\widehat{\mathfrak{R}}_S(\mathcal{F}) = \mathbb{E}\left[\sup_{f \in \mathcal{F}} \frac{1}{m} \sum_{i=1}^{m} \sigma_i f(x_i)\right] ,$$

where the expectation is taken over the m independent Rademacher random variables $\sigma_i \sim \mathbf{Unif}(\{+1, -1\})$. The *Rademacher complexity* of $\mathcal{F}$ is defined as the expectation of $\widehat{\mathfrak{R}}_S(\mathcal{F})$ over the draw of a sample S of size m:

$$\mathfrak{R}_m(\mathcal{F}) = \mathop{\mathbb{E}}_{S \sim D^m}\left[\widehat{\mathfrak{R}}_S(\mathcal{F})\right] .$$

Rademacher complexity bounds can be directly used to derive data-dependent generalization bounds for a variety of learning tasks [11,24,32]. Since the derivation of these learning bounds from Rademacher complexity bounds is now standard and depends on the learning task, we will not provide them here explicitly. Instead, we will discuss multiple applications of our techniques in an extended version of this paper, which will also contain explicit generalization bounds for several set-ups relevant to practical applications.

3   Classes of Rational Functions

In this section, we introduce three different classes of rational functions described in terms of distinct quantities. These quantities, such as the number of states of a WFA representation, the norm of the rational function, or that of its Hankel matrix, control the complexity of the classes of rational functions in distinct ways and each class admits distinct benefits in the analysis of learning with WFAs.

3.1   The Class $\mathcal{A}_{n,p,r}$

We start by considering the case where each rational function is given by a fixed WFA representation. Our learning bounds would then naturally depend on the number of states and the weights of the WFA representations.

Fix an integer $n > 0$ and let $\mathcal{A}_n$ denote the set of all WFAs with n states. Note that any $A \in \mathcal{A}_n$ is identified by the $d = n(kn + 2)$ parameters required to specify its initial, final, and transition weights. Thus, we can identify $\mathcal{A}_n$ with the vector space $\mathbb{R}^d$ by suitably defining addition and scalar multiplication. In particular, given $A, A' \in \mathcal{A}_n$ and $c \in \mathbb{R}$, we define:

$$A + A' = \langle \boldsymbol{\alpha}, \boldsymbol{\beta}, \{\mathbf{A}_a\} \rangle + \langle \boldsymbol{\alpha}', \boldsymbol{\beta}', \{\mathbf{A}'_a\} \rangle = \langle \boldsymbol{\alpha} + \boldsymbol{\alpha}', \boldsymbol{\beta} + \boldsymbol{\beta}', \{\mathbf{A}_a + \mathbf{A}'_a\} \rangle$$
$$cA = c\langle \boldsymbol{\alpha}, \boldsymbol{\beta}, \{\mathbf{A}_a\} \rangle = \langle c\boldsymbol{\alpha}, c\boldsymbol{\beta}, \{c\mathbf{A}_a\} \rangle \ .$$

We can view $\mathcal{A}_n$ as a normed vector space by endowing it with any norm from the following family. Let $p, q \in [1, +\infty]$ be Hölder conjugates, i.e. $p^{-1} + q^{-1} = 1$. It is easy to check that the following defines a norm on $\mathcal{A}_n$:

$$\|A\|_{p,q} = \max \left\{ \|\boldsymbol{\alpha}\|_p, \|\boldsymbol{\beta}\|_q, \max_a \|\mathbf{A}_a\|_q \right\} \ ,$$

where $\|\mathbf{A}\|_q$ denotes the matrix norm induced by the corresponding vector norm, that is $\|\mathbf{A}\|_q = \sup_{\|\mathbf{v}\|_q = 1} \|\mathbf{A}\mathbf{v}\|_q$. Given $p \in [1, +\infty]$ and $q = 1/(1 - 1/p)$, we denote by $\mathcal{A}_{n,p,r}$ the set of all WFAs A with n states and $\|A\|_{p,q} \leq r$. Thus, $\mathcal{A}_{n,p,r}$ is the ball of radius r at the origin in the normed vector space $(\mathcal{A}_n, \| \cdot \|_{p,q})$.

3.2   The Class $\mathcal{R}_{p,r}$

Next, we consider an alternative quantity measuring the complexity of rational functions that is *independent* of any WFA representation: their norm.

Given $p \in [1, \infty]$ and $f \colon \Sigma^* \to \mathbb{R}$ we use $\|f\|_p$ to denote the p-norm of f given by

$$\|f\|_p = \left[\sum_{x \in \Sigma^*} |f(x)|^p \right]^{\frac{1}{p}} \ ,$$

which in the case $p = \infty$ amounts to $\|f\|_\infty = \sup_{x \in \Sigma^*} |f(x)|$.

Let $\mathcal{R}_p$ denote the class of rational functions with finite p-norm: $f \in \mathcal{R}_p$ if and only if f is rational and $\|f\|_p < +\infty$. Given some $r > 0$ we also define $\mathcal{R}_{p,r}$, the class of functions with p-norm bounded by r:

$$\mathcal{R}_{p,r} = \{f \colon \Sigma^* \to \mathbb{R} \mid f \text{ rational and } \|f\|_p \leq r\} \ .$$

Note that this definition is independent of the WFA used to represent f.

3.3   The Class $\mathcal{H}_{p,r}$

Here, we introduce a third class of rational functions described via their Hankel matrices, a quantity that is also independent of their WFA representations. To do so, we represent a function f using its Hankel matrix $\mathbf{H}_f$, interpret this matrix as a linear operator $\mathbf{H}_f : \mathbb{R}^{\Sigma^*} \to \mathbb{R}^{\Sigma^*}$ on the free vector space $\mathbb{R}^{\Sigma^*}$, and consider the Schatten p-norm of $\mathbf{H}_f$ as a measure of the complexity of f.

We now proceed to make this more precise. We identify a function $g \colon \Sigma^* \to \mathbb{R}$ with an infinite vector $g \in \mathbb{R}^{\Sigma^*}$. It follows from the definition of a Hankel matrix that we can interpret $\mathbf{H}_f$ as an operator given by

$$(\mathbf{H}_f g)(x) = \sum_{y \in \Sigma^*} f(xy)g(y) \ .$$

Note the similarity of the operation $g \mapsto \mathbf{H}_f g$ with a convolution between f and g. The following result of [10] shows that $\|f\|_1 < \infty$ is a sufficient condition for this operation to be defined.

Lemma 1. *Let $p \in [1, +\infty]$. Assume that $f \colon \Sigma^* \to \mathbb{R}$ satisfies the condition $\|f\|_1 < \infty$. Then, $\|g\|_p < \infty$ implies $\|\mathbf{H}_f g\|_p < \infty$.*

This shows that for $f \in \mathcal{R}_1$ the operator $\mathbf{H}_f \colon \mathcal{R}_p \to \mathcal{R}_p$ is *bounded* for every $p \in [1, +\infty]$. By the Theorem of Fliess, the matrix $\mathbf{H}_f$ has finite rank when f is rational. Thus, this implies (by considering the case $p = 2$) that the bi-infinite matrix $\mathbf{H}_f$ admits a singular value decomposition whenever $f \in \mathcal{R}_1$. In that case, it makes sense to define the Schatten–Hankel p-norm of $f \in \mathcal{R}_1$ as $\|f\|_{\mathrm{H},p} = \|(\mathfrak{s}_1, \ldots, \mathfrak{s}_n)\|_p$, where $\mathfrak{s}_i = \mathfrak{s}_i(\mathbf{H}_f)$ is the ith singular value of $\mathbf{H}_f$ and $\mathrm{rank}(\mathbf{H}_f) = n$. That is, the Schatten–Hankel p-norm of f is exactly the Schatten p-norm of $\mathbf{H}_f$.

Using this notation, we can define several classes of rational functions. For a given $p \in [1, +\infty]$, we denote by $\mathcal{H}_p$ the class of rational functions with $\|f\|_{\mathrm{H},p} < \infty$ and, for any $r > 0$, by $\mathcal{H}_{p,r}$ the class of rational functions with $\|f\|_{\mathrm{H},p} \leq r$.

3.4   Overview of Results

In addition to proving general bounds on the Rademacher complexity of the three classes just described, we will also highlight their application in some important special cases.

Here, we briefly discuss these special cases, stress different properties of the classes of WFAs to which these results apply, and mention several well-known sub-families within each class. We also briefly touch upon the problem of deciding the membership of a given WFA in any of the particular classes defined above.

- $\mathcal{A}_{n,p,r}$ in the case $r = 1$ (Corollary 1): note that for $r = 1$ and $p = 1$, $\mathcal{A}_{n,p,r}$ includes all DFAs and PFAs since for these classes of automata $\boldsymbol{\alpha}$ is either an indicator vector or a probability distribution over states, hence $\|\boldsymbol{\alpha}\|_1 = 1$; $\boldsymbol{\beta}$ has all its entries in $[0, 1]$ since it consists of accept/reject labels or stopping probabilities, hence $\|\boldsymbol{\beta}\|_\infty \leq 1$; and, for any $a \in \Sigma$ and any $i \in [1, n]$, the inequality $\sum_j |\mathbf{A}_a(i,j)| \leq 1$ holds since the transitions can reach at most one state per symbol, or represent a probability distribution over next states, hence $\|\mathbf{A}_a\|_\infty \leq 1$.

- $\mathcal{R}_{p,r}$ in the cases $p = 1$ and $p = 2$ (Corollaries 3 and 2): we note here that PFAs with stopping probabilities are contained in $\mathcal{R}_1$, while there are PFAs without stopping probabilities in $\mathcal{R}_2 \setminus \mathcal{R}_1$. In general, given a WFA, membership in $\mathcal{R}_{1,r}$ is semi-decidable [5], while membership in $\mathcal{R}_{2,r}$ can be decided in polynomial time [15].
- $\mathcal{H}_{p,r}$ in the cases $p = 1$ and $p = 2$ (Corollaries 5 and 4): as mentioned above, membership in $\mathcal{R}_1$ is sufficient to show membership in $\mathcal{H}_p$ for all $1 \leq p \leq \infty$. Assuming membership in $\mathcal{H}_\infty$, it is possible to decide membership in $\mathcal{H}_{p,r}$ in polynomial time [10].

4   Rademacher Complexity of $\mathcal{A}_{n,p,r}$

In this section, we present an upper bound on the Rademacher complexity of the class of WFAs $\mathcal{A}_{n,p,r}$. To bound $\mathfrak{R}_m(\mathcal{A}_{n,p,r})$, we will use an argument based on covering numbers. We first introduce some notation, then state our general bound and related corollaries, and finally prove the main result of this section.

Let $S = (x_1, \ldots, x_m) \in (\Sigma^*)^m$ be a sample of m strings with maximum length $L_S = \max_i |x_i|$. The expectation of this quantity over a sample of m strings drawn i.i.d. from some fixed distribution D will be denoted by $L_m = \mathbb{E}_{S \sim D^m}[L_S]$. It is interesting at this point to note that L_m appears in our bound and introduces a dependency on the distribution D which will exhibit different growth rates depending on the behavior of the tails of D. For example, it is well known that if the random variable $|x|$ for $x \sim D$ is sub-Gaussian,[1] then $L_m = O(\sqrt{\log m})$. Similarly, if the tail of D is sub-exponential, then $L_m = O(\log m)$ and if the tail is a power-law with exponent $s + 1$, $s > 0$, then $L_m = O(m^{1/s})$. Note that in the latter case the distribution of $|x|$ has finite variance if and only if $s > 1$.

Theorem 1. *The following inequality holds for every sample $S \in (\Sigma^*)^m$:*

$$\widehat{\mathfrak{R}}_S(\mathcal{A}_{n,p,r}) \leq \inf_{\eta > 0} \left(\eta + r^{L_S + 2} \sqrt{\frac{2n(kn+2) \log \left(2r + \frac{r^{L_S+2}(L_S+2)}{\eta} \right)}{m}} \right).$$

By considering the case $r = 1$ and choosing $\eta = (L_S + 2)/m$ we obtain the following corollary.

---

[1] Recall that a non-negative random variable X is sub-Gaussian if $\mathbb{P}[X > k] \leq \exp(-\Omega(k^2))$, sub-exponential if $\mathbb{P}[X > k] \leq \exp(-\Omega(k))$, and follows a power-law with exponent $(s + 1)$ if $\mathbb{P}[X > k] \leq O(1/k^{s+1})$.

Corollary 1. *For any $m \geq 1$ and $n \geq 1$ the following inequality holds:*

$$\mathfrak{R}_m(\mathcal{A}_{n,p,1}) \leq \sqrt{\frac{2n(kn+2)\log(m+2)}{m}} + \frac{L_m + 2}{m} .$$

4.1   Proof of Theorem 1

We begin the proof by recalling several well-known facts and definitions related to covering numbers (see e.g. [16]). Let $V \subset \mathbb{R}^m$ be a set of vectors and $S = (x_1, \ldots, x_m) \in (\Sigma^*)^m$ a sample of size m. Given a WFA A, we define $A(S) \in \mathbb{R}^m$ by $A(S) = (A(x_1), \ldots, A(x_m)) \in \mathbb{R}^m$. We say that V is an (ℓ_1, η)-cover for S with respect to $\mathcal{A}_{n,p,r}$ if for every $A \in \mathcal{A}_{n,p,r}$ there exists some $\mathbf{v} \in V$ such that

$$\frac{1}{m}\|\mathbf{v} - A(S)\|_1 = \frac{1}{m}\sum_{i=1}^m |\mathbf{v}_i - A(x_i)| \leq \eta .$$

The ℓ_1-covering number of S at level η with respect to $\mathcal{A}_{n,p,r}$ is defined as follows:

$$\mathcal{N}_1(\eta, \mathcal{A}_{n,p,r}, S) = \min\{|V| : V \subset \mathbb{R}^m \text{ is an } (\ell_1, \eta)\text{-cover for } S \text{ w.r.t. } \mathcal{A}_{n,p,r}?\} .$$

A typical analysis based on covering numbers would now proceed to obtain a bound on the growth of $\mathcal{N}_1(\eta, \mathcal{A}_{n,p,r}, S)$ in terms of the number of strings m in S. Our analysis requires a slightly finer approach where the size of S is characterized by m and L_S. Thus, we also define for every integer $L \geq 0$ the following covering number

$$\mathcal{N}_1(\eta, \mathcal{A}_{n,p,r}, m, L) = \max_{S \in (\Sigma^{\leq L})^m} \mathcal{N}_1(\eta, \mathcal{A}_{n,p,r}, S) .$$

The first step in the proof of Theorem 1 is to bound $\mathcal{N}_1(\eta, \mathcal{A}_{n,p,r}, m, L)$. In order to derive such a bound, we will make use of the following technical results.

Lemma 2 (Corollary 4.3 in [35]). *A ball of radius $R > 0$ in a real d-dimensional Banach space can be covered by $R^d(2 + 1/\rho)^d$ balls of radius $\rho > 0$.*

Lemma 3. *Let $A, B \in \mathcal{A}_{n,p,r}$. Then the following hold for any $x \in \Sigma^*$:*

1. $|A(x)| \leq r^{|x|+2}$,
2. $|A(x) - B(x)| \leq r^{|x|+1}(|x| + 2)\|A - B\|_{p,q}$.

Proof. The first bound follows from applying Hölder's inequality and the sub-multiplicativity of the norms in the definition of $\|A\|_{p,q}$ to (1). The second bound was proven in [8]. $\quad\square$

Combining these lemmas yields the following bound on the covering number $\mathcal{N}_1(\eta, \mathcal{A}_{n,p,r}, m, L)$.

Lemma 4.

$$\mathcal{N}_1(\eta, \mathcal{A}_{n,p,r}, m, L) \leq r^{n(kn+2)}\left(2 + \frac{r^{L+1}(L+2)}{\eta}\right)^{n(kn+2)} .$$

Proof. Let $d = n(kn+2)$. By Lemma 2 and Lemma 3, for any $\rho > 0$, there exists a finite set $\mathcal{C}_\rho \subset \mathcal{A}_{n,p,r}$ with $|\mathcal{C}_\rho| \leq r^d(2 + 1/\rho)^d$ such that: for every $A \in \mathcal{A}_{n,p,r}$ there exists $B \in \mathcal{C}_\rho$ satisfying $|A(x) - B(x)| \leq r^{|x|+1}(|x| + 2)\rho$ for every $x \in \Sigma^*$. Thus, taking $\rho = \eta/(r^{L+1}(L + 2))$ we see that for every $S \in (\Sigma^{\leq L})^m$ the set $V = \{B(S) : B \in \mathcal{C}_\rho\} \subset \mathbb{R}^m$ is an η-cover for S with respect to $\mathcal{A}_{n,p,r}$.   □

The last step of the proof relies on the following well-known result due to Massart.

Lemma 5 (Massart [28]). *Given a finite set of vectors* $V = \{\mathbf{v}_1, \ldots, \mathbf{v}_N\} \subset \mathbb{R}^m$, *the following holds*

$$\frac{1}{m}\mathbb{E}\left[\max_{\mathbf{v} \in V}\langle \boldsymbol{\sigma}, \mathbf{v}\rangle\right] \leq \left(\max_{\mathbf{v} \in V}\|\mathbf{v}\|_2\right)\frac{\sqrt{2\log(N)}}{m} ,$$

where the expectation is over the vector $\boldsymbol{\sigma} = (\sigma_1, \ldots, \sigma_m)$ *whose entries are independent Rademacher random variables* $\sigma_i \sim \mathbf{Unif}(\{+1, -1\})$.

Fix $\eta > 0$ and let $V_{S,\eta}$ be an (ℓ_1, η)-cover for S with respect to $\mathcal{A}_{n,p,r}$. By Massart's lemma, we can write

$$\widehat{\mathfrak{R}}_S(\mathcal{A}_{n,p,r}) \leq \eta + \left(\max_{\mathbf{v} \in V_{S,\eta}}\|\mathbf{v}\|_2\right)\frac{\sqrt{2\log|V_{S,\eta}|}}{m} . \tag{2}$$

Since $|A(x_i)| \leq r^{L_S+2}$ by Lemma 3, we can restrict the search for (ℓ_1, η)-covers for S to sets $V_{S,\eta} \subset \mathbb{R}^m$ where all $\mathbf{v} \in V_{S,\eta}$ must satisfy $\|\mathbf{v}\|_\infty \leq r^{L_S+2}$. By construction, such a covering satisfies $\max_{\mathbf{v} \in V_{S,\eta}}\|\mathbf{v}\|_2 \leq r^{L_S+2}\sqrt{m}$. Finally, plugging in the bound for $|V_{S,\eta}|$ given by Lemma 4 into (2) and taking the infimum over all $\eta > 0$ yields the desired result.   □

5   Rademacher Complexity of $\mathcal{R}_{p,r}$

In this section, we study the complexity of rational functions from a different perspective. Instead of analyzing their complexity in terms of the parameters of WFAs computing them, we consider an intrinsic associated quantity: their norm. We present upper bounds on the Rademacher complexity of the classes of rational functions $\mathcal{R}_{p,r}$ for any $p \in [1, +\infty]$ and $r > 0$.

It will be convenient for our analysis to identify a rational function $f \in \mathcal{R}_{p,r}$ with an infinite-dimensional vector $\mathbf{f} \in \mathbb{R}^{\Sigma^*}$ with $\|\mathbf{f}\|_p \leq r$. That is, $\mathbf{f}$ is an infinite vector indexed by strings in Σ^* whose xth entry is $\mathbf{f}_x = f(x)$. An important observation is that using this notation, for any given $x \in \Sigma^*$, we can write $f(x)$ as the inner product $\langle \mathbf{f}, \mathbf{e}_x\rangle$, where $\mathbf{e}_x \in \mathbb{R}^{\Sigma^*}$ is the indicator vector corresponding to string x.

Theorem 2. *Let* $p^{-1} + q^{-1} = 1$. *Let* $S = (x_1, \ldots, x_m)$ *be a sample of* m *strings. Then, the following holds for any* $r > 0$:

$$\widehat{\mathfrak{R}}_S(\mathcal{R}_{p,r}) = \frac{r}{m}\mathbb{E}\left[\left\|\sum_{i=1}^m \sigma_i \mathbf{e}_{x_i}\right\|_q\right] ,$$

where the expectation is over the m independent Rademacher random variables $\sigma_i \sim \textbf{Unif}(\{+1, -1\})$.

Proof. In view of the notation just introduced described, we can write

$$\widehat{\mathfrak{R}}_S(\mathcal{R}_{p,r}) = \mathbb{E}\left[\sup_{f \in \mathcal{R}_{p,r}} \frac{1}{m} \sum_{i=1}^{m} \langle \textbf{f}, \sigma_i \textbf{e}_{x_i} \rangle \right] = \frac{1}{m} \mathbb{E}\left[\sup_{f \in \mathcal{R}_{p,r}} \left\langle \textbf{f}, \sum_{i=1}^{m} \sigma_i \textbf{e}_{x_i} \right\rangle \right]$$

$$= \frac{r}{m} \mathbb{E}\left[\left\| \sum_{i=1}^{m} \sigma_i \textbf{e}_{x_i} \right\|_q \right] ,$$

where the last inequality holds by definition of the dual norm.   □

The next corollaries give non-trivial bounds on the Rademacher complexity in the case $p = 1$ and the case $p = 2$.

Corollary 2. *For any $m \geq 1$ and any $r > 0$, the following inequalities hold:*

$$\frac{r}{\sqrt{2m}} \leq \mathfrak{R}_m(\mathcal{R}_{2,r}) \leq \frac{r}{\sqrt{m}}.$$

Proof. The upper bound follows directly from Theorem 2 and Jensen's inequality:

$$\mathbb{E}\left[\left\| \sum_{i=1}^{m} \sigma_i \textbf{e}_{x_i} \right\|_2 \right] \leq \sqrt{\mathbb{E}\left[\left\| \sum_{i=1}^{m} \sigma_i \textbf{e}_{x_i} \right\|_2^2 \right]} = \sqrt{m} .$$

The lower bound is obtained using Khintchine–Kahane's inequality (see appendix of [32]):

$$\mathbb{E}\left[\left\| \sum_{i=1}^{m} \sigma_i \textbf{e}_{x_i} \right\|_2 \right]^2 \geq \frac{1}{2} \mathbb{E}\left[\left\| \sum_{i=1}^{m} \sigma_i \textbf{e}_{x_i} \right\|_2^2 \right] = \frac{m}{2},$$

which completes the proof.   □

The following definitions will be needed to present our next corollary. Given a sample $S = (x_1, \ldots, x_m)$ and a string $x \in \Sigma^*$ we denote by $s_x = |\{i : x_i = x\}|$ the number of times x appears in S. Let $M_S = \max_{s \in \Sigma^*} s_x$. Given a probability distribution D over Σ^* we also define $M_m = \mathbb{E}_{S \sim D^m}[M_S]$. Note that M_m is the expected maximum number of collisions (repeated strings) in a sample of size m drawn from D, and that we have the straightforward bounds $1 \leq M_S \leq m$.

Corollary 3. *For any $m \geq 1$ and any $r > 0$, the following upper bound holds:*

$$\mathfrak{R}_m(\mathcal{R}_{1,r}) \leq \frac{r\sqrt{2M_m \log(2m)}}{m}.$$

Proof. Let $S = (x_1, \ldots, x_m)$ be a sample with m strings. For any $x \in \Sigma^*$ define the vector $\mathbf{v}_x \in \mathbb{R}^m$ given by $\mathbf{v}_x(i) = \mathbb{I}_{x_i = x}$. Let V be the set of vectors $\mathbf{v}_x$ which are not identically zero, and note we have $|V| \leq m$. Also note that by construction we have $\max_{\mathbf{v}_x \in V} \|\mathbf{v}_x\|_2 = \sqrt{M_S}$. Now, by Theorem 2 we have

$$\widehat{\mathfrak{R}}_S(\mathcal{R}_{1,r}) = \frac{r}{m} \mathbb{E}\left[\left\|\sum_{i=1}^m \sigma_i \mathbf{e}_{x_i}\right\|_\infty\right] = \frac{r}{m} \mathbb{E}\left[\max_{\mathbf{v}_x \in V \cup (-V)} \langle \boldsymbol{\sigma}, \mathbf{v}_x \rangle\right] .$$

Therefore, using Massart's Lemma we get

$$\widehat{\mathfrak{R}}_S(\mathcal{R}_{1,r}) \leq \frac{r \sqrt{2 M_S \log(2m)}}{m} .$$

The result now follows from taking the expectation over S and using Jensen's inequality to see that $\mathbb{E}[\sqrt{M_S}] \leq \sqrt{M_m}$. □

Note in this case we cannot rely on the Khintchine–Kahane inequality to obtain lower bounds on $\mathfrak{R}_m(\mathcal{R}_{1,r})$ because there is no version of this inequality for the case $q = \infty$.

6 Rademacher Complexity of $\mathcal{H}_{p,r}$

In this section, we present our last set of upper bounds on the Rademacher complexity of WFAs. Here, we characterize the complexity of WFAs in terms of the spectral properties of their Hankel matrix.

The Hankel matrix of a function $f : \Sigma^* \to \mathbb{R}$ is the bi-infinite matrix $\mathbf{H}_f \in \mathbb{R}^{\Sigma^* \times \Sigma^*}$ whose entries are defined by $\mathbf{H}_f(u, v) = f(uv)$. Note that any string $x \in \Sigma^*$ admits $|x| + 1$ decompositions $x = uv$ into a prefix $u \in \Sigma^*$ and a suffix $v \in \Sigma^*$. Thus, $\mathbf{H}_f$ contains a high degree of redundancy: for any $x \in \Sigma^*$, $f(x)$ is the value of at least $|x| + 1$ entries of $\mathbf{H}_f$ and we can write $f(x) = \mathbf{e}_u^\top \mathbf{H}_f \mathbf{e}_v$ for any decomposition $x = uv$.

Let $\mathfrak{s}_i(\mathbf{M})$ denote the ith singular value of a matrix $\mathbf{M}$. For $1 \leq p \leq \infty$, let $\|\mathbf{M}\|_{\mathrm{S},p}$ denote the p-Schatten norm of $\mathbf{M}$ defined by $\|\mathbf{M}\|_{\mathrm{S},p} = \left[\sum_{i \geq 1} \mathfrak{s}_i(\mathbf{M})^p\right]^{\frac{1}{p}}$.

Theorem 3. *Let $p, q \geq 1$ with $p^{-1} + q^{-1} = 1$ and let $S = (x_1, \ldots, x_m)$ be a sample of m strings in Σ^*. For any decomposition $x_i = u_i v_i$ of the strings in S and any $r > 0$, the following inequality holds:*

$$\widehat{\mathfrak{R}}_S(\mathcal{H}_{p,r}) \leq \frac{r}{m} \mathbb{E}\left[\left\|\sum_{i=1}^m \sigma_i \mathbf{e}_{u_i} \mathbf{e}_{v_i}^\top\right\|_{\mathrm{S},q}\right] .$$

Proof. For any $1 \leq i \leq m$, let $x_i = u_i v_i$ be an arbitrary decomposition and let $\mathbf{R}$ denote $\mathbf{R} = \sum_{i=1}^m \sigma_i \mathbf{e}_{u_i} \mathbf{e}_{v_i}^\top$. Then, in view of the identity $f(x_i) = \mathbf{e}_{u_i}^\top \mathbf{H}_f \mathbf{e}_{v_i}$, we can write

$$\widehat{\mathfrak{R}}_S(\mathcal{H}_{p,r}) = \mathbb{E}\left[\sup_{f \in \mathcal{H}_{p,r}} \frac{1}{m} \sum_{i=1}^m \sigma_i \mathbf{e}_{u_i}^\top \mathbf{H}_f \mathbf{e}_{v_i}\right]$$

$$= \frac{1}{m} \mathbb{E}\left[\sup_{f \in \mathcal{H}_{p,r}} \sum_{i=1}^m \mathrm{Tr}\left(\sigma_i \mathbf{e}_{v_i} \mathbf{e}_{u_i}^\top \mathbf{H}_f\right)\right] = \frac{1}{m} \mathbb{E}\left[\sup_{f \in \mathcal{H}_{p,r}} \langle \mathbf{R}, \mathbf{H}_f \rangle\right] .$$

Then, by von Neumann's trace inequality [29] and Hölder's inequality, the following holds:

$$\mathbb{E}\left[\sup_{f\in\mathcal{H}_{p,r}}\langle\mathbf{R},\mathbf{H}_f\rangle\right]\leq\mathbb{E}\left[\sup_{f\in\mathcal{H}_{p,r}}\sum_{j\geq1}\mathfrak{s}_j(\mathbf{R})\cdot\mathfrak{s}_j(\mathbf{H}_f)\right]$$

$$\leq\mathbb{E}\left[\sup_{f\in\mathcal{H}_{p,r}}\|\mathbf{R}\|_{\mathrm{S},q}\|\mathbf{H}_f\|_{\mathrm{S},p}\right]=r\,\mathbb{E}\left[\|\mathbf{R}\|_{\mathrm{S},q}\right],$$

which completes the proof.   □

Note that, in this last result, the equality condition for von Neumann's inequality cannot be used to obtain a lower bound on $\widehat{\mathfrak{R}}_S(\mathcal{H}_{p,r})$ since it requires the simultaneous diagonalizability of the two matrices involved, which is difficult to control in the case of Hankel matrices.

As in the previous sections, we now proceed to derive specialized versions of the bound of Theorem 3 for the cases $p=1$ and $p=2$. First, note that the corresponding q-Schatten norms have given names: $\|\mathbf{R}\|_{\mathrm{S},2}=\|\mathbf{R}\|_F$ is the Frobenius norm, and $\|\mathbf{R}\|_{\mathrm{S},\infty}=\|\mathbf{R}\|_{\mathrm{op}}$ is the operator norm.

Corollary 4. *For any $m\geq1$ and any $r>0$, the Rademacher complexity of $\mathcal{H}_{2,r}$ can be bounded as follows:*

$$\mathfrak{R}_m(\mathcal{H}_{2,r})\leq\frac{r}{\sqrt{m}}.$$

Proof. In view of Theorem 3 and using Jensen's inequality, we can write

$$\mathfrak{R}_m(\mathcal{H}_{2,r})\leq\frac{r}{m}\mathbb{E}\left[\|\mathbf{R}\|_F\right]\leq\frac{r}{m}\sqrt{\mathbb{E}\left[\|\mathbf{R}\|_F^2\right]}$$

$$=\frac{r}{m}\sqrt{\mathbb{E}\left[\sum_{i,j=1}^{m}\sigma_i\sigma_j\langle\mathbf{e}_{u_i}\mathbf{e}_{v_i}^\top,\mathbf{e}_{u_j}\mathbf{e}_{v_j}^\top\rangle\right]}$$

$$=\frac{r}{m}\sqrt{\mathbb{E}\left[\sum_{i=1}^{m}\langle\mathbf{e}_{u_i}\mathbf{e}_{v_i}^\top,\mathbf{e}_{u_i}\mathbf{e}_{v_i}^\top\rangle\right]}=\frac{r}{\sqrt{m}},$$

which concludes the proof.   □

We now introduce a combinatorial number depending on S and the decomposition selected for each string x_i. Let $U_S=\max_{u\in\Sigma^*}|\{i\colon u_i=u\}|$ and $V_S=\max_{v\in\Sigma^*}|\{i\colon v_i=v\}|$. Then, we define $W_S=\min\max\{U_S,V_S\}$, where then minimum is taken over all possible decompositions of the strings in S. If S is sampled from a distribution D, we also define $W_m=\mathbb{E}_{S\sim D^m}[W_S]$. It is easy to show that we have the bounds $1\leq W_S\leq m$. Indeed, for the case $W_S=m$ consider a sample with m copies of the empty string, and for the case $W_S=1$ consider a sample with m different strings of length m. The following result can be stated using this definition.

Corollary 5. *There exists a universal constant $C > 0$ such that for any $m \geq 1$ and any $r > 0$, the following inequality holds:*

$$\mathfrak{R}_m(\mathcal{H}_{1,r}) \leq \frac{Cr\left(\log(m+1) + \sqrt{W_m \log(m+1)}\right)}{m} \, .$$

Proof. First, note that by Corollary 7.3.2 of [34] applied to the random matrix $\mathbf{R}$, the following inequality holds:

$$\mathbb{E}[\|\mathbf{R}\|_{\mathrm{op}}] \leq C\left(\log(m+1) + \sqrt{\mu \log(m+1)}\right) \, ,$$

where $\mu = \max\{\|\sum_i \mathbf{e}_{u_i} \mathbf{e}_{u_i}^\top\|_{\mathrm{op}}, \|\sum_i \mathbf{e}_{v_i} \mathbf{e}_{v_i}^\top\|_{\mathrm{op}}\}$ and $C > 0$ is a constant. Next, observe that $\mathbf{D} = \sum_i \mathbf{e}_{u_i} \mathbf{e}_{u_i}^\top \in \mathbb{R}^{\Sigma^* \times \Sigma^*}$ is a diagonal matrix with $\mathbf{D}(u,u) = \sum_i \mathbb{I}_{u=u_i}$. Thus, $\|\mathbf{D}\|_{\mathrm{op}} = \max_u \mathbf{D}(u,u) = \max_{u \in \Sigma^*} |\{i \colon u_i = u\}| = U_S$. Similarly, we have $\|\sum_i \mathbf{e}_{v_i} \mathbf{e}_{v_i}^\top\|_{\mathrm{op}} = V_S$. Thus, since the decomposition of the strings in S is arbitrary, we can choose it such that $\mu = W_S$. In addition, Jensen's inequality implies $\mathbb{E}_S[\sqrt{W_S}] \leq \sqrt{W_m}$. Applying Theorem 3 now yields the desired bound. ∎

7 Conclusion

We introduced three general classes of WFAs described via different natural quantities and for each, proved upper bounds on their Rademacher complexity. An interesting property of these bounds is the appearance of different combinatorial parameters tying the sample to the convergence rate, whose nature depends on the way chosen to measure the complexity of the hypotheses: the length of the longest string L_S for $\mathcal{A}_{n,p,r}$; the maximum number of collisions M_S for $\mathcal{R}_{p,r}$; and, the minimum number of prefix or suffix collisions over all possible splits W_S for $\mathcal{H}_{p,r}$.

Another important feature of our bounds for the classes $\mathcal{H}_{p,r}$ is that they depend on spectral properties of Hankel matrices, which are commonly used in spectral learning algorithms for WFAs [8,21]. We hope to exploit this connection in the future to provide more refined analyses of these learning algorithms. Our results can also be used to improve some aspects of existing spectral learning algorithms. For example, it might be possible to use the analysis in Theorem 3 for deriving strategies to help choose which prefixes and suffixes to consider in algorithms working with finite sub-blocks of an infinite Hankel matrix. This is a problem of practical relevance when working with large amounts of data which require balancing trade-offs between computation and accuracy [6].

Acknowledgments. This work was partly funded by the NSF award IIS-1117591 and NSERC.

References

1. Abe, N., Warmuth, M.K.: On the computational complexity of approximating distributions by probabilistic automata. Machine Learning (1992)
2. Albert, J., Kari, J.: Digital image compression. In: Handbook of weighted automata. Springer (2009)
3. Baier, C., Größer, M., Ciesinski, F.: Model checking linear-time properties of probabilistic systems. In: Handbook of Weighted automata. Springer (2009)
4. Bailly, R., Denis, F., Ralaivola, L.: Grammatical inference as a principal component analysis problem. In: ICML (2009)
5. Bailly, R., Denis, F.: Absolute convergence of rational series is semi-decidable. Inf. Comput. (2011)
6. Balle, B., Carreras, X., Luque, F., Quattoni, A.: Spectral learning of weighted automata: A forward-backward perspective. Machine Learning (2014)
7. Balle, B., Hamilton, W., Pineau, J.: Methods of moments for learning stochastic languages: unified presentation and empirical comparison. In: ICML (2014)
8. Balle, B., Mohri, M.: Spectral learning of general weighted automata via constrained matrix completion. In: NIPS (2012)
9. Balle, B., Mohri, M.: Learning weighted automata. In: CAI (2015)
10. Balle, B., Panangaden, P., Precup, D.: A canonical form for weighted automata and applications to approximate minimization. In: Logic in Computer Science (LICS) (2015)
11. Bartlett, P.L., Mendelson, S.: Rademacher and gaussian complexities: risk bounds and structural results. In: Helmbold, D.P., Williamson, B. (eds.) COLT 2001 and EuroCOLT 2001. LNCS (LNAI), vol. 2111, pp. 224–240. Springer, Heidelberg (2001)
12. Berstel, J., Reutenauer, C.: Noncommutative rational series with applications. Cambridge University Press (2011)
13. Boots, B., Siddiqi, S., Gordon, G.: Closing the learning-planning loop with predictive state representations. In: RSS (2009)
14. Carlyle, J.W., Paz, A.: Realizations by stochastic finite automata. J. Comput. Syst. Sci. **5**(1) (1971)
15. Cortes, C., Mohri, M., Rastogi, A.: Lp distance and equivalence of probabilistic automata. International Journal of Foundations of Computer Science (2007)
16. Devroye, L., Lugosi, G.: Combinatorial methods in density estimation. Springer (2001)
17. Eilenberg, S.: Automata, Languages and Machines, vol. A. Academic Press (1974)
18. Fliess, M.: Matrices de Hankel. Journal de Mathématiques Pures et Appliquées **53** (1974)
19. de Gispert, A., Iglesias, G., Blackwood, G., Banga, E., Byrne, W.: Hierarchical phrase-based translation with weighted finite-state transducers and shallow-n grammars. Computational Linguistics (2010)
20. Hamilton, W.L., Fard, M.M., Pineau, J.: Modelling sparse dynamical systems with compressed predictive state representations. In: ICML (2013)
21. Hsu, D., Kakade, S.M., Zhang, T.: A spectral algorithm for learning hidden Markov models. In: COLT (2009)
22. Ishigami, Y., Tani, S.: Vc-dimensions of finite automata and commutative finite automata with k letters and n states. Discrete Applied Mathematics (1997)
23. Knight, K., May, J.: Applications of weighted automata in natural language processing. In: Handbook of Weighted Automata. Springer (2009)

24. Koltchinskii, V., Panchenko, D.: Rademacher processes and bounding the risk of function learning. In: High Dimensional Probability II, pp. 443–459. Birkhäuser (2000)
25. Kuich, W., Salomaa, A.: Semirings, automata, languages. In: EATCS. Monographs on Theoretical Computer Science, vol. 5. Springer-Verlag, Berlin-New York (1986)
26. Kulesza, A., Jiang, N., Singh, S.: Low-rank spectral learning with weighted loss functions. In: AISTATS (2015)
27. Kulesza, A., Rao, N.R., Singh, S.: Low-rank spectral learning. In: AISTATS (2014)
28. Massart, P.: Some applications of concentration inequalities to statistics. In: Annales de la Faculté des Sciences de Toulouse (2000)
29. Mirsky, L.: A trace inequality of John von Neumann. Monatshefte für Mathematik (1975)
30. Mohri, M.: Weighted automata algorithms. In: Handbook of Weighted Automata. Monographs in Theoretical Computer Science, pp. 213–254. Springer (2009)
31. Mohri, M., Pereira, F.C.N., Riley, M.: Speech recognition with weighted finite-state transducers. In: Handbook on Speech Processing and Speech Comm. Springer (2008)
32. Mohri, M., Rostamizadeh, A., Talwalkar, A.: Foundations of machine learning. MIT press (2012)
33. Salomaa, A., Soittola, M.: Automata-Theoretic Aspects of Formal Power Series. Springer-Verlag, New York (1978)
34. Tropp, J.A.: An Introduction to Matrix Concentration Inequalities (2015). ArXiv abs/1501.01571
35. Vershynin, R.: Lectures in Geometrical Functional Analysis. Preprint (2009)

Multi-task and Lifelong Learning of Kernels

Anastasia Pentina[1]([✉]) and Shai Ben-David[2]

[1] Institute of Science and Technology Austria, 3400 Klosterneuburg, Austria
apentina@ist.ac.at
[2] School of Computer Science, University of Waterloo, Waterloo, ON, Canada
shai@uwaterloo.ca

Abstract. We consider a problem of learning kernels for use in SVM classification in the multi-task and lifelong scenarios and provide generalization bounds on the error of a large margin classifier. Our results show that, under mild conditions on the family of kernels used for learning, solving several related tasks simultaneously is beneficial over single task learning. In particular, as the number of observed tasks grows, assuming that in the considered family of kernels there exists one that yields low approximation error on all tasks, the overhead associated with learning such a kernel vanishes and the complexity converges to that of learning when this good kernel is given to the learner.

Keywords: Multi-task learning · Lifelong learning · Kernel learning

1 Introduction

State-of-the-art machine learning algorithms are able to solve many problems sufficiently well. However, both theoretical and experimental studies have shown that in order to achieve solutions of reasonable quality they need an access to extensive amounts of training data. In contrast, humans are known to be able to learn concepts from just a few examples. A possible explanation may lie in the fact that humans are able to reuse the knowledge they have gained from previously learned tasks for solving a new one, while traditional machine learning algorithms solve tasks in isolation. This observation motivates an alternative, transfer learning approach. It is based on idea of transferring information between related learning tasks in order to improve performance.

There are various formal frameworks for transfer learning, modeling different learning scenarios. In this work we focus on two of them: the multi-task and the lifelong settings. In the multi-task scenario, the learner faces a fixed set of learning tasks simultaneously and its goal is to perform well on all of them. In the lifelong learning setting, the learner encounters a stream of tasks and its goal is to perform well on new, yet unobserved tasks.

For any transfer learning scenario to make sense (that is, to benefit from the multiplicity of tasks), there must be some kind of relatedness between the tasks. A common way to model such task relationships is through the assumption that there exists some data representation under which learning each of the tasks

© Springer International Publishing Switzerland 2015
K. Chaudhuri et al. (Eds.): ALT 2015, LNAI 9355, pp. 194–208, 2015.
DOI: 10.1007/978-3-319-24486-0_13

is relatively easy. The corresponding transfer learning methods aim at learning such a representation.

In this work we focus on the case of large-margin learning of kernels. We consider sets of tasks and families of kernels and analyze the sample complexity of finding a kernel in a kernel family that allows low expected error on average over the set of tasks (in the multi-task scenario), or in expectation with respect to some unknown task-generating probability distribution (in the lifelong scenario). We provide generalization bounds for empirical risk minimization learners for both settings. Under the assumption that the considered kernel family has finite pseudodimension, we show that by learning several tasks simultaneously the learner is guaranteed to have low estimation error with fewer training samples per task (compared to solving them independently). In particular, if there exists a kernel with low approximation error for all tasks, then, as the number of observed tasks grows, the problem of learning any specific task with respect to a family of kernels converges to learning when the learner knows a good kernel in advance - the multiplicity of tasks relieves the overhead associated with learning a kernel. Our assumption on finite pseudodimension of the kernel family is satisfied in many practical cases, like families of Gaussian kernels with a learned covariance matrix, and linear and convex combinations of a finite set of kernels (see [4]). We also show that this is the case for families of all sparse combinations of kernels from a large "dictionary" of kernels.

1.1   Related Previous Work

Multi-task and Lifelong Learning. A method for learning a common feature representation for linear predictors in the multi-task scenario was proposed in [9]. A similar idea was also used by [10] and extended to the lifelong scenario by [11]. A natural extension of representation learning approach was proposed for kernel methods in [12,13], where the authors described a method for learning a kernel that is shared between tasks as a combination of some base kernels using maximum entropy discrimination approach. A similar approach, with additional constraints on sparsity of kernel combinations, was used by [17]. These ideas were later generalized to the case, when related tasks may use slightly different kernel combinations [14,18], and successfully used in practical applications [15,16].

Despite intuitive attractiveness of the possibility of automatically learning a suitable feature representation compared to learning with a fixed, perhaps high-dimensional or just irrelevant set of features, relatively little is known about its theoretical justifications. A seminal systematic theoretical study of the multi-task/lifelong learning settings was done by Baxter in [6]. There the author provided sample complexity bounds for both scenarios under the assumption that the tasks share a common optimal hypothesis class. The possible advantages of these approaches according to Baxter's results depend on the behavior of complexity terms, which, however, due to the generality of the formulation, often can not be inferred easily given a particular setting. Therefore, studying more specific scenarios by using more intuitive complexity measures may lead to better understanding of the possible benefits of the multi-task/lifelong settings, even if, in some sense, they can be viewed as particular cases of Baxter's result. Along that line, Maurer in [19] proved that learning a common low-dimensional representation in the case of lifelong learning of linear least-squares regression tasks is beneficial.

Multiple Kernel Learning. The problem of multiple kernel learning in the single-task scenario has been theoretically analyzed using different techniques. By using covering numbers, Srebro et al in [4] have shown generalization bounds with additive dependence on the pseudodimension of the kernel family. Another bound with multiplicative dependence on the pseudodimension was presented in [3], where the authors used Rademacher chaos complexity measure. Both results have a form $O(\sqrt{d/m})$, where d is the pseudodimension of the kernel family and m is the sample size. By carefully analyzing the growth rate of the Rademacher complexity in the case of the linear combinations of finitely many kernels with l_p constraint on the weights, Cortes et al in [2] have improved the above results. In particular, in the case of l_1 constraints, the bound from [4] has a form $O(\sqrt{k/m})$, where k in the total number of kernels, while the bound from [2] is $O(\sqrt{\log(k)/m})$. The fast rate analysis of the linear combinations of kernels using local Rademacher complexities was performed by Kloft et al in [1].

In this work we utilize techniques from [4]. It allows us to formulate results that hold for any kernel family with finite pseudodimension and not only for the case of linear combinations, though at the price of potentially suboptimal dependence on the number of kernels in the latter case. Moreover, additive dependence on the pseudodimension is especially appealing for the analysis of the multi-task and lifelong scenarios, as it allows obtaining bounds where that additional complexity term vanishes as the number of tasks grows and therefore these bounds clearly show possible advantages of transfer learning.

We start by describing the formal set up and preliminaries in Section 2.1,2.2 and providing a list of known kernel families with finite pseudodimensions, including our new result for sparse linear combinations, in 2.3. In Section 3 we provide the proof of the generalization bound for the multi-task case and extend it to the lifelong setting in Section 4. We conclude by discussion in Section 5.

2   Preliminaries

2.1   Formal Setup

Throughout the paper we denote the input space by X and the output space by $Y = \{-1, 1\}$. We assume that the learner (both in the multi-task and the lifelong learning scenarios) has an access to n tasks represented by the corresponding training sets $\mathbf{z_1}, \ldots, \mathbf{z_n} \in (X \times Y)^m$, where each $\mathbf{z_i} = \{(x_{i1}, y_{i1}), \ldots, (x_{im}, y_{im})\}$ consists of m i.i.d. samples from some unknown task-specific data distribution P_i over $Z = X \times Y$. In addition we assume that the learner is given a family $\mathcal{K}$ of kernel functions[1] defined on $X \times X$ and uses the corresponding set of linear

---

[1] A function $K : X \times X \to \mathbb{R}$ is called a kernel, if there exist a Hilbert space $\mathcal{H}$ and a mapping $\phi : X \to \mathcal{H}$ such that $K(x, x') = \langle \phi(x), \phi(x') \rangle$ for all $x, x' \in X$.

predictors for learning. Formally, for every kernel $K \in \mathcal{K}$ we define $\mathcal{F}_K$ to be such set:

$$\mathcal{F}_K \stackrel{\text{def}}{=} \{h : x \mapsto \langle w, \phi(x) \rangle \mid \|w\| \leq 1, K(x, x') = \langle \phi(x), \phi(x') \rangle \} \tag{1}$$

and $\mathbb{H}$ to be the union of them: $\mathbb{H} = \cup_{K \in \mathcal{K}} \mathcal{F}_K$.

In the multi-task scenario the data distributions $P_1, \ldots, P_n$ are assumed to be fixed and the goal of the learner is to identify a kernel $K \in \mathcal{K}$ that performs well on all of them. Therefore we would like to bound the difference between the expected error rate over the tasks:

$$er(\mathcal{F}_K) = \frac{1}{n} \sum_{i=1}^{n} \inf_{h \in \mathcal{F}_K} \mathbf{E}_{(x,y) \sim P_i} [\![yh(x) < 0]\!] \tag{2}$$

and the corresponding empirical margin error rate:

$$\widehat{er}_z^\gamma(\mathcal{F}_K) = \frac{1}{n} \sum_{i=1}^{n} \inf_{h \in \mathcal{F}_K} \frac{1}{m} \sum_{j=1}^{m} [\![y_{ij} h(x_{ij}) < \gamma]\!]. \tag{3}$$

Alternatively the learner may be interested in identifying a particular predictor for every task. If we define $\mathcal{F}_K^n = \{\mathbf{h} = (h_1, \ldots, h_n) : h_i \in \mathcal{F}_K \ \forall i = 1 \ldots n\}$ and $\mathbb{H}^n = \cup_K \mathcal{F}_K^n$, then it means finding some $\mathbf{h} \in \mathbb{H}^n$ with low generalization error:

$$er(\mathbf{h}) = \frac{1}{n} \sum_{i=1}^{n} \mathbf{E}_{(x,y) \sim P_i} [\![yh_i(x) < 0]\!] \tag{4}$$

based on its empirical margin performance:

$$\widehat{er}_z^\gamma(\mathbf{h}) = \frac{1}{n} \sum_{i=1}^{n} \frac{1}{m} \sum_{j=1}^{m} [\![y_{ij} h_i(x_{ij}) < \gamma]\!]. \tag{5}$$

However, due to the following inequality, it is enough to bound the probability of large estimation error for the second case and a bound for the first one will follow immediately:

$$Pr \left\{ z \in Z^{(n,m)} \ \exists \ K \in \mathcal{K} : er(\mathcal{F}_K) > \widehat{er}_z^\gamma(\mathcal{F}_K) + \epsilon \right\} \leq$$
$$Pr \left\{ z \in Z^{(n,m)} \ \exists \ \mathbf{h} \in \mathbb{H}^n : er(\mathbf{h}) > \widehat{er}_z^\gamma(\mathbf{h}) + \epsilon \right\}.$$

For the lifelong learning scenario we adopt the notion of task environment proposed in [6] and assume that there exists a set of possible data distributions (i.e. tasks) $\mathcal{P}$ and that the observed tasks are sampled from it i.i.d. according to some unknown distribution Q. The goal of the learner is to find a kernel $K \in \mathcal{K}$ that would work well on future, yet unobserved tasks from the environment $(\mathcal{P}, Q)$. Therefore we would like to bound the probability of large deviations between the expected error rate on new tasks, given by:

$$er(\mathcal{F}_K) = \mathbb{E}_{P \sim Q} \inf_{h \in \mathcal{F}_K} \mathbb{E}_{(x,y) \sim P} [\![h(x)y < 0]\!], \tag{6}$$

and the corresponding empirical margin error rate $\widehat{er}_z^{\gamma}(\mathcal{F}_K)$.

In order to obtain the generalization bounds in both cases we employ the technique of covering numbers.

2.2   Covering Numbers and Pseudodimensions

In this subsection we describe the types of covering numbers we will need and establish their connections to pseudodimensions of kernel families.

Definition 1. *A subset* $\tilde{A} \subset A$ *is called an* ϵ-*cover of* A *with respect to a distance measure* d, *if for every* $a \in A$ *there exists a* $\tilde{a} \in \tilde{A}$ *such that* $d(a, \tilde{a}) < \epsilon$. *The covering number* $N_d(A, \epsilon)$ *is the size of the smallest* ϵ-*cover of* A.

To derive bounds for the multi-task setting we will use covers of $\mathbb{H}^n$ with respect to ℓ_{∞} metric associated with a sample $\mathbf{x} \in X^{(n,m)}$:

$$d_{\infty}^{\mathbf{x}}(\mathbf{h}, \tilde{\mathbf{h}}) = \max_{i=1\dots n} \max_{j=1\dots m} |h_i(x_{ij}) - \tilde{h}_i(x_{ij})| < \epsilon. \tag{7}$$

The corresponding uniform covering number $N_{(n,m)}(\mathbb{H}^n, \epsilon)$ is given by considering all possible samples $\mathbf{x} \in X^{(n,m)}$:

$$N_{(n,m)}(\mathbb{H}^n, \epsilon) = \max_{\mathbf{x} \in X^{(n,m)}} N_{d_{\infty}^{\mathbf{x}}}(\mathbb{H}^n, \epsilon). \tag{8}$$

In contrast, for the lifelong learning scenario we will need covers of the kernel family $\mathcal{K}$ with respect to a probability distribution. For any probability distribution P over $X \times Y$, we denote its projection on X by P_X and define the following distance between the kernels:

$$D_P(K, \tilde{K}) = \max\{\max_{h \in \mathcal{F}_K} \min_{h' \in \mathcal{F}_{\tilde{K}}} \underset{x \sim P_X}{\mathbb{E}} |h(x) - h'(x)|, \max_{h' \in \mathcal{F}_{\tilde{K}}} \min_{h \in \mathcal{F}_K} \underset{x \sim P_X}{\mathbb{E}} |h(x) - h'(x)|\}. \tag{9}$$

Similarly, for any set of n distributions $\mathbf{P} = (P_1, \dots, P_n)$ we define:

$$D_{\mathbf{P}}(K, \tilde{K}) = \max_{i=1\dots n} D_{P_i}(K, \tilde{K}). \tag{10}$$

The minimal size of the corresponding ϵ-cover of a set of kernels $\mathcal{K}$ we will denote by $N_{D_{\mathbf{P}}}(\mathcal{K}, \epsilon)$ and the corresponding uniform covering number by by $N_{(D,n)}(\mathcal{K}, \epsilon) = \max_{(P_1,\dots,P_n)} N_{D_{\mathbf{P}}}(\mathcal{K}, \epsilon)$.

In order to make the guarantees given by the generalization bounds, that we provide, more intuitively appealing we state them using a natural measure of complexity of kernel families, namely, pseudodimension [4]:

Definition 2. *The class* $\mathcal{K}$ *pseudo-shatters the set of* n *pairs of points* $(x_1, x_1'), \dots, (x_n, x_n')$ *if there exist thresholds* $t_1, \dots, t_n$ *such that for any* $b_1, \dots, b_n \in \{-1, +1\}$ *there exists* $K \in \mathcal{K}$ *such that* $\text{sign}(K(x_i, x_i') - t_i) = b_i$. *The pseudodimension* $d_{\phi}(\mathcal{K})$ *is the largest* n *such that there exists a set of* n *pairs pseudo-shattered by* $\mathcal{K}$.

To do so we develop upper bounds on the covering numbers we use in terms of the pseudodimension of the kernel family $\mathcal{K}$. First, we prove the result for $N_{(n,m)}(\mathbb{H}^n, \epsilon)$ that will be used in the multi-task setting:

Lemma 1. *For any set $\mathcal{K}$ of kernels bounded by B ($K(x,x) \leq B$ for all $K \in \mathcal{K}$ and all x) with pseudodimension d_ϕ the following inequality holds:*

$$N_{(n,m)}(\mathbb{H}^n, \epsilon) \leq 2^n \left(\frac{4en^2m^3B}{\epsilon^2 d_\phi}\right)^{d_\phi} \left(\frac{16mB}{\epsilon^2}\right)^{\frac{64Bn}{\epsilon^2} \log\left(\frac{\epsilon\epsilon m}{8\sqrt{B}}\right)}.$$

In order to prove this result, we first introduce some additional notation. For a sample $\mathbf{x} = (x_1, \ldots, x_m) \in X^m$ we define l_∞ distance between two functions:

$$d_\infty^{\mathbf{x}}(f_1, f_2) = \max_{i=1 \ldots m} |f_1(x_i) - f_2(x_i)|. \tag{11}$$

Then the corresponding uniform covering number is:

$$N_m(\mathcal{F}, \epsilon) = \sup_{\mathbf{x} \in X^m} N_{d_\infty^{\mathbf{x}}}(\mathcal{F}, \epsilon) \tag{12}$$

We also define l_∞ distance between kernels with respect to a sample $\mathbf{x} = (\mathbf{x_1}, \ldots, \mathbf{x_n}) \in X^{(n,m)}$ with the corresponding uniform covering number:

$$D_\infty^{\mathbf{x}}(K, \hat{K}) = \max_i |K_{\mathbf{x_i}} - \hat{K}_{\mathbf{x_i}}|_\infty, \quad N_{(n,m)}(\mathcal{K}, \epsilon) = \sup_{\mathbf{x} \in X^{(n,m)}} N_{D_\infty^{\mathbf{x}}}(\mathcal{K}, \epsilon).$$

In contrast, in [4] the distance between two kernels is defined based on a single sample $\mathbf{x} = (x_1, \ldots, x_m)$ of size m:

$$D_\infty^{\mathbf{x}}(K, \hat{K}) = |K_{\mathbf{x}} - \hat{K}_{\mathbf{x}}|_\infty \tag{13}$$

and the corresponding covering number is $N_m(\mathcal{K}, \epsilon)$. Note that this definition is in strong relation with ours: $N_{(n,m)}(\mathcal{K}, \epsilon) \leq N_{mn}(\mathcal{K}, \epsilon)$, and therefore, by Lemma 3 in [4]:

$$N_{(n,m)}(\mathcal{K}, \epsilon) \leq N_{nm}(\mathcal{K}, \epsilon) \leq \left(\frac{en^2m^2B}{\epsilon d_\phi}\right)^{d_\phi} \tag{14}$$

for any kernel family $\mathcal{K}$ bounded by B with pseudodimension d_ϕ. Now we can prove Lemma 1:

Proof (of lemma 1). Fix $\mathbf{x} = (\mathbf{x_1}, \ldots, \mathbf{x_n}) \in X^{(n,m)}$. Define $\epsilon_K = \epsilon^2/4m$ and $\epsilon_F = \epsilon/2$. Let $\tilde{\mathcal{K}}$ be an ϵ_K-net of $\mathcal{K}$ with respect to $D_\infty^{\mathbf{x}}$. For every $\tilde{K} \in \tilde{\mathcal{K}}$ and every $i = 1 \ldots n$ let $\tilde{\mathcal{F}}_{\tilde{K}}^i$ be an ϵ_F-net of $\tilde{\mathcal{F}}_{\tilde{K}}$ with respect to $d_\infty^{\mathbf{x_i}}$. Now fix some $f \in \mathbb{H}^n$. Then there exists a kernel K such that $f = (f_1, \ldots, f_n) \in \mathcal{F}_K^n$. Therefore there exists a kernel $\tilde{K} \in \tilde{\mathcal{K}}$ such that $|K_{\mathbf{x_i}} - \tilde{K}_{\mathbf{x_i}}|_\infty < \epsilon_K$ for every i. By Lemma 1 in [4] $f_i(x_i) = K_{\mathbf{x_i}}^{1/2} w_i$ for some unit norm vector w_i for every i. Therefore for $\tilde{f}_i(\mathbf{x_i}) \stackrel{\text{def}}{=} \tilde{K}_{\mathbf{x_i}}^{1/2} w_i \in \mathcal{F}_{\tilde{K}}$ we obtain that:

$$d_\infty^{\mathbf{x_i}}(f_i, \tilde{f}_i) = \max_j |f_i(x_{ij}) - \tilde{f}_i(x_{ij})| \leq \|f_i(\mathbf{x_i}) - \tilde{f}_i(\mathbf{x_i})\| =$$

$$\|K_{\mathbf{x_i}}^{1/2} w_i - \tilde{K}_{\mathbf{x_i}}^{1/2} w_i\| \leq \sqrt{m|K_{\mathbf{x_i}} - \tilde{K}_{\mathbf{x_i}}|_\infty} \leq \sqrt{m\epsilon_K}.$$

In addition, for every $\tilde{f}_i \in \mathcal{F}_{\tilde{K}}$ there exists $\tilde{\tilde{f}}_i \in \tilde{\mathcal{F}}_{\tilde{K}}^i$ such that $d_\infty^{\mathbf{x}_i}(\tilde{f}_i, \tilde{\tilde{f}}_i) < \epsilon_F$.
Finally, if we define $\tilde{\tilde{f}} = (\tilde{\tilde{f}}_1, \ldots, \tilde{\tilde{f}}_n) \in \tilde{\mathcal{F}}_{\tilde{K}}^1 \times \cdots \times \tilde{\mathcal{F}}_{\tilde{K}}^n$, we obtain:

$$d_\infty^{\mathbf{x}}(f, \tilde{\tilde{f}}) = \max_i d_\infty^{\mathbf{x}_i}(f_i, \tilde{\tilde{f}}_i) \leq \max_i (d_\infty^{\mathbf{x}_i}(f_i, \tilde{f}_i) + d_\infty^{\mathbf{x}_i}(\tilde{f}_i, \tilde{\tilde{f}}_i)) < \sqrt{m\epsilon_K} + \epsilon_F = \epsilon.$$

The above shows that $\tilde{\mathcal{F}}_{\mathcal{K}} = \cup_{\tilde{K} \in \tilde{\mathcal{K}}} \tilde{\mathcal{F}}_{\tilde{K}}^1 \times \cdots \times \tilde{\mathcal{F}}_{\tilde{K}}^n$ is an ϵ-net of $\mathbb{H}^n$ with respect to $\mathbf{x}$. Now the statement follows from (14) and the fact that for any $\mathcal{F}_K$ with bounded by B kernel K ([4,8]):

$$N_m(\mathcal{F}_K, \epsilon) \leq 2\left(4mB/\epsilon^2\right)^{\frac{16B}{\epsilon^2} \log_2\left(\frac{\epsilon e m}{4\sqrt{B}}\right)} \tag{15}$$

$\square$

Analogously we develop an upper bound on the covering number $N_{(D,n)}(\mathcal{K}, \epsilon)$, which we will use for the lifelong learning scenario:

Lemma 2. *There exists a constant C such that for any kernel family $\mathcal{K}$ bounded by B with pseudodimension d_ϕ:*

$$N_{(D,n)}(\mathcal{K}, \epsilon) \leq \left(Cn^5 d_\phi^5 \left(\sqrt{B}/\epsilon\right)^{17}\right)^{d_\phi}. \tag{16}$$

The proof of this result is based on the following lemma that connects sample-based and distribution-based covers of kernel families (for the proof see Appendix A):

Lemma 3. *For any probability distribution P over $X \times Y$ and any B-bounded set of kernels $\mathcal{K}$ with pseudo-dimension d_ϕ there exists a sample $\mathbf{x}$ of size $m = cd_\phi^2 B^{5/2}/\epsilon^5$ for some constant c, such that for every $K, \tilde{K}$ if $D_1^{\mathbf{x}}(K, \tilde{K}) < \epsilon/2$, then $D_P(K, \tilde{K}) < \epsilon$ (where $D_1^{\mathbf{x}}$ is the same as D_P, but all expectations over P are substituted by empirical averages over $\mathbf{x}$).*

Proof (of lemma 2). Fix some set of probability distributions $\mathbf{P} = (P_1, \ldots, P_n)$. For every P_i denote a sample described by Lemma 3 by $\mathbf{x}_i$. Let $\tilde{\mathcal{K}}$ be an $\epsilon/2n$-cover of $\mathcal{K}$ with respect to $D_1^{\mathbf{x}}$, where $\mathbf{x} = (\mathbf{x}_1, \ldots, \mathbf{x}_n) \in X^{mn}$ and $m = cd_\phi^2 B^{5/2}/\epsilon^5$. Then the following chain of inequalities holds:

$$\max_{h \in \mathcal{F}_K} \min_{h' \in \mathcal{F}_{\tilde{K}}} \frac{1}{mn} \sum_{i=1}^n \sum_{j=1}^m |h(x_{ij}) - h'(x_{ij})| \leq \max_h \min_{h'} \|h(\mathbf{x}) - h'(\mathbf{x})\| \leq$$

$$\max_w \|K_{\mathbf{x}}^{\frac{1}{2}} w - \tilde{K}_{\mathbf{x}}^{\frac{1}{2}} w\| \leq \|K_{\mathbf{x}}^{\frac{1}{2}} - \tilde{K}_{\mathbf{x}}^{\frac{1}{2}}\|_2 \leq \sqrt{\|K_{\mathbf{x}} - \tilde{K}_{\mathbf{x}}\|_2} \leq \sqrt{mn|K_{\mathbf{x}} - \tilde{K}_{\mathbf{x}}|_\infty}.$$

Consequently, by Lemma 3 in [4]:

$$|\tilde{\mathcal{K}}| \leq N(\epsilon/2n, \mathcal{K}, D_1^{\mathbf{x}}) \leq \left(\frac{4em^3 n^5 B}{\epsilon^2 d_\phi}\right)^{d_\phi} = \left(Cn^5 d_\phi^5 \left(\sqrt{B}/\epsilon\right)^{17}\right)^{d_\phi}. \tag{17}$$

It is left to show that $\tilde{\mathcal{K}}$ is an ϵ-cover of $\mathcal{K}$ with respect to $D_{\mathbf{P}}$. By definition, for every $K \in \mathcal{K}$ there exists $\tilde{K} \in \tilde{\mathcal{K}}$ such that $D_1^{\times}(K, \tilde{K}) < \epsilon/2n$. Therefore for every $i = 1 \ldots n$:

$$\max_{h \in \mathcal{F}_K} \min_{h' \in \mathcal{F}_{\tilde{K}}} \frac{1}{m} \sum_{j=1}^{m} |h(x_{ij}) - h'(x_{ij})| \leq \max_{h \in \mathcal{F}_K} \min_{h' \in \mathcal{F}_{\tilde{K}}} \frac{n}{mn} \sum_{i,j} |h(x_{ij}) - h'(x_{ij})| < \frac{\epsilon}{2}.$$

Consequently, by Lemma 3, $D_{P_i}(K, \tilde{K}) < \epsilon$ for all $i = 1 \ldots n$. □

2.3 Pseudodimensions of Various Families of Kernels

In [4] the authors have shown the upper bounds on the pseudodimensions of some families of kernels:

- convex or linear combinations of k kernels have pseudodimension at most k
- Gaussian families with learned covariance matrix in $\mathbb{R}^{\ell}$ have $d_{\phi} \leq \ell(\ell+1)/2$
- Gaussian families with learned low-rank covariance have $d_{\phi} \leq kl \log_2(8ekl)$, where k is the maximum rank of the covariance matrix

Here we extend their analysis to the case of sparse combinations of kernels.

Lemma 4. *Let $K_1, \ldots, K_N$ be N kernels and let $\mathcal{K} = \{\sum_{i=1}^{N} w_i K_i : \sum_{i=1}^{N} w_i = 1 \text{ and } \sum_{i=1}^{N} [w_i \neq 0] \leq k\}$. Then:*

$$d_{\phi}(\mathcal{K}) \leq 2k \log(k) + 2k \log(4eN) \tag{18}$$

Proof. For every kernel K define a function $B_K : X \times X \times \mathbb{R} \to \{-1, 1\}$:

$$B_K(x, \bar{x}, t) = sign(K(x, \bar{x}) - t) \tag{19}$$

and denote a set of such functions for all $K \in \mathcal{K}$ by $\mathcal{B}$. Then $d_{\phi}(\mathcal{K}) = VCdim(\mathcal{B})$.

For every index set $1 \leq i_1 < \cdots < i_k \leq N$ define $\mathcal{K}_i$ to be a set of all linear combinations of $K_{i_1}, \ldots, K_{i_k}$. Then: $\mathcal{K} = \cup_i \mathcal{K}_i$ and $d_{\phi}(\mathcal{K}_i) \leq k$. Moreover, there are $\binom{N}{k} \leq \left(\frac{Ne}{k}\right)^k$ of possible sets of indices i. Therefore $\mathcal{B}$ can also be seen as a union of at most $\left(\frac{Ne}{k}\right)^k$ sets with VC-dimension at most k. VC-dimension of a union of r classes of VC-dimension at most d is at most $4d \log(2d) + 2 \log(r)$. The statement of the lemma is obtained by setting $r = \left(\frac{Ne}{k}\right)^k$ and $d = k$. □

3 Multi-task Kernel Learning

We start with formulating the result using covering number $N_{(n,m)}(\mathbb{H}^n, \epsilon)$:

Theorem 1. *For any $\epsilon > 0$, if $m > 2/\epsilon^2$, we have that:*

$$Pr\{\exists \mathbf{h} \in \mathbb{H}^n : er(\mathbf{h}) > \hat{er}_z^{\gamma}(\mathbf{h}) + \epsilon\} \leq 2N_{(n,2m)}(\mathbb{H}^n, \gamma/2) \exp\left(-\frac{nm\epsilon^2}{8}\right). \tag{20}$$

Proof. We utilize the standard 3-steps procedure (see Theorem 10.1 in [8]). If we denote:

$$Q = \left\{ z \in Z^{(n,m)} : \exists \mathbf{h} \in \mathbb{H}^n : \; er(\mathbf{h}) > \widehat{er}_z^\gamma(\mathbf{h}) + \epsilon \right\}$$

$$R = \left\{ z = (r,s) \in Z^{(n,m)} \times Z^{(n,m)} : \exists \mathbf{h} \in \mathbb{H}^n : \; \widehat{er}_s(\mathbf{h}) > \widehat{er}_r^\gamma(\mathbf{h}) + \epsilon/2 \right\},$$

then according to the symmetrization argument $Pr(Q) \leq 2Pr(R)$. Therefore, instead of bounding the probability of Q, we can bound the probability of R.

Next, we define Γ_{2m} to be a set of permutations σ on the set $\{(1,1), \ldots, (n, 2m)\}$ such that $\{\sigma(i,j), \sigma(i, m+j)\} = \{(i,j), (i, m+j)\}$ for every $1 \leq i \leq n$ and $1 \leq j \leq m$. Then $Pr(R) \leq \max_{z \in Z^{(2m,n)}} Pr_\sigma(\sigma z \in R)$.

Now we proceed with the last step - reduction to a finite class. Fix $z \in Z^{(n,2m)}$ and the corresponding $\mathbf{x} = (x_{ij}) \in X^{(n,2m)}$. Let T be a $\gamma/2$-cover of $\mathbb{H}^n$ with respect to $d_\infty^{\mathbf{x}}$ and fix $\sigma z \in R$. By definition there exists $\mathbf{h} \in \mathbb{H}^n$ such that $\widehat{er}_s(\mathbf{h}) > \widehat{er}_r^\gamma(\mathbf{h}) + \epsilon/2$, where $(r,s) = \sigma z$. We can rewrite it as:

$$\frac{1}{n} \sum_{i=1}^n \frac{1}{m} \sum_{j=m+1}^{2m} [\![h_i(x_{\sigma(ij)}) y_{\sigma(ij)} < 0]\!] > \frac{1}{n} \sum_{i=1}^n \frac{1}{m} \sum_{j=1}^m [\![h_i(x_{\sigma(ij)}) y_{\sigma(ij)} < \gamma]\!] + \epsilon/2.$$

If we denote by $\tilde{\mathbf{h}}$ the function in the cover T corresponding to $\mathbf{h}$, then the following inequalities hold:

if $\tilde{h}_i(x_{ij}) y_{ij} < \dfrac{\gamma}{2}$, then $h_i(x_{ij}) y_{ij} < \gamma$; if $h_i(x_{ij}) y_{ij} < 0$, then $\tilde{h}_i(x_{ij}) y_{ij} < \dfrac{\gamma}{2}$.

By combining them with the previous inequality we obtain that:

$$\frac{1}{n} \sum_{i=1}^n \frac{1}{m} \sum_{j=m+1}^{2m} [\![\tilde{h}_i(x_{\sigma(ij)}) y_{\sigma(ij)} < \frac{\gamma}{2}]\!] > \frac{1}{n} \sum_{i=1}^n \frac{1}{m} \sum_{j=1}^m [\![\tilde{h}_i(x_{\sigma(ij)}) y_{\sigma(ij)} < \frac{\gamma}{2}]\!] + \frac{\epsilon}{2}.$$

Now, if we define the following indicator: $v(\tilde{\mathbf{h}}, i, j) = [\![\tilde{h}_i(x_{ij}) y_{ij} < \gamma/2]\!]$, then:

$$Pr_\sigma \{\sigma z \in R\} \leq Pr_\sigma \left\{ \exists \tilde{\mathbf{h}} \in T : \frac{1}{n} \sum_{i=1}^n \frac{1}{m} \sum_{j=1}^m (v(\tilde{\mathbf{h}}, \sigma(i, m+j)) - v(\tilde{\mathbf{h}}, \sigma(i,j))) > \frac{\epsilon}{2} \right\}$$

$$\leq |T| \max_{\tilde{\mathbf{h}} \in T} Pr_\beta \left\{ \frac{1}{n} \sum_{i=1}^n \frac{1}{m} \sum_{j=1}^m |v(\tilde{\mathbf{h}}, i, m+j) - v(\tilde{\mathbf{h}}, i, j)| \beta_{ij} > \epsilon/2 \right\} = (*),$$

where β_{ij} are independent random variables uniformly distributed over $\{-1, 1\}$. Then $\{|v(\tilde{\mathbf{h}}, i, m+j) - v(\tilde{\mathbf{h}}, i, j)| \beta_{ij}\}$ are nm independent random variables that take values between -1 and 1 and have zero mean. Therefore by Hoeffding's inequality:

$$(*) \leq |T| \exp\left(-\frac{2(nm)^2 \epsilon^2/4}{mn \cdot 4} \right) = |T| \exp\left(-\frac{nm\epsilon^2}{8} \right).$$

By noting that $|T| \leq N_{(n,2m)}(\mathbb{H}^n, \gamma/2)$, we conclude the proof of Theorem 1. $\quad\square$

By using the same technique as for proving Theorem 1, we can obtain a lower bound on the difference between the empirical error rate $\widehat{er}_z^\gamma(\mathbf{h})$ and the expected error rate with double margin:

$$er^{2\gamma}(\mathbf{h}) = \frac{1}{n} \sum_{i=1}^{n} \mathbf{E}_{(x,y)\sim P_i} [\![yh_i(x) < 2\gamma]\!]. \tag{21}$$

Theorem 2. *For any $\epsilon > 0$, if $m > 2/\epsilon^2$, the following holds:*

$$Pr\{\exists \mathbf{h} \in \mathbb{H}^n : er^{2\gamma}(\mathbf{h}) < \widehat{er}_z^\gamma(\mathbf{h}) - \epsilon\} \leq 2N_{(n,2m)}(\mathbb{H}^n, \gamma/2) \exp\left(-\frac{nm\epsilon^2}{8}\right). \tag{22}$$

Now, by combining Theorems 1, 2 and Lemma 1 we can state the final result for the multi-task scenario in terms of pseudodimensions:

Theorem 3. *For any probability distributions $P_1, \ldots, P_n$ over $X \times \{-1, +1\}$, any kernel family $\mathcal{K}$, bounded by B with pseudodimension d_ϕ, and any fixed $\gamma > 0$, for any $\epsilon > 0$, if $m > 2/\epsilon^2$, then, for a sample z generated by $\Pi_{i=1}^n(P_i)^m$:*

$$Pr\left\{\forall \mathbf{h} \in \mathbb{H}^n \ er^{2\gamma}(\mathbf{h}) + \epsilon \geq \widehat{er}_z^\gamma(\mathbf{h}) \geq er(\mathbf{h}) - \epsilon\right\} \geq 1 - \delta, \tag{23}$$

where

$$\epsilon = \sqrt{8\frac{\frac{2\log 2 - \log \delta}{n} + \log 2 + \frac{d_\phi}{n} \log \frac{128en^2m^3B}{\gamma^2 d_\phi} + \frac{256B}{\gamma^2} \log \frac{\gamma em}{8\sqrt{B}} \log \frac{128mB}{\gamma^2}}{m}}. \tag{24}$$

Discussion: The most significant implications of this result are for the case where there exists some kernel $K \in \mathcal{K}$ that has low approximation error for each of the tasks P_i (this is what makes the tasks "related" and, therefore, the multi-task approach advantageous). In such a case, the kernel that minimizes the average error over the set of tasks is a useful kernel for each of these tasks.

1. Maybe the first point to note about the above generalization result is that as the number of tasks (n) grows, while the number of examples per task (m) remains constant, the error bound behaves like the bound needed to learn with respect to a single kernel. That is, if a learner wishes to learn some specific task P_i, and all the learner knows is that in the big family of kernels $\mathcal{K}$, there exists some useful kernel K for P_i that is also good on average over the other tasks, then the training samples from the other tasks allow the learner of P_i to learn as if he had access to a specific good kernel K.
2. Another worthwhile consequence of the above theorem is that it shows the usefulness of an empirical risk minimization approach. Namely,

   Corollary 1. *Let $\widehat{h}$ be a minimizer, over $\mathbb{H}^n$, of the empirical γ-margin loss, $\widehat{er}_z^\gamma(\mathbf{h})$. Then for any $\mathbf{h}^* \in \mathbb{H}^n$ (and in particular for a minimizer over $\mathbb{H}^n$ of the true 2γ-loss $er^{2\gamma}(\mathbf{h})$):*

   $$er(\widehat{h}) \leq er^{2\gamma}(h^*) + 2\epsilon.$$

Proof. The result is implied by the following chain of inequalities:

$$er(\widehat{h}) - \epsilon \leq_1 \widehat{er}^\gamma(\widehat{h}) \leq_2 \widehat{er}^\gamma(h^*) \leq_3 er^{2\gamma}(h^*) + \epsilon$$

where $(\leq_1)$ and $(\leq_3)$ follow from the above theorem and $(\leq_2)$ follows from the definition of an empirical risk minimizer. $\qquad\square$

4   Lifelong Kernel Learning

In this section we generalize the results of the previous section to the case of lifelong learning in two steps. First, note that by using the same arguments as for proving Theorem 1 we can obtain a bound on the difference between $\widehat{er}_z^{2\gamma}(\mathcal{F}_K)$ and:

$$\widehat{er}_{\mathbf{P}}^\gamma(\mathcal{F}_K) = \frac{1}{n} \sum_{i=1}^{n} \inf_{h \in \mathcal{F}_K} \mathbb{E}_{(x,y) \sim P_i}[\![h(x)y < \gamma]\!]. \qquad (25)$$

Therefore the only thing that is left is a bound on the difference between $er(\mathcal{F}_K)$ and $\widehat{er}_{\mathbf{P}}^\gamma(\mathcal{F}_K)$.

We will use the following notation:

$$er_P(\mathcal{F}_K) = \inf_{h \in \mathcal{F}_K} \mathbb{E}_{(x,y) \sim P}[\![h(x)y < 0]\!], \quad er_P^\gamma(\mathcal{F}_K) = \inf_{h \in \mathcal{F}_K} \mathbb{E}_{(x,y) \sim P}[\![h(x)y < \gamma]\!]$$

and proceed in a way analogous to the proof of Theorem 1. First, if we define:

$$Q = \{\mathbf{P} = (P_1, \ldots, P_n) \in \mathcal{P}^n \; \exists \mathcal{F}_K : \; er(\mathcal{F}_K) > \widehat{er}_{\mathbf{P}}^\gamma(\mathcal{F}_K) + \epsilon\}$$
$$R = \{z = (r, s) \in \mathcal{P}^{2n} \; \exists \mathcal{F}_K : \; \widehat{er}_s(\mathcal{F}_K) > \widehat{er}_r^\gamma(\mathcal{F}_K) + \epsilon/2\},$$

then according to the symmetrization argument $Pr(Q) \leq 2Pr(R)$.

Now, if we define Γ_{2n} to be a set of permutations σ on a set $\{1, 2, \ldots, 2n\}$, such that $\{\sigma(i), \sigma(n+i)\} = \{i, n+i\}$ for all $i = 1 \ldots n$, we obtain that $Pr(R) \leq \max_z Pr_\sigma(\sigma z \in R)$, if $n > 2/\epsilon^2$. So, the only thing that is left is reduction to a finite class.

Fix z and denote by $\tilde{\mathcal{K}} \subset \mathcal{K}$ a set of kernels, such that for every $K \in \mathcal{K}$ there exists a $\hat{K} \in \tilde{\mathcal{K}}$ such that:

$$er_{P_i}^\gamma(\mathcal{F}_K) + \epsilon/8 \geq er_{P_i}^{\gamma/2}(\mathcal{F}_{\hat{K}}) \geq er_{P_i}(\mathcal{F}_K) - \epsilon/8 \; \forall i = 1 \ldots n. \qquad (26)$$

Then, if $\mathcal{F}_K$ is such that $\widehat{er}_s(\mathcal{F}_K) > \widehat{er}_r^\gamma(\mathcal{F}_K) + \epsilon/2$, then the corresponding $\hat{K}$ satisfies $\widehat{er}_s^{\gamma/2}(\mathcal{F}_{\hat{K}}) > \widehat{er}_r^{\gamma/2}(\mathcal{F}_{\hat{K}}) + \epsilon/4$. Therefore:

$$Pr_\sigma\{\sigma z \in R\} \leq Pr_\sigma \left\{ \exists K \in \tilde{K} : \frac{1}{n} \sum_{i=1}^{n} (er_{P_{\sigma(n+i)}}^{\gamma/2}(\mathcal{F}_K) - er_{P_{\sigma(i)}}^{\gamma/2}(\mathcal{F}_K)) > \epsilon/4 \right\} \leq$$

$$|\tilde{K}| \max_{K \in \tilde{K}} Pr_\sigma \left\{ \frac{1}{n} \sum_{i=1}^{n} (er_{P_{\sigma(n+i)}}^{\gamma/2}(\mathcal{F}_K) - er_{P_{\sigma(i)}}^{\gamma/2}(\mathcal{F}_K)) > \epsilon/4 \right\} =$$

$$|\tilde{K}| \max_{K \in \tilde{K}} Pr_\beta \left\{ \frac{1}{n} \sum_{i=1}^{n} |er_{P_{n+i}}^{\gamma/2}(\mathcal{F}_K) - er_{P_i}^{\gamma/2}(\mathcal{F}_K)| \beta_i > \epsilon/4 \right\} = (*),$$

where β_i are independent random variables uniformly distributed over $\{-1, +1\}$. As in the previous section, $\{|er_{P_{n+i}}^{\gamma/2}(\mathcal{F}_K) - er_{P_i}^{\gamma/2}(\mathcal{F}_K)|\beta_i\}$ are n independent random variables that take values between -1 and 1 and have zero mean. Therefore by applying Hoeffding's inequality we obtain:

$$(*) \leq |\tilde{\mathcal{K}}| \exp\left(-\frac{2n^2\epsilon^2/16}{4n}\right) = |\tilde{\mathcal{K}}| \exp\left(-\frac{n\epsilon^2}{32}\right). \tag{27}$$

To conclude the proof we need to understand how $|\tilde{\mathcal{K}}|$ behaves. For that we prove the following lemma:

Lemma 5. *For any set of probability distributions* $\mathbf{P} = (P_1, \ldots, P_n)$ *there exists* $\tilde{\mathcal{K}}$ *that satisfies condition of equation* (26) *and* $|\tilde{\mathcal{K}}| \leq N_{(D,n)}(\mathcal{K}, \epsilon\gamma/16)$.

Proof. Fix a set of distributions $\mathbf{P} = (P_1, \ldots, P_n)$ and denote by $\tilde{\mathcal{K}}$ an $\epsilon\gamma/16$-cover of $\mathcal{K}$ with respect to $D_{\mathbf{P}}$. Then $|\tilde{\mathcal{K}}| \leq N_{(D,n)}(\mathcal{K}, \epsilon\gamma/16)$. By definition of a cover for any kernel $K \in \mathcal{K}$ there exists $\tilde{K} \in \tilde{\mathcal{K}}$ such that $D_{\mathbf{P}}(K, \tilde{K}) < \epsilon\gamma/16$. Equivalently, it means that for every $K \in \mathcal{K}$ there exists $\tilde{K} \in \tilde{\mathcal{K}}$ such that the following two conditions hold for every $i = 1 \ldots n$:

$$1. \forall\, h \in \mathcal{F}_K \; \exists h' \in \mathcal{F}_{\tilde{K}} : \; \mathbb{E}_{(x,y)\sim P_i}(|h(x) - h'(x)|) < \frac{\epsilon\gamma}{16}, \tag{28}$$

$$2. \forall\, h' \in \mathcal{F}_{\tilde{K}} \; \exists h \in \mathcal{F}_K : \; \mathbb{E}_{(x,y)\sim P_i}(|h(x) - h'(x)|) < \frac{\epsilon\gamma}{16}. \tag{29}$$

Fix some K and the corresponding kernel $\tilde{K}$ from the cover and take any P_i. By Markov's inequality applied to the first condition we obtain that for every $h \in \mathcal{F}_K$ there exists a $h' \in \mathcal{F}_{\tilde{K}}$ such that $Pr\{x \sim P_i : |h(x) - h'(x)| > \gamma/2\} < \epsilon/8$. Then $er_{P_i}^{\gamma/2}(h') \leq er_{P_i}^{\gamma}(h) + \epsilon/8$. By applying the same argument to the second condition we conclude that for every $h' \in \mathcal{F}_{\tilde{K}}$ there exists a $h \in \mathcal{F}_K$ such that $Pr\{x \sim P_i : |h(x) - h'(x)| > \gamma/2\} < \epsilon/8$. Then $er_{P_i}(h) \leq er_{P_i}^{\gamma/2}(h') + \epsilon/8$. By definition of infimum $\hat{er}_z^{2\gamma}(\mathcal{F}_K)$ for every δ there exists $h \in \mathcal{F}_K$ such that $er_{P_i}(\mathcal{F}_K) + \delta > er_{P_i}^{\gamma}(h) \geq er_{P_i}^{\gamma}(\mathcal{F}_K)$. By above construction for such h there exists $h' \in \mathcal{F}_{\tilde{K}}$ such that $er_{P_i}^{\gamma}(h) \geq er_{P_i}^{\gamma/2}(h') - \epsilon/8 \geq er_{P_i}^{\gamma/2}(\mathcal{F}_{\tilde{K}}) - \epsilon/8$. By combining these inequalities we obtain that for every $\delta > 0$ $er_{P_i}^{\gamma}(\mathcal{F}_K) + \delta > er_{P_i}^{\gamma/2}(\mathcal{F}_{\tilde{K}}) - \epsilon/8$, or, equivalently, $er_{P_i}^{\gamma}(\mathcal{F}_K) \geq er_{P_i}^{\gamma/2}(\mathcal{F}_{\tilde{K}}) - \epsilon/8$. Analogously we can get that $er_{P_i}^{\gamma/2}(\mathcal{F}_{\tilde{K}}) \geq er_{P_i}(\mathcal{F}_K) - \epsilon/8$. So, we obtain condition (26). $\quad\square$

By combining the above Lemma with (27) we obtain the following result (the second inequality can be obtain in a similar manner):

Theorem 4. *For any* $\epsilon > 0$, *if* $n > 2/\epsilon^2$, *the following holds:*

$$Pr\left\{\exists K \in \mathcal{K} : \; er(\mathcal{F}_K) > \hat{er}_{\mathbf{P}}^{\gamma}(\mathcal{F}_K) + \epsilon\right\} \leq 2N_{(D,n)}(\mathcal{K}, \epsilon\gamma/16) \exp\left(-\frac{n\epsilon^2}{32}\right),$$

$$Pr\left\{\exists K \in \mathcal{K} : er^{2\gamma}(\mathcal{F}_K) < \hat{er}_{\mathbf{P}}^{\gamma}(\mathcal{F}_K) - \epsilon\right\} \leq 2N_{(D,n)}(\mathcal{K}, \epsilon\gamma/16) \exp\left(-\frac{n\epsilon^2}{32}\right).$$

Note that by exactly following the proof of Theorem 1 one can obtain that:

$$Pr\left\{\exists K \in \mathcal{K} \; \widehat{er}_{\mathbf{P}}^{\gamma/2}(\mathcal{F}_K) - \widehat{er}_z^{\gamma}(\mathcal{F}_K) > \frac{\epsilon}{2}\right\} < 2N_{(n,2m)}(\mathbb{H}^n, \gamma/4) \exp\left(-\frac{nm\epsilon^2}{32}\right).$$

Therefore, by combining the above result with its equivalent in the opposite direction with Theorem 4 and Lemmas 1 and 2 we obtain the final result for the lifelong kernel learning:

Theorem 5. *For any task environment, any kernel family $\mathcal{K}$, bounded by B with pseudodimension d_ϕ, any fixed $\gamma > 0$ and any $\epsilon > 0$, if $n > 8/\epsilon^2$ and $m > 8/\epsilon^2$, then:*

$$Pr\left\{\forall K \in \mathcal{K} \; er^{2\gamma}(\mathcal{F}_K) + \epsilon \geq \widehat{er}_z^{\gamma}(\mathcal{F}_K) \geq er(\mathcal{F}_K) - \epsilon\right\} \geq 1 - \delta, \qquad (30)$$

where

$$\delta = 2^{n+2}\left(\frac{512en^2m^3B}{\gamma^2 d_\phi}\right)^{d_\phi}\left(\frac{512mB}{\gamma^2}\right)^{\frac{1024Bn}{\gamma^2}\log\left(\frac{e\gamma m}{16\sqrt{B}}\right)}\exp\left(-\frac{nm\epsilon^2}{32}\right) +$$

$$4\left(Cn^5 d_\phi^5\left(\frac{64\sqrt{B}}{\epsilon\gamma}\right)^{17}\right)^{d_\phi}\exp\left(-\frac{n\epsilon^2}{128}\right).$$

Discussion: As for the multi-task case, the most significant implications of this result are for the case where there exists some kernel $K \in \mathcal{K}$ that has low approximation error for all tasks in the environment. In such a case, the kernel that minimizes the average error over the set of observed tasks is a useful kernel for all the tasks.

1. First, note, that the only difference between Theorem 5 and Theorem 3 is the presence of the second term. This additional complexity comes from the fact that for the lifelong learner we are bounding the expected error on new, yet unobserved tasks. Therefore we have to pay additionally for not knowing exactly what these new tasks are going to be.
2. Second, the behavior of the above result is similar to that of Theorem 3 in the limit of infinitely many observed tasks ($n \to \infty$). In this case, the second term vanishes, because by observing large enough amount of tasks the learner gets the full knowledge about the task environment. The first term behaves exactly the same as the one in Theorem 3: its part that depends on d_ϕ vanishes and therefore it converges to the complexity of learning one task as if the learner would know a good kernel in advance.
3. This theorem also shows the usefulness of an empirical risk minimization approach as we can obtain a corollary of exactly the same form as Corollary 1.

5   Conclusions

Multi-task and lifelong learning have been a topic of significant interest of research in recent years and attempts for solving these problems in different directions have been made. Methods of learning kernels in these scenarios have been shown to lead to effective algorithms and became popular in applications. In this work, we have established sample complexity error bounds that justify this approach. Our results show that, under mild conditions on the used family of kernels, by solving multiple tasks jointly the learner can "spread out" the overhead associated with learning a kernel and as the number of observed tasks grows, the complexity converges to that of learning when a good kernel was known in advance. This work constitutes a step forward better understanding of the conditions under which multi-task/lifelong learning is beneficial.

A   Proof of Lemma 3

Define $G = \left\{ g : X \to [0,1] : g(x) = \frac{|h(x) - h'(x)|}{\sqrt{B}} \text{ for some } h, h' \in \cup \mathcal{F}_K \right\}$. Then (using Lemma 2 and 3 in [5] and Theorem 1 in [4]):

$$Pr\left\{ \mathbf{x} \in X^m : \exists K, \tilde{K} : |D_1^{\mathbf{x}}(K, \tilde{K}) - D_P(K, \tilde{K})| > \epsilon/2 \right\} \leq$$

$$Pr\left\{ \mathbf{x} \in X^m : \exists h, h' \in \cup \mathcal{F}_K : \left| \frac{1}{m} \sum_{i=1}^{m} |h(x_i) - h'(x_i)| - \mathop{\mathbb{E}}_{(x,y) \sim P} |h(x) - h'(x)| \right| > \frac{\epsilon}{2} \right\} =$$

$$Pr\left\{ \mathbf{x} \in X^m : \exists g, g' \in G : \left| \frac{1}{m} \sum_{i=1}^{m} g(x_i) - \mathbb{E}_{(x,y) \sim P} g(x) \right| > \epsilon/2\sqrt{B} \right\} \leq$$

$$4 \max_{\mathbf{x}} N(\frac{\epsilon/32}{\sqrt{B}}, G, d_1^{\mathbf{x}}) e^{-\frac{\epsilon^2 m}{512B}} \leq 4 \max_{\mathbf{x}} N(\frac{\epsilon}{64\sqrt{B}}, \cup \mathcal{F}_K/\sqrt{B}, d_1^{\mathbf{x}})^2 e^{-\epsilon^2 m/512B} =$$

$$4 \max_{\mathbf{x}} N(\epsilon/64, \cup \mathcal{F}_K, d_1^{\mathbf{x}})^2 e^{-\epsilon^2 m/512B} \leq 4 \max_{\mathbf{x}} N(\epsilon/64, \cup \mathcal{F}_K, d_\infty^{\mathbf{x}})^2 e^{-\epsilon^2 m/512B} \leq$$

$$4 \cdot 4 \cdot N(\mathcal{K}, \epsilon^2/(64^2 \cdot 4m))^2 \cdot \left(\frac{16mB64^2}{\epsilon^2} \right)^{\frac{2 \cdot 64^3 B}{\epsilon^2} \log\left(\frac{\epsilon \epsilon m}{64 * 8 \sqrt{B}} \right)} e^{-\epsilon^2 m/512B} \leq$$

$$16 \left(\frac{2^{14} e m^3 B}{\epsilon^2 d_\phi} \right)^{2d_\phi} \left(\frac{2^{16} m B}{\epsilon^2} \right)^{\frac{2^{19} B}{\epsilon^2} \log\left(\frac{\epsilon \epsilon m}{2^9 \sqrt{B}} \right)} e^{-\epsilon^2 m/512B} = (**)$$

For big enough m $(**)$ is less than 1, which means that there is a sample $\mathbf{x} \in X^m$ such that for all kernels $K, \tilde{K}$ we have $|D_1^{\mathbf{x}}(K, \tilde{K}) - D_P(K, \tilde{K})| \leq \epsilon/2$. More precisely, m should be bigger than $c d_\phi^2 B^{5/2}/\epsilon^5$ for some constant c.   □

Acknowledgments. This work was in parts funded by the European Research Council under the European Union's Seventh Framework Programme (FP7/2007-2013)/ERC grant agreement no 308036.

References

1. Kloft, M., Blanchard, G.: On the convergence rate of lp-norm multiple kernel learning. Journal of Machine Learning Research (2012)
2. Cortes, C., Mohri, M., Rostamizadeh, A.: Generalization bounds for learning kernels. In: Proceedings of the International Conference on Machine Learning (2010)
3. Ying, Y., Campbell, C.: Generalization bounds for learning the kernel. In: Proceedings of the Workshop on Computational Learning Theory (2009)
4. Srebro, N., Ben-David, S.: Learning bounds for support vector machines with learned kernels. In: Lugosi, G., Simon, H.U. (eds.) COLT 2006. LNCS (LNAI), vol. 4005, pp. 169–183. Springer, Heidelberg (2006)
5. Bartlett, P.L., Kulkarni, S.R., Posner, S.E.: Covering Numbers for Real-Valued Function Classes. IEEE Transactions on Information Theory **43**, 1721–1724 (1997)
6. Baxter, J.: A Model of Inductive Bias Learning. Journal of Artificial Intelligence Research **12**, 149–198 (2000)
7. Evgeniou, T., Pontil, M.: Regularized multi-task learning. In: Proceedings of the International Conference on Knowledge Discovery and Data Mining (2004)
8. Anthony, M., Bartlett, P.L.: Neural Network Learning: Theoretical Foundations. Cambridge University Press (1999)
9. Argyriou, A., Evgeniou, T., Pontil, M.: Convex Multi-task Feature Learning. Machine Learning **73** (2008)
10. Kumar, A., Daumé III, H.: Learning task grouping and overlap in multi-task learning. In: Proceedings of the International Conference on Machine Learning (2012)
11. Eaton, E., Ruvolo, P.L.: ELLA: an efficient lifelong learning algorithm. In: Proceedings of the International Conference on Machine Learning (2013)
12. Jebara, T.: Multi-task feature and kernel selection for SVMs. In: Proceedings of the International Conference on Machine Learning (2004)
13. Jebara, T.: Multitask Sparsity via Maximum Entropy Discrimination. Journal of Machine Learning Research (2011)
14. Gönen, M., Kandemir, M., Kaski, S.: Multitask learning using regularized multiple kernel learning. In: Lu, B.-L., Zhang, L., Kwok, J. (eds.) ICONIP 2011, Part II. LNCS, vol. 7063, pp. 500–509. Springer, Heidelberg (2011)
15. Lampert, C.H., Blaschko, M.B.: A multiple kernel learning approach to joint multi-class object detection. In: Rigoll, G. (ed.) DAGM 2008. LNCS, vol. 5096, pp. 31–40. Springer, Heidelberg (2008)
16. Samek, W., Binder, A., Kawanabe, M.: Multi-task learning via non-sparse multiple kernel learning. In: Computer Analysis of Images and Patterns (2011)
17. Rakotomamonjy, A., Flamary, R., Gasso, G., Canu, S.: lp-lq penalty for sparse linear and sparse multiple kernel multi-task learning. IEEE Transactions on Neural Networks (2011)
18. Zhou, Y., Jin, R., Hoi, S.C.H.: Exclusive lasso for multi-task feature selection. In: Proceedings of the Conference on Uncertainty in Artificial Intelligence (2010)
19. Maurer, A.: Transfer bounds for linear feature learning. Machine Learning **75** (2009)

Permutational Rademacher Complexity
A New Complexity Measure for Transductive Learning

Ilya Tolstikhin[1]([⊠]), Nikita Zhivotovskiy[2,3], and Gilles Blanchard[4]

[1] Max-Planck-Institute for Intelligent Systems, Tübingen, Germany
ilya@tuebingen.mpg.de
[2] Moscow Institute of Physics and Technology, Moscow, Russia
[3] Institute for Information Transmission Problems, Moscow, Russia
nikita.zhivotovskiy@phystech.edu
[4] Department of Mathematics, Universität Potsdam, Potsdam, Germany
gilles.blanchard@math.uni-potsdam.de

Abstract. Transductive learning considers situations when a learner observes m labelled training points and u unlabelled test points with the final goal of giving correct answers for the test points. This paper introduces a new complexity measure for transductive learning called *Permutational Rademacher Complexity* (PRC) and studies its properties. A novel symmetrization inequality is proved, which shows that PRC provides a tighter control over expected suprema of empirical processes compared to what happens in the standard i.i.d. setting. A number of comparison results are also provided, which show the relation between PRC and other popular complexity measures used in statistical learning theory, including Rademacher complexity and Transductive Rademacher Complexity (TRC). We argue that PRC is a more suitable complexity measure for transductive learning. Finally, these results are combined with a standard concentration argument to provide novel data-dependent risk bounds for transductive learning.

Keywords: Transductive learning · Rademacher complexity · Statistical learning theory · Empirical processes · Concentration inequalities

1 Introduction

Rademacher complexities ([14], [2]) play an important role in the widely used concentration-based approach to statistical learning theory [4], which is closely related to the analysis of empirical processes [21]. They measure a complexity of function classes and provide data-dependent risk bounds in the standard i.i.d. framework of inductive learning, thanks to symmetrization and concentration inequalities. Recently, a number of attempts were made to apply this machinery also to the *transductive learning* setting [22]. In particular, the authors of [10] introduced a notion of *transductive Rademacher complexity* and provided an extensive study of its properties, as well as general transductive risk bounds based on this new complexity measure.

© Springer International Publishing Switzerland 2015
K. Chaudhuri et al. (Eds.): ALT 2015, LNAI 9355, pp. 209–223, 2015.
DOI: 10.1007/978-3-319-24486-0_14

In the transductive learning, a learner observes m labelled training points and u unlabelled test points. The goal is to give correct answers on the test points. Transductive learning naturally appears in many modern large-scale applications, including text mining, recommender systems, and computer vision, where often the objects to be classified are available beforehand. There are two different settings of transductive learning, defined by V. Vapnik in his book [22, Chap.8]. The first one assumes that all the objects from the training and test sets are generated i.i.d. from an unknown distribution P. The second one is *distribution free*, and it assumes that the training and test sets are realized by a uniform and random partition of a fixed and finite general population of cardinality $N := m + u$ into two disjoint subsets of cardinalities m and u; moreover, no assumptions are made regarding the underlying source of this general population. The second setting has gained much attention[1] ([22], [9], [7], [10], [8], and [20]), probably due to the fact that any upper risk bound for this setting directly implies a risk bound also for the first setting [22, Theorem8.1]. In essence, the second setting studies uniform deviations of risks computed on two disjoint finite samples. Following Vapnik's discussion in [6, p.458], we would also like to emphasize that the second setting of transductive learning naturally appears as a middle step in proofs of the standard inductive risk bounds, as a result of symmetrization or the so-called *double-sample* trick. This way better transductive risk bounds also translate into better inductive ones.

An important difference between the two settings discussed above lies in the fact that the m elements of the training set in the second setting are interdependent, because they are sampled uniformly *without replacement* from the general population. As a result, the standard techniques developed for inductive learning, including concentration and Rademacher complexities mentioned in the beginning, can not be applied in this setting, since they are heavily based on the i.i.d. assumption. Therefore, it is important to study empirical processes in the setting of sampling without replacement.

Previous Work. A large step in this direction was made in [10], where the authors presented a version of McDiarmid's bounded difference inequality [5] for sampling without replacement together with the Transductive Rademacher Complexity (TRC). As a main application the authors derived an upper bound on the binary test error of a transductive learning algorithm in terms of TRC. However, the analysis of [10] has a number of shortcomings. Most importantly, TRC depends on the unknown labels of the test set. In order to obtain computable risk bounds, the authors resorted to the contraction inequality [15], which is known to be a loose step [17], since it destroys any dependence on the labels.

Another line of work was presented in [20], where variants of Talagrand's concentration inequality were derived for the setting of sampling without replacement. These inequalities were then applied to achieve transductive risk bounds with fast rates of convergence $o(m^{-1/2})$, following a *localized* approach [1]. In contrast, in this work we consider only the worst-case analysis based on the

---

[1]   For the extensive overview of transductive risk bounds we refer the reader to [18].

global complexity measures. An analysis under additional assumptions on the problem at hand, including Mammen-Tsybakov type low noise conditions [4], is an interesting open question and left for future work.

Summary of Our Results. This paper continues the analysis of empirical processes indexed by arbitrary classes of uniformly bounded functions in the setting of sampling without replacement, initiated by [10]. We introduce a new complexity measure called *permutational Rademacher complexity* (PRC) and argue that it captures the nature of this setting very well. Due to space limitations we present the analysis of PRC only for the special case when the training and test sets have the same size $m = u$, which is nonetheless sufficiently illustrative[2].

We prove a novel symmetrization inequality (Theorem 2), which shows that the expected PRC and the expected suprema of empirical processes when sampling without replacement are equivalent up to multiplicative constants. Quite remarkably, the new upper and lower bounds (the latter is often called *desymmetrization inequality*) both hold without any additive terms when $m = u$, in contrast to the standard i.i.d. setting, where an additive term of order $O(m^{-1/2})$ is unavoidable in the lower bound. For TRC even the upper symmetrization inequality [10, Lemma 4] includes an additive term of the order $O(m^{-1/2})$ and no desymmetrization inequality is known. This suggests that PRC may be a more suitable complexity measure for transductive learning. We would also like to note that the proof of our new symmetrization inequality is surprisingly simple, compared to the one presented in [10].

Next we compare PRC with other popular complexity measures used in statistical learning theory. In particular, we provide achievable upper and lower bounds, relating PRC to the conditional Rademacher complexity (Theorem 3). These bounds show that the PRC is upper and lower bounded by the conditional Rademacher complexity up to additive terms of orders $o(m^{-1/2})$ and $O(m^{-1/2})$ respectively, which are achievable (Lemma 1). In addition to this, Theorem 3 also significantly improves bounds on the complexity measure called *maximum discrepancy* presented in [2, Lemma 3]. We also provide a comparison between expected PRC and TRC (Corollary 1), which shows that their values are close up to small multiplicative constants and additive terms of order $O(m^{-1/2})$.

Finally, we apply these results to obtain a new computable data-dependent risk bound for transductive learning based on the PRC (Theorem 5), which holds for any bounded loss functions. We conclude by discussing the advantages of the new risk bound over the previously best known one of [10].

2 Notations

We will use calligraphic symbols to denote sets, with subscripts indicating their cardinalities: $\text{card}(\mathcal{Z}_m) = m$. For any function f we will denote its average value computed on a finite set S by $\bar{f}(S)$. In what follows we will consider an arbitrary

---

[2] All the results presented in this paper are also available for the general $m \neq u$ case, but we defer them to a future extended version of this paper.

space $\mathcal{Z}$ (for instance, a space of input-output pairs) and class F of functions (for instance, loss functions) mapping $\mathcal{Z}$ to $\mathbb{R}$. Most of the proofs are deferred to the last section for improved readability.

Arguably, one of the most popular complexity measures used in statistical learning theory is the Rademacher complexity ([15], [14], [2]):

Definition 1 (Conditional Rademacher complexity). *Fix any subset* $\mathcal{Z}_m = \{Z_1, \ldots, Z_m\} \subseteq \mathcal{Z}$. *The following random quantity is commonly known as a* conditional Rademacher complexity:

$$\hat{R}_m(F, \mathcal{Z}_m) = \mathbb{E}_\epsilon \left[\frac{2}{m} \sup_{f \in F} \sum_{i=1}^m \epsilon_i f(Z_i) \right],$$

where $\epsilon = \{\epsilon_i\}_{i=1}^m$ *are i.i.d. Rademacher signs, taking values* ± 1 *with probabilities* $1/2$. *When the set* $\mathcal{Z}_m$ *is clear from the context we will simply write* $\hat{R}_m(F)$.

As discussed in the introduction, Rademacher complexities play an important role in the analysis of empirical processes and statistical learning theory. However, this measure of complexity was devised mainly for the i.i.d. setting, which is different from our setting of sampling without replacement. The following complexity measure was introduced in [10] to overcome this issue:

Definition 2 (Transductive Rademacher complexity). *Fix any set* $\mathcal{Z}_N = \{Z_1, \ldots, Z_N\} \subseteq \mathcal{Z}$, *positive integers* m, u *such that* $N = m + u$, *and* $p \in [0, \frac{1}{2}]$. *The following quantity is called* Transductive Rademacher complexity *(TRC):*

$$\hat{R}_{m+u}^{td}(F, \mathcal{Z}_N, p) = \left(\frac{1}{m} + \frac{1}{u} \right) \mathbb{E}_\sigma \left[\sup_{f \in F} \sum_{i=1}^N \sigma_i f(Z_i) \right],$$

where $\sigma = \{\sigma_1\}_{i=1}^{m+u}$ *are i.i.d. random variables taking values* ± 1 *with probabilities* p *and* 0 *with probability* $1 - 2p$.

We summarize the importance of these two complexity measures in the analysis of empirical processes when sampling without replacement in the following result:

Theorem 1. *Fix an* N-element subset $\mathcal{Z}_N \subseteq \mathcal{Z}$ *and let* $m < N$ *elements of* $\mathcal{Z}_m$ *be sampled uniformly without replacement from* $\mathcal{Z}_N$. *Also let* m *elements of* $\mathcal{X}_m$ *be sampled uniformly with replacement from* $\mathcal{Z}_N$. *Denote* $\mathcal{Z}_u := \mathcal{Z}_N \setminus \mathcal{Z}_m$ *with* $u := \mathrm{card}(\mathcal{Z}_u) = N - m$. *The following upper bound in terms of the i.i.d. Rademacher complexity was provided in [20]:*

$$\mathbb{E}_{\mathcal{Z}_m} \sup_{f \in F} \left(\bar{f}(\mathcal{Z}_u) - \bar{f}(\mathcal{Z}_m) \right) \leq \mathbb{E}_{\mathcal{X}_m} \left[\hat{R}_m(F, \mathcal{X}_m) \right]. \tag{1}$$

The following bound in terms of TRC was provided in [10]. Assume that functions in F *are uniformly bounded by* B. *Then for* $p_0 := \frac{mu}{N^2}$ *and* $c_0 < 5.05$:

$$\mathbb{E}_{\mathcal{Z}_m} \sup_{f \in F} \left(\bar{f}(\mathcal{Z}_u) - \bar{f}(\mathcal{Z}_m) \right) \leq \hat{R}_{m+u}^{td}(F, \mathcal{Z}_N, p_0) + c_0 B \frac{N \sqrt{\min(m, u)}}{mu}. \tag{2}$$

While (1) did not explicitly appear in [20], it can be immediately derived using [20, Corollary8] and i.i.d. symmetrization of [13, Theorem 2.1].

Finally, we introduce our new complexity measure:

Definition 3 (Permutational Rademacher complexity). *Let $\mathcal{Z}_m \subseteq \mathcal{Z}$ be any fixed set of cardinality m. For any $n \in \{1, \ldots, m-1\}$ the following quantity will be called a* permutational Rademacher complexity (PRC):

$$\hat{Q}_{m,n}(F, \mathcal{Z}_m) = \underset{\mathcal{Z}_n}{\mathbb{E}} \sup_{f \in F} \left(\bar{f}(\mathcal{Z}_k) - \bar{f}(\mathcal{Z}_n)\right),$$

where $\mathcal{Z}_n$ is a random subset of $\mathcal{Z}_m$ containing n elements sampled uniformly without replacement and $\mathcal{Z}_k := \mathcal{Z}_m \setminus \mathcal{Z}_n$. When the set $\mathcal{Z}_m$ is clear from the context we will simply write $\hat{Q}_{m,n}(F)$.

The name PRC is explained by the fact that if m is even then the definitions of $\hat{Q}_{m,m/2}(F)$ and $\hat{R}_m(F)$ are very similar. Indeed, the only difference is that the expectation in the PRC is over the randomly permuted sequence containing *equal number* of " -1 " and " $+1$ ", whereas in Rademacher complexity the average is w.r.t. all the possible sequences of signs. The term "permutation complexity" has already appeared in [16], where it was used to denote a novel complexity measure for a model selection. However, this measure was specific to the i.i.d. setting and *binary* loss. Moreover, the bounds presented in [16] were of the same order as the risk bounds based on the Rademacher complexity with worse constants in the slack term.

3   Symmetrization and Comparison Results

We start with showing a version of the i.i.d. symmetrization inequality (references can be found in [15], [13]) for the setting of sampling without replacement. It shows that the expected supremum of empirical processes in this setting is up to multiplicative constants equivalent to the expected PRC.

Theorem 2. *Fix an N-element subset $\mathcal{Z}_N \subseteq \mathcal{Z}$ and let $m < N$ elements of $\mathcal{Z}_m$ be sampled uniformly without replacement from $\mathcal{Z}_N$. Denote $\mathcal{Z}_u := \mathcal{Z}_N \setminus \mathcal{Z}_m$ with $u := \mathrm{card}(\mathcal{Z}_u) = N - m$. If $m = u$ and m is even then for any $n \in \{1, \ldots, m-1\}$:*

$$\frac{1}{2} \underset{\mathcal{Z}_m}{\mathbb{E}} \left[\hat{Q}_{m,m/2}(F, \mathcal{Z}_m)\right] \leq \underset{\mathcal{Z}_m}{\mathbb{E}} \sup_{f \in F} \left(\bar{f}(\mathcal{Z}_u) - \bar{f}(\mathcal{Z}_m)\right) \leq \underset{\mathcal{Z}_m}{\mathbb{E}} \left[\hat{Q}_{m,n}(F, \mathcal{Z}_m)\right].$$

The inequalities also hold if we include absolute values inside the suprema.

Proof. The proof can be found in Sect. 5.1.

This inequality should be compared to the previously known complexity bounds of Theorem 1. First of all, in contrast to (1) and (2) the new bound provides a two sided control, which shows that PRC is a "correct" complexity measure for our setting. It is also remarkable that the lower bound (commonly known as the *desymmetrization inequality*) does not include any additive terms, since in the standard i.i.d. setting the lower bound holds only up to an additive term of order $O(m^{-1/2})$ [13, Sect.2.1]. Also note that this result does not assume the boundedness of functions in F, which is a necessary assumptions both in (2) and in the i.i.d. desymmetrization inequality.

Next we compare PRC with the conditional Rademacher complexity:

Theorem 3. *Let* $\mathcal{Z}_m \subseteq \mathcal{Z}$ *be any fixed set of even cardinality* m. *Then:*

$$\hat{Q}_{m,m/2}(F, \mathcal{Z}_m) \le \left(1 + \frac{2}{\sqrt{2\pi m} - 2}\right) \hat{R}_m(F, \mathcal{Z}_m). \tag{3}$$

Moreover, if the functions in F *are absolutely bounded by* B *then*

$$\left|\hat{Q}_{m,m/2}(F, \mathcal{Z}_m) - \hat{R}_m(F, \mathcal{Z}_m)\right| \le \frac{2B}{\sqrt{m}}. \tag{4}$$

The results also hold if we include absolute values inside suprema in $\hat{Q}_{m,n}, \hat{R}_m$.

Proof. Conceptually the proof is based on the coupling between a sequence $\{\epsilon_i\}_{i=1}^m$ of i.i.d. Rademacher signs and a uniform random permutation $\{\eta_i\}_{i=1}^m$ of a set containing $m/2$ plus and $m/2$ minus signs. This idea was inspired by the techniques used in [11]. The detailed proof can be found in Sect. 5.2.

Note that a typical order of $\hat{R}_m(F)$ is $O(m^{-1/2})$, thus the multiplicative upper bound (3) can be much tighter than the upper bound of (4). We would also like to note that Theorem 3 significantly improves bounds of Lemma 3 in [2], which relate the so-called *maximal discrepancy* measure of the class F to its Rademacher complexity (for the further discussion we refer to Appendix).

Our next result shows that bounds of Theorem 3 are essentially tight.

Lemma 1. *Let* $\mathcal{Z}_m \subseteq \mathcal{Z}$ *with even* m. *There are two finite classes* F_m' *and* F_m'' *of functions mapping* $\mathcal{Z}$ *to* $\mathbb{R}$ *and absolutely bounded by* 1, *such that:*

$$\hat{Q}_{m,m/2}(F_m', \mathcal{Z}_m) = 0, \quad (2m)^{-1/2} \le \hat{R}_m(F_m', \mathcal{Z}_m) \le 2m^{-1/2}; \tag{5}$$

$$\hat{Q}_{m,m/2}(F_m'', \mathcal{Z}_m) = 1, \quad 1 - \sqrt{\frac{2}{\pi m}} \le \hat{R}_m(F_m'', \mathcal{Z}_m) \le 1 - \frac{4}{5}\sqrt{\frac{2}{\pi m}}. \tag{6}$$

Proof. The proof can be found in Sect. 5.3.

Inequalities (5) simultaneously show that (a) the order $O(m^{-1/2})$ of the additive bound (4) can not be improved, and (b) the multiplicative upper bound (3) can not be reversed. Moreover, it can be shown using (6) that the factor appearing in (3) can not be improved to $1 + o(m^{-1/2})$.

Finally, we compare PRC to the transductive Rademacher complexity:

Lemma 2. *Fix any set* $\mathcal{Z}_N = \{Z_1, \ldots, Z_N\} \subseteq \mathcal{Z}$. *If* $m = u$ *and* $N = m + u$:

$$\hat{R}_N(F, \mathcal{Z}_N) \le \hat{R}_{m+u}^{td}(F, \mathcal{Z}_N, 1/4) \le 2\hat{R}_N(F, \mathcal{Z}_N).$$

Proof. The upper bound was presented in [10, Lemma 1]. For the lower bound, notice that if $p = 1/4$ the i.i.d. signs σ_i presented in Definition 2 have the same distribution as $\epsilon_i \eta_i$, where ϵ_i are i.i.d. Rademacher signs and η_i are i.i.d. Bernoulli random variables with parameters $1/2$. Thus, Jensen's inequality gives:

$$\hat{R}_{m+u}^{td}(F, \mathcal{Z}_N, 1/4) = \frac{4}{N} \mathop{\mathbb{E}}_{(\epsilon, \eta)} \left[\sup_{f \in F} \sum_{i=1}^{m+u} \epsilon_i \eta_i f(Z_i) \right] \geq \frac{4}{N} \mathop{\mathbb{E}}_{\epsilon} \left[\sup_{f \in F} \sum_{i=1}^{m+u} \epsilon_i \frac{1}{2} f(Z_i) \right].$$

Together with Theorems 2 and 3 this result shows that when $m = u$ the PRC can not be much larger than transductive Rademacher complexity:

Corollary 1. *Using notations of Theorem 2, we have:*

$$\mathop{\mathbb{E}}_{\mathcal{Z}_m} \left[\hat{Q}_{m,m/2}(F, \mathcal{Z}_m) \right] \leq \left(2 + \frac{4}{\sqrt{2\pi N} - 2} \right) \hat{R}_{m+u}^{td}(F, \mathcal{Z}_N, 1/4).$$

If functions in F are uniformly bounded by B then we also have a lower bound:

$$\mathop{\mathbb{E}}_{\mathcal{Z}_m} \left[\hat{Q}_{m,m/2}(F, \mathcal{Z}_m) \right] \geq \frac{1}{2} \hat{R}_{m+u}^{td}(F, \mathcal{Z}_N, 1/4) + \frac{2B}{\sqrt{N}}.$$

Proof. Simply notice that $\mathbb{E}_{\mathcal{Z}_m} \left[\sup_{f \in F} \left(\bar{f}(\mathcal{Z}_u) - \bar{f}(\mathcal{Z}_m) \right) \right] = \hat{Q}_{N,m}(F, \mathcal{Z}_N)$.

4   Transductive Risk Bounds

Next we will use the results of Sect. 3 to obtain a new transductive risk bound. First we will shortly describe the setting.

We will consider the second, distribution-free setting of transductive learning described in the introduction. Fix any finite *general population* of input-output pairs $\mathcal{Z}_N = \{(x_i, y_i)\}_{i=1}^N \subseteq \mathcal{X} \times \mathcal{Y}$, where $\mathcal{X}$ and $\mathcal{Y}$ are arbitrary input and output spaces. We make no assumptions regarding underlying source of $\mathcal{Z}_N$. The learner receives the labeled *training set* $\mathcal{Z}_m$ consisting of $m < N$ elements sampled uniformly without replacement from $\mathcal{Z}_N$. The remaining *test set* $\mathcal{Z}_u := \mathcal{Z}_N \setminus \mathcal{Z}_m$ is presented to the learner *without labels* (we will use $\mathcal{X}_u$ to denote the inputs of $\mathcal{Z}_u$). The goal of the learner is to find a predictor in the fixed *hypothesis class* $\mathcal{H}$ based on the training sample $\mathcal{Z}_m$ and unlabelled test points $\mathcal{X}_u$, which has a small test risk measured using bounded *loss function* $\ell \colon \mathcal{Y} \times \mathcal{Y} \to [0, 1]$. For $h \in \mathcal{H}$ and $(x, y) \in \mathcal{Z}_N$ denote $\ell_h(x, y) = \ell(h(x), y)$ and also denote the loss class $L_{\mathcal{H}} = \{\ell_h \colon h \in \mathcal{H}\}$. Then the test and training risks of $h \in \mathcal{H}$ are defined as $\mathrm{err}_u(h) := \overline{\ell_h}(\mathcal{Z}_u)$ and $\mathrm{err}_m(h) := \overline{\ell_h}(\mathcal{Z}_m)$ respectively.

Following risk bound in terms of TRC was presented in [10, Corollary 2]:

Theorem 4 ([10]). *If $m = u$ then with probability at least $1 - \delta$ over the random training set $\mathcal{Z}_m$ any $h \in \mathcal{H}$ satisfies:*

$$\mathrm{err}_u(h) \leq \mathrm{err}_m(h) + \hat{R}_{m+u}^{td}(L_{\mathcal{H}}, \mathcal{Z}_N, 1/4) + 11\sqrt{\frac{2}{N}} + \sqrt{\frac{2N \log(1/\delta)}{(N - 1/2)^2}}. \tag{7}$$

Using results of Sect. 3 we obtain the following risk bound:

Theorem 5. *If $m = u$ and $n \in \{1, \ldots, m-1\}$ then with probability at least $1 - \delta$ over the random training set $\mathcal{Z}_m$ any $h \in \mathcal{H}$ satisfies:*

$$\mathrm{err}_u(h) \leq \mathrm{err}_m(h) + \mathop{\mathbb{E}}_{\mathcal{S}_m} \left[\hat{Q}_{m,n}(L_{\mathcal{H}}, \mathcal{Z}_m) \right] + \sqrt{\frac{2N \log(1/\delta)}{(N - 1/2)^2}}. \tag{8}$$

Moreover, with probability at least $1 - \delta$ any $h \in \mathcal{H}$ satisfies:

$$\mathrm{err}_u(h) \leq \mathrm{err}_m(h) + \hat{Q}_{m,n}(L_{\mathcal{H}}, \mathcal{Z}_m) + 2\sqrt{\frac{2N \log(2/\delta)}{(N - 1/2)^2}}. \tag{9}$$

Proof. The proof can be found in Sect. 5.4.

We conclude by comparing risk bounds of Theorems 5 and 4:

1. First of all, the upper bound of (9) is computable. This bound is based on the concentration argument, which shows that the expected PRC (appearing in (8)) can be nicely estimated using the training set. Meanwhile, the upper bound of (7) depends on the *unknown* labels of the test set through TRC. In order to make it computable the authors of [10] resorted to the contraction inequality, which allows to drop any dependence on the labels for Lipschitz losses, which is known to be a loose step [17].

2. Moreover, we would like to note that for binary loss function TRC (as well as the Rademacher complexity) does not depend on the labels at all. Indeed, this can be shown by writing $\ell_{01}(y, y') = (1 - yy')/2$ for $y, y' \in \{-1, +1\}$ and noting that σ_i and $\sigma_i y$ are identically distributed for σ_i used in Definition 2. This is not true for PRC, which is *sensitive* to the labels even in this setting. As a future work we hope to use this fact for analysis in the low noise setting [4].

3. The slack term appearing in (8) is significantly smaller than the one of (7). For instance, if $\delta = 0.01$ then the latter is 13 times larger. This is caused by the additive term in symmetrization inequality (2). At the same time, Corollary 1 shows that the complexity term appearing in (8) is at most two times larger than TRC, appearing in (7).

4. Comparison result of Theorem 3 shows that the upper bound of (9) is also tighter than the one which can be obtained using (1) and conditional Rademacher complexity.

5. Similar upper bounds (up to extra factor of 2) also hold for the *excess risk* $\mathrm{err}_u(h_m) - \inf_{h \in \mathcal{H}} \mathrm{err}_u(h)$, where h_m minimizes the training risk err_m over $\mathcal{H}$. This can be proved using a similar argument to Theorem 5.

6. Finally, one more application of the concentration argument can simplify the computation of PRC, by estimating the expected value appearing in Definition 3 with only one random partition of $\mathcal{Z}_m$.

5   Full Proofs

5.1   Proof of Theorem 2

Lemma 3. *For $0 < m \leq N$ let $\mathcal{S}_m := \{s_1, \ldots, s_m\}$ be sampled uniformly without replacement from a finite set of real numbers $\mathcal{C} = \{c_1, \ldots, c_N\} \subset \mathbb{R}$. Then:*

$$\mathop{\mathbb{E}}_{\mathcal{S}_m} \left[\frac{1}{m} \sum_{i=1}^{m} s_i \right] = \frac{1}{\binom{N}{m}} \sum_{\mathcal{S}_m \subseteq \mathcal{C}} \frac{1}{m} \sum_{z \in \mathcal{S}_m} z = \frac{1}{m\binom{N}{m}} \sum_{i=1}^{N} \binom{N-1}{m-1} c_i = \frac{1}{N} \sum_{i=1}^{N} c_i.$$

Proof (of Theorem 2). Fix any *positive* integers n and k such that $n + k = m$, which implies $n < m$ and $k < m = u$. Note that Lemma 3 implies:

$$\bar{f}(\mathcal{Z}_u) = \mathop{\mathbb{E}}_{\mathcal{S}_k} \left[\bar{f}(\mathcal{S}_k) \right], \quad \bar{f}(\mathcal{Z}_m) = \mathop{\mathbb{E}}_{\mathcal{S}_n} \left[\bar{f}(\mathcal{S}_n) \right],$$

where $\mathcal{S}_k$ and $\mathcal{S}_n$ are sampled uniformly without replacement from $\mathcal{Z}_u$ and $\mathcal{Z}_m$ respectively. Using Jensen's inequality we get:

$$\mathop{\mathbb{E}}_{\mathcal{Z}_m} \sup_{f \in F} \left(\bar{f}(\mathcal{Z}_u) - \bar{f}(\mathcal{Z}_m) \right) = \mathop{\mathbb{E}}_{\mathcal{Z}_m} \sup_{f \in F} \left(\mathop{\mathbb{E}}_{\mathcal{S}_k} \left[\bar{f}(\mathcal{S}_k) \right] - \mathop{\mathbb{E}}_{\mathcal{S}_n} \left[\bar{f}(\mathcal{S}_n) \right] \right)$$

$$\leq \mathop{\mathbb{E}}_{(\mathcal{Z}_m, \mathcal{S}_k, \mathcal{S}_n)} \sup_{f \in F} \left(\bar{f}(\mathcal{S}_k) - \bar{f}(\mathcal{S}_n) \right). \tag{10}$$

The marginal distribution of $(\mathcal{S}_k, \mathcal{S}_n)$, appearing in (10), can be equivalently described by first sampling $\mathcal{Z}_m$ from $\mathcal{Z}_N$, then $\mathcal{S}_n$ from $\mathcal{Z}_m$ (both times uniformly without replacement), and setting $\mathcal{S}_k := \mathcal{Z}_m \setminus \mathcal{S}_n$ (recall that $n + k = m$). Thus

$$\mathop{\mathbb{E}}_{(\mathcal{Z}_m, \mathcal{S}_k, \mathcal{S}_n)} \sup_{f \in F} \left(\bar{f}(\mathcal{S}_k) - \bar{f}(\mathcal{S}_n) \right) = \mathop{\mathbb{E}}_{\mathcal{Z}_m} \left[\mathop{\mathbb{E}}_{\mathcal{S}_n} \left[\sup_{f \in F} \left(\bar{f}(\mathcal{Z}_m \setminus \mathcal{S}_n) - \bar{f}(\mathcal{S}_n) \right) \Big| \mathcal{Z}_m \right] \right],$$

which completes the proof of the upper bound.

We have shown that for $n \in \{1, \ldots, m-1\}$ and $k := m - n$:

$$\mathop{\mathbb{E}}_{\mathcal{Z}_m} \left[\hat{Q}_{m,n}(F, \mathcal{Z}_m) \right] = \mathop{\mathbb{E}}_{(\mathcal{Z}_k, \mathcal{Z}_n)} \sup_{f \in F} \left(\bar{f}(\mathcal{Z}_k) - \bar{f}(\mathcal{Z}_n) \right), \tag{11}$$

where $\mathcal{Z}_n$ and $\mathcal{Z}_k$ are sampled uniformly without replacement from $\mathcal{Z}_N$ and $\mathcal{Z}_N \setminus \mathcal{Z}_n$ respectively. Let $\mathcal{Z}_{m-n}$ be sampled uniformly without replacement from $\mathcal{Z}_N \setminus (\mathcal{Z}_n \cup \mathcal{Z}_k)$ and let $\mathcal{Z}_{u-k}$ be the remaining $u - k$ elements of $\mathcal{Z}_N$. Using Lemma 3 once again we get:

$$\mathbb{E} \left[\bar{f}(\mathcal{Z}_{m-n}) \big| (\mathcal{Z}_n, \mathcal{Z}_k) \right] = \mathbb{E} \left[\bar{f}(\mathcal{Z}_{u-k}) \big| (\mathcal{Z}_n, \mathcal{Z}_k) \right].$$

We can rewrite the r.h.s. of (11) as:

$$\mathop{\mathbb{E}}_{(\mathcal{Z}_n, \mathcal{Z}_k)} \sup_{f \in F} \left(\bar{f}(\mathcal{Z}_k) - \bar{f}(\mathcal{Z}_n) + \mathbb{E} \left[\bar{f}(\mathcal{Z}_{u-k}) - \bar{f}(\mathcal{Z}_{m-n}) \big| (\mathcal{Z}_n, \mathcal{Z}_k) \right] \right)$$

$$\leq \mathbb{E} \sup_{f \in F} \left(\bar{f}(\mathcal{Z}_k) - \bar{f}(\mathcal{Z}_n) + \bar{f}(\mathcal{Z}_{u-k}) - \bar{f}(\mathcal{Z}_{m-n}) \right),$$

where we have used Jensen's inequality. If we take $n^* = k^* = m/2$ we get

$$\underset{\mathcal{Z}_m}{\mathbb{E}} \left[\hat{Q}_{m,m/2}(F, \mathcal{Z}_m) \right] \leq \mathbb{E} \sup_{f \in F} \left(2\bar{f}(\mathcal{Z}_{k^*} \cup \mathcal{Z}_{u-k^*}) - 2\bar{f}(\mathcal{Z}_{n^*} \cup \mathcal{Z}_{m-n^*}) \right).$$

It is left to notice that the random subsets $\mathcal{Z}_{k^*} \cup \mathcal{Z}_{u-k^*}$ and $\mathcal{Z}_{n^*} \cup \mathcal{Z}_{m-n^*}$ have the same distributions as $\mathcal{Z}_u$ and $\mathcal{Z}_m$.

5.2   Proof of Theorem 3

Let $m = 2 \cdot n$, $\boldsymbol{\epsilon} = \{\epsilon_i\}_{i=1}^m$ be i.i.d. Rademacher signs, and $\boldsymbol{\eta} = \{\eta_i\}_{i=1}^m$ be a uniform random permutation of a set containing n plus and n minus signs. The proof of Theorem 3 is based on the coupling of random variables $\boldsymbol{\epsilon}$ and $\boldsymbol{\eta}$, which is described in Lemma 4. We will need a number of definitions. Consider binary cube $B_m := \{-1, +1\}^m$. Denote $S_m := \{v \in B_m \colon \sum_{i=1}^m v_i = 0\}$, which is a set of all the vectors in B_m having equal number of plus and minus signs. For any $v \in B_m$ denote $\|v\|_1 = \sum_{i=1}^m |v_i|$ and consider the following set:

$$T(v) = \arg\min_{v' \in S_m} \|v - v'\|_1,$$

which consists of the points in S_m closest to v in Hamming metric. For any $v \in B_m$ let $t(v)$ be a random element of $T(v)$, distributed uniformly. We will use $t_i(v)$ to denote i-th coordinate of the vector $t(v)$.

Remark 1. If $v \in S_m$ then $T(v) = \{v\}$. Otherwise, $T(v)$ will clearly contain more than one element of S_m. Namely, it can be shown, that if for some positive integer q it holds that $\sum_{i=1}^m v_i = q$, then q is necessarily even and $T(v)$ consists of all the vectors in S_m which can be obtained by replacing $q/2$ of $+1$ signs in v with -1 signs, and thus in this case $\text{card}(T(v)) = \binom{(m+q)/2}{q/2}$.

Lemma 4 (Coupling). *Assume that $m = 2 \cdot n$. Then the random sequence $t(\boldsymbol{\epsilon})$ has the same distribution as $\boldsymbol{\eta}$.*

Proof. Note that the support of $t(\boldsymbol{\epsilon})$ is equal to S_m. From symmetry it is easy to conclude that the distribution of $t(\boldsymbol{\epsilon})$ is exchangable. This means that it is invariant under permutations and as a consequence uniform on S_m.

Next result is in the core of the multiplicative upper bound (3).

Lemma 5. *Assume that $m = 2 \cdot n$. For any $q \in \{1, \ldots, m\}$ the following holds:*

$$\mathbb{E}[\epsilon_q | t(\boldsymbol{\epsilon})] = \left(1 - 2^{-m} \binom{m}{n} \right) t_q(\boldsymbol{\epsilon}) \geq \left(1 - 2(2\pi m)^{-1/2} \right) t_q(\boldsymbol{\epsilon}).$$

Proof. We will first upper bound $\mathbb{P}\{\epsilon_q \neq t_q(\boldsymbol{\epsilon}) | t(\boldsymbol{\epsilon}) = e\}$, where $e = \{e_i\}_{i=1}^m$ is (w.l.o.g.) a sequence of n plus signs followed by a sequence of n minus signs.

$$\mathbb{P}\{\epsilon_q \neq t_q(\boldsymbol{\epsilon}) | t(\boldsymbol{\epsilon}) = e\} = \frac{\mathbb{P}\{\epsilon_q \neq t_q(\boldsymbol{\epsilon}) \cap t(\boldsymbol{\epsilon}) = e\}}{\mathbb{P}\{t(\boldsymbol{\epsilon}) = e\}}$$

$$= \binom{m}{n} 2^{-m} \sum_s \mathbb{P}\{\epsilon_q \neq t_q(\boldsymbol{\epsilon}) \cap t(\boldsymbol{\epsilon}) = e | \boldsymbol{\epsilon} = s\}, \quad (12)$$

where we have used Lemma 4 and the sum is over all different sequences of m signs $s = \{s_i\}_{i=1}^m$. For any s denote $S(s) = \sum_{j=1}^n s_j$ and consider terms in (12) corresponding to s with $S(s) = 0$, $S(s) > 0$, and $S(s) < 0$:

Case 1: $S(s) = 0$. These terms will be zero, since $t(s) = s$.

Case 2: $S(s) > 0$. This means that s "has more plus signs than it should" and according to Remark 1 the mapping $t(\cdot)$ will replace several of "+1" with "-1". In particular, if $s_q = -1$ then $t_q(s) = s_q$ and thus the corresponding terms will be zero. If $s_q = 1$ and in the same time $e_q = 1$ the event $\{\epsilon_q \neq t_q(\epsilon) \cap t(\epsilon) = e\}$ also can not hold. Moreover, note that identity $e = t(s)$ can hold only if $e \in T(s)$, which necessarily leads to

$$\{j \in \{1, \ldots, m\}: s_j = -1\} \subseteq \{j \in \{1, \ldots, m\}: e_j = -1\}. \tag{13}$$

From this we conclude that if $q \in \{1, \ldots, n\}$ then all the terms corresponding to s with $S(s) > 0$ are zero. We will use $U_q(e)$ to denote the subset of B_m consisting of sequences s, such that (a) $S(s) > 0$, (b) $s_q = 1$, and (c) condition (13) holds. It can be seen that if $s \in U_q(e)$ then:

$$\mathbb{P}\{\epsilon_q \neq t_q(\epsilon) \cap t(\epsilon) = e | \epsilon = s\} = \binom{n + S(s)/2}{S(s)/2}^{-1}.$$

This holds since, according to Remark 1, $t(\epsilon)$ can take exactly $\binom{n+S(s)/2}{S(s)/2}$ different values, while only one of them is equal to e.

Let us compute the cardinality of $U_q(e)$ for $q \in \{n+1, \ldots, m\}$. It is easy to check that condition $S(s) = 2j$ for some positive integer j implies that s has exactly $n - j$ minus signs. Considering the fact that $s_q = 1$ for $s \in U_q(e)$ we have:

$$\text{card}\,(U_q(e)) = \binom{n-1}{n-j}.$$

Combining everything together we have:

$$\sum_{s:\, S(s)>0} \mathbb{P}\{\epsilon_q \neq t_q(\epsilon) \cap t(\epsilon) = e | \epsilon = s\} = \mathbb{1}\{q > n\} \sum_{j=1}^n \frac{\binom{n-1}{n-j}}{\binom{n+j}{j}}.$$

Finally, it is easy to show using induction that:

$$\sum_{j=1}^n \frac{\binom{n-1}{n-j}}{\binom{n+j}{j}} = \frac{1}{2}.$$

Case 3: $S(s) < 0$. We can repeat all the steps of the previous case and get:

$$\sum_{s:\, S(s)<0} \mathbb{P}\{\epsilon_q \neq t_q(\epsilon) \cap t(\epsilon) = e | \epsilon = s\} = \frac{1}{2}\mathbb{1}\{q \leq n\}.$$

Accounting for these three cases in (12) we conclude that

$$\mathbb{P}\{\epsilon_q \neq t_q(\epsilon) | t(\epsilon) = e\} = \frac{1}{2}\binom{m}{n}2^{-m} \leq \frac{1}{\sqrt{2\pi m}},$$

where we have used the upper bound on the binomial coefficient from [19, Corollary2.4]. We can conclude the proof of lemma by writing:

$$\mathbb{E}[\epsilon_q|t(\epsilon)] = t_q(\epsilon)\left(1 - 2\mathbb{P}\{\epsilon_q \neq t_q(\epsilon)|t(\epsilon)\}\right) \geq t_q(\epsilon)\left(1 - 2(2\pi m)^{-1/2}\right).$$

Proof (of Theorem 3). First we prove (3). Let $\mathcal{Z}_m = \{z_1, \ldots, z_m\}$. We can write:

$$\hat{Q}_{m,n}(F) = \mathbb{E}\left[\sup_{f \in F} \frac{2}{m} \sum_{i=1}^{m} t_i(\epsilon)f(z_i)\right] \tag{14}$$

$$\leq \left(1 - 2(2\pi m)^{-1/2}\right)^{-1} \mathbb{E}\left[\sup_{f \in F} \frac{2}{m} \sum_{i=1}^{m} \mathbb{E}[\epsilon_i|t(\epsilon)]f(z_i)\right] \tag{15}$$

$$\leq \left(1 + \frac{2}{\sqrt{2\pi m} - 2}\right) \mathbb{E}\left[\sup_{f \in F} \frac{2}{m} \sum_{i=1}^{m} \epsilon_i f(z_i)\right], \tag{16}$$

where we have used coupling Lemma 4 in (14), Lemma 5 in (15), and Jensen's inequality in (16). This completes the proof of (3).

Next we prove (4). We have:

$$\left|\hat{Q}_{m,n}(F) - \hat{R}_m(F)\right| = \left|\mathbb{E}_{\eta}\left[\sup_{f \in F} \frac{2}{m} \sum_{i=1}^{m} \eta_i f(z_i)\right] - \mathbb{E}_{\epsilon}\left[\sup_{f \in F} \frac{2}{m} \sum_{i=1}^{m} \epsilon_i f(z_i)\right]\right|.$$

Using Lemma 4 and Jensen's inequality we further get:

$$\left|\hat{Q}_{m,n}(F) - \hat{R}_m(F)\right|$$

$$= \left|\mathbb{E}_{\epsilon}\left[\mathbb{E}_{t}\left[\sup_{f \in F} \frac{2}{m} \sum_{i=1}^{m} t_i(\epsilon)f(z_i)\Big|\epsilon\right]\right] - \mathbb{E}_{\epsilon}\left[\sup_{f \in F} \frac{2}{m} \sum_{i=1}^{m} \epsilon_i f(z_i)\right]\right|$$

$$\leq \mathbb{E}_{\epsilon}\left[\mathbb{E}_{t}\left[\left|\sup_{f \in F} \frac{2}{m} \sum_{i=1}^{m} t_i(\epsilon)f(z_i) - \sup_{f \in F} \frac{2}{m} \sum_{i=1}^{m} \epsilon_i f(z_i)\right|\Big|\epsilon\right]\right], \tag{17}$$

where we have, perhaps misleadingly, denoted the conditional expectation with respect to the uniform choice from $T(\epsilon)$ given ϵ using $\mathbb{E}_t[\cdot|\epsilon]$. Next we have:

$$\left|\sup_{f \in F} \frac{2}{m} \sum_{i=1}^{m} t_i(\epsilon)f(z_i) - \sup_{f \in F} \frac{2}{m} \sum_{i=1}^{m} \epsilon_i f(z_i)\right| \leq \left|\sup_{f \in F} \frac{4}{m} \sum_{i \in S(\epsilon,t)} \epsilon_i f(z_i)\right|, \tag{18}$$

where $S(\epsilon, t) \subseteq \{1, \ldots, m\}$ is a subset of indices, s.t. $(t(\epsilon))_i \neq \epsilon_i$ iff $i \in S(\epsilon, t)$. We can continue by writing

$$\left|\sup_{f \in F} \frac{2}{m} \sum_{i=1}^{m} t_i(\epsilon)f(z_i) - \sup_{f \in F} \frac{2}{m} \sum_{i=1}^{m} \epsilon_i f(z_i)\right| \leq \frac{4}{m} \sup_{f \in F} \sum_{i \in S(\epsilon,t)} |f(z_i)|. \tag{19}$$

Note that since functions in F are absolutely bounded by B:

$$\sup_{f \in F} \sum_{i \in S(\epsilon, t)} |f(z_i)| \leq B \cdot \text{card}\left(S(\epsilon, t)\right).$$

Returning to (17) and using Remark 1 we obtain:

$$\left|\hat{Q}_{m,n}(F) - 2\hat{R}_m(F)\right| \leq \frac{4B}{m} \mathbb{E}_\epsilon \left[\mathbb{E}_t \left[\text{card}\left(S(\epsilon, t)\right) |\epsilon\right]\right] = \mathbb{E}_\epsilon \left[\frac{1}{2}\left|\sum_{i=1}^m \epsilon_i\right|\right].$$

Khinchin's inequality [15, Lemma 4.1] together with the best known constant due to [12] gives $\mathbb{E}_\epsilon \left[|\sum_{i=1}^m \epsilon_i|\right] \leq \sqrt{m}$, which completes the proof of (4).

5.3 Proof of Lemma 5

Proof. Let $\mathcal{Z}_m = \{z_1, \dots, z_m\}$. Take F'_m to be a set of two constant functions, $f_1(z) = 1$ and $f_2(z) = 0$ for all $z \in \mathcal{Z}$. Clearly, $\hat{Q}_{m,n}(F'_m) = 0$. In the same time:

$$\mathbb{E}_\epsilon \left[\sup_{f \in F'_m} \frac{2}{m} \sum_{i=1}^m \epsilon_i f(z_i)\right] = \mathbb{E}_\epsilon \left[\max\left\{0, \frac{2}{m}\sum_{i=1}^m \epsilon_i\right\}\right] \leq \mathbb{E}_\epsilon \left[\left|\frac{2}{m}\sum_{i=1}^m \epsilon_i\right|\right] \leq \frac{2}{\sqrt{m}},$$

where we used Khinchin's inequality. Finally, Khinchin's inequality also gives:

$$\mathbb{E}_\epsilon \left[\max\left\{0, \frac{2}{m}\sum_{i=1}^m \epsilon_i\right\}\right] = \frac{1}{2}\mathbb{E}_\epsilon \left[\left|\frac{2}{m}\sum_{i=1}^m \epsilon_i\right|\right] \geq \frac{1}{\sqrt{2m}}.$$

Next, let F''_m contain $\binom{m}{m/2}$ functions, such that their projections on $\mathcal{Z}_m$ recover all the permutations of binary vector containing equal number of 0 and 1. Clearly, in this case $\hat{Q}_{m,n}(F''_m) = 1$. Straightforward calculations show that in the same time $\hat{R}_m(F''_m) = 1 - 2^{-m}\binom{m}{n}$ and we conclude the proof using upper and lower bounds on the binomial coefficient from [19, Corollary 2.4].

5.4 Proof of Theorem 5

The following version of McDiarmid's bounded difference inequality for the setting of sampling without replacement was presented in [10, Lemma 2] and further improved in [8, Theorem 5]:

Theorem 6 ([10], [8]). *Let $\mathcal{Z}_m$ be sampled uniformly without replacement from a fixed set $\mathcal{Z}_{m+u} \subseteq \mathcal{Z}$ of $m+u$ elements. Let $g: \mathcal{Z}^m \to \mathbb{R}$ be a symmetric function s.t. for all $i = 1, \dots, m$ and for all $z_1, \dots, z_m \in \mathcal{Z}$ and $z'_1, \dots, z'_m \in \mathcal{Z}$,*

$$\left|g(z_1, \dots, z_m) - g(z_1, \dots, z_{i-1}, z'_i, z_{i+1}, \dots, z_m)\right| \leq c. \tag{20}$$

Then if $m = u$ with probability not less than $1 - \delta$ the following holds:

$$g \leq \mathbb{E}[g] + \sqrt{\frac{c^2 N^3 \log(1/\delta)}{8(N - 1/2)^2}}.$$

Note that function $\sup_{h \in \mathcal{H}}(\mathrm{err}_h(\mathcal{Z}_u) - \mathrm{err}_h(\mathcal{Z}_m))$ maps $(\mathcal{X} \times \mathcal{Y})^m$ to $\mathbb{R}$ and is of course symmetric. Straightforward calculations show that this function satisfies bounded difference condition (20) with $c = \frac{1}{m} + \frac{1}{u}$ ([10, Inequality 9]). Theorem 6 states that with probability not less than $1 - \delta$:

$$\sup_{h \in \mathcal{H}}(\mathrm{err}_u(h) - \mathrm{err}_m(h)) \leq \mathop{\mathbb{E}}_{\mathcal{S}_m}\left[\sup_{h \in \mathcal{H}}(\mathrm{err}_u(h) - \mathrm{err}_m(h))\right] + \sqrt{\frac{2N\log(1/\delta)}{(N - 1/2)^2}}. \quad (21)$$

Using upper bound of Theorem 2 with $L_{\mathcal{H}}$ in place of F we complete the proof of (8). Next, consider a symmetric function $-\hat{Q}_{m,n}(L_{\mathcal{H}}, \mathcal{Z}_m)$ which also maps $(\mathcal{X} \times \mathcal{Y})^m$ to $\mathbb{R}$. It can be shown again that it satisfies bounded difference condition (20) with $c = \frac{2}{m}$. And thus, Theorem 6 gives that with probability not less than $1 - \delta$:

$$\mathop{\mathbb{E}}_{\mathcal{S}_m}\left[\hat{Q}_{m,n}(L_{\mathcal{H}}, \mathcal{Z}_m)\right] \leq \hat{Q}_{m,n}(L_{\mathcal{H}}, \mathcal{Z}_m) + \sqrt{\frac{2N\log(1/\delta)}{(N - 1/2)^2}}. \quad (22)$$

Using this inequality together with (8) in a union bound we obtain the second inequality of the theorem.

Appendix: Improving Lemma 3 of [2]

Let μ be a probability distribution on $\mathcal{Z}$ and $\mathcal{X}_m := \{X_1, \ldots, X_m\}$ be i.i.d. samples selected according to μ. Maximal discrepancy of F was defined in [2] as:

$$\hat{D}_m(F, \mathcal{X}_m) = \sup_{f \in F}\left(\frac{2}{m}\sum_{i=1}^{m/2} f(X_i) - \frac{2}{m}\sum_{i=m/2+1}^{m} f(X_i)\right).$$

It was shown in [2] that if functions in F are uniformly bounded by 1 then:

$$\frac{1}{2}\mathbb{E}\left[\hat{R}_m(F, \mathcal{X}_m)\right] - 2\sqrt{\frac{2}{m}} \leq \mathbb{E}\left[\hat{D}_m(F, \mathcal{X}_m)\right] \leq \mathbb{E}\left[\hat{R}_m(F, \mathcal{X}_m)\right] + 4\sqrt{\frac{2}{m}}. \quad (23)$$

Since elements in $\mathcal{X}_m$ are i.i.d. the distribution of $\hat{D}_m$ is invariant under their permutations and thus $\mathbb{E}\left[\hat{D}_m(F, \mathcal{X}_m)\right] = \mathbb{E}\left[\hat{Q}_{m,m/2}(F, \mathcal{X}_m)\right]$. Now we can use Theorem 3 to significantly improve bounds in (23):

$$\mathbb{E}\left[\hat{R}_m(F, \mathcal{X}_m)\right] - \frac{2}{\sqrt{m}} \leq \mathbb{E}\left[\hat{D}_m(F, \mathcal{X}_m)\right] \leq \left(1 + \frac{2}{\sqrt{2\pi m} - 2}\right)\mathbb{E}\left[\hat{R}_m(F, \mathcal{X}_m)\right].$$

Acknowledgments. The authors are thankful to Marius Kloft and Ruth Urner for useful discussions and to the anonymous reviewers for their comments. GB acknowledges support of the DFG through the FOR-1735 grant. NZ was supported solely by the Russian Science Foundation grant (project 14-50-00150).

References

1. Bartlett, P., Bousquet, O., Mendelson, S.: Local rademacher complexities. The Annals of Statistics **33**(4), 1497–1537 (2005)
2. Bartlett, P., Mendelson, S.: Rademacher and Gaussian complexities: Risk bounds and structural results. Journal of Machine Learning Research **3**, 463–482 (2001)
3. Blum, A., Langford, J.: PAC-MDL bounds. In: Schölkopf, B., Warmuth, M.K. (eds.) COLT/Kernel 2003. LNCS (LNAI), vol. 2777, pp. 344–357. Springer, Heidelberg (2003)
4. Boucheron, S., Lugosi, G., Bousquet, O.: Theory of classification: a survey of recent advances. ESAIM: Probability and Statistics **9**, 323–375 (2005)
5. Boucheron, S., Lugosi, G., Massart, P.: Concentration Inequalities: A Nonasymptotic Theory of Independence. Oxford University Press (2013)
6. Chapelle, O., Schölkopf, B., Zien, A.: Semi-Supervised Learning. MIT Press (2006)
7. Cortes, C., Mohri, M.: On transductive regression. In: NIPS 2006, pp. 305–312 (2007)
8. Cortes, C., Mohri, M., Pechyony, D., Rastogi, A.: Stability analysis and learning bounds for transductive regression algorithms (2009). CoRR abs/0904.0814
9. Derbeko, P., El-Yaniv, R., Meir, R.: Explicit learning curves for transduction and application to clustering and compression algorithms. Journal of Artificial Intelligence Research **22**(1), 117–142 (2004)
10. El-Yaniv, R., Pechyony, D.: Transductive rademacher complexity and its applications. Journal of Artificial Intelligence Research **35**(1), 193–234 (2009)
11. Gross, D., Nesme, V.: Note on sampling without replacing from a finite collection of matrices (2010). http://arxiv.org/abs/1001.2738v2
12. Haagerup, U.: The best constants in Khinchine inequality. Studia Mathematica **70**(3), 231–283 (1981)
13. Koltchinskii, V.: Oracle inequalities in empirical risk minimization and sparse recovery problems. Springer (2011)
14. Koltchinskii, V., Panchenko, D.: Rademacher processes and bounding the risk of function learning. In: Gine. D.E., Wellner, J. (eds.) High Dimensional Probability, II, pp. 443–457. Birkhauser (1999)
15. Ledoux, M., Talagrand, M.: Probability in Banach Space. Springer-Verlag (1991)
16. Magdon-Ismail, M.: Permutation complexity bound on out-sample error. In: Advances in Neural Information Processing Systems (NIPS 2010), pp. 1531–1539 (2010)
17. Mendelson, S.: Learning without Concentration (2014). CoRR abs/1401.0304
18. Pechyony, D.: Theory and Practice of Transductive Learning. PhD thesis (2008)
19. Stanica, P.: Good lower and upper bounds on binomial coefficients. Journal of Inequalities in Pure and Applied Mathematics **2**(3) (2001)
20. Tolstikhin, I., Blanchard, G., Kloft, M.: Localized complexities for transductive learning. In: COLT 2014, pp. 857–884 (2014)
21. Van der Vaart, A.W., Wellner, J.: Weak Convergence and Empirical Processes: With Applications to Statistics. Springer (2000)
22. Vapnik, V.: Statistical Learning Theory. John Wiley & Sons (1998)

Subsampling in Smoothed Range Spaces

Jeff M. Phillips and Yan Zheng[✉]

University of Utah, Salt Lake City, UT, USA
{jeffp,yanzheng}@cs.utah.edu

Abstract. We consider smoothed versions of geometric range spaces, so an element of the ground set (e.g. a point) can be contained in a range with a non-binary value in $[0, 1]$. Similar notions have been considered for kernels; we extend them to more general types of ranges. We then consider approximations of these range spaces through ε-nets and ε-samples (aka ε-approximations). We characterize when size bounds for ε-samples on kernels can be extended to these more general smoothed range spaces. We also describe new generalizations for ε-nets to these range spaces and show when results from binary range spaces can carry over to these smoothed ones.

1 Introduction

This paper considers traditional sample complexity problems but adapted to when the range space (or function space) smoothes out its boundary. This is important in various scenarios where either the data points or the measuring function is noisy. Similar problems have been considered in specific contexts of functions classes with a $[0, 1]$ range or kernel density estimates. We extend and generalize various of these results, motivated by scenarios like the following.

(S1) Consider maintaining a random sample of noisy spatial data points (say twitter users with geo-coordinates), and we want this sample to include a witness to every large enough event. However, because the data coordinates are noisy we use a kernel density estimate to represent the density. And moreover, we do not want to consider regions with a single or constant number of data points which only occurred due to random variations. In this scenario, how many samples do we need to maintain?

(S2) Next consider a large approximate (say high-dimensional image feature [1]) dataset, where we want to build a linear classifier. Because the features are approximate (say due to feature hashing techniques), we model the classifier boundary to be randomly shifted using Gaussian noise. How many samples from this dataset do we need to obtain a desired generalization bound?

(S3) Finally, consider one of these scenarios in which we are trying to create an informative subset of the enormous full dataset, but have the opportunity to do so in ways more intelligent than randomly sampling. On such a reduced

---

Thanks to supported by NSF CCF-1350888, IIS-1251019, and ACI-1443046.

© Springer International Publishing Switzerland 2015
K. Chaudhuri et al. (Eds.): ALT 2015, LNAI 9355, pp. 224–238, 2015.
DOI: 10.1007/978-3-319-24486-0_15

dataset one may want to train several types of classifiers, or to estimate the density of various subsets. Can we generate a smaller dataset compared to what would be required by random sampling?

The traditional way to study related sample complexity problems is through range spaces (a ground set X, and family of subsets $\mathcal{A}$) and their associated dimension (e.g., VC-dimension [25]). We focus on a smooth extension of range spaces defined on a geometric ground set. Specifically, consider the ground set P to be a subset of points in $\mathbb{R}^d$, and let $\mathcal{A}$ describe subsets defined by some geometric objects, for instance a halfspace or a ball. Points $p \in \mathbb{R}^d$ that are inside the object (e.g., halfspace or ball) are typically assigned a value 1, and those outside a value 0. In our smoothed setting points near the boundary are given a value between 0 and 1, instead of discretely switching from 0 to 1.

In learning theory these smooth range spaces can be characterized by more general notions called P-dimension [22] (or Pseudo dimension) or V-dimension [24] (or "fat" versions of these [2]) and can be used to learn real-valued functions for regression or density estimation, respectively.

In geometry and data structures, these smoothed range spaces are of interest in studying noisy data. Our work extends some recent work [12,21] which examines a special case of our setting that maps to kernel density estimates, and matches or improves on related bounds for non-smoothed versions.

Main Contributions. We next summarize the main contributions in this paper.

- We define a general class of *smoothed range spaces* (Sec 3.1), with application to density estimation and noisy agnostic learning, and we show that these can inherit sample complexity results from *linked* non-smooth range spaces (Corollary 1).
- We define an (ε, τ)-net for a smoothed range space (Sec 3.3). We show how this can inherit sampling complexity bounds from *linked* non-smooth range spaces (Theorem 2), and we relate this to non-agnostic density estimation and hitting set problems.
- We provide discrepancy-based bounds and constructions for ε-samples on smooth range spaces requiring significantly fewer points than uniform sampling approaches (Theorems 4 and 5), and also smaller than discrepancy-based bounds on the linked binary range spaces.

2   Definitions and Background

Recall that we will focus on geometric range spaces $(P, \mathcal{A})$ where the ground set $P \subset \mathbb{R}^d$ and the family of ranges $\mathcal{A}$ are defined by geometric objects. It is common to approximate a range space in one of two ways, as an ε-sample (aka ε-approximation) or an ε-net. An ε-*sample* for a range space $(P, \mathcal{A})$ is a subset $Q \subset P$ such that

$$\max_{A \in \mathcal{A}} \left| \frac{|A \cap P|}{|P|} - \frac{|Q \cap A|}{|Q|} \right| \leq \varepsilon.$$

An ε-*net* of a range space $(P, \mathcal{A})$ is a subset $Q \subset P$ such that

$$\text{for all } A \in \mathcal{A} \text{ such that } \frac{|P \cap A|}{|P|} \geq \varepsilon \text{ then } A \cap Q \neq \emptyset.$$

Given a range space $(P, \mathcal{A})$ where $|P| = m$, then $\pi_{\mathcal{A}}(m)$ describes the maximum number of possible distinct subsets of P defined by some $A \in \mathcal{A}$. If we can bound, $\pi_{\mathcal{A}}(m) \leq C m^{\nu}$ for absolute constant C, then $(P, \mathcal{A})$ is said to have *shatter dimension* ν. For instance the shatter dimension of $\mathcal{H}$ halfspaces in $\mathbb{R}^d$ is d, and for $\mathcal{B}$ balls in $\mathbb{R}^d$ is $d + 1$. For a range space with shatter dimension ν, a random sample of size $O((1/\varepsilon^2)(\nu + \log(1/\delta)))$ is an ε-sample with probability at least $1 - \delta$ [14,25], and a random sample of size $O((\nu/\varepsilon) \log(1/\varepsilon\delta))$ is an ε-net with probability at least $1 - \delta$ [11,18].

An ε-sample Q is sufficient for agnostic learning with generalization error ε, where the best classifier might misclassify some points. An ε-net Q is sufficient for non-agnostic learning with generalization error ε, where the best classifier is assumed to have no error on P.

The size bounds can be made deterministic and slightly improved for certain cases. An ε-sample Q can be made of size $O(1/\varepsilon^{2\nu/(\nu+1)})$ [15] and this bound can be no smaller [16] in the general case. For balls $\mathcal{B}$ in $\mathbb{R}^d$ which have shatter-dimension $\nu = d + 1$, this can be improved to $O(1/\varepsilon^{2d/(d+1)} \log^{d/(d+1)}(1/\varepsilon))$ [4, 16], and the best known lower bound is $O(1/\varepsilon^{2d/(d+1)})$. For axis-aligned rectangles $\mathcal{R}$ in $\mathbb{R}^d$ which have shatter-dimension $\nu = 2d$, this can be improved to $O((1/\varepsilon) \log^{d+1/2}(1/\varepsilon))$ [13].

For ε-nets, the general bound of $O((\nu/\varepsilon) \log(1/\varepsilon))$ can also be made deterministic [15], and for halfspaces in $\mathbb{R}^4$ the size must be at least $\Omega((1/\varepsilon) \log(1/\varepsilon))$ [19]. But for halfspaces in $\mathbb{R}^3$ the size can be $O(1/\varepsilon)$ [10,17], which is tight. By a simple lifting, this also applies for balls in $\mathbb{R}^2$. For other range spaces, such as axis-aligned rectangles in $\mathbb{R}^2$, the size bound is $\Theta((1/\varepsilon) \log \log(1/\varepsilon))$ [3,19].

2.1   Kernels

A *kernel* is a bivariate similarity function $K : \mathbb{R}^d \times \mathbb{R}^d \to \mathbb{R}^+$, which can be normalized so $K(x, x) = 1$ (which we assume through this paper). Examples include ball kernels ($K(x, p) = \{1 \text{ if } \|x - p\| \leq 1 \text{ and } 0 \text{ otherwise}\}$), triangle kernels ($K(x, p) = \max\{0, 1 - \|x - p\|\}$), Epanechnikov kernels ($K(x, p) = \max\{0, 1 - \|x - p\|^2\}$), and Gaussian kernels ($K(x, p) = \exp(-\|x - p\|^2)$), which is reproducing). In this paper we focus on symmetric, shift invariant kernels which depend only on $z = \|x - p\|$, and can be written as a single parameter function $K(x, p) = k(z)$; these can be parameterized by a single bandwidth (or just width) parameter w so $K(x, p) = k_w(\|x - p\|/w)$.

Given a point set $P \subset \mathbb{R}^d$ and a kernel, a *kernel density estimate* KDE_P is the convolution of that point set with K. For any $x \in \mathbb{R}^d$ we define $\text{KDE}_P(x) = \frac{1}{|P|} \sum_{p \in P} K(x, p)$.

A *kernel range space* [12,21] $(P, \mathcal{K})$ is an extension of the combinatorial concept of a range space $(P, \mathcal{A})$ (or to distinguish it we refer to the classic notion as a *binary range space*). It is defined by a point set $P \subset \mathbb{R}^d$ and a kernel K. An element K_x of $\mathcal{K}$ is a kernel $K(x, \cdot)$ applied at point $x \in \mathbb{R}^d$; it assigns a value in $[0, 1]$ to each point $p \in P$ as $K(x, p)$. If we use a ball kernel, then each value is exactly $\{0, 1\}$ and we recover exactly the notion of a binary range space for geometric ranges defined by balls.

The notion of an ε-kernel sample [12] extends the definition of ε-sample. It is a subset $Q \subset P$ such that

$$\max_{x \in \mathbb{R}^d} |\mathrm{KDE}_P(x) - \mathrm{KDE}_Q(x)| \leq \varepsilon.$$

A binary range space $(P, \mathcal{A})$ is *linked* to a kernel range space $(P, \mathcal{K})$ if the set $\{p \in P \mid K(x, p) \geq \tau\}$ is equal to $P \cap A$ for some $A \in \mathcal{A}$, for any threshold value τ. [12] showed that an ε-sample of a linked range space $(P, \mathcal{A})$ is also an ε-kernel sample of a corresponding kernel range space $(P, \mathcal{K})$. Since all range spaces defined by symmetric, shift-invariant kernels are linked to range spaces defined by balls, they inherit all ε-sample bounds, including that random samples of size $O((1/\varepsilon^2)(d + \log(1/\delta)))$ provide an ε-kernel sample with probability at least $1 - \delta$. Then [21] showed that these bounds can be improved through discrepancy-based methods to $O(((1/\varepsilon)\sqrt{\log(1/\varepsilon\delta)})^{2d/(d+2)})$, which is $O((1/\varepsilon)\sqrt{\log(1/\varepsilon\delta)})$ in $\mathbb{R}^2$.

A more general concept has been studied in learning theory on real-valued functions, where a function f as a member of a function class $\mathcal{F}$ describes a mapping from $\mathbb{R}^d$ to $[0, 1]$ (or more generally $\mathbb{R}$). A kernel range space where the linked binary range space has bounded shatter-dimension ν is said to have bounded V-dimension [24] (see [2]) of ν. Given a ground set X, then for $(X, \mathcal{F})$ this describes the largest subset Y of X which can be shattered in the following sense. Choose any value $s \in [0, 1]$ for all points $y \in Y$, and then for each subset of $Z \subset Y$ there exists a function $f \in \mathcal{F}$ so $f(y) > s$ if $y \in Z$ and $f(y) < s$ if $y \notin Z$. The best sample complexity bounds for ensuring Q is an ε-sample of P based on V-dimension are derived from a more general sort of dimension (called a P-dimension [22] where in the shattering definition, each y may have a distinct $s(y)$ value) requires $|Q| = O((1/\varepsilon^2)(\nu + \log(1/\delta)))$ [14]. As we will see, these V-dimension based results are also general enough to apply to the to-be-defined smooth range spaces.

3 New Definitions

In this paper we extend the notion of a kernel range spaces to other *smoothed range spaces* that are "linked" with common range spaces, e.g., halfspaces. These inherent the construction bounds through the linking result of [12], and we show cases where these bounds can also be improved. We also extend the notion of ε-nets to kernels and smoothed range spaces, and showing linking results for these as well.

3.1   Smoothed Range Spaces

Here we will define the primary smoothed combinatorial object we will examine, starting with halfspaces, and then generalizing. Let $\mathcal{H}_w$ denote the family of

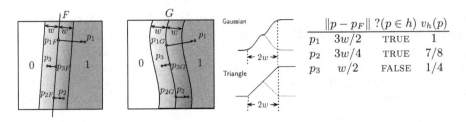

Fig. 1. Illustration of the smoothed halfspace, and smoothed polynomial surface, with function value of three points $\{p_1, p_2, p_3\}$ defined using a triangle kernel.

smoothed halfspaces with width parameter w, and let $(P, \mathcal{H}_w)$ be the associated smoothed range space where $P \subset \mathbb{R}^d$. Given a point $p \in P$, then smoothed halfspace $h \in \mathcal{H}_w$ maps p to a value $v_h(p) \in [0, 1]$ (rather than the traditional $\{0, 1\}$ in a binary range space).

We first describe a specific mapping to the function value $v_h(p)$ that will be sufficient for the development of most of our techniques. Let F be the $(d-1)$-flat defining the boundary of halfspace h. Given a point $p \in \mathbb{R}^d$, let $p_F = \arg\min_{q \in F} \|p - q\|$ describe the point on F closest to p. Now we define

$$v_{h,w}(p) = \begin{cases} 1 & p \in h \text{ and } \|p - p_F\| \geq w \\ \frac{1}{2} + \frac{1}{2}\frac{\|p - p_F\|}{w} & p \in h \text{ and } \|p - p_F\| < w \\ \frac{1}{2} - \frac{1}{2}\frac{\|p - p_F\|}{w} & p \notin h \text{ and } \|p - p_F\| < w \\ 0 & p \notin h \text{ and } \|p - p_F\| \geq w. \end{cases}$$

These points within a slab of width $2w$ surrounding F can take on a value between 0 and 1, where points outside of this slab revert back to the binary values of either 0 or 1.

We can make this more general using a shift-invariant kernel $k(\|p - x\|) = K(p, x)$, where $k_w(\|p - x\|) = k(\|p - x\|/w)$ allows us to parameterize by w. Define $v_{h,w}(p)$ as follows.

$$v_{h,w}(p) = \begin{cases} 1 - \frac{1}{2}k_w(\|p - p_F\|) & p \in h \\ \frac{1}{2}k_w(\|p - p_F\|) & p \notin h. \end{cases}$$

For brevity, we will omit the w and just use $v_h(p)$ when clear. These definitions are equivalent when using the triangle kernel. But for instance we could also use a Epanechnikov kernel or Gaussian kernel. Although the Gaussian kernel does not satisfy the restriction that only points in the width $2w$ slab take non $\{0, 1\}$ values, we can use techniques from [21] to extend to this case as well. This is illustrated in Figure 1. Another property held by this definition which we will exploit is that the slope ς of these kernels is bounded by $\varsigma = O(1/w) = c/w$, for some constant c; the constant $c = 1/2$ for triangle and Gaussian, and $c = 1$ for Epanechnikov.

Finally, we can further generalize this by replacing the flat F at the boundary of h with a polynomial surface G. The point $p_G = \arg\min_{q \in G} \|p - q\|$ replaces

p_F in the above definitions. Then the slab of width $2w$ is replaced with a curved volume in $\mathbb{R}^d$; see Figure 1. For instance, if G defines a circle in $\mathbb{R}^d$, then v_h defines a disc of value 1, then an annulus of width $2w$ where the function value decreases to 0. Alternatively, if G is a single point, then we essentially recover the kernel range space, except that the maximum height is $1/2$ instead of 1. We will prove the key structural results for polynomial curves in Section 5, but otherwise focus on halfspaces to keep the discussion cleaner. The most challenging elements of our results are all contained in the case with F as a $(d-1)$-flat.

3.2   ε-Sample in a Smoothed Range Space

It will be convenient to extend the notion of a kernel density estimate to these smoothed range space. A *smoothed density estimate* SDE$_P$ is defined for any $h \in \mathcal{H}_w$ as

$$\text{SDE}_P(h) = \frac{1}{|P|} \sum_{p \in P} v_h(p).$$

An *ε-sample Q of a smoothed range space* $(P, \mathcal{H}_w)$ is a subset $Q \subset P$ such that

$$\max_{h \in \mathcal{H}_w} |\text{SDE}_P(h) - \text{SDE}_Q(h)| \le \varepsilon.$$

Given such an ε-sample Q, we can then consider a subset $\bar{\mathcal{H}}_w$ of $\mathcal{H}_w$ with bounded integral (perhaps restricted to some domain like a unit cube that contains all of the data P). If we can learn the smooth range $\hat{h} = \arg\max_{h \in \bar{\mathcal{H}}_w} \text{SDE}_Q(h)$, then we know $\text{SDE}_P(h^*) - \text{SDE}_Q(\hat{h}) \le \varepsilon$, where $h^* = \arg\max_{h \in \bar{\mathcal{H}}_w} \text{SDE}_P(h)$, since $\text{SDE}_Q(\hat{h}) \ge \text{SDE}_Q(h^*) \ge \text{SDE}_P(h^*) - \varepsilon$. Thus, such a set Q allows us to learn these more general density estimates with generalization error ε.

We can also learn smoothed classifiers, like scenario (S2) in the introduction, with generalization error ε, by giving points in the negative class a weight of -1; this requires separate $(\varepsilon/2)$-samples for the negative and positive classes.

3.3   (ε, τ)-Net in a Smoothed Range Space

We now generalize the definition of an ε-net. Recall that it is a subset $Q \subset P$ such that Q "hits" all large enough ranges ($|P \cap A|/|P| \ge \varepsilon$). However, the notion of "hitting" is now less well-defined since a point $q \in Q$ may be in a range but with value very close to 0; if a smoothed range space is defined with a Gaussian or other kernel with infinite support, any point q will have a non-zero value for *all* ranges! Hence, we need to introduce another parameter $\tau \in (0, \varepsilon)$, to make the notion of hitting more interesting in this case.

A subset $Q \subset P$ is an *(ε, τ)-net of smoothed range space* $(P, \mathcal{H}_w)$ if for any smoothed range $h \in \mathcal{H}_w$ such that $\text{SDE}_P(h) \ge \varepsilon$, then there exists a point $q \in Q$ such that $v_h(q) \ge \tau$.

The notion of ε-net is closely related to that of hitting sets. A *hitting set* of a binary range space $(P, \mathcal{A})$ is a subset $Q \subset P$ so every $A \in \mathcal{A}$ (not just the large enough ones) contains some $q \in Q$. To extend these notions to the smoothed setting, we again need an extra parameter $\tau \in (0, \varepsilon)$, and also need to only consider large enough smoothed ranges, since there are now an infinite number even if P is finite. A subset $Q \subset P$ is an (ε, τ)-*hitting set of smoothed range space* $(P, \mathcal{H}_w)$ if for any $h \in \mathcal{H}_w$ such that $\mathrm{SDE}_P(h) \geq \varepsilon$, then $\mathrm{SDE}_Q(h) \geq \tau$.

In the binary range space setting, an ε-net Q of a range space $(P, \mathcal{A})$ is sufficient to learn the best classifier on P with generalization error ε in the non-agnostic learning setting, that is assuming a perfect classifier exists on P from $\mathcal{A}$. In the density estimation setting, there is not a notion of a perfect classifier, but if we assume some other properties of the data, the (ε, τ)-net will be sufficient to recover them. For instance, consider (like scenario (S1) in the introduction) that P is a discrete distribution so for some "event" points $p \in P$, there is at least an ε-fraction of the probability distribution describing P at p (e.g., there are more than $\varepsilon|P|$ points very close to p). In this setting, we can recover the location of these points since they will have probability at least τ in the (ε, τ)-net Q.

4 Linking and Properties of (ε, τ)-Nets

First we establish some basic connections between ε-sample, (ε, τ)-net, and (ε, τ)-hitting set in smoothed range spaces. In binary range spaces an ε-sample Q is also an ε-net, and a hitting set is also an ε-net; we show a similar result here up to the covering constant τ.

Lemma 1. *For a smoothed range space $(P, \mathcal{H}_w)$ and $0 < \tau < \varepsilon < 1$, an (ε, τ)-hitting set Q is also an (ε, τ)-net of $(P, \mathcal{H}_w)$.*

Proof. The (ε, τ)-hitting set property establishes for all $h \in \mathcal{H}_w$ with $\mathrm{SDE}_P(h) \geq \varepsilon$, then also $\mathrm{SDE}_Q(h) \geq \tau$. Since $\mathrm{SDE}_Q(h) = \frac{1}{|Q|} \sum_{q \in Q} v_h(q)$ is the average over all points $q \in Q$, then it implies that at least one point also satisfies $v_h(q) \geq \tau$. Thus Q is also an (ε, τ)-net. ☐

In the other direction an (ε, τ)-net is not necessarily an (ε, τ)-hitting set since the (ε, τ)-net Q may satisfy a smoothed range $h \in \mathcal{H}_w$ with a single point $q \in Q$ such that $v_h(q) \geq \tau$, but all others $q' \in Q \setminus \{q\}$ having $v_h(q') \ll \tau$, and thus $\mathrm{SDE}_Q(h) < \tau$.

Theorem 1. *For $0 < \tau < \varepsilon < 1$, an $(\varepsilon - \tau)$-sample Q in smoothed range space $(P, \mathcal{H}_w)$ is an (ε, τ)-hitting set in $(P, \mathcal{H}_w)$, and thus also an (ε, τ)-net of $(P, \mathcal{H}_w)$.*

Proof. Since Q is the $(\varepsilon - \tau)$-sample in the smoothed range space, for any smoothed range $h \in \mathcal{H}_w$ we have $|\mathrm{SDE}_P(h) - \mathrm{SDE}_Q(h)| \leq \varepsilon - \tau$. We consider the upper and lower bound separately.

If $\mathrm{SDE}_P(h) \geq \varepsilon$, when $\mathrm{SDE}_P(h) \geq \mathrm{SDE}_Q(h)$, we have

$$\mathrm{SDE}_Q(h) \geq \mathrm{SDE}_P(h) - (\varepsilon - \tau) \geq \varepsilon - (\varepsilon - \tau) = \tau.$$

And more simply when $\mathrm{SDE}_Q(h) \geq \mathrm{SDE}_P(h)$ and $\mathrm{SDE}_P(h) \geq \varepsilon \geq \tau$, then $\mathrm{SDE}_Q(h) \geq \tau$. Thus in both situations, Q is an (ε, τ)-hitting set of $(P, \mathcal{H}_w)$. And then by Lemma 1 Q is also an (ε, τ)-net of $(P, \mathcal{H}_w)$. ☐

4.1   Relations Between Smoothed Range Spaces and Linked Binary Range Spaces

Consider a smoothed range space $(P, \mathcal{H}_w)$, and for one smoothed range $h \in \mathcal{H}_w$, examine the range boundary F (e.g. a $(d-1)$-flat, or polynomial surface) along with a symmetric, shift invariant kernel K that describes v_h. The *superlevel* set $(v_h)^\tau$ is all points $x \in \mathbb{R}^d$ such that $v_h(x) \geq \tau$. Then recall a smoothed range space $(P, \mathcal{H}_w)$ is *linked* to a binary range space $(P, \mathcal{A})$ if every set $\{p \in P \mid v_h(p) \geq \tau\}$ for any $h \in \mathcal{H}_w$ and any $\tau > 0$, is exactly the same as some range $A \cap P$ for $A \in \mathcal{A}$. For smoothed range spaces defined by halfspaces, then the linked binary range space is also defined by halfspaces. For smoothed range spaces defined by points, mapping to kernel range spaces, then the linked binary range spaces are defined by balls.

Joshi *et al.* [12] established that given a kernel range space $(P, \mathcal{K})$, a linked binary range space $(P, \mathcal{A})$, and an ε-sample Q of $(P, \mathcal{A})$, then Q is also an ε-kernel sample of $(P, \mathcal{K})$. An inspection of the proof reveals the same property holds directly for smoothed range spaces, as the only structural property needed is that all points $p \in P$, as well as all points $q \in Q$, can be sorted in decreasing function value $K(p, x)$, where x is the center of the kernel. For smoothed range space, this can be replaced with sorting by $v_h(p)$.

Corollary 1 ([12]). *Consider a smoothed range space $(P, \mathcal{H}_w)$, a linked binary range space $(P, \mathcal{A})$, and an ε-sample Q of $(P, \mathcal{A})$ with $\varepsilon \in (0, 1)$. Then Q is an ε-sample of $(P, \mathcal{H}_w)$.*

We now establish a similar relationship to (ε, τ)-nets of smoothed range spaces from $(\varepsilon - \tau)$-nets of linked binary range spaces.

Theorem 2. *Consider a smoothed range space $(P, \mathcal{H}_w)$, a linked binary range space $(P, \mathcal{A})$, and an $(\varepsilon - \tau)$-net Q of $(P, \mathcal{A})$ for $0 < \tau < \varepsilon < 1$. Then Q is an (ε, τ)-net of $(P, \mathcal{H}_w)$.*

Proof. Let $|P| = n$. Then since Q is an $(\varepsilon - \tau)$-net of $(P, \mathcal{A})$, for any range $A \in \mathcal{A}$, if $|P \cap A| \geq (\varepsilon - \tau)n$, then $Q \cap A \neq \emptyset$.

Suppose $h \in \mathcal{H}_w$ has $\text{SDE}_P(h) \geq \varepsilon$ and we want to establish that $\text{SDE}_Q(h) \geq \tau$. Let $A \in \mathcal{A}$ be the range such that $(\varepsilon - \tau)n$ points with largest $v_h(p_i)$ values are exactly the points in A. We now partition P into three parts (1) let P_1 be the $(\varepsilon - \tau)n - 1$ points with largest v_h values, (2) let y be the point in P with $(\varepsilon - \tau)n$th largest v_h value, and (3) let P_2 be the remaining $n - n(\varepsilon - \tau)$ points. Thus for every $p_1 \in P_1$ and every $p_2 \in P_2$ we have $v_h(p_2) \leq v_h(y) \leq v_h(p_1) \leq 1$.

Now using our assumption $n \cdot \text{SDE}_P(h) \geq n\varepsilon$ we can decompose the sum

$$n \cdot \text{SDE}_P(h) = \sum_{p_1 \in P_1} v_h(p_1) + v_h(y) + \sum_{p_2 \in P_2} v_h(p_2) \geq n\varepsilon,$$

and hence using upper bounds $v_h(p_1) \leq 1$ and $v_h(p_2) \leq v_h(y)$,

$$v_h(y) \geq n\varepsilon - \sum_{p_1 \in P_1} v_h(p_1) - \sum_{p_2 \in P_2} v_h(p_2)$$

$$\geq n\varepsilon - (n(\varepsilon - \tau) - 1) \cdot 1 - (n - n(\varepsilon - \tau))v_h(y).$$

Solving for $v_h(y)$ we obtain

$$v_h(y) \geq \frac{n\tau + 1}{n - n(\varepsilon - \tau) + 1} \geq \frac{n\tau}{n - n(\varepsilon - \tau)} \geq \frac{n\tau}{n} = \tau.$$

Since $(P, \mathcal{A})$ is linked to $(P, \mathcal{H}_w)$, there exists a range $A \in \mathcal{A}$ that includes precisely $P_1 \cup y$ (or more points with the same $v_h(y)$ value as y). Because Q is an $(\varepsilon - \tau)$-net of $(P, \mathcal{A})$, Q contains at least one of these points, lets call it q. Since all of these points have function value $v_h(p) \geq v_h(y) \geq \tau$, then $v_h(q) \geq \tau$. Hence Q is also an (ε, τ)-net of $(P, \mathcal{H}_w)$, as desired. □

This implies that if $\tau \leq c\varepsilon$ for any constant $c < 1$, then creating an (ε, τ)-net of a smoothed range space, with a known linked binary range space, reduces to computing an ε-net for the linked binary range space. For instance any linked binary range space with shatter-dimension ν has an ε-net of size $O(\frac{\nu}{\varepsilon} \log \frac{1}{\varepsilon})$, including halfspaces in $\mathbb{R}^d$ with $\nu = d$ and balls in $\mathbb{R}^d$ with $\nu = d + 1$; hence there exists $(\varepsilon, \varepsilon/2)$-nets of the same size. For halfspaces in $\mathbb{R}^2$ or $\mathbb{R}^3$ (linked to smoothed halfspaces) and balls in $\mathbb{R}^2$ (linked to kernels), the size can be reduced to $O(1/\varepsilon)$ [10,17,23].

5   Min-cost Matchings Within Cubes

Before we proceed with our construction for smaller ε-samples for smoothed range spaces, we need to prepare some structural results about min-cost matchings. Following some basic ideas from [21], these matchings will be used for discrepancy bounds on smoothed range spaces in Section 6.

In particular, we analyze some properties of the interaction of a min-cost matching M and some basic shapes ([21] considered only balls). Let $P \subset \mathbb{R}^d$ be a set of $2n$ points. A *matching* $M(P)$ is a decomposition of P into n pairs $\{p_i, q_i\}$ where $p_i, q_i \in P$ and each p_i (and q_i) is in exactly one pair. A *min-cost matching* is the matching M that minimizes $\mathsf{cost}_1(M, P) = \sum_{i=1}^n \|p_i - q_i\|$. The min-cost matching can be computed in $O(n^3)$ time by [9] (using an extension of the Hungarian algorithm from the bipartite case). In $\mathbb{R}^2$ it can be calculated in $O(n^{3/2} \log^5 n)$ time [26].

Following [21], again we will base our analysis on a result of [5] which says that if $P \subset [0,1]^d$ (a unit cube) then for d a constant, $\mathsf{cost}_d(M, P) = \sum_{i=1}^n \|p_i - q_i\|^d = O(1)$, where M is the min-cost matching. We make no attempt to optimize constants, and assume d is constant.

One simple consequence, is that if P is contained in a d-dimensional cube of side length ℓ, then $\mathsf{cost}_d(M, P) = \sum_{i=1}^n \|p_i - q_i\|^d = O(\ell^d)$.

We are now interested in interactions with a matching M for P in a d-dimensional cube of side length ℓ $\mathsf{C}_{\ell,d}$ (call this shape an (ℓ, d)-*cube*), and more general objects; in particular C_w a (w, d)-cube and, S_w a slab of width $2w$, both restricted to be within $\mathsf{C}_{\ell,d}$. Now for such an object O_w (which will either be C_w or S_w) and an edge $\{p, q\}$ where line segment $\overline{pq}$ intersects O_w define point p_B (resp. q_B) as the point on segment $\overline{pq}$ inside O_w closest to p (resp. q). Note if p (resp. q) is inside O then $p_B = p$ (resp. $q_B = q$), otherwise it is on the boundary of O_w. For instance, see C_{20w} in Fig 2.

Define the *length* of a matching M restricted to an object $O_w \subset \mathbb{R}^d$ as

$$\rho(O_w, M) = \sum_{(q,p) \in M} \min\left\{ (2w)^d, \|p_B - q_B\|^d \right\}.$$

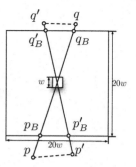

Fig. 2. (T3) edges

Note this differs from a similar definition by [21] since that case did not need to consider when both p and q were both outside of O_w, and did not need the $\min\{(2w)^d, \ldots\}$ term because all objects had diameter 2.

Lemma 2. *Let $P \subset C_{\ell,d}$, where d is constant, and M be its min-cost matching. For any (w,d)-cube $C_w \subset C_{\ell,d}$ we have $\rho(C_w, M) = O(w^d)$.*

Proof. We cannot simply apply the result of [5] since we do not restrict that $P \subset C_w$. We need to consider cases where either p or q or both are outside of C_w. As such, we have three types of edges we consider, based on a cube C_{20w} of side length $20w$ and with center the same as C_w.

(T1) Both endpoints are within C_{20w} of edge length at most $\sqrt{d}20w$.
(T2) One endpoint is in C_w, the other is outside C_{20w}.
(T3) Both endpoints are outside C_{20w}.

For all (T1) edges, the result of Bern and Eppstein can directly bound their contribution to $\rho(C_w, M)$ as $O(w^d)$ (scale to a unit cube, and rescale). For all (T2) edges, we can also bound their contribution to $\rho(C_w, M)$ as $O(w^d)$, by extending an analysis of [21] when both C_w and C_{20w} are similarly proportioned balls. This analysis shows there are $O(1)$ such edges.

We now consider the case of (T3) edges, restricting to those that also intersect C_w. We argue there can be at most $O(1)$ of them. In particular consider two such edges $\{p,q\}$ and $\{p',q'\}$, and their mappings to the boundary of C_{20w} as p_B, q_B, p'_B, q'_B; see Figure 2. If $\|p_B - p'_B\| \le 10w$ and $\|q_B - q'_B\| \le 10w$, then we argue next that this cannot be part of a min-cost matching since $\|p - p'\| + \|q - q'\| < \|p - q\| + \|p' - q'\|$, and it would be better to swap the pairing. Then it follows from the straight-forward net argument below that there can be at most $O(1)$ such pairs.

We first observe that $\|p_B - p'_B\| + \|q_B - q'_B\| \le 10w + 10w < 20w + 20w \le \|p_B - q_B\| + \|p'_B - q'_B\|$. Now we can obtain our desired inequality using that $\|p - q\| = \|p - p_B\| + \|p_B - q_B\| + \|q_B - q\|$ (and similar for $\|p' - q'\|$) and that $\|p - p'\| \le \|p - p_B\| + \|p_B - p'_B\| + \|p'_B - p'\|$ by triangle inequality (and similar for $\|q - q'\|$).

Next we describe the net argument that there can be at most $O(d^2 \cdot 2^{2d}) = O(1)$ such pairs with $\|p_B - p'_B\| > 10w$ and $\|q_B - q'_B\| > 10w$. First place a $5w$-net $\mathcal{N}_f$ on each $(d-1)$-dimensional face f of C_{20w} so that any point $x \in f$ is within $5w$ of some point $\eta \in \mathcal{N}_f$. We can construct $\mathcal{N}_f$ of size $O(2^d)$ with a simple grid. Then let $\mathcal{N} = \bigcup_f \mathcal{N}_f$ as the union of the nets on each face; its size is $O(d \cdot 2^d)$.

Now for any point $p \notin C_{20w}$ let $\eta(p) = \arg\min_{\eta \in N} \|p_B - \eta\|$ be the closest point in N to p_B. If two points p and p' have $\eta(p) = \eta(p')$ then $\|p - p'\| \leq 10w$. Hence there can be at most $O((d \cdot 2^d)^2)$ edges with $\{p, q\}$ mapping to unique $\eta(p)$ and $\eta(q)$ if no other edge $\{p', q'\}$ has $\|p_B - p'_B\| \leq 10w$ and $\|q_B - q'_B\| \leq 10w$.

Concluding, there can be at most $O(d^2 \cdot 2^{2d}) = O(1)$ edges in M of type (T3), and the sum of their contribution to $\rho(C_w, M)$ is at most $O(w^d)$, completing the proof.   □

Lemma 3. *Let $P \subset C_{\ell,d}$, where d is constant, and let M be its min-cost matching. For any width $2w$ slab S_w restricted to $C_{\ell,d}$ we have $\rho(S_w, M) = O(\ell^{d-1}w)$.*

Proof. We can cover the slab S_w with $O((\ell/w)^{d-1})$ (w, d)-cubes. To make this concrete, we cover $C_{\ell,d}$ with $\lceil \ell/w \rceil^d$ cubes on a regular grid. Then in at least one basis direction (the one closest to orthogonal to the normal of F) any column of cubes can intersect S_w in at most 4 cubes. Since there are $\lceil \ell/w \rceil^{d-1}$ such columns, the bound holds. Let C_w be the set of these cubes covering S_w.

Restricted to any one such cube C_w, the contribution of those edges to $\rho(S_w, M)$ is at most $O(w^d)$ by Lemma 2. Now we need to argue that we can just sum the effect of all covering cubes. The concern is that an edge goes through many cubes, only contributing a small amount to each $\rho(C_w, M)$ term, but when the total length is taken to the dth power it is much more. However, since each edge's contribution is capped at $(2w)^2$, we can say that if any edge goes through more than $O(1)$ cubes, its length must be at least w, and its contribution in one such cube is already $\Omega(w)$, so we can simply inflate the effect of each cube towards $\rho(S_w, M)$ by a constant.

In particular, consider any edge $\overline{pq}$ that has $p \in C_w$. Each cube has $3^d - 1$ neighboring cubes, including through vertex incidence. Thus if edge $\overline{pq}$ passes through more than 3^d cubes, q must be in a cube that is not one of C'_w's neighbors. Thus it must have length at least w; and hence its length in at least one cube C'_w must be at least $w/3^d$, with its contribution to $\rho(C'_w, M) > w^d/(3^{d^2})$. Thus we can multiply the effect of each edge in $\rho(C_w, M)$ by $3^{d^2} 2^d = O(1)$ and be sure it is at least as large as the effect of that edge in $\rho(S_w, M)$. Hence

$$\rho(S_w, M) \leq 3^{d^2} 2^d \sum_{C_w \in \mathcal{C}_w} \rho(C_w, M) \leq O(1) \sum_{C_w \in \mathcal{C}_w} O(w^d)$$
$$= O((\ell/w)^{d-1}) \cdot O(w^d) = O(\ell^{d-1}w).   □$$

We can apply the same decomposition as used to prove Lemma 3 to also prove a result for a w-expanded volume G_w around a degree g polynomial surface G. A degree g polynomial surface can intersect a line at most g times, so for some $C_{\ell,d}$ the expanded surface $G_w \cap C_{\ell,d}$ can be intersected by $O(g(\ell/w)^{d-1})$ (w, d)-cubes. Hence we can achieve the following bound.

Corollary 2. *Let $P \subset C_{\ell,d}$, where d is constant, and let M be its min-cost matching. For any volume G_w defined by a polynomial surface of degree g expanded by a width w, restricted to $C_{\ell,d}$ we have $\rho(G_w, M) = O(g\ell^{d-1}w)$.*

6   Constructing ε-Samples for Smoothed Range Spaces

In this section we build on the ideas from [21] and the new min-cost matching results in Section 5 to produce new discrepancy-based ε-sample bounds for smoothed range spaces. The basic construction is as follows. We create a min-cost matching M on P, then for each pair $(p, q) \in M$, we retain one of the two points at random, halving the point set. We repeat this until we reach our desired size. This should not be unfamiliar to readers familiar with discrepancy-based techniques for creating ε-samples of binary range spaces [6,16]. In that literature similar methods exist for creating matchings "with low-crossing number". Each such matching formulation is specific to the particular combinatorial range space one is concerned with. However, in the case of smoothed range spaces, we show that the min-cost matching approach is a universal algorithm. It means that an ε-sample Q for one smoothed range space $(P, \mathcal{H}_w)$ is also an ε-sample for any other smoothed range space $(P, \mathcal{H}'_w)$, perhaps up to some constant factors. We also show how these bounds can sometimes improve upon ε-sample bounds derived from linked range spaces; herein the parameter w will play a critical role.

6.1   Discrepancy for Smoothed Halfspaces

To simplify arguments, we first consider $P \subset \mathbb{R}^2$ extending to $\mathbb{R}^d$ in Section 6.4.

Let $\chi : P \to \{-1, +1\}$ be a coloring of P, and define the *discrepancy* of $(P, \mathcal{H}_w)$ with coloring χ as $\mathsf{disc}_\chi(P, \mathcal{H}_w) = \max_{h \in \mathcal{H}_w} |\sum_{p \in P} \chi(p) v_h(p)|$. Restricted to one smoothed range $h \in \mathcal{H}_w$ this is $\mathsf{disc}_\chi(P, h) = |\sum_{p \in P} \chi(p) v_h(p)|$. We construct a coloring χ using the min-cost matching M of P; for each $\{p_i, q_i\} \in M$ we randomly select one of p_i or q_i to have $\chi(p_i) = +1$, and the other $\chi(q_i) = -1$. We next establish bounds on the discrepancy of this coloring for a ς-bounded smoothed range space $(P, \mathcal{H}_w)$, i.e., where the gradient of v_h is bounded by $\varsigma \leq c_1/w$ for a constant c_1 (see Section 3.1).

For any smoothed range $h \in \mathcal{H}_w$, we can now define a random variable $X_j = \chi(p_j) v_h(p_j) + \chi(q_j) v_h(q_j)$ for each pair $\{p_j, q_j\}$ in the matching M. This allows us to rewrite $\mathsf{disc}_\chi(P, h) = |\sum_j X_j|$. We can also define a variable $\Delta_j = 2|v_h(p_j) - v_h(q_j)|$ such that $X_j \in \{-\Delta_j/2, \Delta_j/2\}$. Now following the key insight from [21] we can bound $\sum_j \Delta_j^2$ using results from Section 5, which shows up in the following Chernoff bound from [8]: Let $\{X_1, X_2, \ldots\}$ be independent random variables with $\mathbf{E}[X_j] = 0$ and $X_j = \{-\Delta_j/2, \Delta_j/2\}$ then

$$\Pr\left[\mathsf{disc}_\chi(P, h) \geq \alpha\right] = \Pr\left[\left|\sum_j X_j\right| \geq \alpha\right] \leq 2 \exp\left(\frac{-2\alpha^2}{\sum_j \Delta_j^2}\right). \tag{1}$$

Lemma 4. *Assume $P \subset \mathbb{R}^2$ is contained in some cube $C_{\ell,2}$ and with min-cost matching M defining χ, and consider a ς-bounded smoothed halfspace $h \in \mathcal{H}_w$ associated with slab S_w. Let $\rho(S_w, M) \leq c_2(\ell w)$ for constant c_2 (see definition of ρ in Section 5). Then $\Pr\left[\mathsf{disc}_\chi(P, h) > C\sqrt{\frac{\ell}{w} \log(2/\delta)}\right] \leq \delta$ for any $\delta > 0$ and constant $C = c_1\sqrt{2c_2}$.*

Proof. Using the gradient of v_h is at most $\varsigma = c_1/w$ and $|v_h(p_j) - v_h(q_j)| \leq \varsigma \max\{2w, \|p_j - q_j\|\}$ we have

$$\sum_j \Delta_j^2 = \sum_j 4(v_h(p_j) - v_h(q_j))^2 \leq 4\varsigma^2 \rho(S_w, M) \leq 4c_1^2/w^2 \cdot c_2\ell w = 4c_1^2 c_2 \ell/w,$$

where the second inequality follows by Lemma 3 which shows that $\rho(S_w, M) = \sum_j \max\{(2w)^2, \|p_j - q_j\|^2\} \leq c_2(\ell w)$.

We now study the random variable $\mathrm{disc}_\chi(P, h) = |\sum_i X_i|$ for a single $h \in \mathcal{H}_w$. Invoking (1) we can bound $\Pr[\mathrm{disc}_\chi(P, h) > \alpha] \leq 2\exp(-\alpha^2/(2c_1^2 c_2 \ell/w))$. Setting $C = c_1\sqrt{2c_2}$ and $\alpha = C\sqrt{\frac{\ell}{w} \log(2/\delta)}$ reveals $\Pr\left[\mathrm{disc}_\chi(P, h) > C\sqrt{\frac{\ell}{w} \log(2/\delta)}\right] \leq \delta$. $\square$

6.2 From a Single Smoothed Halfspace to a Smoothed Range Space

The above theorems imply small discrepancy for a single smoothed halfspace $h \in \mathcal{H}_w$, but this does not yet imply small discrepancy $\mathrm{disc}_\chi(P, \mathcal{H}_w)$, for all choices of smoothed halfspaces simultaneously. And in a smoothed range space, the family $\mathcal{H}_w$ is not finite, since even if the same set of points have $v_h(p) = 1$, $v_h(p) = 0$, or are in the slab S_w, infinitesimal changes of h will change $\mathrm{SDE}_P(h)$. So in order to bound $\mathrm{disc}_\chi(P, \mathcal{H}_w)$, we will show that there are polynomial in n number of smoothed halfspaces that need to be considered, and then apply a union bound across this set. The proof is deferred to the full version.

Theorem 3. *For $P \subset \mathbb{R}^2$ of size n, for $\mathcal{H}_w$, and value $\Psi(n, \delta) = O(\sqrt{\frac{\ell}{w} \log \frac{n}{\delta}})$ for $\delta > 0$, we can choose a coloring χ such that $\Pr[\mathrm{disc}_\chi(P, \mathcal{H}_w) > \Psi(n, \delta)] \leq \delta$.*

6.3 ε-Samples for Smoothed Halfspaces

To transform this discrepancy algorithm to ε-samples, let $f(n) = \mathrm{disc}_\chi(P, \mathcal{H}_w)/n$ be the value of ε in the ε-samples generated by a single coloring of a set of size n. Solving for n in terms of ε, the sample size is $s(\varepsilon) = O(\frac{1}{\varepsilon}\sqrt{\frac{\ell}{w} \log \frac{\ell}{w\varepsilon\delta}})$. We can then apply the *MergeReduce* framework [7]; iteratively apply this random coloring in $O(\log n)$ rounds on disjoint subsets of size $O(s(\varepsilon))$. Using a generalized analysis (c.f., Theorem 3.1 in [20]), we have the same ε-sample size bound.

Theorem 4. *For $P \subset C_{\ell,2} \subset \mathbb{R}^2$, with probability at least $1 - \delta$, we can construct an ε-sample of $(P, \mathcal{H}_w)$ of size $O(\frac{1}{\varepsilon}\sqrt{\frac{\ell}{w} \log \frac{\ell}{w\varepsilon\delta}})$.*

To see that these bounds make rough sense, consider a random point set P in a unit square. Then setting $w = 1/n$ will yield roughly $O(1)$ points in the slab (and should roughly revert to the non-smoothed setting); this leads to $\mathrm{disc}_\chi(P, \mathcal{H}_w) = O(\sqrt{n}\sqrt{\log(n/\delta)})$ and an ε-sample of size $O((1/\varepsilon^2)\sqrt{\log(1/\varepsilon\delta)})$, basically the random sampling bound. But setting $w = \varepsilon$ so about εn points are in the slab (the same amount of error we allow in an ε-sample) yields $\mathsf{disc}_\chi(P, \mathcal{H}_w) = O((1/\sqrt{\varepsilon n}) \cdot \sqrt{\log(n/\delta)})$ and the size of the ε-sample to be $O(\frac{1}{\varepsilon}\sqrt{\log(1/\varepsilon\delta)})$, which is a large improvement over $O(1/\varepsilon^{4/3})$, and the best bound known for non-smoothed range spaces [16].

However, the assumption that $P \subset \mathsf{C}_{\ell,2}$ (although not uncommon [16]) can be restrictive. In the full version we relax the condition for well clustered data.

6.4   Generalization to d Dimensions

We now extend from $\mathbb{R}^2$ to $\mathbb{R}^d$ for $d > 2$. Using results from Section 5 we implicitly get a bound on $\sum_j \Delta_j^d$, but the Chernoff bound we use requires a bound on $\sum_j \Delta_j^2$. As in [21], we can attain a weaker bound using Jensen's inequality over at most n terms

$$\left(\sum_j \frac{1}{n}\Delta_j^2\right)^{d/2} \le \sum_j \frac{1}{n}\left(\Delta_j^2\right)^{d/2} \quad \text{so} \quad \sum_j \Delta_j^2 \le n^{1-2/d}\left(\sum_j \Delta_j^d\right)^{2/d}. \quad (2)$$

Replacing this bound and using $\rho(S_w, M) \le O(\ell^{d-1}w)$ in Lemma 4 and considering $\varsigma = c_1/w$ for some constant c_1 results in the next lemma. Its proof is deferred to the full version.

Lemma 5. *Assume $P \subset \mathbb{R}^d$ is contained in some cube $\mathsf{C}_{\ell,d}$ and with min-cost matching M, and consider a ς-bounded smoothed halfspace $h \in \mathcal{H}_w$ associated with slab S_w. Let $\rho(S_w, M) \le c_2(\ell^{d-1}w)$ for constant c_2. Then $\Pr\left[\mathsf{disc}_\chi(P, h) > Cn^{1/2-1/d}(\ell/w)^{1-1/d}\sqrt{\log(2/\delta)}\right] \le \delta$ for any $\delta > 0$ and $C = \sqrt{2}c_1(c_2)^{1/d}$.*

For all choices of smoothed halfspaces, applying the union bound, the discrepancy is increased by a $\sqrt{\log n}$ factor, with the following probabilistic guarantee,

$$\Pr[\mathsf{disc}_\chi(P, \mathcal{H}_w) > Cn^{1/2-1/d}(\ell/w)^{1-1/d}\sqrt{\log(n/\delta)}] \le \delta.$$

Ultimately, we can extend Theorem 4 to the following.

Theorem 5. *For $P \subset \mathsf{C}_{\ell,d} \subset \mathbb{R}^d$, where d is constant, with probability at least $1 - \delta$, we can construct an ε-sample of $(P, \mathcal{H}_w)$ of size $O\left((\ell/w)^{2(d-1)/(d+2)} \cdot \left(\frac{1}{\varepsilon}\sqrt{\log \frac{\ell}{w\varepsilon\delta}}\right)^{2d/(d+2)}\right)$.*

Note this result addresses scenario (S3) from the introduction where we want to find a small set (the ε-sample) so that it could be much smaller than the d/ε^2 random sampling bound, and allows generalization error $O(\varepsilon)$ for agnostic learning as described in Section 3.2. When ℓ/w is constant, the exponents on $1/\varepsilon$ are also better than those for binary ranges spaces (see Section 2).

References

1. Fitzgibbon, A., Bergamo, A., Torresani, L.: Picodes: learning a compact code for novel-category recognition. In: NIPS (2011)
2. Alon, N., Ben-David, S., Cesa-Bianchi, N., Haussler, D.: Scale-sensitive dimensions, uniform convergence, and learnability. Journal of ACM **44**, 615–631 (1997)
3. Aronov, B., Ezra, E., Sharir, M.: Small size ε-nets for axis-parallel rectangles and boxes. Siam Journal of Computing **39**, 3248–3282 (2010)
4. Beck, J.: Irregularities of distribution I. Acta Mathematica **159**, 1–49 (1987)
5. Bern, M., Eppstein, D.: Worst-case bounds for subadditive geometric graphs. In: SOCG (1993)
6. Chazelle, B.: The Discrepancy Method. Cambridge (2000)
7. Chazelle, B., Matousek, J.: On linear-time deterministic algorithms for optimization problems in fixed dimensions. J. Algorithms **21**, 579–597 (1996)
8. Dubhashi, D.P., Panconesi, A.: Concentration of Measure for the Analysis of Randomized Algorithms. Cambridge (2009)
9. Edmonds, J.: Paths, trees, and flowers. Canadian Journal of Mathematics **17**, 449–467 (1965)
10. Har-Peled, S., Kaplan, H., Sharir, M., Smorodinksy, S.: ε-nets for halfspaces revisited. Technical report (2014). arXiv:1410.3154
11. Haussler, D., Welzl, E.: epsilon-nets and simplex range queries. Disc. & Comp. Geom. **2**, 127–151 (1987)
12. Joshi, S., Kommaraju, R.V., Phillips, J.M., Venkatasubramanian, S.: Comparing distributions and shapes using the kernel distance. In: SOCG (2011)
13. Larsen, K.G.: On range searching in the group model and combinatorial discrepancy. In: FOCS (2011)
14. Li, Y., Long, P.M., Srinivasan, A.: Improved bounds on the samples complexity of learning. J. Comp. and Sys. Sci. **62**, 516–527 (2001)
15. Matoušek, J.: Tight upper bounds for the discrepancy of halfspaces. Discrete & Computational Geometry **13**, 593–601 (1995)
16. Matoušek, J.: Geometric Discrepancy. Springer (1999)
17. Matoušek, J., Seidel, R., Welzl, E.: How to net a lot with little: small ε-nets for disks and halfspaces. In: SOCG (1990)
18. Pach, J., Agarwal, P.K.: Combinatorial geometry. Wiley, Wiley-Interscience series in discrete mathematics and optimization (1995)
19. Pach, J., Tardos, G.: Tight lower bounds for the size of epsilon-nets. Journal of American Mathematical Society **26**, 645–658 (2013)
20. Phillips, J.M.: Algorithms for ε-approximations of terrains. In: Automata, Languages and Programming, pp. 447–458. Springer (2008)
21. Phillips, J.M.: Eps-samples for kernels. In: SODA (2013)
22. Pollard, D.: Emperical processes: theory and applications. In: NSF-CBMS REgional Confernece Series in Probability and Statistics (1990)
23. Pyrga, E., Ray, S.: New existence proofs ε-nets. In: SOCG (2008)
24. Vapnik, V.: Inductive principles of the search for empirical dependencies. In: COLT (1989)
25. Vapnik, V., Chervonenkis, A.: On the uniform convergence of relative frequencies of events to their probabilities. Theory of Probability and its Applications **16**, 264–280 (1971)
26. Varadarajan, K.R.: A divide-and-conquer algorithm for min-cost perfect matching in the plane. In: FOCS (1998)

Information Preserving Dimensionality Reduction

Shrinu Kushagra$^{(\boxtimes)}$ and Shai Ben-David

School of Computer Science, University of Waterloo, Waterloo, ON N2L 3G1, Canada
skushagr@uwaterloo.ca, shai@cs.uwaterloo.ca

Abstract. Dimensionality reduction is a very common preprocessing approach in many machine learning tasks. The goal is to design data representations that on one hand reduce the dimension of the data (therefore allowing faster processing), and on the other hand aim to retain as much task-relevant information as possible. We look at generic dimensionality reduction approaches that do not rely on much task-specific prior knowledge. However, we focus on scenarios in which unlabeled samples are available and can be utilized for evaluating the usefulness of candidate data representations.

We wish to provide some theoretical principles to help explain the success of certain dimensionality reduction techniques in classification prediction tasks, as well as to guide the choice of dimensionality reduction tool and parameters. Our analysis is based on formalizing the often implicit assumption that "similar instances are likely to have similar labels". Our theoretical analysis is supported by experimental results.

1 Introduction

Many successful machine learning tools are essentially heuristic, in the sense that while they work well for many practical tasks, we do not have a solid formal explanation to why that is the case. Various reduction tools fall into that category. For example, while we know that PCA retains as much of the data variance as possible (within the constraints of the linearity of the transformation and upper bounding the dimension of the resulting data representation), it is not explicitly clear as to why this property should suffice for the purpose of classification prediction. We propose a step in the direction of providing such formal explanation, based on some generally conceivable assumptions about the nature of a given classification task.

Most, if not all, of the practical learning algorithms are based on the assumption that similar instances have similar labels. Assuming that the data representation on a given task enjoys such a property (otherwise there is very little chance that learning will succeed), one would like to retain it when changing the representation (such as when applying dimensionality reduction tool).

We propose a notion of metric (distance) retention that is, on one hand, formal and quantifiable, and on the other hand reflects the properties of practically common data representation techniques. We show that common techniques

© Springer International Publishing Switzerland 2015
K. Chaudhuri et al. (Eds.): ALT 2015, LNAI 9355, pp. 239–253, 2015.
DOI: 10.1007/978-3-319-24486-0_16

like Random Projections and PCA have this property. Furthermore, motivated by the availability of abundant unlabeled data in many real world tasks, we show that the proposed notion can be reliably estimated from sufficiently large unlabeled samples.

The second key component in our analysis is a formal tool for quantifying the degree by which a given classification task meets the *"similar instances are likely to have similar labels"* requirement. A notion of *Probabilistic Lipschitzness* (PL) was introduced by Urner et. al [1] to capture this intuition. We propose a new variant of PL, and show that it controls the error rates of Nearest Neighbor learning rules.

We then combine the two notions to show that a distance retaining embedding also preserves PL. That is, we prove that if PL property exists in a higher dimension and an embedding retains distance then the PL property exists in the reduced dimension as well. Hence, we are able to show that an embedding which has the metric retention preserves nearest neighbor learnability. Nearest neighbor learning suffers from the curse of dimensionality as the number of samples needed is exponential in the dimension of the space. We show that if an embedding has metric retention, then we can first use the embedding to reduce the dimension and then do nearest neighbor which leads to exponential sample savings.

Finally, we provide experiments which show how the PL property captures the usefulness of a learned/reduced representation. This also has some interesting implications for some deep feature learning algorithms like RBMs. More concretely, we show that successful features have better PL property.

1.1   Related Work

Dimensionality reduction aims to transform data from high dimensional space to low dimensional space while preserving important properties (like interpoint distances) of the data. Dimensionality reduction techniques are used in many domains of Machine Learning as a preprocessing step. Some of the popular dimensionality reduction techniques include Principal Component Analysis (PCA), Laplacian Eigenmaps, Multidimensional Scaling, Isomap, Neural Autoencoders, K-means based Dimensionality Reduction and many more. [2] and [3] present a nice overview of these techniques.

The task of finding embeddings which preserve inter-point distances for all the pairs of points has been extensively addressed. A notion of *Metric Distortion* is used to quantify the change in the metric due to an embedding. More formally,

Definition 1 (Metric Distortion). *Let f be any embedding from a metric space (X, d) to a metric space (X', d'). Then the distortion of the embedding f is defined as*

$$dist(f) = \max \left\{ \sup_{x,y} \frac{d(x,y)}{d'(f(x), f(y))}, \sup_{x,y} \frac{d'(f(x), f(y))}{d(x,y)} \right\}$$

A common algorithmic task is to find an embedding of a given data set into a "well behaved" metric space (like low dimensional Euclidean space) while minimizing the distortion. A classical and perhaps the most important result in this area is the Johnson-Lindenstrauss Lemma [4]. It says that n points can be embedded in dimension $O(\epsilon^{-2} \log n)$ with distortion $1 + \epsilon$. Moreover, this embedding can be found using a linear map and in randomized polynomial time. The algorithmic version of the Johnson-Lindenstrauss Lemma is stated as Lemma 1 in Section 4.1.

A natural relaxation to metric distortion was considered by Abraham et. al [5]. They introduce the notion of *average distortion* where the goal is to construct embeddings which have small average distortion rather than small maximum distortion. Another relaxation was considered by Chan et. al [6]. They introduce ϵ-*slack distortion*. They construct embeddings such that distortion is small for $1 - \epsilon$ fraction of the pairs of points. Our notion of Metric Retention is very closely related to this notion of ϵ-slack distortion.

Another direction of work which is related to ours is Probabilistic Lipschitzness (PL) [1]. An assumption inherent in many machine learning paradigms is that *close points tend to have the same labels*. To model this property, PL was introduced by Urner et. al [1]. Under PL Assumptions, they showed that Nearest Neighbor has bounded sample complexity [7]. Not only that, they show that PL assumptions lead to sample savings, i.e., faster learning from nicer distributions [7]. They also show that under PL assumptions, proper semi-supervised learning has reduced sample complexity.

2   Metric Retention (MR)

Framework - Given a domain set $X \subseteq \mathbb{R}^N$, a probability distribution P over X and $\mathbb{R}^n$ respectively and an embedding E which maps points from $\mathbb{R}^N$ to $\mathbb{R}^n$. Define $d(x, y) = \|x - y\|$ and $d'(x, y) = \|E(x) - E(y)\|/K$ where K is some positive normalization constant.

Definition 2 (ϵ-slack Distortion [6]). *Given a finite sample S, an embedding E has distortion $D(\epsilon)$ with ϵ-slack if all but an ϵ-fraction of distances have distortion at most $D(\epsilon)$ under E.*

$$d(x, y) \leq d'(x, y) \leq D(\epsilon) d(x, y)$$

Our definition of metric retention is based on Def. 2 but with few important distinctions. We want to give guarantees for the data distribution P and not only a given sample S. Hence, instead of talking about fractions of pairs of points, we talk of probabilities over pairs of points. Also, we don't normalize the distance $d(x, y)$ by a constant K. But rather consider violations of distortion in both directions.

Definition 3 (Metric Retention). *We say that an embedding E has (ψ_1, ψ_2) - metric retention, if for all $\epsilon \in (0, 1)$:*

$$\mathbb{P}\big[d'(x, y) < (1 - \epsilon) d(x, y)\big] \leq \psi_1(\epsilon) \quad and \quad \mathbb{P}\big[d'(x, y) > (1 + \epsilon) d(x, y)\big] \leq \psi_2(\epsilon)$$

where the probability is over the pairs of points (x, y) generated i.i.d by P^2 and over the randomness (if any) in the embedding E.

Example 1 (Comparison with Metric Distortion). *Let P be the probability distribution over $[0,1] \times \{0,2\}$ defined by picking $(x,0)$ with probability $1 - \delta$ uniformly over $[0,1] \times 0$ and picking $(x,2)$ with probability δ uniformly over $[0,1] \times 2$. Let the dimensionality reduction technique be such that it always projects points along the x-axis.*

Then for such a reduction technique, $\psi_1(\epsilon) = 2\delta(1 - \delta)$ and $\psi_2(\epsilon) = 0$. However, with very high probability, the distortion for this technique would be large (> 2) for large datasets.

Example 2 (Comparison with Average Distortion). *Consider a sample $S \subseteq R^2$. $S = \{(1/n,0),(2/n,0),(3/n,0),\ldots,(1,0),(1/2,n)\}$ where n is an odd integer. Let the embedding be such that it always projects points along the x-axis.*

It is easy to see that the average distortion for this embedding is > 4. On the other hand, assuming that the distribution is uniform over the domain S, we get the metric retention functions $\psi_1 = 2n/(n+1)^2$ and $\psi_2 = 0$.

3   Estimating Metric Retention Parameters

Before choosing a dimensionality reduction technique, we would like to estimate how "good" is it at preserving inter-point distances. Hence, we would like to estimate the probabilities in Def. 3 from only a sample S.

This can be done by the following procedure. Let $S = \{(x_1, y_1), \ldots, (x_m, y_m)\}$ be a sample of m pairs of points where each pair is generated independently and identically by the distribution P^2. Now, denote by

$$f_1(\epsilon) = \underset{\text{of points in S}}{\text{Fraction of pairs}} [d' < (1-\epsilon)d] \quad \text{and} \quad f_2(\epsilon) = \underset{\text{of points in S}}{\text{Fraction of pairs}} [d' > (1+\epsilon)d]$$

where S is generated i.i.d by P^2. We will now prove that f_1 and f_2 as defined above are 'close to' to the true probabilities in Def. 3. Before we do that, we need the following result on ϵ-approximations from VC-Theory.

Theorem 1 (Vapnik and Chervonenkis [8]). *Let X be a domain set and D a probability distribution over X. Let H be a class of subsets of X of finite VC-dimension d. Let $\epsilon, \delta \in (0,1)$. Let $S \subseteq X$ be picked i.i.d according to D of size m. If $m > \frac{c}{\epsilon^2}(d \log \frac{d}{\epsilon} + \log \frac{1}{\delta})$, then with probability $1 - \delta$ over the choice of S, we have that $\forall h \in H$*

$$\left| \frac{|h \cap S|}{|S|} - P(h) \right| < \epsilon$$

Theorem 2. *Let f_1, f_2 be as defined above. Let S be an i.i.d sample generated by P^2 of size m. Given $\alpha, \beta \in (0,1)$. If $m > \frac{1}{\alpha^2}(\log \frac{1}{\alpha} + \frac{1}{\beta})$, then with probability at least $1 - \beta$, we have that for all $\epsilon \in (0,1)$*

$$\left| f_1(\epsilon) - \Pr_{x,y \sim P}[d'(x,y) < (1-\epsilon)\,d(x,y)] \right| < \alpha$$

A similar result holds for $f_2(\epsilon)$ as well.

Proof. Let $h_\epsilon = \{(x, y) : d'(x, y) < (1 - \epsilon)d(x, y)\}$. Let $H = \{h_\epsilon : \epsilon \in (0, 1)\}$. It is easy to see that for $\epsilon' < \epsilon$, $h_\epsilon \subseteq h_{\epsilon'}$. Hence, H is a union of hypothesis which are ordered by inclusion. It is easy to see that $\mathrm{VCDim}(H) = 1$. Now, using Theorem 1 gives the desired result.

Observe that such an estimation can be done in an unsupervised manner by just looking at the unlabeled examples and then evaluating the dimensionality reduction technique.

4 Embeddings with Metric Retention

4.1 Random Projections (RP) have MR

The use of random projections for dimensionality reduction is based on a classical and a very important theorem popularly known as the JL-Lemma [4]. It states that m points can be embedded in lower dimension of size $O(\epsilon^{-2} \log m)$ while preserving the Euclidean 2-norm (small distortion). We do not state the actual JL-Lemma here but another useful lemma.

Lemma 1. *Let $v \in R^N$ and let $W \in R^{n,N}$ be a random matrix such that each $W_{i,j}$ is an independent standard normal random variable (mean 0 and variance 1). Then, for every $\epsilon \in (0, 3)$ we have*

$$\Pr\left[\frac{\|Wv\|^2}{\|v\|^2} \leq (1 - \epsilon)n\right] \leq e^{-\epsilon^2 n/6} \quad and \quad \Pr\left[\frac{\|Wv\|^2}{\|v\|^2} \geq (1 + \epsilon)n\right] \leq e^{-\epsilon^2 n/6}$$

Proof. The proof of this lemma can be found in [4] or Lemma 23.3 in [9].

Note that the actual JL-Lemma which we do not state here is a rather straightforward consequence of Lemma 1. We now show that Random Projections have the Metric Retention property. Consider the following Random Projection (RP) algorithm. Construct a matrix $W \in R^{n,N}$ such that each $W_{i,j}$ is an independent normal variable. Now, given any point x, output $RP(x) := \frac{1}{\sqrt{n}}Wx$. Observe that in this framework, $d(x, y) = \|x - y\|$ and $d'(x, y) = \frac{1}{\sqrt{n}}\|Wx - Wy\|$.

Theorem 3. *Let domain be $X \subseteq R^N$ generated i.i.d by some distribution P. Let $RP(x) = \frac{1}{\sqrt{n}}Wx$ take points from R^N to R^n where W is as defined above. Then RP retains distance with $\psi_1(\epsilon) = e^{-\epsilon^2 n/6}$ and $\psi_2(\epsilon) = e^{-\epsilon^2 n/6}$.*

Proof. Using $v = x - y$ in Lemma 1 and $\epsilon \in (0, 1)$, we get that $\Pr[(x, y) : d' \leq (1 - \epsilon)d] \leq \Pr[(x, y) : d' \leq (1 - \epsilon)^{1/2}d] \leq e^{-\epsilon^2 n/6}$. Similarly, $\Pr[(x, y) : d' \geq (1 + \epsilon)d] \leq \Pr[(x, y) : d' \leq (1 + \epsilon)^{1/2}d] \leq e^{-\epsilon^2 n/6}$

Discussion. We see that for a large enough n both the functions ψ_1 and ψ_2 are small. Hence, the fraction of pairs of points which have large distortion is small. Theorem 3 provides a way explaining why JL works in practice.

4.2   PCA has MR

PCA is one of the most popular dimensionality reduction techniques. PCA projects the original N-dimensional data along the n principal directions on an n-dimensional linear subspace. For background and more details on PCA, we refer the reader to [3] or any standard machine learning text.

From the point of view of real-world applications, the important decision here is the choice of n, the dimensionality of the space to which the data should be reduced. Practitioners often choose n such that "most" of the variance of the data is captured. A dimension n which retains 99% of the variance is often considered a good choice. Intuitively, capturing the variance seems equivalent to capturing the relevant variations and information inherent in the data.

Let us assume that PCA projects the data along the n orthogonal unit vectors given by $v_1, \ldots, v_n$. Let $v_{n+1}, \ldots, v_N$ be some other vectors such that $v_1, \ldots, v_N$ form an orthogonal basis for the original N-dimensional space. Then PCA is equivalent to choosing the first n dimensions amongst the N given dimensions in this space. Let the distance functions be the standard Euclidean distances. Hence, in this framework $d(x, y) = \sqrt{(x_1 - y_1)^2 + \ldots + (x_N - y_N)^2}$ and $d'(x, y) = \sqrt{(x_1 - y_1)^2 + \ldots + (x_n - y_n)^2}$.

Theorem 4. *Let the domain $X \subseteq \mathbb{R}^N$ be generated i.i.d by some unknown probability distribution P. Let $var(P_i) < \delta, \forall n + 1 \leq i \leq N$. In addition, let $\Pr[(x, y) : d(x, y) < t] \leq c$ for some constants t, c. Then PCA retains distances with $\psi_1(\epsilon) = c + \frac{8 (N-n)^2 \delta^2}{t^2 \epsilon^2}$ and $\psi_2(\epsilon) = 0$.*

Proof. Please refer to the appendix for the proof of this theorem.

Discussion. We see that $\psi_1(\epsilon) \in O(\delta^2)$ and $\psi_2(\epsilon) = 0$. Hence, as more variance is captured by the n principal directions, the smaller is δ and hence smaller is ψ_1 implying that the metric retention property of PCA improves. Hence, our notion is able to model PCA using similar assumptions and intuitions which are used in practice. In a sense, we also provide justification for something which is used in practice. Note that, the metric distortion of PCA would be large. Since for some of pairs of points the distortion may be large. However, Theorem 4 shows that for most of the pairs of points the distortion is small.

5   Probabilistic Lipschitzness (PL)

The notion of Probabilistic Lipschitzness was introduced by Urner et. al [1]. The PL assumption says that the probability of two close points having different labels is bounded and small. Such a relation is inherent in many Machine Learning paradigms.

In this work, we consider a slightly different definition of PL. We show that the sample complexity bounds for Nearest Neighbor that were obtained using the original definition can also be obtained using our definition. We will refer to our definition of PL as *PL-Conditional* and the original definition as *PL-Unary*.

Definition 4 (PL-Unary [7]). *The labeling function l satisfies ϕ-Probabilistic Lipschitzness w.r.t to the original definition, if for all $\lambda > 0$*

$$\Pr_{x \sim P} \left[\Pr_{y \sim P} [d(x, y) < \lambda \wedge l(x) \neq l(y)] > 0 \right] \leq \phi(\lambda)$$

Definition 5 (PL-Conditional). *We say that the labeling function l satisfies ϕ-PL-Conditional if for all $\lambda > 0$*

$$\Pr_{x, y \sim P} \left[l(x) \neq l(y) \mid d(x, y) < \lambda \right] \leq \phi(\lambda)$$

Throughout the remainder of the section, we bound the sample complexity of Nearest Neighbour learning (m_{NN}) w.r.t the class of all distributions that satisfy PL assumptions and have deterministic labeling functions. ϵ, δ denote the usual accuracy and confidence parameters respectively.

Theorem 5 (Urner and Ben-David [7]). *Let the domain be $X \subseteq [0, 1]^N$ generated i.i.d by some distribution P and labeled by some deterministic function l which satisfies PL-Unary with function ϕ . Then the sample complexity of Nearest Neighbor m_{NN} is upper bounded by*

$$m_{NN}(\epsilon, \delta) \quad \leq \quad \frac{2}{\epsilon \delta e} \cdot \left(\frac{\sqrt{N}}{\phi^{-1}(\epsilon/2)} \right)^N \tag{1}$$

They also showed that PL-Unary leads to faster learning from nicer distributions. Here, we show that similar results can also be obtained under PL-Conditional assumptions.

5.1   Sample Complexity Bounds for Nearest Neighbor under PL-Conditional

Theorem 6. *Let the domain be $X \subseteq [0, 1]^N$ generated i.i.d by some distribution P and labeled by some deterministic function l which satisfies PL-Conditional with function ϕ . Then the sample complexity of Nearest Neighbor m_{NN} is upper bounded by*

$$m_{NN}(\epsilon, \delta) \quad \leq \quad \frac{2}{\epsilon \delta e} \cdot \left(\frac{\sqrt{N}}{\phi^{-1}(\epsilon \delta/2)} \right)^N \tag{2}$$

Proof. Please refer to the appendix for the proof of this theorem.

Faster Learning from Nicer Distributions. We now look at learning rates for distributions which satisfy PL-Conditional with function $\phi(\lambda) = \lambda^a$ for some $a \in \mathbb{N}$. In this case, we see that the upper bound in Thm. 6 evaluates to

$$O\left(\left(\frac{1}{\epsilon} \right)^{\frac{N}{a}+1} \left(\frac{1}{\delta} \right)^{\frac{N}{a}+1} \right) \tag{3}$$

Hence, we see that the nicer the distribution (large a), the faster is the convergence of nearest neighbor.

5.2   Comparison of Convergence Rates for Nearest Neighbor

PL-Conditional vs PL-Unary. First, we want to compare convergence rates for Nearest Neighbor (m_{NN}) under PL-Unary and PL-Conditional assumptions. Thm. 7 which is stated below shows that both the notions are orthogonal to one another. That is, there are examples for which PL-Unary is smaller and there are examples for which PL-Conditional is smaller. Hence, none always leads to faster learning as compared to the other.

Theorem 7. *Let the domain be $X \subseteq [0,1]$. Denote by ϕ_{PLC} the function which parametrizes our definition (Eq. 5) and by ϕ_{PLU} the function which parametrizes the earlier definition (Eq. 4). Then there exists distributions D_1 and D_2 such that $\phi_{PLC} < \phi_{PLU}$ for D_1 and $\phi_{PLU} < \phi_{PLC}$ for D_2.*

PL-Conditional vs Lipschitzness. In our analysis, we assumed that the labeling function is deterministic. The Lipschitz assumption works with the non-deterministic case and hence is not directly related to the present work. In this case, there is conditional probability function over the labels, η defined as, $\eta(x) = \mathbb{P}[y = 1|x]$. Assuming that η satisfies c-Lipschitzness, Thm 19.3 in [9] shows that the sample complexity of nearest neighbor learning is upper bounded by $m_{NN}(\epsilon, \delta) \leq \left(\frac{4c\sqrt{N}}{\epsilon\delta} \right)^{N+1}$, that is the sample complexity evaluates to

$$O\left(\left(\frac{1}{\epsilon}\right)^{N+1} \left(\frac{1}{\delta}\right)^{N+1} \right) \tag{4}$$

Hence, comparing Eqns. 3 and 4, we see that for $a > 1$, PL-Conditional leads to faster learning than the Lipschitz assumption.

5.3   Estimating ϕ PL-Conditional

Given a labelled sample, we would like to evaluate how label homogeneous it is. Probabilistic Lipschitzness provides a way to quantify this. Specifically, we would like to estimate the probability $\mathbb{P}[l(x) \neq l(y)|d(x,y) < \lambda]$ from some sample S.

This can be done by the following procedure. Let $S = \{(x_1, y_1), \ldots, (x_m, y_m)\}$ be a sample of m pairs of points where each pair is generated independently and identically by the distribution P^2. Now, denote by

$$f_1(\lambda) = \underset{\text{of points in S}}{\text{Fraction of pairs}} [l(x) \neq l(y) \cap d < \lambda] \quad \text{and} \quad g_1(\lambda) = \underset{\text{of points in S}}{\text{Fraction of pairs}} [d < \lambda]$$

We will prove that given a large enough sample, f_1, g_1 are good estimates of $\mathbb{P}[l(x) \neq l(y) \cap d(x,y) < \lambda]$ and $\mathbb{P}[d(x,y) < \lambda]$ respectively.

Theorem 8. *Let $f_1(\lambda), g_1(\lambda)$ be as defined above. Let S be an i.i.d sample generated by P^2 of size m. Given $\alpha, \beta \in (0,1)$ and $\lambda > 0$. If $m > \frac{1}{2\alpha^2} \log \frac{4}{\beta}$, then with probability atleast $1 - \beta$, we have that*

$$\left| f_1(\lambda) - \mathbb{P}[l(x) \neq l(y) \cap d(x,y) < \lambda] \right| < \alpha \quad \text{and} \quad \left| g_1(\lambda) - \mathbb{P}[d(x,y) < \lambda] \right| < \alpha$$

Proof. Denote by $PLA(\lambda) = \mathbb{P}[l(x) \neq l(y) \cap d(x,y) < \lambda]$. Using standard concentration bounds (Hoeffding's inequality), we get that

$$\mathbb{P}[|f_1(\lambda) - PLC(\lambda)| > \alpha] \leq 2\exp(-2m\alpha^2)$$

Same bound holds for g_1 as well. Substituting $m > \frac{\log(4/\beta)}{2\alpha^2}$ gives the result.

Theorem 8 forms the basis for our experiments in Section 7. It shows that given a large enough sample with high probability $f_1(\lambda)$ is an accurate estimate of the corresponding probability. Alg. 1 shows a procedure of how to evaluate $f = f_1/g_1$ for different datasets.

One important point to note is that we do not evaluate $f(\lambda)$ for very small values of λ. This is because for very small λ, we expect f_1, g_1 to be very close to zero. In such a case, even a small additive error of α, might imply large multiplicative error.

6   Metric Retention and PL-Conditional

6.1   Metric Retention Preserves PL-Conditional

An embedding which has metric retention property preserves some of the *nice* properties of the distribution as well. Thm. 9 and Lemma 2 together show that if we have PL-Conditional in the original dimension then any dimensionality reduction technique which has metric retention preserves the PL-Conditional property as well. Ex. 3 shows that the same is not true for PL-Unary.

Another way of stating our results is that, Nearest Neighbor has bounded sample complexity in the reduced dimension space as well. There are certain conditions under which Nearest Neighbor has bounded sample complexity. We want to show that under a metric distance retaining reduction, if those conditions are true in the original dimension, then those conditions are true in the reduced dimension as well. PL-Conditional and PL-Unary are such conditions and hence we investigate as to which of these also hold in the reduced dimension.

We break the PL-Conditional definition into two statements. The first is that the probability of the 'and' condition is upper bounded and that the denominator is lower bounded by some function of λ. More formally, we assume that

$$\Pr_{x,y \sim P}[l(x) \neq l(y) \wedge d(x,y) < \lambda] \leq \alpha(\lambda) \text{ and } \Pr_{x,y \sim P}[d(x,y) < \lambda] \geq \beta(\lambda)$$

where $\phi = \frac{\alpha}{\beta}$. We will now show that these quantities are bounded in the reduced dimension space as well.

Theorem 9. *Consider the framework as in the Def. 3. Domain set $X \subseteq \mathbb{R}^N$ generated by some probability distribution P, distance functions d and d', embedding E which has metric retention w.r.t functions ψ_1 and ψ_2 and a labeling function l. If* $\Pr_{x,y \sim P}[l(x) \neq l(y) \wedge d(x,y) < \lambda] \leq \alpha(\lambda)$ *then* $\Pr_{x,y \sim P}[l(x) \neq l(y) \wedge d'(x,y) < \lambda] \leq \alpha_1(\lambda)$ *where $\alpha_1(\lambda) = \psi_1(\epsilon) + \alpha(\frac{\lambda}{1-\epsilon})$ and $\epsilon \in (0,1)$.*

Proof.

$$\Pr_{x,y\sim P}\big[d'(x,y) < \lambda \wedge l(x) \neq l(y)\big]$$

$$\leq \Pr_{x,y\sim P}\Big[d(x,y) < \frac{\lambda}{1-\epsilon} \wedge l(x) \neq l(y)\Big] + \Pr_{x,y\sim P}\big[d'(x,y) < (1-\epsilon)d(x,y)\big]$$

$$\leq \alpha\Big(\frac{\lambda}{1-\epsilon}\Big) + \psi_1(\epsilon) =: \alpha_1(\lambda)$$

Lemma 2. *Consider the framework as in Theorem 9. If* $\Pr_{x,y\sim P}\big[d(x,y) < \lambda\big] \geq \beta(\lambda)$, *then*
$\Pr_{x,y\sim P}\big[d'(x,y) < \lambda\big] \geq \beta_1(\lambda)$ *where* $\beta_1(\lambda) = \beta(\frac{\lambda}{1+\epsilon}) - \psi_2(\epsilon)$ *and* $\epsilon \in (0,1)$.

Proof. The proof is left as an exercise for the reader.

In our discussion in this chapter, one slight detail is missing. Note that Probabilistic Lipschitzness is sensitive to the scale of the data. To obtain bounds for Nearest Neighbor under PL, we needed an implicit assumption on the diameter of the data. This is because we assumed that the domain was $[0,1]^N$. Hence, we should upper bound the diameter in the reduced dimension as well. The next lemma bounds this quantity.

Lemma 3. *Denote by* d_{max} *the diameter of the data in the original dimension. Then,* $\Pr_{x,y\sim P}[d'(x,y) > (1+\epsilon)d_{max}] \leq \psi_2(\epsilon)$.

Proof. $\Pr[d'(x,y) > (1+\epsilon)d_{max}] \leq \Pr[d'(x,y) > (1+\epsilon)d(x,y)] \leq \psi_2(\epsilon)$

Example 3 (Metric Retention does not preserve PL-Unary). *Consider points in 2-dimensional plane. Fix some* $0 < \lambda < 1$. *Let the distribution* P *be such that it generates points* $y = 0$ *and* $x \in (0,\lambda) \cup (\lambda, 2\lambda) \cup \ldots \cup ((R-1)\lambda, R\lambda)$ *with probability* $1-\delta$ *uniformly with label 1 and points* $y = 2$ *and* $x \in \{\lambda, 2\lambda, \ldots, R\lambda\}$ *with probability* δ/R *with label 0. Let the dimensionality reduction technique be such that it always projects points along the* x-*axis.*

In the original representation, it is easy to see that $\phi_{PLU}(\lambda) = 0$. *Also,* $\psi_1(\epsilon) = 2(1-\delta)\delta$ *and* $\psi_2(\epsilon) = 0$. *In the reduced representation,* $\phi_{PLU}(\lambda) = 1$ *as for all points there exists another point of opposite label at distance* $< \lambda$.

6.2   Sample Complexity Benefits

One of the benefits of dimensionality reduction is the computational cost of classification. As an example, as the dimension reduces the computational cost of nearest neighbor also decreases. However, if the embedding has metric retention then we get statistical benefits (sample complexity) as well.

The sample complexity of nearest neighbor is given by $O\big(\frac{\sqrt{N}^N}{(\phi^{-1}(\epsilon\delta/2))^N}\big)$. Intuitively, when an embedding has MR, it leads to a slight worsening in the ϕ function as given by Thm. 9 and Lemma 2. However the exponential dependence on dimension improves which should lead to reduction in sample complexity.

Example 4. *Let α, β be as defined in Thm 9 and Lemma 2. Let $\alpha(\lambda) = \lambda^{2a}$ and $\beta(\lambda) = \lambda^a$ for some $a \in \mathbb{N}$. Let $\psi_1(\epsilon) = \epsilon$ and $\psi_2(\epsilon) = 0$ for all $\epsilon \in (0,1)$ (Note, this is true for PCA).*

Choose an ϵ s.t. $\epsilon \leq \frac{1-2^{-1/a}}{1+2^{-1/a}}$. Then, we have that $\phi_1(\lambda) = \alpha_1(\lambda)/\beta_1(\lambda) \leq 2\lambda^a + \epsilon(1+\epsilon)^a \lambda^{-a} < 3\lambda^a$. Hence, we see that the sample complexity in the original dimension is

$$O\left(\frac{\sqrt{N}^N}{(\epsilon\delta)^{N/a+1}} \right) \text{ which reduces to } O\left(\frac{\sqrt{n}^n}{(\epsilon\delta)^{n/a+1}} \right)$$

Hence, we have an exponential decrease in the number of samples needed.

7   Experiments

We now describe the experiments we ran to validate and compare the different notions we introduced in this work. We ran all our experiments on a standard Linux distribution running Ubuntu 14.04 with 16GBs Main Memory. We have used Matlab and C++ to implement all our algorithms. We implemented our algorithms on two different datasets namely, *MNIST, ISOLET [10]*.

7.1   Probabilistic Lipschitzness Captures Usefulness

How related the labels are to its marginal distribution is type of information which is stored in a representation. PL is a measure which tries to quantify this information. Let us consider any binary classification task. In this case, the "best" representation or the "best" feature learning algorithm would be one which gives all points of label 0 as the same representation (say **0**) and all points of label 1 the same representation (say **1**). For such a representation both PL-Conditional and PL-Unary measures are 0 implying perfect label homogeneity.

Hence, PL has implications for feature learning algorithms as well. The goal is to give experimental evidence supporting the claim that *"Better representations should have better PL"*. Also, better feature learning algorithms should produce features which have better PL.

We present experiments related to PL-Conditional and show how it is able to capture the usefulness of a representation. We need to compute ϕ (Def. 5) which depends on knowledge of the distribution. This is never available in practical situations. Hence, instead of calculating probabilities, we calculate the fraction of pairs of points over which the predicate is satisfied. The procedure is defined in Alg. 1. A formal justification of this approach is provided in Sec. 5.3, Thm. 8.

Deep Belief Network (DBN) have become quite popular and successful as feature learning algorithms especially for vision related tasks (Hinton et. al [11]). We used the DeepBNet Toolbox (Keyvanrad et. al [12]) to train a deep belief network with 3 hidden layers of sizes $500, 500$ and 2000 respectively on the MNIST dataset and obtained the learned features.

Algorithm 1. Evaluating a learned representation

Input: Unlabelled data set $X \subseteq \mathbb{R}^N$

Output: $f(\lambda)$ for different values of λ.

1: Let $S = \{(x_1, y_1), \ldots, (x_m, y_m)\}$ where each pair of points is independently and identically uniformly generated.

2: Let $f(\lambda) = $ Fraction of pairs of points $[\ l(x) \neq l(y) \mid d(x,y) < \lambda\ ]$

3: Compute f for a particular λ on the set S.

4: Repeat 1-3 for different values of λ.

Hence, we get four different learned feature representations. We first compute the f values (Alg. 1). Next, we run three classification algorithms namely, Nearest Neighbor, Regression, Linear SVM on these representations. We then calculate the accuracy difference between the original and the learned representations averaged over these three classification algorithms. This we call the "performance gap". Intuitively, better features will perform better on different classification algorithms and hence should have larger performance gap.

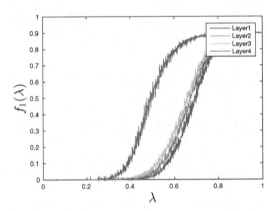

Table 1. Difference of classification accuracy (test) of PCA representation vs Original representation on different datasets averaged over 3 different algorithms

Features	Accuracy Gap(in %)
Layer-1	0.00
Layer-2	3.23
Layer-3	3.76
Layer-4	3.91

Fig. 1. Fraction of points which have different labels given their distance is ($< \lambda$) for the given representation.

Fig. 1 shows the plot of f (Alg. 1) for the features learned from different layers. Table 1 gives the average classification accuracy gap between the different layers (using layer-1 features as the base). The f plot is noisy but the general trend is that it increases with increasing distance (λ). Observe the nice correlation between the f values and the accuracy gap. We observe that as the depth of the network increases, the features computed by DBNs have better PL (more label homogeneous). The observation that deeper layers are more label homogeneous is not new. Figures 3(B) and 4(B) in Hinton et. al [11] visualize a trained autoencoder with 784-1000-500-250-2 layers and obtain similar conclusions for the features of the output layer (dimension 2). However, for higher dimensions (> 3) such a visualization is not possible. Probabilistic Lipschitzness (f plots) provides a way to measure and visualize the same.

8   Conclusion and Future Work

We introduced a new notion to model data embeddings which preserve inter-point distances with high probability, called *Metric Retention* (Def. 3). We showed that common techniques like PCA and Random Projections satisfy this property (Thm. 4). We then showed that an embedding which satisfies metric retention also preserves essential niceness properties of the distribution, namely the PL-Conditional property (Thm. 9). Our experiments showed some correlation between the usefulness of a representation and its PL-Conditional measure.

In future, we want to identify assumptions under which other embeddings such as Neural Autoencoders might retain distance. Another direction is using PL-Conditional or some similar niceness assumption to obtain sample complexity bounds for other popular learning paradigms like Linear Classification, SVMs etc.

References

1. Urner, R., Ben-David, S., Shalev-Shwartz, S.: Access to unlabeled data can speed up prediction time. In: ICML (2011)
2. van der Maaten, L.J., Postma, E.O., van den Herik, H.J.: Dimensionality reduction: A comparative review. Journal of Machine Learning Research **10**(1–41), 66–71 (2009)
3. Ghodsi, A.: Dimensionality reduction a short tutorial. Department of Statistics and Actuarial Science, Univ. of Waterloo, Ontario, Canada (2006)
4. Johnson, W.B., Lindenstrauss, J.: Extensions of lipschitz mappings into a hilbert space. Contemporary Mathematics **26**(189–206), 1 (1984)
5. Abraham, I., Bartal, Y., Neiman, O.: On embedding of finite metric spaces into hilbert space. Tech. Report, Technical report (2006)
6. Chan, T.H.H., Dhamdhere, K., Gupta, A., Kleinberg, J., Slivkins, A.: Metric embeddings with relaxed guarantees. SIAM Journal on Computing **38**(6), 2303–2329 (2009)
7. Urner, R., Ben-David, S.: Probabilistic lipschitzness a niceness assumption for deterministic labels. In: Learning Faster from Easy Data-Workshop@ NIPS (2013)
8. Vapnik, V.N., Chervonenkis, A.Y.: On the uniform convergence of relative frequencies of events to their probabilities. Theory of Probability & Its Applications **16**(2), 264–280 (1971)
9. Shalev-Shwartz, S., Ben-David, S.: Understanding machine learning (2014)
10. Bache, K., Lichman, M.: UCI machine learning repository (2013)
11. Hinton, G.E., Salakhutdinov, R.R.: Reducing the dimensionality of data with neural networks. Science **313**(5786), 504–507 (2006)
12. Keyvanrad, M.A., Homayounpour, M.M.: A brief survey on deep belief networks and introducing a new object oriented matlab toolbox (deebnet) (2014). arXiv preprint arXiv:1408.3264
13. Knuth, D.E.: The art of computer programming: Fundamental algorithms, vol. i (1968)

A   Proof of Theorems

Proof of Thm. 4. We first show that $\mathbb{P}[d - d' > \lambda]$ is small.

$$\Pr_{x,y\sim P}\left[d(x,y) - d'(x,y) \geq \lambda\right] \leq \Pr_{x,y\sim P}\left[\exists i : (x_i - y_i)^2 \geq \frac{\lambda^2}{N-n}\right]$$

$$\leq \sum_{i=n+1}^{N} \Pr_{x,y\sim P}\left[|x_i - \mu_i| + |y_i - \mu_i| \geq \frac{\lambda}{\sqrt{N-n}}\right] \text{ where } \mu_i = \text{mean}(P_i).$$

$$= 2\sum_{i=n+1}^{N} \Pr_{x_i\sim P_i}\left[|x_i - \mu_i| \geq \frac{\lambda}{2\sqrt{N-n}}\right]. \text{ Using Chebyshev's Inequality [13],}$$

$$\leq \sum_{i=n+1}^{N} \frac{8\,(N-n)\,var^2(P_i)}{\lambda^2} \leq \frac{8\,(N-n)^2\,\delta^2}{\lambda^2} \tag{5}$$

Now, using Eqn. 5 we will bound the probability of the actual event.

$$\mathbb{P}[d'(x,y) \leq (1 - \epsilon)d(x,y)] = \mathbb{P}[d(x,y) - d'(x,y) \geq \epsilon d(x,y)]$$

$$\leq \mathbb{P}[d(x,y) - d'(x,y) \geq \epsilon\lambda] + \mathbb{P}[d(x,y) < \lambda] \leq c + \frac{8(N-n)^2\delta^2}{\epsilon^2\lambda^2}$$

Proof of Thm. 6. Partition the domain $\mathbb{X} = [0,1]^N$ into $r = (\sqrt{N}/\lambda)^N$ axis-aligned boxes $C_1,\ldots,C_r$ each of length $\lambda/\sqrt{N}$ and diameter λ for some λ (to be chosen later). For any $x \in [0,1]^N$ denote by $C(x)$ the region in which x lies and by $\pi_S(x)$ the Nearest Neighbor of x in S.

$$Err_P(NN(S)) = \Pr_{x\sim P}\left[l(\pi_S(x)) \neq l(x)\right]$$

$$\leq \Pr_{x\sim P}\left[C(x) \cap S = \phi\right] + \Pr_{x\sim P}\left[C(x) \cap S \neq \phi \wedge l(\pi_S(x)) \neq l(x)\right]$$

$$= \sum_{i:C_i\cap S=\phi} \Pr_{x\sim P}\left[x \in C_i\right] + \sum_{i:C_i\cap S\neq\phi} \Pr_{x\sim P}\left[x \in C_i \wedge l(\pi_S(x)) \neq l(x)\right]$$

$$= \sum_{i:C_i\cap S=\phi} \mathbb{P}[C_i] + \sum_{i:C_i\cap S\neq\phi} \mathbb{P}[C_i] \Pr_{x\sim P_{C_i}}\left[l(\pi_S(x)) \neq l(x)\right] \tag{6}$$

where $\mathbb{P}[C_i] := \Pr_{x\sim P}\left[x \in C_i\right]$ and P_{C_i} denotes the distribution P restricted to the set C_i. Observe that, since $C_i \cap S \neq \phi$, we have that $d(x, \pi_S(x)) \leq \lambda$. Denote $a(x,y) := (l(x) \neq l(y))$. Now consider the expectation over the sample S of the quantity on the extreme right of Equation 6.

$$\text{Exp}_{S\sim P^m}\left[\sum_{i:C_i\cap S\neq\phi} \mathbb{P}[C_i] \Pr_{x\sim P_{C_i}}\left[a(\pi_S(x), x)\right]\right]$$

$$\leq \sum_{i=1}^{r} \mathbb{P}[C_i] \text{Exp}_{S\sim P^m}\left[\Pr_{x\sim P_{C_i}}\left[a(\pi_S(x), x)\right]\right] = \sum_{i=1}^{r} \mathbb{P}[C_i] \Pr_{\substack{S\sim P^m \\ x\sim P_{C_i}}}\left[a(\pi_S(x), x)\right] \tag{7}$$

Denote $\mathbb{P}_S[\pi_{x_j}] := \Pr_{\substack{S \sim P^m \\ x \sim P_{C_i}}}[\pi_S(x) = x_j]$. Now, observe that the labeling function l is independent of the choice of the nearest Neighbor $\pi_S(x)$ which depends only on the distance function d. Hence, we get

$$\Pr_{\substack{S \sim P^m \\ x \sim P_{C_i}}}\left[a(\pi_S(x), x)\right] \leq \sum_{j=1}^{m} \mathbb{P}_S[\pi_{x_j}] \Pr_{\substack{x_1,\ldots,x_m \sim P \\ x \sim P_{C_i}}}\left[a(x_j, x)\right]$$

$$= \sum_{j=1}^{m} \mathbb{P}_S[\pi_{x_j}] \Pr_{\substack{x_j \sim P \\ x \sim P_{C_i}}}\left[l(x_j) \neq l(x)\right] \leq \sum_{j=1}^{m} \mathbb{P}_S[\pi_{x_j}]\phi(\lambda) \leq \phi(\lambda). \text{ Hence,}$$

$$\Exp_{S \sim P^m}\left[\sum_{i:C_i \cap S \neq \phi} \mathbb{P}[C_i] \Pr_{x \sim P_{C_i}}\left[NN(S)(x) \neq l(x)\right]\right] \leq \phi(\lambda) \tag{8}$$

Now, using Equations 6 and 8 together with Lemma 19.2 from [9], we get that

$$\Exp_{S \sim P^m}\left[Err_P(NN(S))\right] \leq \frac{r}{me} + \phi(\lambda). \text{ Now, using Markov's inequality}$$

$$\Pr_{S \sim P^m}\left[Err_P(NN(S)) > \epsilon\right] \leq \frac{1}{me\epsilon}\left(\frac{\sqrt{N}}{\lambda}\right)^N + \frac{\phi(\lambda)}{\epsilon}$$

Using $\lambda = \phi^{-1}(\epsilon\delta/2)$, we get the result of the Theorem.   □

Proof of Thm. 7. We first construct a distribution such that ϕ_{PLC} is small but $\phi_{PLU} = 1$ (Example 5). Next we construct another distribution such that $\phi_{PLC} = 1$ but ϕ_{PLU} is small. (Example 6). This completes the proof of our theorem.

Example 5. *(PL-Conditional but not PL-Unary)* Let λ and $\gamma << \lambda$ be constants. Let $X_k = [k\lambda - \gamma, k\lambda + \gamma]$ for some $k \in \mathbb{N}$. Let the domain $\mathbb{X} = \cup X_k$ for all k such that $X_k \subseteq [0, 1]$. Let the labeling function be such that all points of the form $k\lambda$ where $k \in \mathbb{N}$ are labeled 1 and all the other points are labeled 0. Now, consider the following distribution P over the domain $\mathbb{X}$. All the subintervals are given equal weight of $\frac{1}{n}$. Within the subinterval, the point of the form $k\lambda$ (center) is given a weight of $\frac{1-\gamma}{n}$ and the remaining weight of $\frac{\gamma}{n}$ is spread uniformly over the remaining points.

Now, observe that, given any point x there exists another point y with finite probability such that $d(x, y) < \lambda$ and $l(x) \neq l(y)$. Hence, $\phi_{PLU} = 1$. Standard calculations show that for the above example $\phi_{PLC} = \frac{4\gamma}{\lambda}$.

Example 6. *(PL-Unary but not PL-Conditional)* Let $\lambda << 1$ be some small constant. Let $S = \{\lambda, \ldots, k\lambda\}$ be the maximal set such that $k \in \mathbb{N}$ and $S \subseteq [0, 1]$. Let the domain $\mathbb{X} = S \cup (k + \frac{1}{2}\lambda)$. Let n be the number of points in the domain $\mathbb{X}$. Then $\frac{1}{\lambda} \leq n \leq \frac{1}{\lambda} + 1$. Let P be the uniform distribution over the domain $\mathbb{X}$. Let the labeling function be such that it labels all points in S as 1 and the point $(k + \frac{1}{2}\lambda)$ as 0. It is easy to see that in this case, $\phi_{PLC}(\lambda) = 1$ whereas $\phi_{PLU}(\lambda) = 2\lambda$.

Learning with Deep Cascades

Giulia DeSalvo[1]([✉]), Mehryar Mohri[1,2], and Umar Syed[2]

[1] Courant Institute of Mathematical Sciences, 251 Mercer Street,
New York, NY 10012, USA
desalvo@cims.nyu.edu
[2] Google Research, 111 8th Avenue, New York, NY 10011, USA

Abstract. We introduce a broad learning model formed by cascades of predictors, *Deep Cascades*, that is structured as general decision trees in which leaf predictors or node questions may be members of rich function families. We present new data-dependent theoretical guarantees for learning with Deep Cascades with complex leaf predictors and node questions in terms of the Rademacher complexities of the sub-families composing these sets of predictors and the fraction of sample points reaching each leaf that are correctly classified. These guarantees can guide the design of a variety of different algorithms for deep cascade models and we give a detailed description of two such algorithms. Our second algorithm uses as node and leaf classifiers SVM predictors and we report the results of experiments comparing its performance with that of SVM combined with polynomial kernels.

Keywords: Decision trees · Learning theory · Supervised learning

1 Introduction

Decision trees are learning models commonly used in classification, regression, and clustering applications [6,23]. They can be defined as binary trees augmented with indicator functions at each internal node and assignment functions at each leaf. A sample point is processed by a decision tree by answering questions at each node of a tree until a leaf is reached. The label assignment at that leaf then determines the value returned by the tree for that point.

In standard decision trees, the node questions are selected from a fixed family of functions and similarly for the leaf predictors [6,23]. The complexity of a decision tree directly depends on these two families of functions and the depth of the tree. Thus, in practice, to limit the risk of overfitting, relatively simple families of functions are used: node questions are typically selected to be threshold functions based on the input features, leaf predictors often chosen to be constant functions.

This paper considers a significantly broader learning model formed by cascades of predictors, *Deep Cascades*, structured as a decision tree. In this model, the leaf predictors can be chosen out of a complex hypothesis set H and, similarly, the node questions from a family Q. For some difficult learning tasks,

© Springer International Publishing Switzerland 2015
K. Chaudhuri et al. (Eds.): ALT 2015, LNAI 9355, pp. 254–269, 2015.
DOI: 10.1007/978-3-319-24486-0_17

the flexibility of allowing leaf predictors to be selected from a more complex set H (or node questions from Q) may be needed to achieve a high performance. However, cascades with leaf predictors freely selected from H are likely to be prone to overfitting, even with a relatively large number of training samples. Can we preserve the flexibility of using complex leaf predictors (or node questions) and yet avoid overfitting?

Suppose H can be decomposed as the union of p distinct hypothesis sets $H_1, \ldots, H_p$ with increasing complexity. For example, H_k could be the family of threshold functions based on feature monomials of degree k, or polynomial functions of degree k, or H_k could be the family of linear classifiers based on polynomial kernels of degree k. The simpler form of our theoretical analysis shows that, remarkably, it is possible to choose a leaf predictor function from H_k with a relatively large k while benefitting from strong learning guarantees, so long as the fraction of training sample points reaching that leaf is small compared to the complexity of H_k. Our full analysis provides finer guarantees that we will describe in detail.

We present data-dependent theoretical guarantees for learning with Deep Cascades with leaf predictors chosen from the hypothesis sets H_k and node question functions selected from different hypothesis sets Q_j. Our learning bounds are expressed in terms of the Rademacher complexities of the families of leaf predictors H_k and the families of node questions Q_j. These general guarantees can guide the design of a variety of different algorithms for deep cascade models. We describe in depth two such algorithms for learning deep cascades. Our second algorithm uses as node and leaf classifiers SVM predictors and we report the results of experiments comparing its performance with that of SVM combined with polynomial kernels.

Our theory and algorithm have many connections with the wide literature on decision trees and some more recent publications on cascades of classifiers. They are also related to classification with reject option and to a series of articles about combining decision trees with the SVM algorithm. We briefly discuss some of these connections and highlight our contributions here. A more detailed discussion of the previous work is presented in the full version of the paper [11].

Several types of generalization bounds have been given in the past for decision trees. Mansour and McAllester [19] provided non-trivial generalization bounds for decision trees where the node questions are selected from a single hypothesis set and where the leaves are simply labeled with zero or one. These are special cases of the deep cascades we are considering. As in our analysis, their bounds depend on the actual training sample and the tree structure, but the complexity term of their bound is the size of the tree, while ours are expressed in terms of the empirical Rademacher complexities of the hypothesis sets used. A similar approach was adopted by Nobel [21] who further proved the consistency of pruned trees under certain assumptions. Golea et al. [13] gave generalization bounds in terms of the VC-dimension of the node functions and the number of leaves but the trees analyzed are much less general than the deep cascades.

Lastly, Scott and Nowak [26] presented an analysis of a specific family of decision trees, Dyadic Decision Trees (DDT).

Cascades have been extensively used in object detection starting with the work of Viola and Jones [28] who introduced attentional cascades and combined complex classifiers in a linear structure to create a highly accurate face detector. Their work inspired a number of variants of their training procedures [8,16,22,25]. Most of these papers focus on finding the best trade-off between computational cost and classification accuracy, which differs from our main objective here. Additionally, the deep cascades we consider admit a more general structure than those considered by this previous work.

Since one of our deep cascade algorithms uses SVMs, we also review the related previous work on combining SVMs with decision trees. Bennett and Blue [5] used SVMs as node questions in decision trees. They did not present a theoretical analysis of these models and did not address the issue of overfitting, but they proposed an optimization problem for which they gave a heuristic solution and presented preliminary empirical results. Some of the papers in this area focus on multi-class classification [12,18,27]. However, they partition the feature space in a different way from our cascade models. Other articles attempted to increase SVM's computational speeds by using decision trees [1,2,7,15?], but both the splitting criteria and class assignments are very different from ours.

The layout of the paper is as follows. We introduce the notation adopted throughout the paper and give a formal definition of the family of deep cascades in Section 2. Next, in Section 3, we present data-dependent learning bounds for deep cascades, first in the case of leaf functions taking values in $\{-1, +1\}$, and later in the more general case where they take values in the interval $[-1, +1]$. In Section 4, we describe two binary classification algorithms whose design is guided by these bounds and which benefit from these learning guarantees. We report the results of several experiments using one of these algorithms in Section 5.

2   Preliminaries

Let $\mathcal{X}$ denote the input space. We consider the standard supervised learning scenario where the training and test points are drawn i.i.d. according to some distribution $\mathcal{D}$ over $\mathcal{X} \times \{-1, +1\}$ and denote by $S = ((x_1, y_1), \ldots, (x_m, y_m))$ a training sample of size m drawn according to $\mathcal{D}^m$.

Let $l \geq 1$. For any $k \in [1, l]$, let $\mathcal{S}_k$ denote a family of functions mapping $\mathcal{X}$ to $\{0, 1\}$ and let $\mathcal{H}$ denote a family of p hypothesis sets of functions mapping $\mathcal{X}$ to $[0, 1]$. A *deep cascade* with $l \geq 1$ leaves is a tree of classifiers which, in the most generic view, can be defined by a triplet $(\mathbf{H}, \mathbf{s}, \mathbf{h})$ where

- $\mathbf{H} = (H_1, \ldots, H_l)$ is an element of $\mathbf{H}^l$ which determines, for all k, the hypothesis set H_k used at leaf k;
- $\mathbf{s} \colon \mathcal{X} \times [1, l] \to \{0, 1\}$ is a *leaf selector*, that is $\mathbf{s}(x, k) = 1$ if x is assigned to leaf k, $\mathbf{s}(x, k) = 0$ otherwise; for each k, function $\mathbf{s}(\cdot, k)$ is an element of $\mathcal{S}_k$;

- $\mathbf{h} = (h_1, \ldots, h_l)$, with $h_k \colon \mathcal{X} \to [-1, +1]$ the *leaf classifier* for leaf k, which is an element of the family of functions H_k.

We denote by $\mathcal{H}_k = \{x \mapsto \mathsf{s}(x, k)h_k(x) \colon \mathsf{s}(\cdot, k) \in \mathcal{S}_k, h_k \in H_k\}$ the family composed of products of a k-leaf selector and a k-leaf classifier.

We will later assume, as in standard decision trees, that the leaf selector s can be decomposed into *node questions* (or their complements): for any $x \in \mathcal{X}$ and $k \in [1, p]$, $\mathsf{s}(x, k) = \prod_{j=1}^{d_k} q_j(x)$, where d_k is the depth of node k and where each function $q_j \colon \mathcal{X} \to \{0, 1\}$ is an element of a family Q_j.[1] Yet much of our analysis holds without this assumption.

Each triplet $(\mathbf{H}, \mathsf{s}, \mathbf{h})$ defines a *deep cascade function* $f \colon \mathcal{X} \to [-1, +1]$ as follows:

$$\forall x \in \mathcal{X}, \ f(x) = \sum_{k=1}^{l} \mathsf{s}(x, k)h_k(x). \tag{1}$$

We denote by $\mathcal{T}_l$ the family of all deep cascade functions f with l leaves thereby defined. We also denote by $R(f) = \mathbb{E}_{(x,y)\sim\mathcal{D}}[1_{yf(x)\leq 0}]$ the binary classification error of a function $f \in \mathcal{T}_l$, by $\widehat{R}_S(f) = \mathbb{E}_{(x,y)\sim S}[1_{yf(x)\leq 0}]$ its empirical error and, for any $\rho > 0$, by $\widehat{R}_{S,\rho}(f) = \mathbb{E}_{(x,y)\sim S}[1_{yf(x)\leq\rho}]$ its empirical margin error over a labeled sample S, where the notation $(x, y) \sim S$ means that (x, y) is drawn according to the empirical distribution defined by S. We further denote by $\mathfrak{R}_m(H)$ the Rademacher complexity and by $\widehat{\mathfrak{R}}_S(H)$ the empirical Rademacher complexity of a hypothesis set H [3,14].

3 Data-dependent Learning Guarantees

In this section, we present data-dependent learning guarantees for deep cascades that depend, for each leaf k, on the Rademacher complexity of the family $\mathcal{H}_k$ and on the fraction of the points in the training sample that reach leaf k and that are correctly classified, denoted by m_k^+/m. m_k^+ is thus defined by $m_k^+ = |\{i \colon y_i h_k(x_i) > 0, \mathsf{s}(x_i, k) = 1\}|$. Similarly, the number of points that reach leaf k that are incorrectly classified is denoted by m_k^- and defined by $m_k^- = |\{i \colon y_i h_k(x_i) \leq 0, \mathsf{s}(x_i, k) = 1\}|$.

We first analyse the case where the leaf classifiers h_k take values in $\{-1, +1\}$ (Section 3.1), and next consider the more general case where they take values in $[-1, +1]$ (Section 3.2). In the full version of this paper [11], we further extend our analysis and data-dependent learning guarantees to the setting of multi-class classification.

3.1 Leaf Classifiers Taking Values in $\{-1, +1\}$

The main result of this section is Theorem 1, which provides a data-dependent generalization bound for deep cascade functions in the case where leaf classifiers

---

[1] Each q_j is either a node question q or its complement $\bar{q}$ defined by $\bar{q}(x) = 1$ iff $q(x) = 0$. The family Q_j is assumed symmetric: it contains $\bar{q}$ when it contains q.

take values in $\{-1, +1\}$. The following is a simpler form of that result: with high probability, for all $f \in \mathcal{T}_l$,

$$R(f) \leq \widehat{R}_S(f) + \sum_{k=1}^{l} \min\left(4\widehat{\mathfrak{R}}_S(\mathcal{H}_k), \frac{m_k^+}{m}\right) + \tilde{O}\left(l\sqrt{\frac{\log pl}{m}}\right). \qquad (2)$$

Remarkably, this suggests that a strong learning guarantee holds even when a very complex hypothesis set $\mathcal{H}_k$ is used in a deep cascade model, so long as m_k^+/m, the fraction of the points in the training sample that reach leaf k and are correctly classified, is relatively small. Observe that the result remains remarkable and non-trivial even if we upper bound m_k^+ by m_k, the total number of points reaching leaf k. The fraction of the points in the training sample that reach leaf k and are correctly classified depends on the choice of the cascade. Thus, the bound can provide a quantitative guide for the choice of the best deep cascade. Even for $p = 1$, the result is striking since, while in the worst case the complexity term could be in $O(l\widehat{\mathfrak{R}}_S(\mathcal{H}_1))$, this data-dependent result suggests that it can be substantially less for some deep cascades since we may have $m_k^+/m \ll \widehat{\mathfrak{R}}_S(\mathcal{H}_1)$ for many leaves. Also, note that the dependency of the bound on the number of distinct hypothesis sets p is only logarithmic. In Section 4, we present several algorithms exploiting this generalization bound for deep cascades.

For clarity, we will sometimes use the shorthand $r_k = \widehat{\mathfrak{R}}_S(\mathcal{H}_k)$ for any $k \in [1, l]$. We will assume without loss of generality that the leaves are numbered in order of increasing depth and will denote by $\mathcal{K}$ the set of leaves k whose fraction of correctly classified sample points is greater than $4r_k$: $\mathcal{K} = \{k \in [1, l]: \frac{m_k^+}{m} > 4r_k\}$.

Theorem 1. *Fix $\rho > 0$. Assume that for all $k \in [1, l]$, the functions in H_k take values in $\{-1, +1\}$. Then, for any $\delta > 0$, with probability at least $1 - \delta$ over the choice of a sample S of size $m \geq 1$, the following holds for all $l \geq 1$ and all cascade functions $f \in \mathcal{T}_l$ defined by $(\mathbf{H}, \mathbf{s}, \mathbf{h})$:*

$$R(f) \leq \widehat{R}_S(f) + \sum_{k=1}^{l} \min\left(4\widehat{\mathfrak{R}}_S(\mathcal{H}_k), \frac{m_k^+}{m}\right)$$

$$+ \min_{\substack{\mathcal{L} \subseteq \mathcal{K} \\ |\mathcal{L}| \geq |\mathcal{K}| - \frac{1}{\rho}}} \sum_{k \in \mathcal{L}} \left[\frac{m_k^+}{m} - 4\widehat{\mathfrak{R}}_S(\mathcal{H}_k)\right] + C(m, p, \rho) + \sqrt{\frac{\log \frac{4}{\delta}}{2m}},$$

where $C(m, p, \rho) = \frac{2}{\rho}\sqrt{\frac{\log pl}{m}} + \sqrt{\frac{\log pl}{\rho^2 m} \log\left[\frac{\rho^2 m}{\log pl}\right]} = \tilde{O}\left(\frac{1}{\rho}\sqrt{\frac{\log pl}{m}}\right)$.

Proof. First, observe that the classification error of a deep cascade function $f \in \mathcal{T}_l$ only depends on its sign $\text{sgn}(f)$. Let Δ denote the simplex in $\mathbb{R}^l$ and $\text{int}(\Delta)$ its interior. For any $\boldsymbol{\alpha} \in \text{int}(\Delta)$, define $g_{\boldsymbol{\alpha}}$ by

$$\forall x \in X, \quad g_{\boldsymbol{\alpha}}(x) = \sum_{k=1}^{l} \alpha_k \mathbf{s}(x, k) h_k(x). \qquad (3)$$

Then, $\mathsf{sgn}(f)$ coincides with $\mathsf{sgn}(g_{\boldsymbol{\alpha}})$ since $\mathsf{s}(x,k)$ is non-zero for exactly one value of k. We can therefore analyze $R(g_{\boldsymbol{\alpha}})$ instead of $R(f)$, for any $\boldsymbol{\alpha} \in \mathrm{int}(\Delta)$.

Now, since $g_{\boldsymbol{\alpha}}$ is a convex combination of the functions $x \mapsto \mathsf{s}(x,k)h_k(x)$, we can apply to the set of functions $g_{\boldsymbol{\alpha}}$ the learning guarantees for convex ensembles with multiple hypothesis sets given by [9]:

$$R(f) \le \inf_{\boldsymbol{\alpha} \in \mathrm{int}(\Delta)} \left[\widehat{R}_{S,\rho}(g_{\boldsymbol{\alpha}}) + \frac{4}{\rho} \sum_{k=1}^{l} \alpha_k \widehat{\mathfrak{R}}_S(\mathcal{H}_k) \right] + C(m,p,\rho) + \sqrt{\frac{\log \frac{4}{\delta}}{2m}}. \quad (4)$$

This bound is not explicit and depends on the choice of $\boldsymbol{\alpha} \in \mathrm{int}(\Delta)$. The crux of our proof now consists of removing $\boldsymbol{\alpha}$ and deriving an explicit bound. The first term of the right-hand side of (4) can be re-written as $\inf_{\boldsymbol{\alpha} \in \mathrm{int}(\Delta)} A(\boldsymbol{\alpha})$ with

$$A(\boldsymbol{\alpha}) = \frac{1}{m} \sum_{k=1}^{l} \sum_{\mathsf{s}(x_i,k)=1} 1_{y_i \alpha_k h_k(x_i) < \rho} + \frac{4}{\rho} \sum_{k=1}^{l} \alpha_k \widehat{\mathfrak{R}}_S(\mathcal{H}_k), \quad (5)$$

since $\widehat{R}_{S,\rho}(g_{\boldsymbol{\alpha}}) = \frac{1}{m} \sum_{k=1}^{l} \sum_{\mathsf{s}(x_i,k)=1} 1_{y_i \alpha_k h_k(x_i) < \rho}$. Observe that function A can be decoupled as a sum, $A(\boldsymbol{\alpha}) = \sum_{k=1}^{l} A_k(\alpha_k)$, where

$$A_k(\alpha_k) = \frac{1}{m} \sum_{\mathsf{s}(x_i,k)=1} 1_{y_i \alpha_k h_k(x_i) < \rho} + \frac{4}{\rho} \alpha_k r_k$$

with $r_k = \widehat{\mathfrak{R}}_S(\mathcal{H}_k)$. For any $k \in [1,l]$, $A_k(\alpha_k)$ can be rewritten as follows in terms of m_k^- and m_k^+: $A_k(\alpha_k) = \frac{m_k^-}{m} + \frac{m_k^+}{m} 1_{\alpha_k < \rho} + \frac{4}{\rho} \alpha_k r_k$. This implies $\inf_{\alpha_k > 0} A_k(\alpha_k) = \frac{m_k^-}{m} + \min \left(\frac{m_k^+}{m}, 4r_k \right)$. However, we need to ensure the global condition $\sum_{k=1}^{l} \alpha_k \le 1$. First, we let $l' = \min(|\mathcal{K}|, \frac{1}{\rho})$. For any $\mathcal{J} \subseteq \mathcal{K}$ with $|\mathcal{J}| \le l'$, we choose $\alpha_k = \rho$ for $k \in \mathcal{J}$, $\alpha_k \to 0$ otherwise, which guarantees $\sum_{k=1}^{l} \alpha_k = \rho l' \le 1$ and gives the infimum

$$\inf_{\boldsymbol{\alpha} \in \mathrm{int}(\Delta)} A(\boldsymbol{\alpha}) = \min_{\substack{\mathcal{J} \subseteq \mathcal{K} \\ |\mathcal{J}| \le l'}} \left(4 \sum_{k \in \mathcal{J}} r_k + \sum_{k \in \mathcal{K} - \mathcal{J}} \frac{m_k^+}{m} \right) + \sum_{k=1}^{l} \frac{m_k^-}{m} + \sum_{k \notin \mathcal{K}} \frac{m_k^+}{m}.$$

In order to simplify the bound, observe that the following equalities hold:

$$\min_{\mathcal{J}} \left(4 \sum_{k \in \mathcal{J}} r_k + \sum_{k \in \mathcal{K} - \mathcal{J}} \frac{m_k^+}{m} \right) = \min_{\mathcal{J}} \left(4 \sum_{k \in \mathcal{J}} r_k + \sum_{k \in \mathcal{K} - \mathcal{J}} \frac{m_k^+}{m} + \sum_{k \in \mathcal{K} - \mathcal{J}} 4r_k - \sum_{k \in \mathcal{K} - \mathcal{J}} 4r_k \right)$$

$$= \min_{\mathcal{J}} \left(4 \sum_{k \in \mathcal{K}} r_k + \sum_{k \in \mathcal{K} - \mathcal{J}} \frac{m_k^+}{m} - 4r_k \right) = 4 \sum_{k \in \mathcal{K}} r_k + \min_{\mathcal{J}} \left(\sum_{k \in \mathcal{K} - \mathcal{J}} \frac{m_k^+}{m} - 4r_k \right).$$

By definition, $\sum_{k \in \mathcal{K}} 4r_k + \sum_{k \notin \mathcal{K}} \frac{m_k^+}{m} = \sum_{k=1}^{l} \min \left(4r_k, \frac{m_k^+}{m} \right)$. Now, let $\mathcal{L} = \mathcal{K} - \mathcal{J}$ and since $|\mathcal{J}| \le l'$, $|\mathcal{L}| = |\mathcal{K}| - |\mathcal{J}| \ge |\mathcal{K}| - l' = |\mathcal{K}| - \min(|\mathcal{K}|, \frac{1}{\rho}) = \max(0, |\mathcal{K}| - \frac{1}{\rho})$

thus, $|\mathcal{L}| \geq |\mathcal{K}| - \frac{1}{\rho}$. Finally, we write the bound in the following simpler form:

$$\inf_{\alpha \in \text{int}(\Delta)} A(\alpha) = \sum_{k=1}^{l} \min\left(4r_k, \frac{m_k^+}{m}\right) + \min_{\substack{\mathcal{L} \subseteq \mathcal{K} \\ |\mathcal{L}| \geq |\mathcal{K}| - \frac{1}{\rho}}} \left(\sum_{k \in \mathcal{L}} \frac{m_k^+}{m} - 4r_k\right) + \sum_{k=1}^{l} \frac{m_k^-}{m}.$$

Since $\widehat{R}_S(f) = \sum_{k=1}^{l} \frac{m_k^-}{m}$, this coincides with the bound of the theorem.   □

These learning bounds are not straightforward and cannot be derived from standard Rademacher complexity bounds. A finer analysis is used in the proof to relate deep cascades to convex ensembles with multiple hypothesis sets [9].

We already commented on the simpler form (2) of this generalization bound. Our comments apply a fortiori to this finer version of the bound. Let us add that the theorem also provides new learning guarantees in the special case of decision trees. The result may seem surprising since it suggests that the complexity term depends on m_k^+/m when this ratio is sufficiently small; however, for such nodes, typically the fraction of points m_k/m would also be small, where m_k denotes the number of points at leaf k. At a deeper level, these guarantees suggest that for cascades, the complexity of the hypothesis sets may not be the most critical measure, but rather a balance of those complexities and the fractions of points.

The bound of the theorem can be generalized to hold uniformly for all $\rho > 0$ at the price of an additional term in $O\left(\frac{\log \log_2 \frac{1}{\rho}}{m}\right)$. For $|\mathcal{K}| \leq \frac{1}{\rho}$, choosing $\mathcal{L} = \emptyset$ yields:

$$R(f) \leq \widehat{R}_S(f) + \sum_{k=1}^{l} \min\left(4\widehat{\mathfrak{R}}_S(\mathcal{H}_k), \frac{m_k^+}{m}\right) + C(m, p, \rho) + \sqrt{\frac{\log \frac{4}{\delta}}{2m}}. \quad (6)$$

As mentioned above, these learning bounds can be generalized to hold uniformly over all $\rho > 0$: thus, we can choose $\rho = \frac{1}{|\mathcal{K}|}$ at the price of an additional term in the bound varying only in $O\left(\frac{\log \log_2 |\mathcal{K}|}{m}\right) \leq O\left(\frac{\log \log_2 l}{m}\right)$. This gives the simpler form (2) of the bound of Theorem 1, using $C(m, p, \rho) = C(m, p, \frac{1}{|\mathcal{K}|}) \leq C(m, p, \frac{1}{l})$.

The learning bounds just presented are given in terms of the empirical Rademacher complexities $\widehat{\mathfrak{R}}_S(\mathcal{H}_k)$. To derive more explicit guarantees, we must bound each of these quantities in terms of $\widehat{\mathfrak{R}}_S(H_k)$ and $\widehat{\mathfrak{R}}_S(\mathcal{S}_k)$. The following lemma helps us achieve that.

Lemma 1. *Let G_1 be a family of functions mapping $\mathcal{X}$ to $\{0, 1\}$ and let G_2 be a family of functions mapping $\mathcal{X}$ to $\{-1, +1\}$. Let $G = \{g_1 g_2 \colon g_1 \in G_1, g_2 \in G_2\}$. Then, the empirical Rademacher complexity of G for any sample S of size m can be bounded as follows:*

$$\widehat{\mathfrak{R}}_S(G) \leq \widehat{\mathfrak{R}}_S(G_1) + \widehat{\mathfrak{R}}_S(G_2).$$

Proof. Observe that for $g_1 \in G_1$ and $g_2 \in G_2$, $g_1 g_2 = |g_1 + g_2| - 1$. Since $x \mapsto |x| - 1$ is 1-Lipschitz over $[-1, 2]$, by Talagrand's lemma in [20], the following holds: $\widehat{\Re}_S(G) \leq \widehat{\Re}_S(G_1 + G_2) \leq \widehat{\Re}_S(G_1) + \widehat{\Re}_S(G_2)$. □

Thus, in view of the lemma, for any $k \in [1, p]$, we can use the upper bound $\widehat{\Re}_S(\mathcal{H}_k) \leq \widehat{\Re}_S(H_k) + \widehat{\Re}_S(\mathcal{S}_k)$.

We now assume, as previously discussed, that leaf selectors are defined via node questions $q_j \colon \mathcal{X} \to \{0, 1\}$, with $q_j \in Q_j$. Thus, to derive more explicit guarantees in that case, we need to bound $\widehat{\Re}_S(\mathcal{S}_k)$ in terms of the Rademacher complexities $\widehat{\Re}_S(Q_j)$.

Lemma 2. *Let H_1 and H_2 be two families of functions mapping $\mathcal{X}$ to $\{0, 1\}$ and let $H = \{h_1 h_2 \colon h_1 \in H_1, h_2 \in H_2\}$. Then, the empirical Rademacher complexity of H for any sample S of size m can be bounded as follows:*

$$\widehat{\Re}_S(H) \leq \widehat{\Re}_S(H_1) + \widehat{\Re}_S(H_2).$$

Proof. Observe that for any $h_1 \in H_1$ and $h_2 \in H_2$, we can write $h_1 h_2 = (h_1 + h_2 - 1)1_{h_1 + h_2 - 1 \geq 0} = (h_1 + h_2 - 1)_+$. Since $x \mapsto (x - 1)_+$ is 1-Lipschitz over $[0, 2]$, by Talagrand's lemma in [20], the following holds: $\widehat{\Re}_S(H) \leq \widehat{\Re}_S(H_1 + H_2) \leq \widehat{\Re}_S(H_1) + \widehat{\Re}_S(H_2)$. □

In view of Lemmas 2 and 1, the Rademacher complexities of the hypothesis sets $\mathcal{H}_k$ can be explicitly bounded as follows for any $k \in [1, l]$: $\widehat{\Re}_S(\mathcal{H}_k) \leq \sum_{j=1}^{d_k} \widehat{\Re}_S(Q_j) + \widehat{\Re}_S(H_k)$. Clearly, if the same hypothesis set is used for all node questions, that is $Q_j = Q$ for all j for some Q, then the bound admits the following simpler form: $\widehat{\Re}_S(\mathcal{H}_k) \leq d_k \widehat{\Re}_S(Q) + \widehat{\Re}_S(H_k)$. The Rademacher complexity of the hypothesis sets $\mathcal{H}_k$ can also be bounded in terms of the growth function of H_k and of Q_j (see full paper [11]).

To the best of our knowledge, Lemmas 2 and 1 are both novel and can be used as general tools for the analysis of the Rademacher complexity of other families. In the full version of this paper [11], we also provide a lower bound for the Rademacher complexity of the product of two hypothesis sets as a linear combination of the Rademacher complexity of the two sets. This shows that the upper bounds given by Lemma 2 cannot be significantly improved in general.

3.2 Leaf Classifiers Taking Values in $[-1, +1]$

A similar but somewhat more complex analysis can be given in the case where the leaf classifiers take values in $[-1, +1]$. Define $\rho_k = \min\{y_i h_k(x_i) \colon y_i h_k(x_i) > 0, \mathsf{s}(x_i, k) = 1\}$ as the smallest confidence value over the correctly classified sample points at leaf k. If there are no correctly classified points, then define $\rho_k = 0$. Let $\widetilde{\mathcal{K}} = \left\{ k \in [1, l] \colon \frac{m_k^+}{m} > \frac{4 r_k}{\rho_k} \right\}$ and denote its weighted cardinality as $|\widetilde{\mathcal{K}}|_{\tilde{w}} = \sum_{k=1}^{l} \frac{1}{\rho_k}$. Then, it can be shown that for any $\delta > 0$, for $|\widetilde{\mathcal{K}}|_{\tilde{w}} \leq \frac{1}{\rho}$, the following holds with probability at least $1 - \delta$:

$$R(f) \leq \widehat{R}_S(f) + \sum_{k=1}^{l} \min\left(\frac{4\widehat{\mathfrak{R}}_S(\mathcal{H}_k)}{\rho_k}, \frac{m_k^+}{m}\right) + \widetilde{O}\left(l\sqrt{\frac{\log pl}{m}}\right), \qquad (7)$$

which is the analogue of the learning bound (2) obtained in the case of leaf classifiers taking values in $\{-1, +1\}$. The full proof of this result, as well as that of more refined results, is given in the full version of this paper in [11]. As in the discrete case, to derive an explicit bound, we need to upper bound for all $k \in [1, l]$ the Rademacher complexity $\widehat{\mathfrak{R}}_S(\mathcal{H}_k)$ in terms of those of H_k and Q_j. To do so, we will need a new tool provided by the following lemma.

Lemma 3. *Let H_1 and H_2 be two families of functions mapping $\mathcal{X}$ to $[0, +1]$ and let F_1 and F_2 be two families of functions mapping $\mathcal{X}$ to $[-1, +1]$. Let $H = \{h_1 h_2 \colon h_1 \in H_1, h_2 \in H_2\}$ and let $F = \{f_1 f_2 \colon f_1 \in F_1, f_2 \in F_2\}$. Then, the empirical Rademacher complexities of H and F for any sample S of size m are bounded as follows:*

$$\widehat{\mathfrak{R}}_S(H) \leq \frac{3}{2}\big(\widehat{\mathfrak{R}}_S(H_1) + \widehat{\mathfrak{R}}_S(H_2)\big) \qquad \widehat{\mathfrak{R}}_S(F) \leq 2\big(\widehat{\mathfrak{R}}_S(F_1) + \widehat{\mathfrak{R}}_S(F_2)\big).$$

Proof. Observe that for any $h_1 \in H_1$ and $h_2 \in H_2$, we can write $h_1 h_2 = \frac{1}{4}[(h_1 + h_2)^2 - (h_1 - h_2)^2]$. For bounding the term $(h_1 + h_2)^2$, note that the function $x \mapsto \frac{1}{4}x^2$ is 1-Lipschitz over $[0, 2]$. For the term $(h_1 - h_2)^2$, observe that the function $x \mapsto \frac{1}{4}x^2$ is 1/2-Lipschitz over $[-1, 1]$. Thus, by Talagrand's lemma (see [20]), $\widehat{\mathfrak{R}}_S(H) \leq \widehat{\mathfrak{R}}_S(H_1 + H_2) + \frac{1}{2}\widehat{\mathfrak{R}}_S(H_1 - H_2) \leq \frac{3}{2}\big(\widehat{\mathfrak{R}}_S(H_1) + \widehat{\mathfrak{R}}_S(H_2)\big)$. Similarly, the same equation holds for the product $f_1 f_2$ with $f_1 \in F_1$ and $f_2 \in F_2$, but now the function $x \mapsto \frac{1}{4}x^2$ is 1-Lipschitz over $[-2, 2]$. Thus, by Talagrand's lemma [20], the following holds: $\widehat{\mathfrak{R}}_S(F) \leq \widehat{\mathfrak{R}}_S(F_1 + F_2) + \widehat{\mathfrak{R}}_S(F_1 - F_2) \leq 2\big(\widehat{\mathfrak{R}}_S(F_1) + \widehat{\mathfrak{R}}_S(F_2)\big)$, which completes the proof. $\square$

Lemma 2 and Lemma 3 yield the following explicit bound for $\widehat{\mathfrak{R}}(\mathcal{H}_k)$ for any $k \in [1, l]$: $\widehat{\mathfrak{R}}_S(\mathcal{H}_k) \leq 2\sum_{j=1}^{d_k} \widehat{\mathfrak{R}}_S(Q_j) + 2\widehat{\mathfrak{R}}_S(H_k)$. When the same hypothesis set is used for all node questions, that is $Q_j = Q$ for all j for some Q, then the bound admits the following simpler form: $\widehat{\mathfrak{R}}_S(\mathcal{H}_k) \leq 2d_k\widehat{\mathfrak{R}}_S(Q) + 2\widehat{\mathfrak{R}}_S(H_k)$.

4   Algorithms

There are several algorithms that could be derived from the learning guarantees presented in the previous section. Here, we will describe two algorithms based on the simplest bound (2) of Section 3.1, which we further bound more explicitly by using the results from the previous section:

$$R(f) \leq \widehat{R}_S(f) + \sum_{k=1}^{l} \min\left(4\left[\sum_{j=1}^{d_k} \widehat{\mathfrak{R}}_S(Q_j) + \widehat{\mathfrak{R}}_S(H_k)\right], \frac{m_k^+}{m}\right) + \widetilde{O}\left(l\sqrt{\frac{\log pl}{m}}\right). \quad (8)$$

For both of our algorithms, we fix the topology of the deep cascade to be binary trees where every left child is a leaf as shown by Figure 2. Other more general tree topologies can be considered, which could further improve our results.

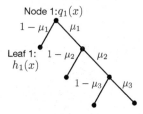

Fig. 1. Tree Topology of deep cascades for DEEPCASCADE and DEEPCASCADESVM Algorithm. The node question at node 1 is denoted by $q_1(x)$ and the leaf classifier at leaf 1 denoted by $h_1(x)$. A μ_k fraction of the points at node k is sent to the right child, and the remaining $(1 - \mu_k)$ fraction of points to the left child. For DEEPCASCADE, all μ_ks are set to be equal: $\mu_k = \mu$ for all k.

4.1   DEEPCASCADE

In this section, we describe a generic algorithm for deep cascades, named DEEP-CASCADE. The algorithm first generates several deep cascades and then chooses the best among them by minimizing the generalization bound (8).

Let $\mathcal{H}$ be a set of p hypothesis sets from which the hypothesis sets H_k are selected. Here, we similarly allow each hypothesis set Q_k to be chosen out of a family of hypothesis sets $\mathcal{Q}$ of cardinality p – it is not hard to see that this affects our learning bound only by the $\log(pl)$ factors being changed into $l \log(p)$ and leaves the main terms we are minimizing unchanged; moreover, since we will be considering cascades with a relatively small depth, say $l \le 4$, the effect will be essentially insignificant.

At any node k, the question q_k is selected so that a μ fraction of the points is sent to the right child. We assume for simplicity that for any node k and any choice μ, there exists a unique node question q_k with that property. For the topology of Figure 2, it is not hard to see that for any k, $\frac{m_k^+}{m}$ is at most μ^{k-1}. The parameter μ controls the fraction of points that proceed deeper into the tree and is introduced in order to find the best trade-off between the complexity

---

Algorithm 1. DEEPCASCADE$(L, \mathcal{M})$

$S_1 \leftarrow S$
for $l \in [1, \dots, L], \mu \in \mathcal{M}, (H_k)_{1 \le k \le l} \subseteq \mathcal{H}, (Q_k)_{1 \le k \le l} \subseteq \mathcal{Q}$ **do**
    for $k = 1$ to l **do**
        $q_k \leftarrow \arg_{q \in Q_k}\{|q^{-1}(1) \cap S_k| = \mu|S_k|\}$
        $S_{k+1} \leftarrow q_k^{-1}(1) \cap S_k$
        $h_k \leftarrow \arg\min_{h \in H_k}\{\hat{R}_{\bar{S}_{k+1}}(h) : \bar{S}_{k+1} = q_k^{-1}(0) \cap S_k\}$
    end for
    $f \leftarrow \sum_{k=1}^{l-1}(\prod_{j=1}^{k-1} q_j)\bar{q}_k h_k + (\prod_{j=1}^{l} q_j)h_l$
    $\mathcal{F} \leftarrow \mathcal{F} \cup \{f\}$
end for
$f^* \leftarrow \arg\min_{f \in \mathcal{F}} R_S(f) + \sum_{k=1}^{l} \min\left(4\left(\sum_{j=1}^{d_k} \widehat{\mathfrak{R}}_S(Q_{f,j}) + \widehat{\mathfrak{R}}_S(H_{f,k})\right), \frac{m_k^+}{m}\right)$
return f^*

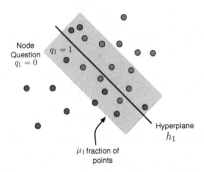

Fig. 2. Illustration of the first step of DEEPCASCADESVM. The hyperplane h_1 is learned using the SVM algorithm over sample points S_1. The node question q_1 equals one in the green area and zero otherwise. The green area contains a μ_1 fraction of the data points that will proceed to the next node.

term and fraction of points at each node. The subsample of the points reaching the internal node k is denoted by S_k and the subsample of those reaching leaf k is denoted by $\bar{S}_{k+1}$, with $|\bar{S}_{k+1}| = m_k$. The leaf classifier h_k is chosen to be the minimizer of the empirical error over $\bar{S}_{k+1}$ since, in this way, it further minimizes bound (8).

Algorithm 1 gives the pseudocode of DEEPCASCADE. The algorithm takes as input the maximum depth L for all the deep cascades generated and the set $\mathcal{M}$ of different fraction values for the parameter μ. For any depth $l \in [1, \ldots, L]$, any $\mu \in \mathcal{M}$, and any sequence of leaf hypothesis sets $(H_k)_{1 \le k \le l} \subseteq \mathcal{H}$ and sequence of node question hypothesis sets $(Q_k)_{1 \le k \le l} \subseteq \mathcal{Q}$, the algorithm defines a new deep cascade function f. At each node k, the question $q_k \in Q_k$ is selected with the μ-property already discussed and the leaf hypothesis $h_k \in H_k$ is selected to minimize the error over the leaf sample. For each f, we denote by $Q_{f,k}$ the question hypothesis set at node k that served to define f and similarly $H_{f,k}$ the hypothesis set at leaf k that was used to define f. The algorithm returns the deep cascade function f^* among these cascade functions f that minimizes the bound (8).

The empirical risk minimization (ERM) method used to determine the leaf classifiers h_k is intractable for some hypothesis sets. In the next section, we present an alternative algorithm using SVMs which can be viewed as an efficient instantiation of this generic algorithm.

4.2   DEEPCASCADESVM

In this section, we describe an algorithm for learning deep cascades that makes use of SVMs and that we named DEEPCASCADESVM. In short, as with DEEP-CASCADE, DEEPCASCADESVM first generates different deep cascade functions, but it uses the SVM algorithm at each node of the cascade and chooses the best among them by minimizing an upper bound of the generalization bound (8).

Algorithm 2. DEEPCASCADESVM($L, \mathcal{M}, \gamma$)

---

for $l \in [1, \ldots, L], (\mu_k)_{1 \le k \le l} \subseteq \mathcal{M}, (\delta_k)_{1 \le k \le l} \subseteq \mathcal{G}$ **do**
  $S_1 \leftarrow S$
  for $k = 1$ to l **do**
    $h_k \leftarrow \text{SVM}(\delta_k, S_k)$
    $q_k \leftarrow \arg_{q \in Q_k} \{|q^{-1}(1) \cap S_k| = \mu_k |S_k|\}$
    $S_{k+1} \leftarrow q_k^{-1}(1) \cap S_k$
  end for
  $f(\cdot) = \sum_{k=1}^{l-1} (\prod_{j=1}^{k-1} q_j) \bar{q}_k h_k + (\prod_{j=1}^{l} q_j) h_l h_k(\cdot)$
  $\mathcal{F} \leftarrow \mathcal{F} \cup \{f\}$
end for

$$f^* \leftarrow \underset{f \in \mathcal{F}}{\arg\min} \, \widehat{R}_S(f) + \sum_{k=1}^{l} \min \left(4\gamma \left[\sum_{j=1}^{d_k} \sqrt{\frac{\mathsf{d}_{f,j} \log\left(\frac{em}{\mathsf{d}_{f,j}}\right)}{m}} + \sqrt{\frac{\mathsf{d}_{f,k} \log\left(\frac{em}{\mathsf{d}_{f,k}}\right)}{m}} \right], \frac{m_k^+}{m} \right)$$

return f^*

The deep cascade functions generated by the algorithm are based on repeatedly using SVMs combined with polynomial kernels of different degree. The leaf hypothesis sets H_k are decision surfaces defined by polynomial kernels. The hypothesis $h_k \in H_k$ is learned via the SVM algorithm with a polynomial kernel degree δ_k on subsample S_k. Note that in the pseudocode of Algorithm 2, we denote this step by $\text{SVM}(\delta_k, S_k)$. The node question hypothesis set Q_k is defined to be the set of indicator functions of $dist(h_k, x) \le c$, where $dist(h_k, x)$ is the Euclidian distance of point x to hyperplane h_k in the feature space. The node question $q_k \in Q_k$ is chosen to be the indicator function of $dist(h_k, x) \le c_k$ where c_k is such that $|q_k(1)^{-1} \cap S_k| = \mu_k |S_k|$, meaning the number of points in S_k that are within a distance c_k to hyperplane h_k equals $\mu_k |S_k|$. In other words, after learning the hyperplane via the SVM algorithm on subsample S_k, the algorithm

1. extracts a μ_k fraction of points closest to the hyperplane;
2. on the next node in the cascade, retrains on this extracted subsample using the SVM algorithm with a polynomial kernel of another degree.

We extract the fraction of points closest to the hyperplane because these points can be harder to classify correctly. Hence, these points will proceed deeper into the cascade in hope to find a better trade-off between the complexity and the fraction of correctly classified points.

The algorithm generates several cascades functions with a given depth $l \in [1, \ldots, L]$. For any depth $l \in [1, \ldots, L]$, any sequence of fraction values $(\mu_k)_{1 \le k \le l} \subseteq \mathcal{M}$ and sequence of degree values $(\delta_k)_{1 \le k \le l} \subseteq \mathcal{G}$, the algorithm defines a new deep cascade function f. At each node k, the question $q_k \in Q_k$ and leaf hypothesis $h_k \in H_k$ are selected as already discussed. Similarly, as before, for each f, we denote by $Q_{f,k}$ the question hypothesis set at node k that served to define f and similarly $H_{f,k}$ the hypothesis set at leaf k that was used to define f. The best cascade f^* is chosen by minimizing an upper bound of the generalization bound (8). More precisely, we first bound the Rademacher complexity in terms of the VC-dimension of the hypothesis set:

$$\sum_{j=1}^{d_k} \widehat{\mathfrak{R}}_S(Q_{f,j}) + \widehat{\mathfrak{R}}_S(H_{f,k}) \leq \sum_{j=1}^{d_k} \sqrt{\frac{\mathsf{d}_{f,j} \log(\frac{em}{\mathsf{d}_{f,j}})}{m}} + \sqrt{\frac{\mathsf{d}_{f,k} \log(\frac{em}{\mathsf{d}_{f,k}})}{m}},$$

where $\mathsf{d}_{f,k}$ is the VC-dimension of $H_{f,k}$ and where we used the fact that $\text{VCdim}(Q_{f,j}) \leq \text{Pdim}(H_{f,j}) = \text{VCdim}(H_{f,j}) = \mathsf{d}_{f,j}$. Then, we rescale the complexity term by a parameter γ, which we will determine by cross-validation. Thus, for a given γ, we chose the deep cascade with the smallest value of the generalization bound :

$$R(f) \leq \widehat{R}_S(f) + \sum_{k=1}^{l} \min\left(4\gamma \left[\sum_{j=1}^{d_k} \sqrt{\frac{\mathsf{d}_{f,j} \log(\frac{em}{\mathsf{d}_{f,j}})}{m}} + \sqrt{\frac{\mathsf{d}_{f,k} \log(\frac{em}{\mathsf{d}_{f,k}})}{m}}\right], \frac{m_k^+}{m}\right). \tag{9}$$

DEEPCASCADESVM can be seen as a tractable version of the generic DEEPCASCADE algorithm with some minor differences in the following ways. Instead of choosing h_k to be the minimizer of the empirical error as done in DEEPCASCADE, the DEEPCASCADESVM chooses the h_k that minimizes a surrogate loss (hinge loss) of the empirical error by using the SVM algorithm. In fact, the γ parameter is introduced because the hinge loss used in the SVM algorithm needs to be re-scaled. Note that h_k is learned via the SVM algorithm on S_k and not on $\bar{S}_{k+1}$, namely the points that reach leaf k, as in the DEEPCASCADE algorithm. One could retrain SVM on the points reaching the leaf to be consistent with the first algorithm, but this typically will not change the hypothesis h_k. The generic node question q_k of DEEPCASCADE are picked to be the distance to the classification hyperplane h_k for a given fraction μ_k of points in DEEPCASCADESVM algorithm. Technically, in the DEEPCASCADE, the μ fractions are the same, but this was done to simplify the exposition of the DEEPCASCADE algorithm. DEEPCASCADE minimizes exactly bound (8), while DEEPCASCADESVM minimizes an upper bound in terms of the VC-dimension.

5   Experiments

This section reports the results of some preliminary experiments with the DEEPCASCADESVM algorithm on several UC Irvine data sets. Since DEEPCASCADESVM uses only polynomial kernels as predictors, we similarly compared our results with those achieved by the SVM algorithm with polynomial kernels over the set $\mathcal{G}$ of polynomial degrees. Of course, a similar set of experiments can be carried out by using both Gaussian kernels or other kernels, which we plan to do in the future.

For our experiments, we used five different data sets from UC Irvine's data repository, http://archive.ics.uci.edu/ml/datasets.html: breastcancer, german (numeric), ionosphere, splice, and a1a. Table 1 gives the sample size and the

Table 1. Results for DEEPCASCADESVM algorithm. The table reports the average test error and standard deviation for DEEPCASCADESVM(γ^*) and for the SVM algorithm. For each data set, the table also indicates the sample size, the number of features, and the depth of the cascade.

Dataset	Number of Examples	Number of Features	SVM Algorithm	DEEPCASCADESVM	Cascade Depth
breastcancer	683	10	0.0426 ± 0.0117	0.0353 ± 0.00975	4
german	1,000	24	0.297 ± 0.0193	0.256 ± 0.0324	4
splice	1,000	60	0.205 ± 0.0134	0.175 ± 0.0152	3
ionosphere	351	34	0.0971 ± 0.0167	0.117 ± 0.0229	4
a1a	1,000	123	0.195 ± 0.0217	0.209 ± 0.0233	2

number of features for each of these data sets. For each of them, we randomly divided the set into five folds and ran the algorithm five times using a different assignment of folds to the training set, validation set, and test set. For each $j \in \{0, 1, 2, 3, 4\}$, the sample points from the fold j was used for testing, the fold $j + 1 \pmod 5$ used for validation, and the remaining sample points used for training.

The following are the parameters used for DEEPCASCADESVM: the maximum tree depth was set to $L = 4$, the set of fraction values was selected to be $\mathcal{M} = \{\frac{i}{10} : i = 1, \cdots, 10\}$ and the set of polynomial degrees $\mathcal{G} = \{1, \ldots, 4\}$. The regularization parameter $C_\delta \in \{10^i : i = -3, \cdots, 2\}$ of SVMs was selected via cross-validation for each polynomial degree $\delta \in \mathcal{G}$. To avoid a grid search at each node, for cascades, the regularization parameter C_{δ_k} for SVMs at node k was simply defined to be $\sqrt{\frac{m_k}{m}} C_\delta$ when using a polynomial degree δ_k.

For each value of the parameter $\gamma \in \{10^i : i = -2, \ldots, 0\}$, we generated several deep cascades and then chose the one that minimized the bound (9). Thus, for each γ, there was a corresponding deep cascade f_γ^*. The parameter γ was chosen via cross-validation. More precisely, we chose the best γ^* by finding the deep cascade $f_{\gamma^*}^*$ that had the smallest validation error among the deep cascade functions f_γ^*. We report the average test error of the deep cascade $f_{\gamma^*}^*$ in Table 1. For SVMs, we report the test errors for the polynomial degree and regularization parameter with the smallest validation error.

The results of Table 1 show that DEEPCASCADESVM outperforms SVMs for three out of the five data sets: breastcancer, german, and splice. The german and splice results are statistically significant at the 5% level using a one-sided paired t-test while breastcancer result is not statistically significant. For the remaining two data sets where SVMs outperforms DEEPCASCADESVM, the a1a result is statistically significant at the 5% level while it is not statistically significant for the ionosphere data set.

Overall, the results demonstrate the benefits of DEEPCASCADESVM in several data sets. Note also that SVMs can be viewed as a special instance of the deep cascades with depth one. It is conceivable of course that for some data sets such simpler cascades would provide a better performance. There are several components in our algorithm that could be optimized more effectively to further improve performance. This includes optimizing over the regularization parameter C at each node of the cascade, testing polynomial degrees higher than 4, or searching over larger sets of μ fraction values and γ values. Yet, even with this rudimentary implementation of an algorithm that minimizes the simplest form of our bound (8), it is striking that it outperforms SVMs for several of the data sets and finds a comparable accuracy for the remaining data sets. More extensive experiments with other variants of the algorithms would be interesting to investigate in the future.

6   Conclusion

We presented two algorithms for learning Deep Cascades, a broad family of hier-archical models which offer the flexibility of selecting node or leaf functions from unions of complex hypothesis sets. We further reported the results of experi-ments demonstrating the performance for one of our algorithms using different data sets. Our algorithms benefit from data-dependent learning guarantees we derived, which are expressed in terms of the Rademacher complexities of the sub-families composing these sets of predictors and the fraction of sample points correctly classified at each leaf. Our theoretical analysis is general and can help guide the design of many other algorithms: different sub-families of leaf or node questions can be chosen and alternative cascade topologies and parameters can be selected. For the design of our algorithms, we used a simpler version of our guarantees. Finer algorithms could be devised to more closely exploit the quan-tities appearing in our learning bounds, which could further improve prediction accuracy.

Acknowledgments. We thank Vitaly Kuznetsov and Andrés Muñoz Medina for com-ments on an earlier draft of this paper. This work was partly funded by the NSF award IIS-1117591 and an NSF Graduate Research Fellowship.

References

1. Arreola, K., Fehr, J., Burkhardt, H.: Fast support vector machine classification using linear SVMs. In: ICPR (2006)
2. Arreola, K., Fehr, J., Burkhardt, H.: Fast support vector machine classification of very large datasets. In: GfKl Conference (2007)
3. Bartlett, P., Mendelson, S.: Rademacher and Gaussian complexities: Risk bounds and structural results. JMLR (2002)
4. Bengio, S., Weston, J., Weston, D.: Label embedding trees for large multi-class tasks. In: NIPS, Vancouver, Canada (2010)
5. Bennet, K., Blue, J.: A support vector machine approach to decision trees. In: IJCNN, Anchorage, Alaska (1998)
6. Breiman, L., Friedman, J., Olshen, R., Stone, C.: Classification and Regression Trees. Wadsworth and Brooks, Monterey (1984)
7. Chang, F., Guo, C., Lin, X., Lu, C.: Tree decomposition for large-scale SVM problems. JMLR (2010)
8. Chen, M., Xu, Z., Kedem, D., Chapelle, O.: Classifier cascade for minimizing feature evaluation cost. In: AISTATS, La Palma, Canary Islands (2012)

9. Cortes, C., Mohri, M., Syed, U.: Deep boosting. In: ICML (2014)
10. Deng, J., Satheesh, S., Berg, A., Fei-Fei, L.: Fast and balanced: efficient label tree learning for large scale object recognition. In: NIPS (2011)
11. DeSalvo, G., Mohri, M., Syed, U.: Learning with Deep Cascades. arXiv (2015)
12. Dong, G., Chen, J.: Study on support vector machine based decision tree and application. In: ICNC-FSKD, Jinan, China (2008)
13. Golea, M., Bartlett, P., Lee, W., Mason, L.: Generalization in decision trees and DNF: does size matter? In: NIPS (1997)
14. Koltchinskii, V., Panchenko, D.: Empirical margin distributions and bounding the generalization error of combined classifiers. Annals of Statistics **30** (2002)
15. Kumar, A., Gopal, M.: A hybrid SVM based decision tree. JPR (2010)
16. Lefakis, L., Fleuret, F.: Joint cascade optimization using a product of boosted classifiers. In: NIPS (2010)
17. Littman, M., Li, L., Walsh, T.: Knows what it knows: a framework for self-aware learning. In: ICML (2008)
18. Madjarov, G., Gjorgjevikj, D.: Hybrid decision tree architecture utilizing local SVMs for multi-label classification. In: Corchado, E., Snášel, V., Abraham, A., Woźniak, M., Graña, M., Cho, S.-B. (eds.) HAIS 2012, Part II. LNCS, vol. 7209, pp. 1–12. Springer, Heidelberg (2012)
19. Mansour, Y., McAllester, D.: Generalization bounds for decision trees. In: COLT (2000)
20. Mohri, M., Rostamizadeh, R., Talwalkar, A.: Foundations of Machine Learning. The MIT Press (2012)
21. Nobel, A.: Analysis of a complexity based pruning scheme for classification trees. IEEE Trans. Inf. Theory (2002)
22. Pujara, J., Daume, H., Getoor, L.: Using classifier cascades for scalable e-mail classification. In: CEAS (2011)
23. Quinlan, J.: Induction of decision trees. Machine Learning **1**(1), 81–106 (1986)
24. Rodriguez-Lujan, I., Cruz, C., Huerta, R.: Hierarchical linear SVM. JPR (2012)
25. Saberian, M., Vasconcelos, N.: Boosting classifier cascades. In: NIPS, Canada (2010)
26. Scott, C., Nowak, R.: On adaptive properties of decision trees. In: NIPS, Canada (2005)
27. Takahashi, F., Abe, S.: Decision tree based multiclass SVMs. In: ICONIP (2002)
28. Viola, P., Jones, M.: Robust real-time face detection. IJCV (2004)
29. Wang, J., Saligrama, V.: Local supervised learning through space partitioning. In: NIPS (2012)
30. Xu, Z., Kusner, M., Weinberger, K., Chen, M.: Cost-sensitive tree of classifiers. In: ICML, Altanta, USA (2013)

Bounds on the Minimax Rate for Estimating a Prior over a VC Class from Independent Learning Tasks

Liu Yang[1]([⊠]), Steve Hanneke[2]([⊠]), and Jaime Carbonell[3]

[1] IBM T.J. Watson Research Center, Yorktown Heights, NY, USA
{liuy,jgc}@cs.cmu.edu
[2] Princeton, NJ, USA
steve.hanneke@gmail.com
[3] Carnegie Mellon University, Pittsburgh, PA, USA

Abstract. We study the optimal rates of convergence for estimating a prior distribution over a VC class from a sequence of independent data sets respectively labeled by independent target functions sampled from the prior. We specifically derive upper and lower bounds on the optimal rates under a smoothness condition on the correct prior, with the number of samples per data set equal the VC dimension. These results have implications for the improvements achievable via transfer learning. We additionally extend this setting to real-valued function, where we establish consistency of an estimator for the prior, and discuss an additional application to a preference elicitation problem in algorithmic economics.

1 Introduction

In the *transfer learning* setting, we are presented with a sequence of learning problems, each with some respective target concept we are tasked with learning. The key question in transfer learning is how to leverage our access to past learning problems in order to improve performance on learning problems we will be presented with in the future.

Among the several proposed models for transfer learning, one particularly appealing model supposes the learning problems are independent and identically distributed, with unknown distribution, and the advantage of transfer learning then comes from the ability to estimate this shared distribution based on the data from past learning problems [2,11]. For instance, when customizing a speech recognition system to a particular speaker's voice, we might expect the first few people would need to speak many words or phrases in order for the system to accurately identify the nuances. However, after performing this for many different people, if the software has access to those past training sessions when customizing itself to a new user, it should have identified important properties of the speech patterns, such as the common patterns within each of the major dialects or accents, and other such information about the *distribution* of speech patterns within the user population. It should then be able to leverage this information to reduce the number of words or phrases the next user needs to speak in

© Springer International Publishing Switzerland 2015
K. Chaudhuri et al. (Eds.): ALT 2015, LNAI 9355, pp. 270–284, 2015.
DOI: 10.1007/978-3-319-24486-0_18

order to train the system, for instance by first trying to identify the individual's dialect, then presenting phrases that differentiate common subpatterns within that dialect, and so forth.

In analyzing the benefits of transfer learning in such a setting, one important question to ask is how quickly we can estimate the distribution from which the learning problems are sampled. In recent work, [11] have shown that under mild conditions on the family of possible distributions, if the target concepts reside in a known VC class, then it is possible to estimate this distribution using only a bounded number of training samples per task: specifically, a number of samples equal the VC dimension. However, that work left open the question of quantifying the *rate* of convergence of the estimate, in terms of the number of tasks. This rate of convergence can have a direct impact on how much benefit we gain from transfer learning when we are faced with only a finite sequence of learning problems. As such, it is certainly desirable to derive tight characterizations of this rate of convergence.

The present work continues that of [11], bounding the rate of convergence for estimating this distribution, under a smoothness condition on the distribution. We derive a generic upper bound, which holds regardless of the VC class the target concepts reside in. The proof of this result builds on that earlier work, but requires several interesting innovations to make the rate of convergence explicit, and to dramatically improve the upper bound implicit in the proofs of those earlier results. We further derive a nontrivial lower bound that holds for certain constructed scenarios, which illustrates a lower limit on how good of a general upper bound we might hope for in results expressed only in terms of the number of tasks, the smoothness conditions, and the VC dimension.

We additionally include an extension of the results of [11] to the setting of real-valued functions, establishing consistency (at a uniform rate) for an estimator of a prior over any VC subgraph class. In addition to the application to transfer learning, analogous to the original work of [11], we also discuss an application of this result to a preference elicitation problem in algorithmic economics, in which we are tasked with allocating items to a sequence of customers to approximately maximize the customers' satisfaction, while permitted access to the customer valuation functions only via value queries.

2   The Setting

Let $(\mathcal{X}, \mathcal{B}_{\mathcal{X}})$ be a measurable space [7] (where $\mathcal{X}$ is called the *instance space*), and let $\mathcal{D}$ be a distribution on $\mathcal{X}$ (called the *data distribution*). Let $\mathbb{C}$ be a VC class of measurable classifiers $h : \mathcal{X} \rightarrow \{-1, +1\}$ (called the *concept space*), and denote by d the VC dimension of $\mathbb{C}$ [9]. We suppose $\mathbb{C}$ is equipped with its Borel σ-algebra $\mathcal{B}$ induced by the pseudo-metric $\rho(h, g) = \mathcal{D}(\{x \in \mathcal{X} : h(x) \neq g(x)\})$. Though our results can be formulated for general $\mathcal{D}$ (with somewhat more complicated theorem statements), to simplify the statement of results we suppose ρ is actually a *metric*.

For any two probability measures μ_1, μ_2 on a measurable space $(\Omega, \mathcal{F})$, define the total variation distance

$$\|\mu_1 - \mu_2\| = \sup_{A \in \mathcal{F}} \mu_1(A) - \mu_2(A).$$

For a set function μ on a *finite* measurable space $(\Omega, \mathcal{F})$, we abbreviate $\mu(\omega) = \mu(\{\omega\})$, $\forall \omega \in \Omega$. Let $\Pi_\Theta = \{\pi_\theta : \theta \in \Theta\}$ be a family of probability measures on $\mathbb{C}$ (called *priors*), where Θ is an arbitrary index set (called the *parameter space*). We suppose there exists a probability measure π_0 on $\mathbb{C}$ (the *reference measure*) such that every π_θ is absolutely continuous with respect to π_0, and therefore has a density function f_θ given by the Radon-Nikodym derivative $\frac{d\pi_\theta}{d\pi_0}$ [7].

We consider the following type of estimation problem. There is a collection of $\mathbb{C}$-valued random variables $\{h_{t\theta}^* : t \in \mathbb{N}, \theta \in \Theta\}$, where for any fixed $\theta \in \Theta$ the $\{h_{t\theta}^*\}_{t=1}^\infty$ variables are i.i.d. with distribution π_θ. For each $\theta \in \Theta$, there is a sequence $\mathcal{Z}^t(\theta) = \{(X_{t1}, Y_{t1}(\theta)), (X_{t2}, Y_{t2}(\theta)), \ldots\}$, where $\{X_{ti}\}_{t,i \in \mathbb{N}}$ are i.i.d. $\mathcal{D}$, and for each $t, i \in \mathbb{N}$, $Y_{ti}(\theta) = h_{t\theta}^*(X_{ti})$. We additionally denote by $\mathcal{Z}_k^t(\theta) = \{(X_{t1}, Y_{t1}(\theta)), \ldots, (X_{tk}, Y_{tk}(\theta))\}$ the first k elements of $\mathcal{Z}^t(\theta)$, for any $k \in \mathbb{N}$, and similarly $\mathbb{X}_{tk} = \{X_{t1}, \ldots, X_{tk}\}$ and $\mathbb{Y}_{tk}(\theta) = \{Y_{t1}(\theta), \ldots, Y_{tk}(\theta)\}$. Following the terminology used in the transfer learning literature, we refer to the collection of variables associated with each t collectively as the tth *task*. We will be concerned with sequences of estimators $\hat{\theta}_{T\theta} = \hat{\theta}_T(\mathcal{Z}_k^1(\theta), \ldots, \mathcal{Z}_k^T(\theta))$, for $T \in \mathbb{N}$, which are based on only a bounded number k of samples per task, among the first T tasks. Our main results specifically study the case of d samples per task. For any such estimator, we measure the *risk* as $\mathbb{E}\left[\|\pi_{\hat{\theta}_{T\theta_\star}} - \pi_{\theta_\star}\|\right]$, and will be particularly interested in upper-bounding the worst-case risk $\sup_{\theta_\star \in \Theta} \mathbb{E}\left[\|\pi_{\hat{\theta}_{T\theta_\star}} - \pi_{\theta_\star}\|\right]$ as a function of T, and lower-bounding the minimum possible value of this worst-case risk over all possible $\hat{\theta}_T$ estimators (called the *minimax risk*).

In previous work, [11] showed that, if Π_Θ is a totally bounded family, then even with only d number of samples per task, the minimax risk (as a function of the number of tasks T) converges to zero. In fact, that work also proved this is not necessarily the case in general for any number of samples less than d. However, the actual rates of convergence were not explicitly derived in that work, and indeed the upper bounds on the rates of convergence implicit in that analysis may often have fairly complicated dependences on $\mathbb{C}$, Π_Θ, and $\mathcal{D}$, and furthermore often provide only very slow rates of convergence.

To derive explicit bounds on the rates of convergence, in the present work we specifically focus on families of *smooth* densities. The motivation for involving a notion of smoothness in characterizing rates of convergence is clear if we consider the extreme case in which Π_Θ contains two priors π_1 and π_2, with $\pi_1(\{h\}) = \pi_2(\{g\}) = 1$, where $\rho(h, g)$ is a very small but nonzero value; in this case, if we have only a small number of samples per task, we would require many tasks (on the order of $1/\rho(h, g)$) to observe any data points carrying any information that would distinguish between these two priors (namely, points x with $h(x) \neq g(x)$); yet $\|\pi_1 - \pi_2\| = 1$, so that we have a slow rate of convergence (at least initially). A total boundedness condition on Π_Θ would limit the number of such pairs present in Π_Θ, so that for instance we cannot have arbitrarily close h and g, but less extreme variants of this can lead to slow asymptotic rates of convergence as well. Specifically, in the present work we consider the following notion of smoothness. For $L \in (0, \infty)$ and $\alpha \in (0, 1]$, a function $f : \mathbb{C} \to \mathbb{R}$ is (L, α)-Hölder smooth if

$$\forall h, g \in \mathbb{C}, |f(h) - f(g)| \leq L\rho(h, g)^\alpha.$$

3   An Upper Bound

We now have the following theorem, holding for an arbitrary VC class $\mathbb{C}$ and data distribution $\mathcal{D}$; it is the main result of this work.

Theorem 1. *For Π_Θ any class of priors on $\mathbb{C}$ having (L, α)-Hölder smooth densities $\{f_\theta : \theta \in \Theta\}$, for any $T \in \mathbb{N}$, there exists an estimator $\hat{\theta}_{T\theta} = \hat{\theta}_T(\mathcal{Z}_d^1(\theta), \ldots, \mathcal{Z}_d^T(\theta))$ such that*

$$\sup_{\theta_* \in \Theta} \mathbb{E}\|\pi_{\hat{\theta}_T} - \pi_{\theta_*}\| = \tilde{O}\left(LT^{-\frac{\alpha^2}{2(d+2\alpha)(\alpha+2(d+1))}}\right).$$

Proof. By the standard PAC analysis [3,8], for any $\gamma > 0$, with probability greater than $1 - \gamma$, a sample of $k = O((d/\gamma)\log(1/\gamma))$ random points will partition $\mathbb{C}$ into regions of width less than γ (under $L_1(\mathcal{D})$). For brevity, we omit the t subscripts and superscripts on quantities such as $\mathcal{Z}_k^t(\theta)$ throughout the following analysis, since the claims hold for any arbitrary value of t.

For any $\theta \in \Theta$, let π'_θ denote a (conditional on $X_1, \ldots, X_k$) distribution defined as follows. Let f'_θ denote the (conditional on $X_1, \ldots, X_k$) density function of π'_θ with respect to π_0, and for any $g \in \mathbb{C}$, let $f'_\theta(g) = \frac{\pi_\theta(\{h \in \mathbb{C}: \forall i \leq k, h(X_i) = g(X_i)\})}{\pi_0(\{h \in \mathbb{C}: \forall i \leq k, h(X_i) = g(X_i)\})}$ (or 0 if $\pi_0(\{h \in \mathbb{C} : \forall i \leq k, h(X_i) = g(X_i)\}) = 0$). In other words, π'_θ has the same probability mass as π_θ for each of the equivalence classes induced by $X_1, \ldots, X_k$, but conditioned on the equivalence class, simply has a constant-density distribution over that equivalence class. Note that every $h \in \mathbb{C}$ has $f'_\theta(h)$ between the smallest and largest values of $f_\theta(g)$ among $g \in \mathbb{C}$ with $\forall i \leq k, g(X_i) = h(X_i)$; therefore, by the smoothness condition, on the event (of probability greater than $1 - \gamma$) that each of these regions has diameter less than γ, we have $\forall h \in \mathbb{C}, |f_\theta(h) - f'_\theta(h)| < L\gamma^\alpha$. On this event, for any $\theta, \theta' \in \Theta$,

$$\|\pi_\theta - \pi_{\theta'}\| = (1/2)\int |f_\theta - f_{\theta'}|\mathrm{d}\pi_0 < L\gamma^\alpha + (1/2)\int |f'_\theta - f'_{\theta'}|\mathrm{d}\pi_0.$$

Furthermore, since the regions that define f'_θ and $f'_{\theta'}$ are the same (namely, the partition induced by $X_1, \ldots, X_k$), we have

$$(1/2)\int |f'_\theta - f'_{\theta'}|\mathrm{d}\pi_0 = (1/2)\sum_{y_1, \ldots, y_k \in \{-1, +1\}} |\pi_\theta(\{h \in \mathbb{C} : \forall i \leq k, h(X_i) = y_i\})$$

$$- \pi_{\theta'}(\{h \in \mathbb{C} : \forall i \leq k, h(X_i) = y_i\})|$$

$$= \|\mathbb{P}_{\mathbb{Y}_k(\theta)|\mathbb{X}_k} - \mathbb{P}_{\mathbb{Y}_k(\theta')|\mathbb{X}_k}\|.$$

Thus, we have that with probability at least $1 - \gamma$,

$$\|\pi_\theta - \pi_{\theta'}\| < L\gamma^\alpha + \|\mathbb{P}_{\mathbb{Y}_k(\theta)|\mathbb{X}_k} - \mathbb{P}_{\mathbb{Y}_k(\theta')|\mathbb{X}_k}\|.$$

Following analogous to the inductive argument of [11], suppose $I \subseteq \{1, \ldots, k\}$, fix $\bar{x}_I \in \mathcal{X}^{|I|}$ and $\bar{y}_I \in \{-1, +1\}^{|I|}$. Then the $\tilde{y}_I \in \{-1, +1\}^{|I|}$ for which $\|\bar{y}_I - \tilde{y}_I\|_1$ is minimal, subject to the constraint that no $h \in \mathbb{C}$ has $h(\bar{x}_I) = \tilde{y}_I$, has $(1/2)\|\bar{y}_I - \tilde{y}_I\|_1 \le d + 1$; also, for any $i \in I$ with $\bar{y}_i \ne \tilde{y}_i$, letting $\bar{y}'_j = \bar{y}_j$ for $j \in I \setminus \{i\}$ and $\bar{y}'_i = \tilde{y}_i$, we have

$$\mathbb{P}_{\mathbb{Y}_I(\theta)|\mathbb{X}_I}(\bar{y}_I|\bar{x}_I) = \mathbb{P}_{\mathbb{Y}_{I\setminus\{i\}}(\theta)|\mathbb{X}_{I\setminus\{i\}}}(\bar{y}_{I\setminus\{i\}}|\bar{x}_{I\setminus\{i\}}) - \mathbb{P}_{\mathbb{Y}_I(\theta)|\mathbb{X}_I}(\bar{y}'_I|\bar{x}_I),$$

and similarly for θ', so that

$$\begin{aligned}
&|\mathbb{P}_{\mathbb{Y}_I(\theta)|\mathbb{X}_I}(\bar{y}_I|\bar{x}_I) - \mathbb{P}_{\mathbb{Y}_I(\theta')|\mathbb{X}_I}(\bar{y}_I|\bar{x}_I)| \\
&\le |\mathbb{P}_{\mathbb{Y}_{I\setminus\{i\}}(\theta)|\mathbb{X}_{I\setminus\{i\}}}(\bar{y}_{I\setminus\{i\}}|\bar{x}_{I\setminus\{i\}}) - \mathbb{P}_{\mathbb{Y}_{I\setminus\{i\}}(\theta')|\mathbb{X}_{I\setminus\{i\}}}(\bar{y}_{I\setminus\{i\}}|\bar{x}_{I\setminus\{i\}})| \\
&\quad + |\mathbb{P}_{\mathbb{Y}_I(\theta)|\mathbb{X}_I}(\bar{y}'_I|\bar{x}_I) - \mathbb{P}_{\mathbb{Y}_I(\theta')|\mathbb{X}_I}(\bar{y}'_I|\bar{x}_I)|.
\end{aligned}$$

Now consider that these two terms inductively define a binary tree. Every time the tree branches left once, it arrives at a difference of probabilities for a set I of one less element than that of its parent. Every time the tree branches right once, it arrives at a difference of probabilities for a $\bar{y}_I$ one closer to an unrealized $\tilde{y}_I$ than that of its parent. Say we stop branching the tree upon reaching a set I and a $\bar{y}_I$ such that either $\bar{y}_I$ is an unrealized labeling, or $|I| = d$. Thus, we can bound the original (root node) difference of probabilities by the sum of the differences of probabilities for the leaf nodes with $|I| = d$. Any path in the tree can branch left at most $k - d$ times (total) before reaching a set I with only d elements, and can branch right at most $d + 1$ times in a row before reaching a $\bar{y}_I$ such that both probabilities are zero, so that the difference is zero. So the depth of any leaf node with $|I| = d$ is at most $(k - d)d$. Furthermore, at any level of the tree, from left to right the nodes have strictly decreasing $|I|$ values, so that the maximum width of the tree is at most $k - d$. So the total number of leaf nodes with $|I| = d$ is at most $(k - d)^2 d$. Thus, for any $\bar{y} \in \{-1, +1\}^k$ and $\bar{x} \in \mathcal{X}^k$,

$$\begin{aligned}
&|\mathbb{P}_{\mathbb{Y}_k(\theta)|\mathbb{X}_k}(\bar{y}|\bar{x}) - \mathbb{P}_{\mathbb{Y}_k(\theta')|\mathbb{X}_k}(\bar{y}|\bar{x})| \\
&\le (k - d)^2 d \cdot \max_{\bar{y}^d \in \{-1,+1\}^d} \max_{D \in \{1,\ldots,k\}^d} |\mathbb{P}_{\mathbb{Y}_d(\theta)|\mathbb{X}_d}(\bar{y}^d|\bar{x}_D) - \mathbb{P}_{\mathbb{Y}_d(\theta')|\mathbb{X}_d}(\bar{y}^d|\bar{x}_D)|.
\end{aligned}$$

Since

$$\|\mathbb{P}_{\mathbb{Y}_k(\theta)|\mathbb{X}_k} - \mathbb{P}_{\mathbb{Y}_k(\theta')|\mathbb{X}_k}\| = (1/2) \sum_{\bar{y}^k \in \{-1,+1\}^k} |\mathbb{P}_{\mathbb{Y}_k(\theta)|\mathbb{X}_k}(\bar{y}^k) - \mathbb{P}_{\mathbb{Y}_k(\theta')|\mathbb{X}_k}(\bar{y}^k)|,$$

and by Sauer's Lemma this is at most

$$(ek)^d \max_{\bar{y}^k \in \{-1,+1\}^k} |\mathbb{P}_{\mathbb{Y}_k(\theta)|\mathbb{X}_k}(\bar{y}^k) - \mathbb{P}_{\mathbb{Y}_k(\theta')|\mathbb{X}_k}(\bar{y}^k)|,$$

we have that

$$
\begin{aligned}
&\|\mathbb{P}_{\mathbb{Y}_k(\theta)|\mathbb{X}_k} - \mathbb{P}_{\mathbb{Y}_k(\theta')|\mathbb{X}_k}\| \\
&\leq (ek)^d k^2 d \max_{\bar{y}^d \in \{-1,+1\}^d} \max_{D \in \{1,\dots,k\}^d} |\mathbb{P}_{\mathbb{Y}_d(\theta)|\mathbb{X}_D}(\bar{y}^d) - \mathbb{P}_{\mathbb{Y}_d(\theta')|\mathbb{X}_D}(\bar{y}^d)|.
\end{aligned}
$$

Thus, we have that

$$
\begin{aligned}
\|\pi_\theta - \pi_{\theta'}\| &= \mathbb{E}\|\pi_\theta - \pi_{\theta'}\| \\
&< \gamma + L\gamma^\alpha + (ek)^d k^2 d \mathbb{E}\left[\max_{\bar{y}^d \in \{-1,+1\}^d} \max_{D \in \{1,\dots,k\}^d} \mathbb{P}_{\mathbb{Y}_d(\theta)|\mathbb{X}_D}(\bar{y}^d) - \mathbb{P}_{\mathbb{Y}_d(\theta')|\mathbb{X}_D}(\bar{y}^d)|\right].
\end{aligned}
$$

Note that

$$
\begin{aligned}
&\mathbb{E}\left[\max_{\bar{y}^d \in \{-1,+1\}^d} \max_{D \in \{1,\dots,k\}^d} |\mathbb{P}_{\mathbb{Y}_d(\theta)|\mathbb{X}_D}(\bar{y}^d) - \mathbb{P}_{\mathbb{Y}_d(\theta')|\mathbb{X}_D}(\bar{y}^d)|\right] \\
&\leq \sum_{\bar{y}^d \in \{-1,+1\}^d} \sum_{D \in \{1,\dots,k\}^d} \mathbb{E}\left[|\mathbb{P}_{\mathbb{Y}_d(\theta)|\mathbb{X}_D}(\bar{y}^d) - \mathbb{P}_{\mathbb{Y}_d(\theta')|\mathbb{X}_D}(\bar{y}^d)|\right] \\
&\leq (2k)^d \max_{\bar{y}^d \in \{-1,+1\}^d} \max_{D \in \{1,\dots,k\}^d} \mathbb{E}\left[|\mathbb{P}_{\mathbb{Y}_d(\theta)|\mathbb{X}_D}(\bar{y}^d) - \mathbb{P}_{\mathbb{Y}_d(\theta')|\mathbb{X}_D}(\bar{y}^d)|\right],
\end{aligned}
$$

and by exchangeability, this last line equals

$$
(2k)^d \max_{\bar{y}^d \in \{-1,+1\}^d} \mathbb{E}\left[|\mathbb{P}_{\mathbb{Y}_d(\theta)|\mathbb{X}_d}(\bar{y}^d) - \mathbb{P}_{\mathbb{Y}_d(\theta')|\mathbb{X}_d}(\bar{y}^d)|\right].
$$

[11] showed that $\mathbb{E}\left[|\mathbb{P}_{\mathbb{Y}_d(\theta)|\mathbb{X}_d}(\bar{y}^d) - \mathbb{P}_{\mathbb{Y}_d(\theta')|\mathbb{X}_d}(\bar{y}^d)|\right] \leq 4\sqrt{\|\mathbb{P}_{\mathcal{Z}_d(\theta)} - \mathbb{P}_{\mathcal{Z}_d(\theta')}\|}$, so that in total we have $\|\pi_\theta - \pi_{\theta'}\| < (L+1)\gamma^\alpha + 4(2ek)^{2d+2}\sqrt{\|\mathbb{P}_{\mathcal{Z}_d(\theta)} - \mathbb{P}_{\mathcal{Z}_d(\theta')}\|}$. Plugging in the value of $k = c(d/\gamma)\log(1/\gamma)$, this is

$$
(L+1)\gamma^\alpha + 4\left(2ec\frac{d}{\gamma}\log\left(\frac{1}{\gamma}\right)\right)^{2d+2}\sqrt{\|\mathbb{P}_{\mathcal{Z}_d(\theta)} - \mathbb{P}_{\mathcal{Z}_d(\theta')}\|}.
$$

Thus, it suffices to bound the rate of convergence (in total variation distance) of some estimator of $\mathbb{P}_{\mathcal{Z}_d(\theta_\star)}$. If $N(\varepsilon)$ is the ε-covering number of $\{\mathbb{P}_{\mathcal{Z}_d(\theta)} : \theta \in \Theta\}$, then taking $\hat{\theta}_{T\theta_\star}$ as the minimum distance skeleton estimate of [5,13] achieves expected total variation distance ε from $\mathbb{P}_{\mathcal{Z}_d(\theta_\star)}$, for some $T = O((1/\varepsilon^2)\log N(\varepsilon/4))$. We can partition $\mathbb{C}$ into $O((L/\varepsilon)^{d/\alpha})$ cells of diameter $O((\varepsilon/L)^{1/\alpha})$, and set a constant density value within each cell, on an $O(\varepsilon)$-grid of density values, and every prior with (L,α)-Hölder smooth density will have density within ε of some density so-constructed; there are then at most $(1/\varepsilon)^{O((L/\varepsilon)^{d/\alpha})}$ such densities, so this bounds the covering numbers of Π_Θ. Furthermore, the covering number of Π_Θ upper bounds $N(\varepsilon)$ [11], so that $N(\varepsilon) \leq (1/\varepsilon)^{O((L/\varepsilon)^{d/\alpha})}$.

Solving $T = O(\varepsilon^{-2}(L/\varepsilon)^{d/\alpha}\log(1/\varepsilon))$ for ε, we have $\varepsilon = O\left(L\left(\frac{\log(TL)}{T}\right)^{\frac{\alpha}{d+2\alpha}}\right)$.

So this bounds the rate of convergence for $\mathbb{E}\|\mathbb{P}_{\mathcal{Z}_d(\hat{\theta}_T)} - \mathbb{P}_{\mathcal{Z}_d(\theta_\star)}\|$, for $\hat{\theta}_T$ the minimum distance skeleton estimate. Plugging this rate into the bound on the priors, combined with Jensen's inequality, we have

$$\mathbb{E}\|\pi_{\hat{\theta}_T} - \pi_{\theta_*}\| < (L+1)\gamma^\alpha + 4\left(2ec\frac{d}{\gamma}\log\left(\frac{1}{\gamma}\right)\right)^{2d+2} \times O\left(L\left(\frac{\log(TL)}{T}\right)^{\frac{\alpha}{2d+4\alpha}}\right).$$

This holds for any $\gamma > 0$, so minimizing this expression over $\gamma > 0$ yields a bound on the rate. For instance, with $\gamma = \tilde{O}\left(T^{-\frac{\alpha}{2(d+2\alpha)(\alpha+2(d+1))}}\right)$, we have

$$\mathbb{E}\|\pi_{\hat{\theta}_T} - \pi_{\theta_*}\| = \tilde{O}\left(LT^{-\frac{\alpha^2}{2(d+2\alpha)(\alpha+2(d+1))}}\right).$$

$\square$

4   A Minimax Lower Bound

One natural quesiton is whether Theorem 1 can generally be improved. While we expect this to be true for some fixed VC classes (e.g., those of finite size), and in any case we expect that some of the constant factors in the exponent may be improvable, it is not at this time clear whether the general form of $T^{-\Theta(\alpha^2/(d+\alpha)^2)}$ is sometimes optimal. One way to investigate this question is to construct specific spaces $\mathbb{C}$ and distributions $\mathcal{D}$ for which a lower bound can be obtained. In particular, we are generally interested in exhibiting lower bounds that are worse than those that apply to the usual problem of density estimation based on direct access to the $h_{t\theta_*}^*$ values (see Theorem 3 below).

Here we present a lower bound that is interesting for this reason. However, although larger than the optimal rate for methods with direct access to the target concepts, it is still far from matching the upper bound above, so that the question of tightness remains open. Specifically, we have the following result.

Theorem 2. *For any integer $d \geq 1$, any $L > 0, \alpha \in (0,1]$, there is a value $C(d,L,\alpha) \in (0,\infty)$ such that, for any $T \in \mathbb{N}$, there exists an instance space $\mathcal{X}$, a concept space $\mathbb{C}$ of VC dimension d, a distribution $\mathcal{D}$ over $\mathcal{X}$, and a distribution π_0 over $\mathbb{C}$ such that, for Π_Θ a set of distributions over $\mathbb{C}$ with (L,α)-Hölder smooth density functions with respect to π_0, any estimator $\hat{\theta}_T = \hat{\theta}_T(\mathcal{Z}_d^1(\theta_*), \ldots, \mathcal{Z}_d^T(\theta_*))$ has*

$$\sup_{\theta_* \in \Theta} \mathbb{E}\left[\|\pi_{\hat{\theta}_T} - \pi_{\theta_*}\|\right] \geq C(d,L,\alpha)T^{-\frac{\alpha}{2(d+\alpha)}}.$$

Proof. (Sketch) We proceed by a reduction from the task of determining the bias of a coin from among two given possibilities. Specifically, fix any $\gamma \in (0,1/2)$, $n \in \mathbb{N}$, and let $B_1(p), \ldots, B_n(p)$ be i.i.d Bernoulli(p) random variables, for each $p \in [0,1]$; then it is known that, for any (possibly nondeterministic) decision rule $\hat{p}_n : \{0,1\}^n \to \{(1+\gamma)/2, (1-\gamma)/2\}$,

$$\frac{1}{2}\sum_{p\in\{(1+\gamma)/2,(1-\gamma)/2\}} \mathbb{P}(\hat{p}_n(B_1(p),\ldots,B_n(p)) \neq p)$$

$$\geq (1/32)\cdot\exp\left\{-128\gamma^2 n/3\right\}. \quad (1)$$

This easily follows from the results of [1], combined with a result of [6] bounding the KL divergence (see also [10])

To use this result, we construct a learning problem as follows. Fix some $m \in \mathbb{N}$ with $m \geq d$, let $\mathcal{X} = \{1, \ldots, m\}$, and let $\mathbb{C}$ be the space of all classifiers $h : \mathcal{X} \to \{-1, +1\}$ such that $|\{x \in \mathcal{X} : h(x) = +1\}| \leq d$. Clearly the VC dimension of $\mathbb{C}$ is d. Define the distribution $\mathcal{D}$ as uniform over $\mathcal{X}$. Finally, we specify a family of (L, α)-Hölder smooth priors, parameterized by $\Theta = \{-1, +1\}^{\binom{m}{d}}$, as follows. Let $\gamma_m = (L/2)(1/m)^\alpha$. First, enumerate the $\binom{m}{d}$ distinct d-sized subsets of $\{1, \ldots, m\}$ as $\mathcal{X}_1, \mathcal{X}_2, \ldots, \mathcal{X}_{\binom{m}{d}}$. Define the reference distribution π_0 by the property that, for any $h \in \mathbb{C}$, letting $q = |\{x : h(x) = +1\}|$, $\pi_0(\{h\}) = (\frac{1}{2})^d \binom{m-q}{d-q}/\binom{m}{d}$. For any $\mathbf{b} = (b_1, \ldots, b_{\binom{m}{d}}) \in \{-1, 1\}^{\binom{m}{d}}$, define the prior $\pi_{\mathbf{b}}$ as the distribution of a random variable $h_{\mathbf{b}}$ specified by the following generative model. Let $i^* \sim \text{Uniform}(\{1, \ldots, \binom{m}{d}\})$, let $C_{\mathbf{b}}(i^*) \sim \text{Bernoulli}((1 + \gamma_m b_{i^*})/2)$; finally, $h_{\mathbf{b}} \sim \text{Uniform}(\{h \in \mathbb{C} : \{x : h(x) = +1\} \subseteq \mathcal{X}_{i^*}, \text{Parity}(|\{x : h(x) = +1\}|) = C_{\mathbf{b}}(i^*)\})$, where $\text{Parity}(n)$ is 1 if n is odd, or 0 if n is even. We will refer to the variables in this generative model below. For any $h \in \mathbb{C}$, letting $H = \{x : h(x) = +1\}$ and $q = |H|$, we can equivalently express
$$\pi_{\mathbf{b}}(\{h\}) = (\tfrac{1}{2})^d \binom{m}{d}^{-1} \sum_{i=1}^{\binom{m}{d}} \mathbb{1}[H \subseteq \mathcal{X}_i](1 + \gamma_m b_i)^{\text{Parity}(q)}(1 - \gamma_m b_i)^{1 - \text{Parity}(q)}.$$
From this explicit representation, it is clear that, letting $f_{\mathbf{b}} = \frac{d\pi_{\mathbf{b}}}{d\pi_0}$, we have $f_{\mathbf{b}}(h) \in [1 - \gamma_m, 1 + \gamma_m]$ for all $h \in \mathbb{C}$. The fact that $f_{\mathbf{b}}$ is Hölder smooth follows from this, since every distinct $h, g \in \mathbb{C}$ have $\mathcal{D}(\{x : h(x) \neq g(x)\}) \geq 1/m = (2\gamma_m/L)^{1/\alpha}$.

Next we set up the reduction as follows. For any estimator $\hat{\pi}_T = \hat{\pi}_T(\mathcal{Z}_d^1(\theta_\star), \ldots, \mathcal{Z}_d^T(\theta_\star))$, and each $i \in \{1, \ldots, \binom{m}{d}\}$, let h_i be the classifier with $\{x : h_i(x) = +1\} = \mathcal{X}_i$; also, if $\hat{\pi}_T(\{h_i\}) > (\frac{1}{2})^d/\binom{m}{d}$, let $\hat{b}_i = 2\text{Parity}(d) - 1$, and otherwise $\hat{b}_i = 1 - 2\text{Parity}(d)$. We use these $\hat{b}_i$ values to estimate the original b_i values. Specifically, let $\hat{p}_i = (1 + \gamma_m \hat{b}_i)/2$ and $p_i = (1 + \gamma_m b_i)/2$, where $\mathbf{b} = \theta_\star$. Then

$$\|\hat{\pi}_T - \pi_{\theta_\star}\| \geq (1/2) \sum_{i=1}^{\binom{m}{d}} |\hat{\pi}_T(\{h_i\}) - \pi_{\theta_\star}(\{h_i\})|$$

$$\geq (1/2) \sum_{i=1}^{\binom{m}{d}} \frac{\gamma_m}{2^d \binom{m}{d}} |\hat{b}_i - b_i|/2 = (1/2) \sum_{i=1}^{\binom{m}{d}} \frac{1}{2^d \binom{m}{d}} |\hat{p}_i - p_i|.$$

Thus, we have reduced from the problem of deciding the biases of these $\binom{m}{d}$ independent Bernoulli random variables. To complete the proof, it suffices to lower bound the expectation of the right side for an *arbitrary* estimator.

Toward this end, we in fact study an even easier problem. Specifically, consider an estimator $\hat{q}_i = \hat{q}_i(\mathcal{Z}_d^1(\theta_\star), \ldots, \mathcal{Z}_d^T(\theta_\star), i_1^*, \ldots, i_T^*)$, where i_t^* is the i^* random variable in the generative model that defines $h_{t\theta_\star}^*$; that is, $i_t^* \sim \text{Uniform}(\{1, \ldots, \binom{m}{d}\})$, $C_t \sim \text{Bernoulli}((1 + \gamma_m b_{i_t^*})/2)$, and $h_{t\theta_\star}^* \sim \text{Uniform}(\{h \in \mathbb{C} : \{x : h(x) = +1\} \subseteq \mathcal{X}_{i_t^*}, \text{Parity}(|\{x : h(x) = +1\}|) = C_t\})$, where the i_t^* are independent across t, as are the C_t and $h_{t\theta_\star}^*$. Clearly the $\hat{p}_i$ from above can be viewed as an estimator of this type, which simply ignores the knowledge of i_t^*. The knowledge of these i_t^* variables simplifies the analysis, since given $\{i_t^* : t \leq T\}$, the data can be partitioned into $\binom{m}{d}$ disjoint sets, $\{\{\mathcal{Z}_d^t(\theta_\star) : i_t^* = i\} : i = 1, \ldots, \binom{m}{d}\}$, and we can use only the set $\{\mathcal{Z}_d^t(\theta_\star) : i_t^* = i\}$ to estimate p_i. Furthermore, we can use only the subset of these for which $\mathbb{X}_{td} = \mathcal{X}_i$, since otherwise we have zero information about the value of Parity($|\{x : h_{t\theta_\star}^*(x) = +1\}|$). That is, given $i_t^* = i$, any $\mathcal{Z}_d^t(\theta_\star)$ is conditionally independent from every b_j for $j \neq i$, and is even conditionally independent from b_i when $\mathbb{X}_{td}$ is not completely contained in $\mathcal{X}_i$; specifically, in this case, regardless of b_i, the conditional distribution of $\mathbb{Y}_{td}(\theta_\star)$ given $i_t^* = i$ and given $\mathbb{X}_{td}$ is a product distribution, which deterministically assigns label -1 to those $Y_{tk}(\theta_\star)$ with $X_{tk} \notin \mathcal{X}_i$, and gives uniform random values to the subset of $\mathbb{Y}_{td}(\theta_\star)$ with their respective $X_{tk} \in \mathcal{X}_i$. Finally, letting $r_t = \text{Parity}(|\{k \leq d : Y_{tk}(\theta_\star) = +1\}|)$, we note that given $i_t^* = i$, $\mathbb{X}_{td} = \mathcal{X}_i$, and the value r_t, b_i is conditionally independent from $\mathcal{Z}_d^t(\theta_\star)$. Thus, the set of values $C_{iT}(\theta_\star) = \{r_t : i_t^* = i, \mathbb{X}_{td} = \mathcal{X}_i\}$ is a sufficient statistic for b_i (hence for p_i). Recall that, when $i_t^* = i$ and $\mathbb{X}_{td} = \mathcal{X}_i$, the value of r_t is equal to C_t, a Bernoulli(p_i) random variable. Thus, we neither lose nor gain anything (in terms of risk) by restricting ourselves to estimators $\hat{q}_i$ of the type $\hat{q}_i = \hat{q}_i(\mathcal{Z}_d^1(\theta_\star), \ldots, \mathcal{Z}_d^T(\theta_\star), i_1^*, \ldots, i_T^*) = \hat{q}_i'(C_{iT}(\theta_\star))$, for some $\hat{q}_i'$ [7]: that is, estimators that are a function of the $N_{iT}(\theta_\star) = |C_{iT}(\theta_\star)|$ Bernoulli(p_i) random variables, which we should note are conditionally i.i.d. given $N_{iT}(\theta_\star)$.

Thus, by (1), for any $n \leq T$,

$$\frac{1}{2} \sum_{b_i \in \{-1,+1\}} \mathbb{E}\left[|\hat{q}_i - p_i| \,\Big|\, N_{iT}(\theta_\star) = n\right] = \frac{1}{2} \sum_{b_i \in \{-1,+1\}} \gamma_m \mathbb{P}\left(\hat{q}_i \neq p_i \,\Big|\, N_{iT}(\theta_\star) = n\right)$$
$$\geq (\gamma_m/32) \cdot \exp\left\{-128\gamma_m^2 N_i/3\right\}.$$

Also, $\forall i$, $\mathbb{E}[N_i] = \frac{d!(1/m)^d}{\binom{m}{d}}T \leq (d/m)^{2d}T = d^{2d}(2\gamma_m/L)^{2d/\alpha}T$. Thus, Jensen's inequality, linearity of expectation, and the law of total expectation imply

$$\frac{1}{2} \sum_{b_i \in \{-1,+1\}} \mathbb{E}\left[|\hat{q}_i - p_i|\right] \geq (\gamma_m/32) \cdot \exp\left\{-43(2/L)^{2d/\alpha}d^{2d}\gamma_m^{2+2d/\alpha}T\right\}.$$

Thus, by linearity of the expectation,

$$\left(\frac{1}{2}\right)^{\binom{m}{d}} \sum_{b \in \{-1,+1\}^{\binom{m}{d}}} \mathbb{E}\left[\sum_{i=1}^{\binom{m}{d}} \frac{1}{2^d\binom{m}{d}}|\hat{q}_i - p_i|\right] = \sum_{i=1}^{\binom{m}{d}} \frac{1}{2^d\binom{m}{d}} \frac{1}{2} \sum_{b_i \in \{-1,+1\}} \mathbb{E}\left[|\hat{q}_i - p_i|\right]$$
$$\geq (\gamma_m/(32 \cdot 2^d)) \cdot \exp\left\{-43(2/L)^{2d/\alpha}d^{2d}\gamma_m^{2+2d/\alpha}T\right\}.$$

In particular, taking $m = \left\lceil (L/2)^{1/\alpha} \left(43(2/L)^{2d/\alpha}d^{2d}T\right)^{\frac{1}{2(d+\alpha)}} \right\rceil$, we have $\gamma_m = \Theta\left(\left(43(2/L)^{2d/\alpha}d^{2d}T\right)^{-\frac{\alpha}{2(d+\alpha)}} \right)$, so that

$$\left(\frac{1}{2}\right)^{\binom{m}{d}} \sum_{\mathbf{b}\in\{-1,+1\}^{\binom{m}{d}}} \mathbb{E}\left[\sum_{i=1}^{\binom{m}{d}} \frac{1}{2^d\binom{m}{d}}|\hat{q}_i - p_i| \right] = \Omega\left(2^{-d}\left(43(2/L)^{2d/\alpha}d^{2d}T\right)^{-\frac{\alpha}{2(d+\alpha)}} \right).$$

In particular, this implies there exists some $\mathbf{b}$ for which

$$\mathbb{E}\left[\sum_{i=1}^{\binom{m}{d}} \frac{1}{2^d\binom{m}{d}}|\hat{q}_i - p_i| \right] = \Omega\left(2^{-d}\left(43(2/L)^{2d/\alpha}d^{2d}T\right)^{-\frac{\alpha}{2(d+\alpha)}} \right).$$

Applying this lower bound to the estimator $\hat{p}_i$ above yields the result. □

It is natural to wonder how these rates might potentially improve if we allow $\hat{\theta}_T$ to depend on more than d samples per data set. To establish limits on such improvements, we note that in the extreme case of allowing the estimator to depend on the full $\mathcal{Z}^t(\theta_*)$ data sets, we may recover the known results lower bounding the risk of density estimation from i.i.d. samples from a smooth density, as indicated by the following result.

Theorem 3. *For any integer $d \geq 1$, there exists an instance space $\mathcal{X}$, a concept space $\mathbb{C}$ of VC dimension d, a distribution $\mathcal{D}$ over $\mathcal{X}$, and a distribution π_0 over $\mathbb{C}$ such that, for Π_Θ the set of distributions over $\mathbb{C}$ with (L,α)-Hölder smooth density functions with respect to π_0, any sequence of estimators, $\hat{\theta}_T = \hat{\theta}_T(\mathcal{Z}^1(\theta_*),\ldots,\mathcal{Z}^T(\theta_*))$ $(T = 1, 2, \ldots)$, has*

$$\sup_{\theta_*\in\Theta} \mathbb{E}\left[\|\pi_{\hat{\theta}_T} - \pi_{\theta_*}\| \right] = \Omega\left(T^{-\frac{\alpha}{d+2\alpha}} \right).$$

The proof is a simple reduction from the problem of estimating π_{θ_*} based on direct access to $h^*_{1\theta_*},\ldots,h^*_{T\theta_*}$, which is essentially equivalent to the standard model of density estimation, and indeed the lower bound in Theorem 3 is a well-known result for density estimation from T i.i.d. samples from a Hölder smooth density in a d-dimensional space [5].

5   Real-Valued Functions and an Application in Algorithmic Economics

In this section, we present results generalizing the analysis of [11] to classes of real-valued functions. We also present an application of this generalization to a preference elicitation problem.

5.1   Consistent Estimation of Priors over Real-Valued Functions at a Bounded Rate

In this section, we let $\mathcal{B}$ denote a σ-algebra on $\mathcal{X} \times \mathbb{R}$, and again let $\mathcal{B}_\mathcal{X}$ denote the corresponding σ-algebra on $\mathcal{X}$. Also, for measurable functions $h, g : \mathcal{X} \to \mathbb{R}$, let $\rho(h, g) = \int |h - g| \mathrm{d}P_X$, where P_X is a distribution over $\mathcal{X}$. Let $\mathcal{F}$ be a class of functions $\mathcal{X} \to \mathbb{R}$ with Borel σ-algebra $\mathcal{B}_\mathcal{F}$ induced by ρ. Let Θ be a set, and for each $\theta \in \Theta$, let π_θ denote a probability measure on $(\mathcal{F}, \mathcal{B}_\mathcal{F})$. We suppose $\{\pi_\theta : \theta \in \Theta\}$ is totally bounded in total variation distance, and that $\mathcal{F}$ is a uniformly bounded VC subgraph class with pseudodimension d. We also suppose ρ is a *metric* when restricted to $\mathcal{F}$.

As above, let $\{X_{ti}\}_{t,i\in\mathbb{N}}$ be i.i.d. P_X random variables. For each $\theta \in \Theta$, let $\{h^*_{t\theta}\}_{t\in\mathbb{N}}$ be i.i.d. π_θ random variables, independent from $\{X_{ti}\}_{t,i\in\mathbb{N}}$. For each $t \in \mathbb{N}$ and $\theta \in \Theta$, let $Y_{ti}(\theta) = h^*_{t\theta}(X_{ti})$ for $i \in \mathbb{N}$, and let $\mathcal{Z}^t(\theta) = \{(X_{t1}, Y_{t1}(\theta)), (X_{t2}, Y_{t2}(\theta)), \ldots\}$; for each $k \in \mathbb{N}$, define $\mathcal{Z}^t_k(\theta) = \{(X_{t1}, Y_{t1}(\theta)), \ldots, (X_{tk}, Y_{tk}(\theta))\}$, $\mathbb{X}_{tk} = \{X_{t1}, \ldots, X_{tk}\}$, and $\mathbb{Y}_{tk}(\theta) = \{Y_{t1}(\theta), \ldots, Y_{tk}(\theta)\}$.

We have the following result. The proof parallels that of [11] (who studied the special case of binary functions), with a few important twists (in particular, a significantly different approach in the analogue of their Lemma 3). Due to space restrictions, the formal details are omitted; we refer the interested reader to the full version of this article online [12].

Theorem 4. *There exists an estimator* $\hat{\theta}_{T\theta_\star} = \hat{\theta}_T(\mathcal{Z}^1_d(\theta_\star), \ldots, \mathcal{Z}^T_d(\theta_\star))$, *and functions* $R : \mathbb{N}_0 \times (0, 1] \to [0, \infty)$ *and* $\delta : \mathbb{N}_0 \times (0, 1] \to [0, 1]$ *such that, for any* $\alpha > 0$, $\lim_{T\to\infty} R(T, \alpha) = \lim_{T\to\infty} \delta(T, \alpha) = 0$ *and for any* $T \in \mathbb{N}_0$ *and* $\theta_\star \in \Theta$,

$$\mathbb{P}\left(\|\pi_{\hat{\theta}_{T\theta_\star}} - \pi_{\theta_\star}\| > R(T, \alpha) \right) \leq \delta(T, \alpha) \leq \alpha.$$

5.2   Maximizing Customer Satisfaction in Combinatorial Auctions

Theorem 4 has a clear application in the context of transfer learning, following analogous arguments to those given in the special case of binary classification by [11]. In addition to that application, we can also use Theorem 4 in the context of the following problem in algorithmic economics, where the objective is to serve a sequence of customers so as to maximize their satisfaction.

Consider an online travel agency, where customers go to the site with some idea of what type of travel they are interested in; the site then poses a series of questions to each customer, and identifies a travel package that best suits their desires, budget, and dates. There are many options of travel packages, with options on location, site-seeing tours, hotel and room quality, etc. Because of this, serving the needs of an *arbitrary* customer might be a lengthy process, requiring many detailed questions. Fortunately, the stream of customers is typically not a worst-case sequence, and in particular obeys many statistical regularities: in particular, it is not too far from reality to think of the customers as being independent and identically distributed samples. With this assumption in mind, it becomes desirable to identify some of these statistical regularities so that we can pose the questions that are typically most relevant, and thereby more quickly identify the travel package that best suits the needs of the typical customer. One straightforward way to do this is to directly *estimate* the distribution of customer value functions, and optimize the questioning system to minimize the expected number of questions needed to find a suitable travel package.

One can model this problem in the style of Bayesian combinatorial auctions, in which each customer has a value function for each possible bundle of items. However, it is slightly different, in that we do not assume the distribution of customers is known, but rather are interested in estimating this distribution; the obtained estimate can then be used in combination with methods based on Bayesian decision theory. In contrast to the literature on Bayesian auctions (and subjectivist Bayesian decision theory in general), this technique is able to maintain general guarantees on performance that hold under an objective interpretation of the problem, rather than merely guarantees holding under an arbitrary assumed prior belief. This general idea is sometimes referred to as *Empirical Bayesian* decision theory in the machine learning and statistics literatures. The ideal result for an Empirical Bayesian algorithm is to be competitive with the corresponding Bayesian methods based on the *actual* distribution of the data (assuming the data are random, with an unknown distribution); that is, although the Empirical Bayesian methods only operate with a data-based estimate of the distribution, the aim is to perform nearly as well as methods based on the true (unobservable) distribution. In this work, we present results of this type, in the context of an abstraction of the aforementioned online travel agency problem, where the measure of performance is the expected number of questions to find a suitable package.

The specific application we are interested in here may be expressed abstractly as a kind of combinatorial auction with preference elicitation. Specifically, we suppose there is a collection of items on a menu, and each possible bundle of items has an associated fixed price. There is a stream of customers, each with a valuation function that provides a value for each possible bundle of items. The objective is to serve each customer a bundle of items that nearly-maximizes his or her surplus value (value minus price). However, we are not permitted direct observation of the customer valuation functions; rather, we may query for the value of any given bundle of items; this is referred to as a *value query* in the literature on preference elicitation in combinatorial auctions (see Chapter 14 of [4], [14]). The objective is to achieve this near-maximal surplus guarantee, while making only a small number of queries per customer. We suppose the customer valuation function are sampled i.i.d. according to an unknown distribution over a known (but arbitrary) class of real-valued functions having finite pseudo-dimension. Reasoning that knowledge of this distribution should allow one to make a smaller number of value queries per customer, we are interested in estimating this unknown distribution, so that as we serve more and more customers, the number of queries per customer required to identify a near-optimal bundle should decrease. In this context, we in fact prove that in the limit, the expected number of queries per customer converges to the number required of a method having direct knowledge of the true distribution of valuation functions.

Formally, suppose there is a menu of n items $[n] = \{1, \ldots, n\}$, and each bundle $B \subseteq [n]$ has an associated price $p(B) \geq 0$. Suppose also there is a sequence of customers, each with a valuation function $v_t : 2^{[n]} \to \mathbb{R}$. We suppose these v_t functions are i.i.d. samples. We can then calculate the satisfaction function for each customer as $s_t(x)$, where $x \in \{0,1\}^n$, and $s_t(x) = v_t(B_x) - p(B_x)$, where $B_x \subseteq [n]$ contains element $i \in [n]$ iff $x_i = 1$.

Now suppose we are able to ask each customer a number of questions before serving up a bundle $B_{\hat{x}_t}$ to that customer. More specifically, we are able to ask for the value $s_t(x)$ for any $x \in \{0,1\}^n$. This is referred to as a *value query* in the literature on preference elicitation in combinatorial auctions (see Chapter 14 of [4], [14]). We are interested in asking as few questions as possible, while satisfying the guarantee that $\mathbb{E}[s_t(\hat{x}_t) - \max_x s_t(x)] \leq \varepsilon$.

Now suppose, for every π and ε, we have a method $A(\pi, \varepsilon)$ such that, given that π is the actual distribution of the s_t functions, $A(\pi, \varepsilon)$ guarantees that the $\hat{x}_t$ value it selects has $\mathbb{E}[\max_x s_t(x) - s_t(\hat{x}_t)] \leq \varepsilon$; also let $\hat{N}_t(\pi, \varepsilon)$ denote the actual (random) number of queries the method $A(\pi, \varepsilon)$ would ask for the s_t function, and let $Q(\pi, \varepsilon) = \mathbb{E}[\hat{N}_t(\pi, \varepsilon)]$. We suppose the method never queries any $s_t(x)$ value twice for a given t, so that its number of queries for any given t is bounded.

Also suppose $\mathcal{F}$ is a VC subgraph class of functions mapping $\mathcal{X} = \{0,1\}^n$ into $[-1, 1]$ with pseudodimension d, and that $\{\pi_\theta : \theta \in \Theta\}$ is a known totally bounded family of distributions over $\mathcal{F}$ such that the s_t functions have distribution π_{θ_*} for some unknown $\theta_* \in \Theta$. For any $\theta \in \Theta$ and $\gamma > 0$, let $B(\theta, \gamma) = \{\theta' \in \Theta : \|\pi_\theta - \pi_{\theta'}\| \leq \gamma\}$.

Suppose, in addition to A, we have another method $A'(\varepsilon)$ that is not π-dependent, but still provides the ε-correctness guarantee, and makes a bounded number of queries (e.g., in the worst case, we could consider querying all 2^n points, but in most cases there are more clever π-independent methods that use far fewer queries, such as $O(1/\varepsilon^2)$). Consider the method described in Algorithm 1; the quantities $\hat{\theta}_{T\theta_*}$, $R(T, \alpha)$, and $\delta(T, \alpha)$ from Theorem 4 are here considered with respect P_X taken as the uniform distribution on $\{0,1\}^n$.

The following theorem indicates that Algorithm 1 is correct, and furthermore that the long-run average number of queries is not much worse than that of a method that has direct knowledge of π_{θ_*}. The proof of this result parallels that of [11] for the transfer learning setting, but is included here for completeness.

Theorem 5. *In Algorithm 1, $\forall t \leq T, \mathbb{E}[\max_x s_t(x) - s_t(\hat{x}_t)] \leq \varepsilon$. Furthermore, if $S_T(\varepsilon)$ is the total number of queries made by the method, then*

$$\limsup_{T \to \infty} \frac{\mathbb{E}[S_T(\varepsilon)]}{T} \leq Q(\pi_{\theta_*}, \varepsilon/4) + d.$$

Proof. By Theorem 4, for any $t \leq T$, if $R(t-1, \varepsilon/2) \leq \varepsilon/8$, then with probability at least $1 - \varepsilon/2$, $\|\pi_{\theta_*} - \pi_{\hat{\theta}_{(t-1)\theta_*}}\| \leq R(t - 1, \varepsilon/2)$, so that a triangle inequality

Algorithm 1. An algorithm for sequentially maximizing expected customer satisfaction.

---

for $t = 1, 2, \ldots, T$ **do**

  Pick points $X_{t1}, X_{t2}, \ldots, X_{td}$ uniformly at random from $\{0, 1\}^n$

  if $R(t - 1, \varepsilon/2) > \varepsilon/8$ **then**

    Run $A'(\varepsilon)$

    Take $\hat{x}_t$ as the returned value

  else

    Let $\breve{\theta}_{t\theta_\star} \in B\left(\hat{\theta}_{(t-1)\theta_\star}, R(t-1, \varepsilon/2)\right)$ be such that

    $Q(\pi_{\breve{\theta}_{t\theta_\star}}, \varepsilon/4) \leq \min\limits_{\theta \in B(\hat{\theta}_{(t-1)\theta_\star}, R(t-1,\varepsilon/2))} Q(\pi_\theta, \varepsilon/4) + \frac{1}{t}$

    Run $A(\pi_{\breve{\theta}_{t\theta_\star}}, \varepsilon/4)$ and let $\hat{x}_t$ be its return value

  end if

end for

---

implies $\|\pi_{\theta_\star} - \pi_{\breve{\theta}_{t\theta_\star}}\| \leq 2R(t - 1, \varepsilon/2) \leq \varepsilon/4$. Thus, $\mathbb{E}\left[\max_x s_t(x) - s_t(\hat{x}_t)\right] \leq \varepsilon/2 + \mathbb{E}\left[\mathbb{E}\left[\max_x s_t(x) - s_t(\hat{x}_t)\big|\breve{\theta}_{t\theta_\star}\right]\mathbb{1}\left[\|\pi_{\breve{\theta}_{t\theta_\star}} - \pi_{\theta_\star}\| \leq \varepsilon/2\right]\right]$. For $\theta \in \Theta$, let $\hat{x}_{t\theta}$ denote the point x that would be returned by $A(\pi_{\breve{\theta}_{t\theta_\star}}, \varepsilon/4)$ when queries are answered by some $s_{t\theta} \sim \pi_\theta$ instead of s_t (and supposing $s_t = s_{t\theta_\star}$). If $\|\pi_{\breve{\theta}_{t\theta_\star}} - \pi_{\theta_\star}\| \leq \varepsilon/4$, then

$$\mathbb{E}\left[\max_x s_t(x) - s_t(\hat{x}_t)\Big|\breve{\theta}_{t\theta_\star}\right] = \mathbb{E}\left[\max_x s_{t\theta_\star}(x) - s_{t\theta_\star}(\hat{x}_t)\Big|\breve{\theta}_{t\theta_\star}\right]$$

$$\leq \mathbb{E}\left[\max_x s_{t\breve{\theta}_{t\theta_\star}}(x) - s_{t\breve{\theta}_{t\theta_\star}}(\hat{x}_{t\breve{\theta}_{t\theta_\star}})\Big|\breve{\theta}_{t\theta_\star}\right] + \|\pi_{\breve{\theta}_{t\theta_\star}} - \pi_{\theta_\star}\| \leq \varepsilon/4 + \varepsilon/4 = \varepsilon/2.$$

Plugging into the above bound, we have $\mathbb{E}\left[\max_x s_t(x) - s_t(\hat{x}_t)\right] \leq \varepsilon$.

For the result on $S_T(\varepsilon)$, first note that $R(t - 1, \varepsilon/2) > \varepsilon/8$ only finitely many times (due to $R(t, \alpha) = o(1)$), so that we can ignore those values of t in the asymptotic calculation (as the number of queries is always bounded), and rely on the correctness guarantee of A'. For the remaining values t, let N_t denote the number of queries made by $A(\pi_{\breve{\theta}_{t\theta_\star}}, \varepsilon/4)$. then $\limsup\limits_{T \to \infty} \frac{\mathbb{E}[S_T(\varepsilon)]}{T} \leq d + \limsup\limits_{T \to \infty} \sum_{t=1}^T \frac{\mathbb{E}[N_t]}{T}$. Since $\lim\limits_{T \to \infty} \frac{1}{T}\sum_{t=1}^T \mathbb{E}\left[N_t\mathbb{1}[\|\pi_{\hat{\theta}_{(t-1)\theta_\star}} - \pi_{\theta_\star}\| > R(t - 1, \varepsilon/2)]\right] \leq \lim\limits_{T \to \infty} \frac{1}{T}\sum_{t=1}^T 2^n\mathbb{P}\left(\|\pi_{\hat{\theta}_{(t-1)\theta_\star}} - \pi_{\theta_\star}\| > R(t - 1, \varepsilon/2)\right) \leq 2^n \lim\limits_{T \to \infty} \frac{1}{T}\sum_{t=1}^T \delta(t - 1, \varepsilon/2) = 0$, we have $\limsup\limits_{T \to \infty} \sum_{t=1}^T \frac{\mathbb{E}[N_t]}{T} = \limsup\limits_{T \to \infty} \frac{1}{T}\sum_{t=1}^T \mathbb{E}\left[N_t\mathbb{1}[\|\pi_{\hat{\theta}_{(t-1)\theta_\star}} - \pi_{\theta_\star}\| \leq R(t-1, \varepsilon/2)]\right]$. For $t \leq T$, let $N_t(\breve{\theta}_{t\theta_\star})$ denote the number of queries $A(\pi_{\breve{\theta}_{t\theta_\star}}, \varepsilon/4)$ would make if queries were answered with $s_{t\breve{\theta}_{t\theta_\star}}$ instead of s_t. On the event $\|\pi_{\hat{\theta}_{(t-1)\theta_\star}} - \pi_{\theta_\star}\| \leq R(t-1, \varepsilon/2)$, we have $\mathbb{E}\left[N_t\big|\breve{\theta}_{t\theta_\star}\right] \leq \mathbb{E}\left[N_t(\breve{\theta}_{t\theta_\star})\big|\breve{\theta}_{t\theta_\star}\right] + 2R(t - 1, \varepsilon/2) = Q(\pi_{\breve{\theta}_{t\theta_\star}}, \varepsilon/4) + 2R(t-1, \varepsilon/2) \leq Q(\pi_{\theta_\star}, \varepsilon/4) + 2R(t-1, \varepsilon/2) + 1/t$. Therefore, $\limsup\limits_{T \to \infty} \frac{1}{T}\sum_{t=1}^T \mathbb{E}\left[N_t\mathbb{1}[\|\pi_{\hat{\theta}_{(t-1)\theta_\star}} - \pi_{\theta_\star}\| \leq R(t-1, \varepsilon/2)]\right] \leq Q(\pi_{\theta_\star}, \varepsilon/4) + \limsup\limits_{T \to \infty} \frac{1}{T}\sum_{t=1}^T 2R(t - 1, \varepsilon/2) + 1/t = Q(\pi_{\theta_\star}, \varepsilon/4).$ $\qquad\square$

In many cases, this result will even continue to hold with an infinite number of goods $(n = \infty)$, since Theorem 4 has no dependence on the cardinality of $\mathcal{X}$.

6 Open Problems

There are several interesting questions that remain open at this time. Can either the lower bound or upper bound be improved in general? If, instead of d samples per task, we instead use $m \geq d$ samples, how does the minimax risk vary with m? Related to this, what is the optimal value of m to optimize the rate of convergence as a function of mT, the total number of samples? More generally, if an estimator is permitted to use N total samples, taken from however many tasks it wishes, what is the optimal rate of convergence as a function of N?

References

1. Bar-Yossef, Z.: Sampling lower bounds via information theory. In: Proceedings of the 35th Annual ACM Symposium on the Theory of Computing, pp. 335–344 (2003)
2. Baxter, J.: A Bayesian/information theoretic model of learning to learn via multiple task sampling. Machine Learning **28**, 7–39 (1997)
3. Blumer, A., Ehrenfeucht, A., Haussler, D., Warmuth, M.: Learnability and the Vapnik-Chervonenkis dimension. Journal of the Association for Computing Machinery **36**(4), 929–965 (1989)
4. Cramton, P., Shoham, Y., Steinberg, R.: Combinatorial Auctions. The MIT Press (2006)
5. Devroye, L., Lugosi, G.: Combinatorial Methods in Density Estimation. Springer, New York (2001)
6. Poland, J., Hutter, M.: MDL convergence speed for Bernoulli sequences. Statistics and Computing **16**, 161–175 (2006)
7. Schervish, M.J.: Theory of Statistics. Springer, New York (1995)
8. Vapnik, V.: Estimation of Dependencies Based on Empirical Data. Springer-Verlag, New York (1982)
9. Vapnik, V., Chervonenkis, A.: On the uniform convergence of relative frequencies of events to their probabilities. Theory of Probability and its Applications **16**, 264–280 (1971)
10. Wald, A.: Sequential tests of statistical hypotheses. The Annals of Mathematical Statistics **16**(2), 117–186 (1945)
11. Yang, L., Hanneke, S., Carbonell, J.: A theory of transfer learning with applications to active learning. Machine Learning **90**(2), 161–189 (2013)
12. Yang, L., Hanneke, S., Carbonell, J.: Bounds on the minimax rate for estimating a prior over a vc class from independent learning tasks. arXiv:1505.05231 (2015)
13. Yatracos, Y.G.: Rates of convergence of minimum distance estimators and Kolmogorov's entropy. The Annals of Statistics **13**, 768–774 (1985)
14. Zinkevich, M., Blum, A., Sandholm, T.: On polynomial-time preference elicitation with value queries. In: Proceedings of the 4th ACM Conference on Electronic Commerce, pp. 175–185 (2003)

Online learning, Stochastic Optimization

Scale-Free Algorithms for Online Linear Optimization

Francesco Orabona and Dávid Pál[✉]

Yahoo Labs, 11th Floor, 229 West 43rd Street, New York, NY 10036, USA
francesco@orabona.com, dpal@yahoo-inc.com

Abstract. We design algorithms for online linear optimization that have optimal regret and at the same time do not need to know any upper or lower bounds on the norm of the loss vectors. We achieve adaptiveness to norms of loss vectors by scale invariance, i.e., our algorithms make exactly the same decisions if the sequence of loss vectors is multiplied by any positive constant. Our algorithms work for any decision set, bounded or unbounded. For unbounded decisions sets, these are the first truly adaptive algorithms for online linear optimization.

1 Introduction

Online Linear Optimization (OLO) is a problem where an algorithm repeatedly chooses a point w_t from a convex decision set K, observes an arbitrary, or even adversarially chosen, loss vector ℓ_t and suffers loss $\langle \ell_t, w_t \rangle$. The goal of the algorithm is to have a small cumulative loss. Performance of an algorithm is evaluated by the so-called regret, which is the difference of cumulative losses of the algorithm and of the (hypothetical) strategy that would choose in every round the same best point in hindsight.

OLO is a fundamental problem in machine learning [3,18]. Many learning problems can be directly phrased as OLO, e.g., learning with expert advice [2,9,21], online combinatorial optimization [7]. Other problems can be reduced to OLO, e.g. online convex optimization [18, Chapter 2], online classification and regression [3, Chapters 11 and 12], multi-armed problems [3, Chapter 6], and batch and stochastic optimization of convex functions [12]. Hence, a result in OLO immediately implies other results in all these domains.

The adversarial choice of the loss vectors received by the algorithm is what makes the OLO problem challenging. In particular, if an OLO algorithm commits to an upper bound on the norm of future loss vectors, its regret can be made arbitrarily large through an adversarial strategy that produces loss vectors with norms that exceed the upper bound.

For this reason, most of the existing OLO algorithms receive as an input—or explicitly assume—an upper bound B on the norm of the loss vectors. The input B is often disguised as the learning rate, the regularization parameter, or the parameter of strong convexity of the regularizer. Examples of such algorithms include the Hedge algorithm or online projected gradient descent with fixed learning rate. However, these algorithms have two obvious drawbacks.

© Springer International Publishing Switzerland 2015
K. Chaudhuri et al. (Eds.): ALT 2015, LNAI 9355, pp. 287–301, 2015.
DOI: 10.1007/978-3-319-24486-0_19

Table 1. Selected results for OLO. Best results in each column are in bold.

Algorithm	Decisions Set(s)	Regularizer(s)	Scale-Free
HEDGE [5]	Probability Simplex	Negative Entropy	No
GIGA [23]	Any Bounded	$\frac{1}{2}\|w\|_2^2$	No
RDA [22]	**Any**	**Any Strongly Convex**	No
FTRL-PROXIMAL [10, 11]	Any Bounded	$\frac{1}{2}\|w\|_2^2$+ any convex func.	**Yes**
ADAGRAD MD [4]	Any Bounded	$\frac{1}{2}\|w\|_2^2$+ any convex func.	**Yes**
ADAGRAD FTRL [4]	**Any**	$\frac{1}{2}\|w\|_2^2$+ any convex func.	No
ADAHEDGE [15]	Probability Simplex	Negative Entropy	**Yes**
OPTIMISTIC MD [14]	$\sup_{u,v \in K} \mathcal{B}_f(u, v) < \infty$	**Any Strongly Convex**	**Yes**
NAG [16]	$\{u : \max_t \langle \ell_t, u \rangle \leq C\}$	$\frac{1}{2}\|w\|_2^2$	Partially[1]
SCALE INVARIANT ALGORITHMS [13]	**Any**	$\frac{1}{2}\|w\|_p^2$+ any convex func. $1 < p \leq 2$	Partially[1]
ADAFTRL [this paper]	Any Bounded	**Any Strongly Convex**	**Yes**
SOLO FTRL [this paper]	**Any**	**Any Strongly Convex**	**Yes**

First, they do not come with any regret guarantee for sequences of loss vectors with norms exceeding B. Second, on sequences where the norm of loss vectors is bounded by $b \ll B$, these algorithms fail to have an optimal regret guarantee that depends on b rather than on B.

There is a clear practical need to design algorithms that adapt automatically to norms of the loss vectors. A natural, yet overlooked, design method to achieve this type of adaptivity is by insisting to have a **scale-free** algorithm. That is, the sequence of decisions of the algorithm does not change if the sequence of loss vectors is multiplied by a positive constant.

A summary of algorithms for OLO is presented in Table 1. While the scale-free property has been looked at in the expert setting, in the general OLO setting this issue has been largely ignored. In particular, the AdaHedge [15] algorithm, for prediction with expert advice, is specifically designed to be scale-free. A notable exception in the OLO literature is the discussion of the "off-by-one" issue in [10], where it is explained that even the popular AdaGrad algorithm [4] is not completely adaptive; see also our discussion in Section 4. In particular, existing scale-free algorithms cover only some norms/regularizers and *only* bounded decision sets. The case of **unbounded decision sets**, practically the most interesting one for machine learning applications, remains completely unsolved.

Rather than trying to design strategies for a particular form of loss vectors and/or decision sets, in this paper we explicitly focus on the scale-free property. Regret of scale-free algorithms is proportional to the scale of the losses, ensuring optimal linear dependency on the maximum norm of the loss vectors.

The contribution of this paper is twofold. First, in Section 3 we show that the analysis and design of AdaHedge can be generalized to the OLO scenario

---

[1] These algorithms attempt to produce an invariant sequence of predictions $\langle w_t, \ell_t \rangle$, rather than a sequence of invariant w_t.

and to any strongly convex regularizer, in an algorithm we call ADAFTRL, providing a new and rather interesting way to adapt the learning rates to have scale-free algorithms. Second, in Section 4 we propose a new and simple algorithm, SOLO FTRL, that is scale-free and is the **first** scale-free online algorithm for unbounded sets with a non-vacuous regret bound. Both algorithms are instances of Follow The Regularized Leader (FTRL) with an adaptive learning rate. Moreover, our algorithms show that scale-free algorithms can be obtained in a "native" and simple way, i.e. without using "doubling tricks" that attempt to fix poorly designed algorithms rather than directly solving the problem.

For both algorithms, we prove that for bounded decision sets the regret after T rounds is at most $O(\sqrt{\sum_{t=1}^{T} \|\ell_t\|_*^2})$. We show that the $\sqrt{\sum_{t=1}^{T} \|\ell_t\|_*^2}$ term is necessary by proving a $\Omega(D\sqrt{\sum_{t=1}^{T} \|\ell_t\|_*^2})$ lower bound on the regret of any algorithm for OLO for any decision set with diameter D with respect to the primal norm $\|\cdot\|$. For the SOLO FTRL algorithm, we prove an $O(\max_{t=1,2,\ldots,T} \|\ell_t\|_* \sqrt{T})$ regret bound for any unbounded decision set.

Our algorithms are also **any-time**, i.e., do not need to know the number of rounds in advance and our regret bounds hold for all time steps simultaneously.

2   Notation and Preliminaries

Let V be a finite-dimensional real vector space equipped with a norm $\|\cdot\|$. We denote by V^* its dual vector space. The bi-linear map associated with (V^*, V) is denoted by $\langle\cdot,\cdot\rangle : V^* \times V \to \mathbb{R}$. The dual norm of $\|\cdot\|$ is $\|\cdot\|_*$.

In OLO, in each round $t = 1, 2, \ldots$, the algorithm chooses a point w_t in the decision set $K \subseteq V$ and then the algorithm observes a loss vector $\ell_t \in V^*$. The instantaneous loss of the algorithm in round t is $\langle\ell_t, w_t\rangle$. The cumulative loss of the algorithm after T rounds is $\sum_{t=1}^{T}\langle\ell_t, w_t\rangle$. The regret of the algorithm with respect to a point $u \in K$ is

$$\text{Regret}_T(u) = \sum_{t=1}^{T}\langle\ell_t, w_t\rangle - \sum_{t=1}^{T}\langle\ell_t, u\rangle,$$

and the regret with respect to the best point is $\text{Regret}_T = \sup_{u \in K} \text{Regret}_T(u)$. We assume that K is a non-empty closed convex subset of V. Sometimes we will assume that K is also bounded. We denote by D its diameter with respect to $\|\cdot\|$, i.e. $D = \sup_{u,v \in K} \|u - v\|$. If K is unbounded, $D = +\infty$.

Convex Analysis. The *Bregman divergence* of a convex differentiable function f is defined as $\mathcal{B}_f(u,v) = f(u) - f(v) - \langle\nabla f(v), u - v\rangle$. Note that $\mathcal{B}_f(u,v) \geq 0$ for any u, v which follows directly from the definition of convexity of f.

The *Fenchel conjugate* of a function $f : K \to \mathbb{R}$ is the function $f^* : V^* \to \mathbb{R} \cup \{+\infty\}$ defined as $f^*(\ell) = \sup_{w \in K} (\langle\ell, w\rangle - f(w))$. The Fenchel conjugate of any function is convex (since it is a supremum of affine functions) and satisfies for all $w \in K$ and all $\ell \in V^*$ the *Fenchel-Young inequality* $f(w) + f^*(\ell) \geq \langle\ell, w\rangle$.

Algorithm 1. FTRL WITH VARYING REGULARIZER

Require: Sequence of regularizers $\{R_t\}_{t=1}^{\infty}$
1: Initialize $L_0 \leftarrow 0$
2: **for** $t = 1, 2, 3, \ldots$ **do**
3:     $w_t \leftarrow \operatorname{argmin}_{w \in K} (\langle L_{t-1}, w \rangle + R_t(w))$
4:     Predict w_t
5:     Observe $\ell_t \in V^*$
6:     $L_t \leftarrow L_{t-1} + \ell_t$
7: **end for**

Monotonicity of Fenchel conjugates follows easily from the definition: If $f, g :$ $K \to \mathbb{R}$ satisfy $f(w) \leq g(w)$ for all $w \in K$ then $f^*(\ell) \geq g^*(\ell)$ for every $\ell \in V^*$.

Given $\lambda > 0$, a function $f : K \to \mathbb{R}$ is called λ-*strongly convex* with respect to a norm $\| \cdot \|$ if and only if, for all $x, y \in K$,

$$f(y) \geq f(x) + \langle \nabla f(x), y - x \rangle + \frac{\lambda}{2} \|x - y\|^2 ,$$

where $\nabla f(x)$ is any subgradient of f at point x.

The following proposition relates the range of values of a strongly convex function to the diameter of its domain. The proof can be found in Appendix A.

Proposition 1 (Diameter vs. Range). *Let $K \subseteq V$ be a non-empty bounded closed convex subset. Let $D = \sup_{u,v \in K} \|u - v\|$ be its diameter with respect to $\| \cdot \|$. Let $f : K \to \mathbb{R}$ be a non-negative lower semi-continuous function that is 1-strongly convex with respect to $\| \cdot \|$. Then, $D \leq \sqrt{8 \sup_{v \in K} f(v)}$.*

Fenchel conjugates and strongly convex functions have certain nice properties, which we list in Proposition 2 below.

Proposition 2 (Fenchel Conjugates of Strongly Convex Functions). *Let $K \subseteq V$ be a non-empty closed convex set with diameter $D := \sup_{u,v \in K} \|u - v\|$. Let $\lambda > 0$, and let $f : K \to \mathbb{R}$ be a lower semi-continuous function that is λ-strongly convex with respect to $\| \cdot \|$. The Fenchel conjugate of f satisfies:*

1. *f^* is finite everywhere and differentiable.*
2. *$\nabla f^*(\ell) = \operatorname{argmin}_{w \in K} (f(w) - \langle \ell, w \rangle)$*
3. *For any $\ell \in V^*$, $f^*(\ell) + f(\nabla f^*(\ell)) = \langle \ell, \nabla f^*(\ell) \rangle$.*
4. *f^* is $\frac{1}{\lambda}$-strongly smooth i.e. for any $x, y \in V^*$, $\mathcal{B}_{f^*}(x, y) \leq \frac{1}{2\lambda} \|x - y\|_*^2$.*
5. *f^* has $\frac{1}{\lambda}$-Lipschitz continuous gradients i.e. $\|\nabla f^*(x) - \nabla f^*(y)\| \leq \frac{1}{\lambda} \|x - y\|_*$ for any $x, y \in V^*$.*
6. *$\mathcal{B}_{f^*}(x, y) \leq D\|x - y\|_*$ for any $x, y \in V^*$.*
7. *$\|\nabla f^*(x) - \nabla f^*(y)\| \leq D$ for any $x, y \in V^*$.*
8. *For any $c > 0$, $(cf(\cdot))^* = cf^*(\cdot/c)$.*

Except for properties 6 and 7, the proofs can be found in [17]. Property 6 is proven in Appendix A. Property 7 trivially follows from property 2.

Generic FTRL with Varying Regularizer. Our scale-free online learning algorithms are versions of the FOLLOW THE REGULARIZED LEADER (FTRL) algorithm with varying regularizers, presented as Algorithm 1. The following lemma bounds its regret.

Lemma 1 (Lemma 1 in [13]). *For any sequence $\{R_t\}_{t=1}^{\infty}$ of strongly convex lower semi-continuous regularizers, regret of Algorithm 1 is upper bounded as*

$$\text{Regret}_T(u) \leq R_{T+1}(u) + R_1^*(0) + \sum_{t=1}^{T} \mathcal{B}_{R_t^*}(-L_t, -L_{t-1}) - R_t^*(-L_t) + R_{t+1}^*(-L_t) .$$

The lemma allows data dependent regularizers. That is, R_t can depend on the past loss vectors $\ell_1, \ell_2, \ldots, \ell_{t-1}$.

3 AdaFTRL

In this section we generalize the AdaHedge algorithm [15] to the OLO setting, showing that it retains its scale-free property. The analysis is very general and based on general properties of strongly convex functions, rather than specific properties of the entropic regularizer like in AdaHedge.

Assume that K is bounded and that $R(w)$ is a strongly convex lower semi-continuous function bounded from above. We instantiate Algorithm 1 with the sequence of regularizers

$$R_t(w) = \Delta_{t-1}R(w) \quad \text{where} \quad \Delta_t = \sum_{i=1}^{t} \Delta_{i-1} \mathcal{B}_{R^*}\left(-\frac{L_i}{\Delta_{i-1}}, -\frac{L_{i-1}}{\Delta_{i-1}}\right) . \tag{1}$$

The sequence $\{\Delta_t\}_{t=0}^{\infty}$ is non-negative and non-decreasing. Also, Δ_t as a function of $\{\ell_s\}_{s=1}^{t}$ is positive homogenous of degree one, making the algorithm scale-free.

If $\Delta_{i-1} = 0$, we define $\Delta_{i-1}\mathcal{B}_{R^*}(\frac{-L_i}{\Delta_{i-1}}, \frac{-L_{i-1}}{\Delta_{i-1}})$ as $\lim_{a\to0^+} a\mathcal{B}_{R^*}(\frac{-L_i}{a}, \frac{-L_{i-1}}{a})$ which always exists and is finite; see Appendix B. Similarly, when $\Delta_{t-1} = 0$, we define $w_t = \text{argmin}_{w\in K}\langle L_{t-1}, w\rangle$ where ties among minimizers are broken by taking the one with the smallest value of $R(w)$, which is unique due to strong convexity; this is the same as $w_t = \lim_{a\to0^+} \text{argmin}_{w\in K}(\langle L_{t-1}, w\rangle + aR(w))$.

Our main result is an $O(\sqrt{\sum_{t=1}^{T}\|\ell_t\|_*^2})$ upper bound on the regret of the algorithm after T rounds, without the need to know before hand an upper bound on $\|\ell_t\|_*$. We prove the theorem in Section 3.1.

Theorem 1 (Regret Bound). *Suppose $K \subseteq V$ is a non-empty bounded closed convex subset. Let $D = \sup_{x,y\in K}\|x - y\|$ be its diameter with respect to a norm $\|\cdot\|$. Suppose that the regularizer $R : K \to \mathbb{R}$ is a non-negative lower semi-continuous function that is λ-strongly convex with respect to $\|\cdot\|$ and is bounded from above. The regret of AdaFTRL satisfies*

$$\text{Regret}_T(u) \leq \sqrt{3}\max\left\{D, \frac{1}{\sqrt{2\lambda}}\right\}\sqrt{\sum_{t=1}^{T}\|\ell_t\|_*^2\,(1 + R(u))} .$$

The regret bound can be optimized by choosing the optimal multiple of the regularizer. Namely, we choose regularizer of the form $\lambda f(w)$ where $f(w)$ is 1-strongly convex and optimize over λ. The result of the optimization is the following corollary. Its proof can be found in Appendix C.

Corollary 1 (Regret Bound). *Suppose $K \subseteq V$ is a non-empty bounded closed convex subset. Suppose $f : K \to \mathbb{R}$ is a non-negative lower semi-continuous function that is 1-strongly convex with respect to $\|\cdot\|$ and is bounded from above. The regret of AdaFTRL with regularizer*

$$R(w) = \frac{f(w)}{16 \cdot \sup_{v \in K} f(v)} \qquad satisfies \qquad \text{Regret}_T \leq 5.3 \sqrt{\sup_{v \in K} f(v) \sum_{t=1}^{T} \|\ell_t\|_*^2} \, .$$

3.1   Proof of Regret Bound for AdaFTRL

Lemma 2 (Initial Regret Bound). *AdaFTRL, for any $u \in K$ and any $u \geq 0$, satisfies $\text{Regret}_T(u) \leq (1 + R(u)) \Delta_T$.*

Proof. Let $R_t(w) = \Delta_{t-1} R(w)$. Since R is non-negative, $\{R_t\}_{t=1}^{\infty}$ is non-decreasing. Hence, $R_t^*(\ell) \geq R_{t+1}^*(\ell)$ for every $\ell \in V^*$ and thus $R_t^*(-L_t) - R_{t+1}^*(-L_t) \geq 0$. So, by Lemma 1,

$$\text{Regret}_T(u) \leq R_{T+1}(u) + R_1^*(0) + \sum_{t=1}^{T} \mathcal{B}_{R_t^*}(-L_t, -L_{t-1}) \, . \tag{2}$$

Since, $\mathcal{B}_{R_t^*}(u, v) = \Delta_{t-1} \mathcal{B}_{R^*}(\frac{u}{\Delta_{t-1}}, \frac{v}{\Delta_{t-1}})$ by definition of Bregman divergence and Part 8 of Proposition 2, we have $\sum_{t=1}^{T} \mathcal{B}_{R_t^*}(-L_t, -L_{t-1}) = \Delta_T$.

Lemma 3 (Recurrence). *Let $D = \sup_{u,v \in K} \|u - v\|$ be the diameter of K. The sequence $\{\Delta_t\}_{t=1}^{\infty}$ generated by AdaFTRL satisfies for any $t \geq 1$,*

$$\Delta_t \leq \Delta_{t-1} + \min \left\{ D\|\ell_t\|_*, \frac{\|\ell_t\|_*^2}{2\lambda \Delta_{t-1}} \right\} \, .$$

Proof. The inequality results from strong convexity of $R_t(w)$ and Proposition 2.

Lemma 4 (Solution of the Recurrence). *Let D be the diameter of K. The sequence $\{\Delta_t\}_{t=0}^{\infty}$ generated by AdaFTRL satisfies for any $T \geq 0$,*

$$\Delta_T \leq \sqrt{3} \max \left\{ D, \frac{1}{\sqrt{2\lambda}} \right\} \sqrt{\sum_{t=1}^{T} \|\ell_t\|_*^2} \, .$$

Proof of the Lemma 4 is deferred to Appendix C. Theorem 1 follows from Lemmas 2 and 4.

4   SOLO FTRL

The closest algorithm to a scale-free one in the OLO literature is the AdaGrad algorithm [4]. It uses a regularizer on each coordinate of the form

$$R_t(w) = R(w) \left(\delta + \sqrt{\sum_{s=1}^{t-1} \|\ell_s\|_*^2} \right).$$

This kind of regularizer would yield a scale-free algorithm *only* for $\delta = 0$. Unfortunately, the regret bound in [4] becomes vacuous for such setting in the unbounded case. In fact, it requires δ to be greater than $\|\ell_t\|_*$ for all time steps t, requiring knowledge of the future (see Theorem 5 in [4]). In other words, despite of its name, AdaGrad is not fully adaptive to the norm of the loss vectors. Identical considerations hold for the FTRL-Proximal in [10,11]: the scale-free setting of the learning rate is valid only in the bounded case.

One simple approach would be to use a doubling trick on δ in order to estimate on the fly the maximum norm of the losses. Note that a naive strategy would still fail because the initial value of δ should be data-dependent in order to have a scale-free algorithm. Moreover, we would have to upper bound the regret in all the rounds where the norm of the current loss is bigger than the estimate. Finally, the algorithm would depend on an additional parameter, the "doubling" power. Hence, even guaranteeing a regret bound[2], such strategy would give the feeling that FTRL needs to be "fixed" in order to obtain a scale-free algorithm.

In the following, we propose a much simpler and better approach. We propose to use Algorithm 1 with the regularizer

$$R_t(w) = R(w) \sqrt{\sum_{s=1}^{t-1} \|\ell_s\|_*^2} ,$$

where $R : K \to \mathbb{R}$ is any strongly convex function. Through a refined analysis, we show that the regularizer suffices to obtain an optimal regret bound for any decision set, bounded or unbounded. We call such variant SCALE-FREE ONLINE LINEAR OPTIMIZATION FTRL algorithm (SOLO FTRL). Our main result is the following Theorem, which is proven in Section 4.1.

Theorem 2 (Regret of SOLO FTRL). *Suppose $K \subseteq V$ is a non-empty closed convex subset. Let $D = \sup_{u,v \in K} \|u - v\|$ be its diameter with respect to a norm $\| \cdot \|$. Suppose that the regularizer $R : K \to \mathbb{R}$ is a non-negative lower semi-continuous function that is λ-strongly convex with respect to $\| \cdot \|$. The regret of SOLO FTRL satisfies*

$$\mathrm{Regret}_T(u) \le \left(R(u) + \frac{2.75}{\lambda} \right) \sqrt{\sum_{t=1}^{T} \|\ell_t\|_*^2} + 3.5 \min \left\{ \frac{\sqrt{T-1}}{\lambda}, D \right\} \max_{t \le T} \|\ell_t\|_*.$$

---

[2] For lack of space, we cannot include the regret bound for the doubling trick version. It would be exactly the same as in Theorem 2, following a similar analysis, but with the additional parameter of the doubling power.

When K is bounded, we can choose the optimal multiple of the regularizer. We choose $R(w) = \lambda f(w)$ where f is a 1-strongly convex function and optimize λ. The result of the optimization is Corollary 2; the proof is in Appendix D. It is similar to Corollary 1 for AdaFTRL. The scaling however is different in the two corollaries. In Corollary 1, $\lambda \sim 1/(\sup_{v \in K} f(v))$ while in Corollary 2 we have $\lambda \sim 1/\sqrt{\sup_{v \in K} f(v)}$.

Corollary 2 (Regret Bound for Bounded Decision Sets). *Suppose $K \subseteq V$ is a non-empty bounded closed convex subset. Suppose that $f : K \to \mathbb{R}$ is a non-negative lower semi-continuous function that is 1-strongly convex with respect to $\|\cdot\|$. SOLO FTRL with regularizer*

$$R(w) = \frac{f(w)\sqrt{2.75}}{\sqrt{\sup_{v \in K} f(v)}} \quad \textit{satisfies} \quad \mathrm{Regret}_T \leq 13.3 \sqrt{\sup_{v \in K} f(v) \sum_{t=1}^{T} \|\ell_t\|_*^2} \,.$$

4.1   Proof of Regret Bound for SOLO FTRL

The proof of Theorem 2 relies on an inequality (Lemma 5). Related and weaker inequalities were proved by [1] and [6]. The main property of this inequality is that on the right-hand side C does *not* multiply the $\sqrt{\sum_{t=1}^{T} a_t^2}$ term. We will also use the well-known technical Lemma 6.

Lemma 5 (Useful Inequality). *Let $C, a_1, a_2, \ldots, a_T \geq 0$. Then,*

$$\sum_{t=1}^{T} \min\left\{ a_t^2 / \sqrt{\sum_{s=1}^{t-1} a_s^2}, \ Ca_t \right\} \leq 3.5C \max_{t=1,2,\ldots,T} a_t + 3.5 \sqrt{\sum_{t=1}^{T} a_t^2} \,.$$

Proof. Without loss of generality, we can assume that $a_t > 0$ for all t. Since otherwise we can remove all $a_t = 0$ without affecting either side of the inequality. Let $M_t = \max\{a_1, a_2, \ldots, a_t\}$ and $M_0 = 0$. We prove that for any $\alpha > 1$

$$\min\left\{ \frac{a_t^2}{\sqrt{\sum_{s=1}^{t-1} a_s^2}}, \ Ca_t \right\} \leq 2\sqrt{1+\alpha^2} \left(\sqrt{\sum_{s=1}^{t} a_s^2} - \sqrt{\sum_{s=1}^{t-1} a_s^2} \right) + \frac{C\alpha(M_t - M_{t-1})}{\alpha - 1}$$

from which the inequality follows by summing over $t = 1, 2, \ldots, T$ and choosing $\alpha = \sqrt{2}$. The inequality follows by case analysis. If $a_t^2 \leq \alpha^2 \sum_{s=1}^{t-1} a_s^2$, we have

$$\min\left\{ \frac{a_t^2}{\sqrt{\sum_{s=1}^{t-1} a_s^2}}, \ Ca_t \right\} \leq \frac{a_t^2}{\sqrt{\sum_{s=1}^{t-1} a_s^2}} = \frac{a_t^2}{\sqrt{\frac{1}{1+\alpha^2}\left(\alpha^2 \sum_{s=1}^{t-1} a_s^2 + \sum_{s=1}^{t-1} a_s^2 \right)}}$$

$$\leq \frac{a_t^2 \sqrt{1+\alpha^2}}{\sqrt{a_t^2 + \sum_{s=1}^{t-1} a_s^2}} = \frac{a_t^2 \sqrt{1+\alpha^2}}{\sqrt{\sum_{s=1}^{t} a_s^2}} \leq 2\sqrt{1+\alpha^2} \left(\sqrt{\sum_{s=1}^{t} a_s^2} - \sqrt{\sum_{s=1}^{t-1} a_s^2} \right)$$

where we have used $x^2/\sqrt{x^2+y^2} \le 2(\sqrt{x^2+y^2}-\sqrt{y^2})$ in the last step. On the other hand, if $a_t^2 > \alpha^2 \sum_{t=1}^{t-1} a_s^2$, we have

$$\min\left\{\frac{a_t^2}{\sqrt{\sum_{s=1}^{t-1} a_s^2}}, Ca_t\right\} \le Ca_t = C\frac{\alpha a_t - a_t}{\alpha - 1} \le \frac{C}{\alpha - 1}\left(\alpha a_t - \alpha\sqrt{\sum_{s=1}^{t-1} a_s^2}\right)$$

$$= \frac{C\alpha}{\alpha - 1}\left(a_t - \sqrt{\sum_{s=1}^{t-1} a_s^2}\right) \le \frac{C\alpha}{\alpha - 1}(a_t - M_{t-1}) = \frac{C\alpha}{\alpha - 1}(M_t - M_{t-1})$$

where we have used that $a_t = M_t$ and $\sqrt{\sum_{s=1}^{t-1} a_s^2} \ge M_{t-1}$.

Lemma 6 (Lemma 3.5 in [1]). *Let $a_1, a_2, \ldots, a_T$ be non-negative real numbers. If $a_1 > 0$ then,*

$$\sum_{t=1}^{T} a_t / \sqrt{\sum_{s=1}^{t} a_s} \le 2\sqrt{\sum_{t=1}^{T} a_t} \ .$$

Proof (Proof of Theorem 2). Let $\eta_t = \frac{1}{\sqrt{\sum_{s=1}^{t-1} \|\ell_s\|_*^2}}$, hence $R_t(w) = \frac{1}{\eta_t}R(w)$. We assume without loss of generality that $\|\ell_t\|_* > 0$ for all t, since otherwise we can remove all rounds t where $\ell_t = 0$ without affecting regret and the predictions of the algorithm on the remaining rounds. By Lemma 1,

$$\text{Regret}_T(u) \le \frac{1}{\eta_{T+1}}R(u) + \sum_{t=1}^{T}\left(\mathcal{B}_{R_t^*}(-L_t, -L_{t-1}) - R_t^*(-L_t) + R_{t+1}^*(-L_t)\right) \ .$$

We upper bound the terms of the sum in two different ways. First, by Proposition 2, we have

$$\mathcal{B}_{R_t^*}(-L_t, -L_{t-1}) - R_t^*(-L_t) + R_{t+1}^*(-L_t) \le \mathcal{B}_{R_t^*}(-L_t, -L_{t-1}) \le \frac{\eta_t\|\ell_t\|_*^2}{2\lambda} \ .$$

Second, we have

$$\mathcal{B}_{R_t^*}(-L_t, -L_{t-1}) - R_t^*(-L_t) + R_{t+1}^*(-L_t)$$

$$= \mathcal{B}_{R_{t+1}^*}(-L_t, -L_{t-1}) + R_{t+1}^*(-L_{t-1}) - R_t^*(-L_{t-1})$$

$$\quad + \langle \nabla R_t^*(-L_{t-1}) - \nabla R_{t+1}^*(-L_{t-1}), \ell_t\rangle$$

$$\le \frac{1}{2\lambda}\eta_{t+1}\|\ell_t\|_*^2 + \|\nabla R_t^*(-L_{t-1}) - \nabla R_{t+1}^*(-L_{t-1})\| \cdot \|\ell_t\|_*$$

$$= \frac{1}{2\lambda}\eta_{t+1}\|\ell_t\|_*^2 + \|\nabla R^*(-\eta_t L_{t-1}) - \nabla R^*(-\eta_{t+1}L_{t-1})\| \cdot \|\ell_t\|_*$$

$$\le \frac{\eta_{t+1}\|\ell_t\|_*^2}{2\lambda} + \min\left\{\frac{1}{\lambda}\|L_{t-1}\|_* (\eta_t - \eta_{t+1}), D\right\}\|\ell_t\|_* \ ,$$

where in the first inequality we have used the fact that $R_{t+1}^*(-L_{t-1}) \le R_t^*(-L_{t-1})$, Hölder's inequality, and Proposition 2. In the second inequality we have used properties 5 and 7 of Proposition 2. Using the definition of η_{t+1} we have

$$\frac{\|L_{t-1}\|_*(\eta_t - \eta_{t+1})}{\lambda} \leq \frac{\|L_{t-1}\|_*}{\lambda\sqrt{\sum_{i=1}^{t-1}\|\ell_i\|_*^2}} \leq \frac{\sum_{i=1}^{t-1}\|\ell_i\|_*}{\lambda\sqrt{\sum_{i=1}^{t-1}\|\ell_i\|_*^2}} \leq \frac{\sqrt{t-1}}{\lambda} \leq \frac{\sqrt{T-1}}{\lambda}.$$

Denoting by $H = \min\left\{\frac{\sqrt{T-1}}{\lambda}, D\right\}$ we have

$$\text{Regret}_T(u) \leq \frac{1}{\eta_{T+1}}R(u) + \sum_{t=1}^{T}\min\left\{\frac{\eta_t\|\ell_t\|_*^2}{2\lambda}, \ H\|\ell_t\|_* + \frac{\eta_{t+1}\|\ell_t\|_*^2}{2\lambda}\right\}$$

$$\leq \frac{1}{\eta_{T+1}}R(u) + \frac{1}{2\lambda}\sum_{t=1}^{T}\eta_{t+1}\|\ell_t\|_*^2 + \frac{1}{2\lambda}\sum_{t=1}^{T}\min\left\{\eta_t\|\ell_t\|_*^2, \ 2\lambda H\|\ell_t\|_*\right\}$$

$$= \frac{1}{\eta_{T+1}}R(u) + \frac{1}{2\lambda}\sum_{t=1}^{T}\frac{\|\ell_t\|_*^2}{\sqrt{\sum_{s=1}^{t}\|\ell_t\|_*^2}} + \frac{1}{2\lambda}\sum_{t=1}^{T}\min\left\{\frac{\|\ell_t\|_*^2}{\sqrt{\sum_{s=1}^{t-1}\|\ell_s\|_*^2}}, \ 2\lambda H\|\ell_t\|_*\right\}.$$

We bound each of the three terms separately. By definition of η_{T+1}, the first term is $\frac{1}{\eta_{T+1}}R(u) = R(u)\sqrt{\sum_{t=1}^{T}\|\ell_t\|_*^2}$. We upper bound the second term using Lemma 6 as

$$\frac{1}{2\lambda}\sum_{t=1}^{T}\frac{\|\ell_t\|_*^2}{\sqrt{\sum_{s=1}^{t}\|\ell_t\|_*^2}} \leq \frac{1}{\lambda}\sqrt{\sum_{t=1}^{T}\|\ell_t\|_*^2}.$$

Finally, by Lemma 5 we upper bound the third term as

$$\frac{1}{2\lambda}\sum_{t=1}^{T}\min\left\{\frac{\|\ell_t\|_*^2}{\sqrt{\sum_{s=1}^{t-1}\|\ell_s\|_*^2}}, \ 2\lambda\|\ell_t\|_*H\right\} \leq 3.5H\max_{t\leq T}\|\ell_t\|_* + \frac{1.75}{\lambda}\sqrt{\sum_{t=1}^{T}\|\ell_t\|_*^2}.$$

Putting everything together gives the stated bound.

5   Lower Bound

We show a lower bound on the worst-case regret of any algorithm for OLO. The proof is a standard probabilistic argument, which we present in Appendix E.

Theorem 3 (Lower Bound). *Let $K \subseteq V$ be any non-empty bounded closed convex subset. Let $D = \sup_{u,v \in K}\|u - v\|$ be the diameter of K. Let A be any (possibly randomized) algorithm for OLO on K. Let T be any non-negative integer and let $a_1, a_2, \ldots, a_T$ be any non-negative real numbers. There exists a sequence of vectors $\ell_1, \ell_2, \ldots, \ell_T$ in the dual vector space V^* such that $\|\ell_1\|_* = a_1, \|\ell_2\|_* = a_2, \ldots, \|\ell_T\|_* = a_T$ and the regret of algorithm A satisfies*

$$\text{Regret}_T \geq \frac{D}{\sqrt{8}}\sqrt{\sum_{t=1}^{T}\|\ell_t\|_*^2}. \tag{3}$$

The upper bounds on the regret, which we have proved for our algorithms, have the same dependency on the norms of loss vectors. However, a gap remains between the lower bound and the upper bounds.

Our upper bounds are of the form $O(\sqrt{\sup_{v \in K} f(v) \sum_{t=1}^{T} \|\ell_t\|_*^2})$ where f is any 1-strongly convex function with respect to $\|\cdot\|$. The same upper bound is also achieved by FTRL with a constant learning rate when the number of rounds T and $\sum_{t=1}^{T} \|\ell_t\|_*^2$ is known upfront [18, Chapter2]. The lower bound is $\Omega(D\sqrt{\sum_{t=1}^{T} \|\ell_t\|_*^2})$.

The gap between D and $\sqrt{\sup_{v \in K} f(v)}$ can be substantial. For example, if K is the probability simplex in $\mathbb{R}^d$ and $f(w) = \ln(d) + \sum_{i=1}^{d} w_i \ln w_i$ is the shifted negative entropy, the $\|\cdot\|_1$-diameter of K is 2, f is non-negative and 1-strongly convex w.r.t. $\|\cdot\|_1$, but $\sup_{v \in K} f(v) = \ln(d)$. On the other hand, if the norm $\|\cdot\|_2 = \sqrt{\langle\cdot,\cdot\rangle}$ arises from an inner product $\langle\cdot,\cdot\rangle$, the lower bound matches the upper bounds within a constant factor. The reason is that for any K with $\|\cdot\|_2$-diameter D, the function $f(w) = \frac{1}{2}\|w - w_0\|_2^2$, where w_0 is an arbitrary point in K, is 1-strongly convex w.r.t. $\|\cdot\|_2$ and satisfies that $\sqrt{\sup_{v \in K} f(v)} \leq D$. This leads to the following open problem (posed also in [8]):

> Given a bounded convex set K and a norm $\|\cdot\|$, construct a non-negative function $f : K \to \mathbb{R}$ that is 1-strongly convex with respect to $\|\cdot\|$ and minimizes $\sup_{v \in K} f(v)$.

As shown in [19], the existence of f with small $\sup_{v \in K} f(v)$ is equivalent to the existence of an algorithm for OLO with $\widetilde{O}(\sqrt{T \sup_{v \in K} f(v)})$ regret assuming $\|\ell_t\|_* \leq 1$. The $\widetilde{O}$ notation hides a polylogarithmic factor in T.

6   Per-Coordinate Learning

An interesting class of algorithms proposed in [11] and [4] are based on the so-called per-coordinate learning rates. As shown in [20], our algorithms, or in fact any algorithm for OLO, can be used with per-coordinate learning rates as well.

Abstractly, we assume that the decision set is a Cartesian product $K = K_1 \times K_2 \times \cdots \times K_d$ of a finite number of convex sets. On each factor K_i, $i = 1, 2, \ldots, d$, we can run any OLO algorithm separately and we denote by $\text{Regret}_T^{(i)}(u_i)$ its regret with respect to $u_i \in K_i$. The overall regret with respect to any $u = (u_1, u_2, \ldots, u_d) \in K$ can be written as

$$\text{Regret}_T(u) = \sum_{i=1}^{d} \text{Regret}_T^{(i)}(u_i) .$$

If the algorithm for each factor is scale-free, the overall algorithm is clearly scale-free as well. Using ADAFTRL or SOLO FTRL for each factor K_i, we generalize and improve existing regret bounds [4,11] for algorithms with per-coordinate learning rates.

References

1. Auer, P., Cesa-Bianchi, N., Gentile, C.: Adaptive and self-confident on-line learning algorithms. Journal of Computer and System Sciences **64**(1), 48–75 (2002)
2. Cesa-Bianchi, N., Freund, Y., Haussler, D., Helmbold, D.P., Schapire, R.E., Warmuth, M.K.: How to use expert advice. J. ACM **44**(3), 427–485 (1997)
3. Cesa-Bianchi, N., Lugosi, G.: Prediction, Learning, and Games. Cambridge University Press, Cambridge (2006)
4. Duchi, J., Hazan, E., Singer, Y.: Adaptive subgradient methods for online learning and stochastic optimization. J. Mach. Learn. Res. **12**, 2121–2159 (2011)
5. Freund, Y., Schapire, R.E.: Large margin classification using the perceptron algorithm. Mach. Learn. **37**(3), 277–296 (1999)
6. Jaksch, T., Ortner, R., Auer, P.: Near-optimal regret bounds for reinforcement learning. J. Mach. Learn. Res. **11**, 1563–1600 (2010)
7. Koolen, W.M., Warmuth, M.K., Kivinen, J.: Hedging structured concepts. In: Proc. of COLT, pp. 93–105 (2010)
8. Kwon, J., Mertikopoulos, P.: A continuous-time approach to online optimization, February 2014. arXiv:1401.6956
9. Littlestone, N., Warmuth, M.K.: The weighted majority algorithm. Information and Computation **108**(2), 212–261 (1994)
10. McMahan, H.B.: Analysis techniques for adaptive online learning (2014). arXiv:1403.3465
11. McMahan, H.B., Streeter, J.M.: Adaptive bound optimization for online convex optimization. In: Proc. of COLT, pp. 244–256 (2010)
12. Nemirovski, A., Yudin, D.B.: Problem complexity and method efficiency in optimization. Wiley (1983)
13. Orabona, F., Crammer, K., Cesa-Bianchi, N.: A generalized online mirror descent with applications to classification and regression. Mach. Learn. **99**, 411–435 (2014)
14. Rakhlin, A., Sridharan, K.: Optimization, learning, and games with predictable sequences. In: Advances in Neural Information Processing Systems, vol. 26 (2013)
15. de Rooij, S., van Erven, T., Grünwald, P.D., Koolen, W.M.: Follow the leader if you can, hedge if you must. J. Mach. Learn. Res. **15**, 1281–1316 (2014)
16. Ross, S., Mineiro, P., Langford, J.: Normalized online learning. In: Proc. of UAI (2013)
17. Shalev-Shwartz, S.: Online Learning: Theory, Algorithms, and Applications. Ph.D. thesis, Hebrew University, Jerusalem (2007)
18. Shalev-Shwartz, S.: Online learning and online convex optimization. Foundations and Trends in Machine Learning **4**(2), 107–194 (2011)
19. Srebro, N., Sridharan, K., Tewari, A.: On the universality of online mirror descent. In: Advances in Neural Information Processing Systems (2011)
20. Streeter, M., McMahan, H.B.: Less regret via online conditioning (2010). arXiv:1002.4862
21. Vovk, V.: A game of prediction with expert advice. Journal of Computer and System Sciences **56**, 153–173 (1998)
22. Xiao, L.: Dual averaging methods for regularized stochastic learning and online optimization. J. Mach. Learn. Res. **11**, 2543–2596 (2010)
23. Zinkevich, M.: Online convex programming and generalized infinitesimal gradient ascent. In: Proc. of ICML, pp. 928–936 (2003)

A   Proofs for Preliminaries

Proof (Proof of Proposition 1). Let $S = \sup_{u \in K} f(u)$ and $v^* = \operatorname{argmin}_{v \in K} f(v)$. The minimizer v^* is guaranteed to exist by lower semi-continuity of f and compactness of K. Optimality condition for v^* and 1-strong convexity of f imply that for any $u \in K$,

$$S \geq f(u) - f(v^*) \geq f(u) - f(v^*) - \langle \nabla f(v^*), u - v^* \rangle \geq \frac{1}{2} \|u - v^*\|^2 .$$

In other words, $\|u - v^*\| \leq \sqrt{2S}$. By triangle inequality,

$$D = \sup_{u,v \in K} \|u - v\| \leq \sup_{u,v \in K} (\|u - v^*\| + \|v^* - v\|) \leq 2\sqrt{2S} = \sqrt{8S} .$$

Proof (Proof of Property 6 of Proposition 2). To bound $\mathcal{B}_{f^*}(x, y)$ we add a non-negative divergence term $\mathcal{B}_{f^*}(y, x)$.

$$\mathcal{B}_{f^*}(x, y) \leq \mathcal{B}_{f^*}(x, y) + \mathcal{B}_{f^*}(y, x) = \langle x - y, \nabla f^*(x) - \nabla f^*(y) \rangle$$
$$\leq \|x - y\|_* \cdot \|\nabla f^*(x) - \nabla f^*(y)\| \leq D\|x - y\|_* ,$$

where we have used Hölder's inequality and Part 7 of the Proposition.

B   Limits

Lemma 7. *Let K be a non-empty bounded closed convex subset of a finite dimensional normed real vector space $(V, \|\cdot\|)$. Let $R : K \to \mathbb{R}$ be a strongly convex lower semi-continuous function bounded from above. Then, for any $x, y \in V^*$,*

$$\lim_{a \to 0^+} a\mathcal{B}_{R^*}(x/a, y/a) = \langle x, u - v \rangle$$

where

$$u = \lim_{a \to 0^+} \operatorname*{argmin}_{w \in K} (aR(w) - \langle x, w \rangle) \quad and \quad v = \lim_{a \to 0^+} \operatorname*{argmin}_{w \in K} (aR(w) - \langle y, w \rangle) .$$

Proof. Using Part 3 of Proposition 2 we can write the divergence

$$a\mathcal{B}_{R^*}(x/a, y/a) = aR^*(x/a) - aR^*(y/a) - \langle x - y, \nabla R^*(y/a) \rangle$$
$$= a\left[\langle x/a, \nabla R^*(x/a) \rangle - R(\nabla R^*(x/a))\right]$$
$$\quad - a\left[\langle y/a, \nabla R^*(y/a) \rangle - R(\nabla R^*(y/a))\right] - \langle x - y, \nabla R^*(y/a) \rangle$$
$$= \langle x, \nabla R^*(x/a) - \nabla R^*(y/a) \rangle - aR(\nabla R^*(x/a)) + aR(\nabla R^*(y/a)) .$$

Part 2 of Proposition 2 implies that

$$u = \lim_{a \to 0^+} \nabla R^*(x/a) = \lim_{a \to 0^+} \operatorname*{argmin}_{w \in K} (aR(w) - \langle x, w \rangle) ,$$
$$v = \lim_{a \to 0^+} \nabla R^*(y/a) = \lim_{a \to 0^+} \operatorname*{argmin}_{w \in K} (aR(w) - \langle y, w \rangle) .$$

The limits on the right exist because of compactness of K. They are simply the minimizers $u = \operatorname{argmin}_{w \in K} -\langle x, w \rangle$ and $v = \operatorname{argmin}_{w \in K} -\langle y, w \rangle$ where ties in argmin are broken according to smaller value of $R(w)$.

By assumption $R(w)$ is upper bounded. It is also lower bounded, since it is defined on a compact set and it is lower semi-continuous. Thus,

$$\lim_{a \to 0^+} a \mathcal{B}_{R^*}(x/a, y/a)$$
$$= \lim_{a \to 0^+} \langle x, \nabla R^*(x/a) - \nabla R^*(y/a) \rangle - aR(\nabla R^*(x/a)) + aR(\nabla R^*(y/a))$$
$$= \lim_{a \to 0^+} \langle x, \nabla R^*(x/a) - \nabla R^*(y/a) \rangle = \langle x, u - v \rangle .$$

C   Proofs for AdaFTRL

Proof (Proof of Corollary 1). Let $S = \sup_{v \in K} f(v)$. Theorem 1 applied to the regularizer $R(w) = \frac{c}{S} f(w)$ and Proposition 1 gives

$$\mathrm{Regret}_T \leq \sqrt{3}(1 + c) \max\left\{ \sqrt{8}, \frac{1}{\sqrt{2c}} \right\} \sqrt{S \sum_{t=1}^{T} \|\ell_t\|_*^2} .$$

It remains to find the minimum of $g(c) = \sqrt{3}(1 + c) \max\{\sqrt{8}, 1/\sqrt{2c}\}$. The function g is strictly convex on $(0, \infty)$ and has minimum at $c = 1/16$ and $g(\frac{1}{16}) = \sqrt{3}(1 + \frac{1}{16})\sqrt{8} \leq 5.3$.

Proof (Proof of Lemma 4). Let $a_t = \|\ell_t\|_* \max\{D, 1/\sqrt{2\lambda}\}$. The statement of the lemma is equivalent to $\Delta_T \leq \sqrt{3 \sum_{t=1}^{T} a_t^2}$ which we prove by induction on T. The base case $T = 0$ is trivial. For $T \geq 1$, we have

$$\Delta_T \leq \Delta_{T-1} + \min\left\{ a_T, \frac{a_T^2}{\Delta_{T-1}} \right\} \leq \sqrt{3 \sum_{t=1}^{T-1} a_t^2} + \min\left\{ a_T, \frac{a_T^2}{\sqrt{3 \sum_{t=1}^{T-1} a_t^2}} \right\}$$

where the first inequality follows from Lemma 3, and the second inequality from the induction hypothesis and the fact that $f(x) = x + \min\{a_T, a_T^2/x\}$ is an increasing function of x. It remains to prove that

$$\sqrt{3 \sum_{t=1}^{T-1} a_t^2} + \min\left\{ a_T, \frac{a_T^2}{\sqrt{3 \sum_{t=1}^{T-1} a_t^2}} \right\} \leq \sqrt{3 \sum_{t=1}^{T} a_t^2} .$$

Dividing through by a_T and making substitution $z = \frac{\sqrt{\sum_{t=1}^{T-1} a_t^2}}{a_T}$, leads to

$$z\sqrt{3} + \min\left\{ 1, \frac{1}{z\sqrt{3}} \right\} \leq \sqrt{3 + 3z^2}$$

which can be easily checked by considering separately the cases $z \in [0, \frac{1}{\sqrt{3}})$ and $z \in [\frac{1}{\sqrt{3}}, \infty)$.

D   Proofs for SOLO FTRL

Proof (Proof of Corollary 2). Let $S = \sup_{v \in K} f(v)$. Theorem 2 applied to the regularizer $R(w) = \frac{c}{\sqrt{S}} f(w)$, together with Proposition 1 and a crude bound $\max_{t=1,2,\dots,T} \|\ell_t\|_* \leq \sqrt{\sum_{t=1}^{T} \|\ell_t\|_*^2}$, give

$$\text{Regret}_T \leq \left(c + \frac{2.75}{c} + 3.5\sqrt{8} \right) \sqrt{S \sum_{t=1}^{T} \|\ell_t\|_*^2} \,.$$

We choose c by minimizing $g(c) = c + \frac{2.75}{c} + 3.5\sqrt{8}$. Clearly, $g(c)$ has minimum at $c = \sqrt{2.75}$ and has minimal value $g(\sqrt{2.75}) = 2\sqrt{2.75} + 3.5\sqrt{8} \leq 13.3$.

E   Lower Bound Proof

Proof (Proof of Theorem 3). Pick $x, y \in K$ such that $\|x - y\| = D$. This is possible since K is compact. Since $\|x - y\| = \sup\{\langle \ell, x - y \rangle : \ell \in V^*, \|\ell\|_* = 1\}$ and the set $\{\ell \in V^* : \|\ell\|_* = 1\}$ is compact, there exists $\ell \in V^*$ such that

$$\|\ell\|_* = 1 \qquad \text{and} \qquad \langle \ell, x - y \rangle = \|x - y\| = D \,.$$

Let $Z_1, Z_2, \dots, Z_T$ be i.i.d. Rademacher variables, that is, $\Pr[Z_t = +1] = \Pr[Z_t = -1] = 1/2$. Let $\ell_t = Z_t a_t \ell$. Clearly, $\|\ell_t\|_* = a_t$. The lemma will be proved if we show that (3) holds with positive probability. We show a stronger statement that the inequality holds in expectation, i.e. $\mathbf{E}[\text{Regret}_T] \geq \frac{D}{\sqrt{8}} \sqrt{\sum_{t=1}^{T} a_t^2}$. Indeed,

$$\mathbf{E}[\text{Regret}_T] \geq \mathbf{E}\left[\sum_{t=1}^{T} \langle \ell_t, w_t \rangle \right] - \mathbf{E}\left[\min_{u \in \{x,y\}} \sum_{t=1}^{T} \langle \ell_t, u \rangle \right]$$

$$= \mathbf{E}\left[\sum_{t=1}^{T} Z_t a_t \langle \ell, w_t \rangle \right] + \mathbf{E}\left[\max_{u \in \{x,y\}} \sum_{t=1}^{T} -Z_t a_t \langle \ell, u \rangle \right]$$

$$= \mathbf{E}\left[\max_{u \in \{x,y\}} \sum_{t=1}^{T} -Z_t a_t \langle \ell, u \rangle \right] = \mathbf{E}\left[\max_{u \in \{x,y\}} \sum_{t=1}^{T} Z_t a_t \langle \ell, u \rangle \right]$$

$$= \frac{1}{2} \mathbf{E}\left[\sum_{t=1}^{T} Z_t a_t \langle \ell, x + y \rangle \right] + \frac{1}{2} \mathbf{E}\left[\left| \sum_{t=1}^{T} Z_t a_t \langle \ell, x - y \rangle \right| \right]$$

$$= \frac{D}{2} \mathbf{E}\left[\left| \sum_{t=1}^{T} Z_t a_t \right| \right] \geq \frac{D}{\sqrt{8}} \sqrt{\sum_{t=1}^{T} a_t^2}$$

where we used that $\mathbf{E}[Z_t] = 0$, the fact that distributions of Z_t and $-Z_t$ are the same, the formula $\max\{a, b\} = (a + b)/2 + |a - b|/2$, and Khinchin's inequality in the last step (Lemma A.9 in [3]).

Online Learning in Markov Decision Processes with Continuous Actions

Yi-Te Hong[(✉)] and Chi-Jen Lu

Institute of Information Science, Academia Sinica, Taipei, Taiwan
{ted0504,cjlu}@iis.sinica.edu.tw

Abstract. We consider the problem of online learning in a Markov decision process (MDP) with finite states but continuous actions. This generalizes both the traditional problem of learning an MDP with finite actions and states, as well as the so-called continuum-armed bandit problem which has continuous actions but with no state involved. Based on previous works for these two problems, we propose a new algorithm for our problem, which dynamically discretizes the action spaces and learns to play strategies over these discretized actions that evolve over time. Our algorithm is able to achieve a T-step regret of about the order of $T^{\frac{d+1}{d+2}}$ with high probability, where d is a newly defined near-optimality dimension we introduce to capture the hardness of learning the MDP.

1 Introduction

We consider the problem of online learning in a Markov decision process (MDP), in which an agent needs to repeatedly play an action and receive a reward in the following way, for some number T of steps. At step t, the agent is in some state s_t and needs to choose some action a_t to play. After playing the action, the agent immediately receives a one-step reward $r(s_t, a_t)$ randomly drawn from some distribution over $[0,1]$ with unknown mean $\bar{r}(s_t, a_t)$, and then goes into some state s_{t+1} according to some unknown transition probability $p(s_{t+1}|s_t, a_t)$. A natural goal of the agent is to maximize her (or his) total reward accumulated through T steps, or equivalently to minimize her regret, defined as the gap between her total reward and that of the best fixed policy in hindsight. Following [3], we consider the so-called bandit setting, in which the only information the agent gets to know is the one-step rewards she receives as well as the states she visits. For this problem in the finite setting, with a finite state space of size S as well as a finite action space of size A, the best regret upper bound currently known is about the order of $DS\sqrt{AT}$ (ignoring a logarithmic factor), achieved by [3], where D is the diameter of the MDP. In this paper, we generalize this problem to the infinite setting, in which the action space is now allowed to be infinite.

Note that traditional reinforcement learning (RL) algorithms, in both the offline and the online models, were mostly designed for the finite setting, and there has been much theory developed to support them. However, many real-world applications of RL today have to deal with state and action spaces which

© Springer International Publishing Switzerland 2015
K. Chaudhuri et al. (Eds.): ALT 2015, LNAI 9355, pp. 302–316, 2015.
DOI: 10.1007/978-3-319-24486-0_20

are huge or even infinite (see for example the survey paper by [5]). In this regime, those traditional algorithms no longer seem appropriate and very little theory seems to be known. Aiming for better theoretical understanding, we start by considering MDP's which still have finite states but can have infinite actions that satisfy some smoothness property. More precisely, we require that with respect to the actions, the rewards as well as the transition probabilities only change smoothly. One natural approach to handle this infinite actions case is to reduce it to the finite case by discretizing the action space into a finite one. That is, one partitions the action space into finite parts, represents each part by an action in it, and feeds the resulting discretized MDP to a traditional RL algorithm. However, without prior knowledge of the MDP, such an approach may be problematic. For example, if the discretization is too coarse, one can not distinguish a large region of actions which have rather different effects. On the other hand, if the discretization is too fine, one will waste time and memory to collect and store statistics for a large number of actions with only marginal benefits. To remedy this, it would be better to use an adaptive discretization scheme, which can discretize the action space in an online and data driven way, with regions of better actions having higher resolutions.

Such issues in fact have been studied before, for the so called continuum-armed bandit problem, which has no state involved and can be seen as a special case of our MDP problem. More precisely, the continuum-armed bandit problem is a generalization of the finite-armed bandit problem, from the setting with finite arms (actions) to the setting with continuous arms. As there are infinitely many arms, it seems hopeless to achieve nontrivial regret bound unless the reward function behaves somewhat nicely and satisfies some additional properties. One property commonly assumed is that the expected reward is a "smooth" function of the arms (for example, satisfying some Lipschitz condition). With such an assumption, [4] showed that for a generic metric space, a regret about the order of $T^{\frac{d+1}{d+2}}$ can be achieved, where d is the so-called zooming dimension of the expected reward function, which captures in some way the hardness of the learning problem. Later, [2] and [6] achieved regret bounds of a similar form, but according to a different dimension d called near-optimality dimension. As it turns out, the continuum-armed bandit problem is much harder than the finite-armed one. For the latter, it is known that logarithmic regret is achievable (see e.g. [1]), while for the former, the regret bounds in typical cases grow as some positive power of T.

To solve our MDP problem with continuous actions, we would like to build on the work for the the continuum-armed bandit problem. A natural approach, which we follow, is to associate each state with a continuum-armed bandit, run for example the algorithm of [2] to adaptively discretize the action space for each state separately, and call for example the algorithm of [3] on the MDP with the discretized action sets. However, there are some issues arising from this approach. First, the MDP considered in [3] has a fixed action set while we have action sets changing over time, so it is not clear if we can still follow their regret analysis. Second, it is not clear if the discretization rule of [2] works for our MDP case involving states. This is because for a continuum-armed bandit, a region of actions is worthy of further discretization if it contains a good action, which can be determined solely by its own reward distribution. On the other hand, in an MDP, whether an action is good or not depends on others—it is good if it can be coupled with actions for other states to form a good policy. Third, the regret achieved by [3] depends on the number of actions, and it is not clear if we will produce too many actions when we keep discretizing our action space using the discretization rule of [2]. Finally, the algorithm of [3] divides time steps into episodes and its regret depends on the number of episodes. As we now have a large number of actions, produced by our discretization, which actually grows with time, it is not clear if the criterion used by [3] to stop an episode still works here, without resulting in too many episodes.

Fortunately, we are able to resolve the issues discussed above by coming up with a criterion for stopping an episode as well as a criterion for further discretization, which can work well together. As a result, we obtain a new algorithm for our problem, which can achieve a regret of about the order of

$$D^{\frac{d+3}{d+2}} S^{\frac{2}{d+2}} T^{\frac{d+1}{d+2}}$$

with high probability, where d here is a new dimension we introduce to capture the hardness of the MDP learning problem. Note that for the case with actions from an Euclidean space of dimension $\bar{d}$, a simple uniform discretization can achieve a regret bound of a similar form, but with $\bar{d}$ in place of d. However, as shown later in Example 1, our dimension d can be much smaller than $\bar{d}$, which allows our algorithm to achieve a much smaller regret. Moreover, our work can be seen as a generalization of two different lines of works: that for learning finite MDP's and that for the continuum-armed bandits. In fact, with finite actions and no need for discretization, our algorithm and regret bound become similar to those in [3], while with $S = D = 1$ for continuum-armed bandits, our algorithm and regret bound become similar to those in [2]. On the other hand, our result is incomparable to that of [7] which considered the setting of finite actions but continuous states using uniform discretization.

Our regret analysis conceptually can be seen as following that for the finite-armed bandits as well as its generalization to the continuum-armed bandits. Recall that in [1,2], the arms are divided into good and bad ones according to their reward distributions, and the regret is guaranteed by showing that each bad arm is only played for a small number of steps and the number of bad arms is small. To follow this approach in our MDP setting, by seeing policies as arms, we would like to divide the policies into good and bad ones according to their average gains. However, even though one can still show that each bad policy is only played for a small number of steps, it turns out that the number of bad policies can be too large for us in general. Instead, we take a slightly different route, by showing that when counting the total number of steps played by all the bad policies together, the number is indeed small. That is, although there are a huge number of potential bad policies, most of them actually will not be selected by our algorithm to play.

2   Preliminaries

2.1   Problem Statement

In a Markov Decision Process (MDP) $M = (\mathcal{S}, \mathcal{X}, p, r)$, with a set $\mathcal{S}$ of states and a set $\mathcal{X}$ of actions, an agent when taking an action $a \in \mathcal{X}$ in a state $s \in \mathcal{S}$ will reach a state $s' \in \mathcal{S}$ with probability $p(s'|s, a)$ and receive a reward $r(s, a) \in [0, 1]$ drawn from some distribution with mean $\bar{r}(s, a)$. Given such an MDP, a standard task in reinforcement learning is to learn a good policy $\pi : \mathcal{S} \to \mathcal{X}$ with a good average gain, defined as

$$\rho(\pi) = \lim_{T \to \infty} \frac{1}{T} \sum_{t=1}^{T} \bar{r}(s_t, \pi(s_t)),$$

where s_t denotes the state reached at time step t when following the stationary policy π.[1] As in [3], we consider the following online learning problem in the bandit setting. In the setting, the distributions of p and r of the MDP are unknown to the agent, and the only information she can receive at each step is the state she is in as well the reward she receives. The agent will play in the MDP for some number T of steps, and the goal is to maximize her cumulative reward $\sum_{t=1}^{T} r(s_t, a_t)$, where s_t is the state she is in and a_t is action she takes at step t. Note that here the agent is allowed to use different policies at different time steps. A standard way to measure the performance of an agent is to compare her cumulative reward with that of the best offline policy π^*. Here, the same offline policy π^* has to be used to select the action at each step, and its cumulative reward is $\sum_{t=1}^{T} r(s_t, \pi^*(s_t))$, which can be easily shown to approach $\rho^* T$ with high probability as T grows, where we denote $\rho^* = \rho(\pi^*)$. Following [3], we define the T-step regret as

$$\text{REGRET}_T = \rho^* T - \sum_{t=1}^{T} r(s_t, a_t),$$

which is the measure we want to minimize.

While previous works focused on the finite setting with finite $\mathcal{S}$ and $\mathcal{X}$, here we generalize the setting to allow a continuous action space $\mathcal{X}$. More precisely, we assume that $S = |\mathcal{S}| < \infty$ and that the continuous action space $\mathcal{X}$ is compact with some distance measure between any two points $a, a' \in \mathcal{X}$, denoted as $\|a - a'\|$. Following [2] and [3], we will make some assumptions, as discussed next.

2.2   Assumptions

First, we assume that the one-step reward and the state transition probability are both smooth with respect to the action.

---

[1] More precisely, we start from state s_1, and when in state s_t at step t, we take the action $\pi(s_t)$ and go to the next state s_{t+1}, which happens with probability $p(s_{t+1}|s_t, \pi(s_t))$.

Assumption 1. *For any $s, s' \in \mathcal{S}$, both $p(s'|s, a)$ and $\bar{r}(s, a)$ are 1-Lipschitz[2] continuous functions with respect to a. That is, for any $a, a' \in \mathcal{X}$,*

$$|p(s'|s, a) - p(s'|s, a')| \leq \|a - a'\| \text{ and } |\bar{r}(s, a) - \bar{r}(s, a')| \leq \|a - a'\|.$$

Next, we assume that the MDP has a finite diameter.

Assumption 2. *There is a parameter $D < \infty$ such that for any states $s, s' \in \mathcal{S}$, there is a policy $\pi : \mathcal{S} \to \mathcal{X}$ such that the expected number of steps for reaching s' from s following π is at most D.*

Finally, as in [2], we assume that the action space $\mathcal{X}$ is equipped with a tree of coverings, which is an infinite binary tree such that each node represents some subset $\mathcal{A}$ of $\mathcal{X}$ and the subsets represented by its children form a partition of $\mathcal{A}$. Note that the leaf nodes of any subtree of the tree form a partition of $\mathcal{X}$. Moreover, we assume that the tree satisfies the following condition.

Assumption 3. *There are real numbers $\nu_1 > 0$, $\nu_2 > 0$ and $0 < \beta < 1$ such that any node representing some $\mathcal{A} \subseteq \mathcal{X}$ at depth ℓ of the tree*

- *has diameter at most $\nu_1 \beta^\ell$ (i.e., $\forall a, a' \in \mathcal{A}, \|a - a'\| \leq \nu_1 \beta^\ell$), and*
- *contains a ball of radius at least $\nu_2 \beta^\ell$ (i.e., $\exists a, \{a' : \|a - a'\| \leq \nu_2 \beta^\ell\} \subseteq \mathcal{A}$).*

2.3   Near-optimal Region and Near-Optimality Dimension

As in [2], we also need the notion of near-optimal region and near-optimality dimension for our MDP setting. For an MDP with optimal gain ρ^* and for $\varepsilon \in (0, 1)$, let us define the ε-optimal region as

$$\Pi_\varepsilon = \left\{\pi \in \mathcal{X}^S : \rho^* - \rho(\pi) \leq \varepsilon\right\}.$$

One possible way to generalize the near-optimality dimension from the $\mathcal{X}$-arm case of [2] to our MDP setting is to consider the logarithm of the packing number of Π_ε by small balls of $\mathcal{X}^S$. However, the dimension defined in this way may usually depend on the number of states of the MDP, which we would like to avoid. Instead, we consider a different definition. For a state s, let $\Pi_\varepsilon^s \subseteq \mathcal{X}$ be the projection of Π_ε in the s-th dimension. Note that Π_ε^s represents a region of actions for state s which can be coupled with actions for other states to form an ε-optimal policy. Let $\mathcal{N}(\Pi_\varepsilon^s, \varepsilon')$ denote the maximal number of disjoint balls of radius ε' in $\mathcal{X}$ that can be packed inside of Π_ε^s. Then we define our near-optimality dimension as follows.

Definition 1. *For $c > 0$, let the c-optimality dimension of an MDP be the smallest real number $d \geq 0$ such that for some constants $c_0, \varepsilon_0 > 0$,*

$$\max_{s \in \mathcal{S}} \mathcal{N}(\Pi_{c\varepsilon}^s, \varepsilon) \leq c_0 c \varepsilon^{-d} \text{ for any } \varepsilon \leq \varepsilon_0.$$

---

[2] In fact, we only need $\bar{r}(s, a)$ to be "weakly-Lipschitz" as in [2]; we use the stronger assumption here just to simplify our presentation.

Let us remark that in the definition above, we allow the bound on the packing number $\mathcal{N}(\Pi^s_{c\varepsilon}, \varepsilon)$ to depend on c, which will make our regret slightly larger. We choose to use such a definition because we will use the bound with c of the order of D, the diameter of the MDP, and it would seem cheating if our definition does not include such a dependency.[3] Note that in the $\mathcal{X}$-armed bandit case, one can take $D = 1$ as there is no state involved, and in that case, our definition coincides with that of [2]. Moreover, as in [2], our dimension in the MDP case can be much smaller than the Euclidean dimension, when having an Euclidian action space. A simple example is the following.

Example 1. Consider the MDP with action space $\mathcal{X} = [0,1]^{\bar{d}}$, expected one-step rewards $\bar{r}(s,a) = 1 - \|a\|_2$, and transition probabilities $p(s'|s,a) \geq q$ for some $q > 0$, for any s, a, s'. Then for any s, $\Pi^s_{c\varepsilon} \subseteq \{a \in \mathcal{X} : \|a\|_2 \leq c\varepsilon/q\}$ and $\mathcal{N}(\Pi^s_{c\varepsilon}, \varepsilon) \leq (c\varepsilon/q)^{\bar{d}}/\varepsilon^{\bar{d}} = (c/q)^{\bar{d}}$. Thus, the c-optimality dimension is 0, even when given an arbitrarily large $\bar{d}$.

2.4 Notation

To make our presentation cleaner, we will treat the numbers ν_1, ν_2, β as constants, and focus on the parameters D, S, d, and T. We will use the standard asymptotic notations $O(\cdot)$, $\Omega(\cdot)$, $\Theta(\cdot)$, and $o(\cdot)$ to hide the factors of those constants, and we will use the notation $\tilde{O}(\cdot)$ which is similar to $O(\cdot)$ but further hides a logarithmic factor of T.

3 Our Algorithm

We would like to extend the UCRL2 algorithm of [3] from the case with finite action spaces to our general case with continuous action spaces. One naive approach is to choose a fixed finite subset of actions and use the UCRL2 algorithm on this subset, but this does not seem to yield a good enough result. Instead, we borrow the idea from [2], which allows us to partition the action spaces dynamically and use different finite action sets at different time steps. Note that for each part of a partition, we will use an arbitrary action in that part to represent it, and we will simply refer to a part of actions as an action. Following [3], we divide the time steps into episodes, and during each episode, we play a fixed policy over some finite action sets at every step, but the main difference is that we allow different finite action sets for different episodes. More precisely, during each episode, we associate each state s with a tree $\mathcal{T}^s$, which corresponds to a subtree of the tree of covering given in Assumption 3. Note that the leaves of $\mathcal{T}^s$ form a partition of $\mathcal{X}$, and we let them constitute the finite action set of state s for that episode (recall that we represent each part of actions by an action in it). The key in designing

---

[3] The linear dependency of course is not the only plausible one; we use it here simply as a demonstration. If one alternatively assumes that $\max_{s \in \mathcal{S}} \mathcal{N}(\Pi^s_{c\varepsilon}, \varepsilon) \leq c_0 c^f \varepsilon^{-d}$ for some f, then one can check that the subsequent analysis still follows similarly and the corresponding regret bound becomes $\tilde{O}(D^{\frac{f+d+2}{d+2}} S^{\frac{2}{d+2}} T^{\frac{d+1}{d+2}})$.

Algorithm 1. Modified UCRL2 for continuous action space $\mathcal{X}$

1: Input: state space $\mathcal{S}$, continuous action space $\mathcal{X}$, and diameter D of the MDP; confidence parameter δ; number of steps T.

2: Set $\mathcal{M}_1$ to contain all possible models with action sets $A_1^s = \{T^s\} = \{\mathcal{X}\}, \forall s \in S$.

3: Set $t = 0$ and observe the first state s_1.

4: **for** episodes $k = 1, 2, ...,$ **do**

5:     Set $t_k \leftarrow t + 1$.

6:     Get optimistic policy π_k and its optimistic average gain ρ_k^* from $\mathcal{M}_k$.

7:     **repeat**

8:         Set $t \leftarrow t + 1$.

9:         Play action $\pi_k(s_t)$, receive reward r_t, and observe next state s_{t+1}.

10:         Update empirical estimates according to (1) and (2).

11:     **until** $\left(\sum_{\tau=t_k}^{t} \frac{\rho_k^* - r_\tau}{t - t_k + 1} \geq c_1 D \sqrt{\frac{\log(T/\delta)}{t - t_k + 1}} \right)$

12:     **for** $s \in S$ **do**

13:         **for** a in current action set A_k^s satisfying the discretization criterion (5) **do**

14:             Split the node $a \in T^s$ to have two children a_1 and a_2.

15:             Set the confidence regions of each (s, a_i) according to (6) and (7).

16:         Set the next action set A_{k+1}^s of state s to be the leaves of T^s.

17:     Set $\mathcal{M}_{k+1}$ to contain all models falling within the confidence regions.

---

our algorithm then is to decide when to stop an episode and how to update these trees after each episode. For convenience, we will refer to a node a of a tree T_s as a node (s, a), and we will refer to the number of steps that state s is visited and action a is played as the number of visits to the node (s, a).

Our algorithm is summarized in Algorithm 1, and for simplicity of presentation, let us assume that the number of steps T is known to our algorithm.[4] Initially, we let every tree T_s contain only one node $\mathcal{X}$, and let the action set A_1^s contain only that node. For any such a node (s, a), we let the confidence region $\mathcal{R}(s, a)$ for $\bar{r}(s, a)$ be the whole interval $[0, 1]$, and let the confidence region $\mathcal{P}(s, a)$ for $p(\cdot|s, a)$ contain all possible probability distributions. When entering episode $k \geq 1$, we will first compute an optimistic policy π_k from the set $\mathcal{M}_k$ of plausible models. Then at each step of the episode, we play this policy, update the empirical estimates after receiving the reward, and then decide if we should stop the current episode and start a new one. Moreover, when the episode stops, we will decide whether or not to refine the current action sets by splitting some nodes, and then update the confidence regions as well as the set of plausible models. Next, we elaborate these points in more details.

3.1   Computing an Optimistic Policy

Following [3], we can combine all MDPs from $\mathcal{M}_k$ into a single MDP and then apply the algorithm of extended value iteration. From this, we can obtain a

---

[4] It is not hard to check that this assumption can be easily lifted by slightly modifying our algorithm, with the factor $\log(T/\delta)$ in Step 11 replaced by $\log(t/\delta)$, and by computing the failure probability in a more careful way, as done by [3].

near-optimistic policy π_k as well as a near-optimistic model M_k, and let ρ_k^* be the average gain of π_k with respect to M_k. Note that according to [3], we can have π_k arbitrarily close to an optimistic one so that the additional regret resulting from this approximation is small enough. Therefore, to make our presentation cleaner, we will assume that π_k is actually the optimistic policy, so that $\rho^* \le \rho_k^*$ whenever $M \in M_k$. Then during episode k, our algorithm will execute this fixed policy π_k for every step.

3.2   Updating the Empirical Estimates

At step t in episode k, our algorithm in state s_t will play the action $a_t = \pi_k(s_t)$ and suppose the next state is s_{t+1}. Then, after receiving a reward $r_t = r(s_t, a_t)$, our algorithm will update the empirical estimates (initially set to zero) for $\bar{r}(s_t, a_t)$ and $p(s_{t+1}|s_t, a_t)$, respectively, as

$$\gamma(s_t, a_t) = \sum_{\tau=1}^{t} \frac{r_\tau \mathbf{1}_{\{s_\tau = s_t, a_\tau = a_t\}}}{n(s_t, a_t)} \text{ and} \tag{1}$$

$$\phi(s_{t+1}|s_t, a_t) = \sum_{\tau=1}^{t} \frac{\mathbf{1}_{\{s_\tau = s_t, a_\tau = a_t, s_{\tau+1} = s_{t+1}\}}}{n(s_t, a_t)}, \tag{2}$$

where $n(s, a) = \sum_{\tau=1}^{t} \mathbf{1}_{\{s_\tau = s, a_\tau = a\}}$ is the number of steps till now that the node (s, a) has been visited.

3.3   Stopping Rule for Each Episode

For an episode k, let t_k be the first step in episode k. Our algorithm terminates episode k when the estimated optimistic gain ρ_k^* of π_k no longer appears accurate as there is a gap from it to the empirical estimate of $\rho_k = \rho(\pi_k)$, which suggests that π_k may not be a good policy as we originally believed. More precisely, we stop the k-th episode at step t once

$$\sum_{\tau=t_k}^{t} \frac{\rho_k^* - r_\tau}{t - t_k + 1} \ge c_1 \alpha_{t,k}, \text{ with } \alpha_{t,k} = D\sqrt{\frac{\log(T/\delta)}{t - t_k + 1}}, \tag{3}$$

where c_1 is a large enough fixed constant. This stopping condition provides the following guarantee on the length of the episode.

Proposition 1. *For any k with $M \in M_k$, the number of steps in episode k is $\Theta(D^2 \log(T/\delta)/(\rho_k^* - \rho_k)^2)$ with probability at least $1 - \delta/T^3$.*

To prove the proposition, we need the following, which can be derived from [3, page 1577] by observing that the average of r_t approaches ρ_k and using Azuma inequality as well as Chernoff bound.

Lemma 1. *Let G be any set of episodes and let N_G be the total number of steps over the episodes in G. Then with probability at least $1 - \delta/T^3$, we have*

$$\left| \sum_{k \in G} \sum_{t \in I_k} (\rho_k - r_t) \right| \leq O\left(D\sqrt{N_G \log(T/\delta)} \right) + D|G|.$$

We state the lemma in a form more general than needed by the proposition, since we will need this form later. With this lemma, we can now prove the proposition.

Proof (of Proposition1). Let $t = t_{k+1} - 2$ so that episode k has not stopped at step t. Then

$$\rho_k^* - \rho_k = \sum_{\tau = t_k}^{t} \frac{\rho_k^* - r_\tau}{t - t_k + 1} + \sum_{\tau = t_k}^{t} \frac{r_\tau - \rho_k}{t - t_k + 1}, \tag{4}$$

where the first sum in (4) is at most $c_1 \alpha_{t,k}$ since the stopping criterion has not yet been met. The second sum in (4) is at most $c_1 \alpha_{t,k}$ with probability $1 - \delta/T^3$ according to Lemma 1, for a large enough constant c_1. As a result, with probability $1 - \delta/T^3$, we have

$$\rho_k^* - \rho_k = \Theta(\alpha_{t,k}) = \Theta\left(D\sqrt{\frac{\log(T/\delta)}{t - t_k + 1}} \right),$$

which implies that $t_{k+1} - t_k = \Theta(t - t_k + 1) = \Theta(D^2 \log(T/\delta)/(\rho_k^* - \rho_k)^2)$. □

Let us remark that we will need both the upper bound and the lower bound given by the proposition. This is because we do not want a bad policy to be executed for too many steps, and we also do not want to switch policies too often and have too many episodes, both of which will be needed in our regret analysis later. Note that an alternative criterion following that of [3] is to stop an episode as soon as some node has been visited during the episode for a number steps enough to cut its confidence substantially (say by a half). However, this alternative may be too aggressive and it seems to result in too many episodes for us. Finally, note that the stopping rule requires the knowledge of the diameter parameter D. If this knowledge is not available, we can use an alternative rule with D omitted from (3). One can check that the same upper bound in the proposition still holds but the lower bound now becomes smaller by a D^2 factor, which results in a similar regret bound as that in the next section, but with a slightly worse dependence on D.

3.4   Splitting Nodes

Immediately after the termination of episode k, our algorithm examines whether or not it should move on to a "higher resolution" of the action spaces for episode $k + 1$. We use the following criterion: we split any leaf node (s, a) at depth ℓ if it has been visited $n(s, a)$ times with

$$c_2 \sqrt{\frac{S \log(T/\delta)}{n(s, a)}} \leq \nu_1 \beta^\ell, \tag{5}$$

where ν_1 and β are the parameters given in Assumption 3, and c_2 is a large enough fixed constant. For a large enough c_2, we immediately have the following by Assumption 1 and standard large deviation bounds, as was done by [3, proof of Lemma 17 in Appendix C.1].

Proposition 2. *When a node* (s, a) *at depth* ℓ *is split, we have with probability* $1 - \delta/T^3$ *that for any action* a' *in the node, both* $|\bar{r}(s, a') - \gamma(s, a)|$ *and* $\|p(\cdot|s, a') - \phi(\cdot|s, a)\|_1$ *are at most* $2\nu_1\beta^\ell$.

When we split a node (s, a), which corresponds to a node a in tree $\mathcal{T}^s$, we append two nodes a_1 and a_2 as children of a in $\mathcal{T}^s$, where a_1 and a_2 are the children of a in the tree of covering in Assumption 3 (recall that each node a_i is actually associated with a subset of actions and we represent the subset by an action in it). For each of these two new nodes, we set its confidence regions as

$$\mathcal{R}(s, a_i) = \left\{ \tilde{r}(s, a_i) : |\tilde{r}(s, a_i) - \gamma(s, a)| \leq 2\nu_1\beta^\ell \right\}, \tag{6}$$

$$\mathcal{P}(s, a_i) = \left\{ \tilde{p}(\cdot|s, a_i) : \|\tilde{p}(\cdot|s, a_i) - \phi(\cdot|s, a)\|_1 \leq 2\nu_1\beta^\ell \right\}. \tag{7}$$

After all the splits, we let the new action set A_{k+1}^s be the leaves of the new tree $\mathcal{T}^s$. Then we update $\mathcal{M}_{k+1}$, the set of plausible models for the next episode, to contain all the models with reward $\tilde{r}(s, a) \in \mathcal{R}(s, a)$ and state transition probability $\tilde{p}(\cdot|s, a) \in \mathcal{P}(s, a)$ for any state s and action $a \in A_{k+1}^s$. If no node has been split in episode k, we have $A_{k+1}^s = A_k^s$ for every s, and in this case, we choose not to update the confidence regions, and note that in this case we have $\mathcal{M}_{k+1} = \mathcal{M}_k$.[5] With this update rule, one can easily show the following, by induction on k.

Proposition 3. *Consider any episode* k *such that* $M \in \mathcal{M}_k$. *Then any leaf node* (s, a) *at depth* ℓ *of* $\mathcal{T}_s$ *must have both* $|\hat{r}(s, a) - \bar{r}(s, a)|$ *and* $\|\hat{p}(\cdot|s, a) - p(\cdot|s, a)\|_1$ *upper bounded by* $4\nu_1\beta^{\ell-1}$, *where* $\hat{r}(s, a)$ *and* $\hat{p}(\cdot|s, a)$ *are the reward and the probability distribution of the optimistic model* M_k *we obtain.*

4   Regret Analysis

In this section, we analyze the regret of our algorithm. Our main result is the following.

Theorem 1. *Consider any MDP with* $S \geq 2$ *states which satisfies the assumptions in Section 2 and has diameter* D *and* c-*optimality dimension* d, *with* $c = 9D(\nu_1/\nu_2)\beta^{-2}$. *Then for* $T \geq 4DS^2 \geq 16$, *the* T-*step regret achieved by our algorithm in such an MDP is with high probability at most*

$$\tilde{O}\left(D^{\frac{d+3}{d+2}} S^{\frac{2}{d+2}} T^{\frac{d+1}{d+2}} \right).$$

---

[5] This turns out to work fine for us although a slightly better result could be achieved by updating the confidence regions to refine the models as done in [3].

Let us make some remarks about the theorem. First, the assumption $T \geq 4DS^2 \geq 16$ can be seen as without loss of generality, because for the problem, one typically assumes that T is large enough and dominates other parameters, and the focus is usually on minimizing the regret in terms of T. Next, although a uniform discretization of $\mathcal{X}$, when $\mathcal{X} \subseteq \mathbb{R}^{\bar{d}}$, can achieve a regret bound of a similar form, but with $\bar{d}$ in place of d, we know from Example 1 that there are cases with d much smaller than $\bar{d}$, in which cases Theorem 1 gives a much smaller regret bound. Finally, observe the curious property of our regret bound that as d grows, the factors involving D and S actually becomes smaller, although the overall regret bound still increases. We have not yet fully understand this phenomenon and we do not know if this is the nature of the problem or a feature of our algorithm. One attempt to understand this is that even though the number of actions involved in ε-optimal policies grows in the order of ε^{-d}, which increases as d increases, by choosing ε properly, in the form of $D^{c_1} S^{c_2} / T^{c_3}$ for some $c_1, c_2, c_3 > 0$, the factors involving D and S indeed decrease as d increases. This can be seen when we minimize the bound (10) in our regret analysis later.

Before proceeding to prove the theorem, let us introduce some notation. Let m denote the number of episodes taken by our algorithm, and let $[m] = \{1, \ldots, m\}$ be the set of episodes. For an episode $k \in [m]$, let $\rho_k = \rho(\pi_k)$, let $I_k = \{t_k, \ldots, t_{k+1} - 1\}$ be the set of steps in episode k, and for $t \in I_k$, let $r_t = r(s_t, \pi_k(s_t))$ be the reward received at step t. Recall that $\delta \in (0,1)$ is the confidence parameter, which bounds the failure probability of our algorithm.

Our proof of the theorem conceptually follows a standard regret analysis of the well-known UCB algorithm for the multi-armed bandit (MAB) problem. That is, by seeing a policy in MDP as an arm in MAB, we will separate policies into good and bad ones and show that bad policies will not be executed for too many steps. Before we can do that, some preprocessing work is needed.

First, following [3], we can focus on the set

$$G = \{k \in [m] : M \in \mathcal{M}_k\}$$

of good episodes, as the regret can be expressed as

$$\sum_{k \in [m]} \sum_{t \in I_k} (\rho^* - r_t) = \sum_{k \in G} \sum_{t \in I_k} (\rho^* - r_t) + \sum_{k \in [m] \setminus G} \sum_{t \in I_k} (\rho^* - r_t), \qquad (8)$$

where the last sum can be bounded by the following.

Lemma 2. *With probability* $1 - \delta/4$, *we have*

$$\sum_{k \in [m] \setminus G} \sum_{t \in I_k} (\rho^* - r_t) \leq \sqrt{T}.$$

We omit the proof as it can be easily derived from [3, Section4.2], which is based on showing that each $\mathcal{M}_k$ fails to include M with a small probability, according to Proposition 2.

The first sum in the righthand side of (8) can be further decomposed as

$$\sum_{k \in G} \sum_{t \in I_k} (\rho^* - \rho_k) + \sum_{k \in G} \sum_{t \in I_k} (\rho_k - r_t), \tag{9}$$

where the last sum in (9) can be upper bounded by $O(D\sqrt{T \log(T/\delta)}) + Dm$ with probability $1 - \delta/T^3 \geq 1 - \delta/4$, using Lemma 1 with the trivial upper bounds $N_G \leq T$ and $|G| \leq m$.

Now let us turn to bound the first sum in (9), which is the critical part in our regret analysis. As discussed before, we would like to divide the policies into good and bad ones, and show that bad policies will not be played for too many steps. For that, let us partition the episodes into buckets $B_1, B_2, \ldots$, with the b-th bucket defined as

$$B_b = \left\{ k \in G : 9D\nu_1 \beta^{b-1} < \rho_k^* - \rho_k \leq 9D\nu_1 \beta^{b-2} \right\},$$

where recall that ρ_k^* is the optimistic average gain calculated for the policy π_k with respect to the optimistic model M_k. Note that for any $k \in B_b$, we have $M \in \mathcal{M}_k$, which implies that

$$\rho^* - \rho_k \leq \rho_k^* - \rho_k \leq 9D\nu_1 \beta^{b-2}.$$

We see policies π_k with $k \in B_b$ as good policies, for b larger than some threshold H to be determined later, since their average gain is close enough to the optimal ρ^*, and we see the remaining ones as bad policies. Then our key lemma is the following, to be proved in Subsection 4.1, which allows us to bound the number of steps taken by bad policies.

Lemma 3. *For any b, the total number of steps taken by all episodes in B_b is at most $\tilde{O}(DS^2 \beta^{-b(d+2)}) + O(|B_b| \beta^{-b})$ with probability $1 - 2\delta/T^2$.*

According to this lemma together with a union bound, we know that the contribution of bad policies to the first sum in (9) is

$$\sum_{b \leq H} \sum_{k \in B_b} \sum_{t \in I_k} (\rho^* - \rho_k) \leq \sum_{b \leq H} \left(\tilde{O}\left(DS^2 \beta^{-b(d+2)} \right) + O\left(|B_b| \beta^{-b} \right) \right) \cdot 9D\nu_1 \beta^{b-2}$$

$$\leq \tilde{O}\left(D^2 S^2 \beta^{-H(d+1)} \right) + O(Dm)$$

with probability $1 - 2\delta/T \geq 1 - \delta/4$. On the other hand, the contribution from good policies can be easily bounded as

$$\sum_{b > H} \sum_{k \in B_b} \sum_{t \in I_k} (\rho^* - \rho_k) \leq T \cdot 9D\nu_1 \beta^{H-1} \leq O\left(DT\beta^H \right).$$

Consequently, the first sum in (9) can be bounded as

$$\sum_{k \in G} \sum_{t \in I_k} (\rho^* - \rho_k) \leq \tilde{O}\left(D^2 S^2 \beta^{-H(d+1)} \right) + O(Dm) + O\left(DT\beta^H \right) \tag{10}$$

with probability $1 - \delta/4$. By choosing H such that $\beta^H = (DS^2/T)^{\frac{1}{d+2}}$ for the bound above, and then by combining all the bounds obtained so far, we can conclude that with probability $1 - 3\delta/4$, the regret of our algorithm is at most

$$\tilde{O}\left(D^{\frac{d+3}{d+2}} S^{\frac{2}{d+2}} T^{\frac{d+1}{d+2}}\right) + O(Dm).$$

Finally, Theorem 1 follows from the following, which provides a bound for m, the number of episodes. We prove the lemma in Subsection 4.2.

Lemma 4. *With probability $1 - \delta/4$, we have*

$$m \leq \tilde{O}\left((DS^2)^{\frac{1}{d+2}} T^{\frac{d+1}{d+2}}\right).$$

4.1 Proof of Lemma 3

Fix any b, and let N denote the total number of steps taken by all the episodes in B_b, which is the number we want to bound. Our approach is the following. First, according to the definition of B_b, we have

$$N \cdot 9D\nu_1 \beta^{b-1} \leq \sum_{k \in B_b} \sum_{t \in I_k} (\rho_k^* - \rho_k). \tag{11}$$

On the other hand, we also have

$$\sum_{k \in B_b} \sum_{t \in I_k} (\rho_k^* - \rho_k) = \sum_{k \in B_b} \sum_{t \in I_k} (\rho_k^* - r_t) + \sum_{k \in B_b} \sum_{t \in I_k} (r_t - \rho_k)$$

$$\leq \sum_{k \in B_b} \sum_{t \in I_k} (\rho_k^* - r_t) + O\left(D\sqrt{N \log(T/\delta)}\right) + D|B_b| \tag{12}$$

with probability $1 - \delta/T^3$ by Lemma 1. The key is to show that the sum in (12) is also small with high probability, so that by combining it with the bound in (11), we can conclude that N is small with high probability.

To bound the sum in (12), we rely on the following, which can be derived easily from [3, Sections 4.3.1 & 4.3.2], so we omit the proof.

Lemma 5. *Let $v(s,a)$ denote the number of steps over episodes in B_b that visit the node (s,a). Let $\sigma(s,a)$ be an upper bound for both $|\hat{r}(s,a) - \bar{r}(s,a)|$ and $\|\hat{p}(\cdot|s,a) - p(\cdot|s,a)\|_1$, where $\hat{r}(s,a)$ and $\hat{p}(\cdot|s,a)$ are the reward and the probability of the optimistic model M_k we obtain. Then with probability $1 - \delta/T^3$,*

$$\sum_{k \in B_b} \sum_{t \in I_k} (\rho_k^* - r_t) \leq 2D \sum_{s,a} v(s,a)\sigma(s,a) + O\left(D\sqrt{N \log(T/\delta)}\right) + D|B_b|.$$

Note that in our case, we have by Proposition 3 that $\sigma(s,a) \leq 4\nu_1 \beta^{\ell-1}$ for any node (s,a) at depth ℓ, which is small when ℓ is large. To bound the sum $\sum_{s,a} v(s,a)\sigma(s,a)$ in the lemma, let us decompose it into two parts as $V_1 + V_2$, where V_1 is the sum over (s,a) in depth at least b and V_2 is the sum over the rest. It is easy to see that $V_1 \leq N \cdot 4\nu_1 \beta^{b-1}$. To bound V_2, we rely on the following two propositions, which we will prove later.

Proposition 4. *For any $\ell \leq b$, any node in depth ℓ with probability $1 - \delta/T^3$ is visited for at most $\tilde{O}(S\beta^{-2b})$ steps during all the episodes in B_b.*

Proposition 5. *For any $\ell \leq b$, the number of different nodes in depth ℓ visited during all the episodes in B_b is at most $\tilde{O}(DS\beta^{-\ell d})$.*

According to these two propositions, together with a union bound, we have with probability $1 - \delta/T^2$ that

$$V_2 \leq \sum_{\ell \leq b} 4\nu_1 \beta^{\ell-1} \cdot \tilde{O}\left(S\beta^{-2b}\right) \cdot \tilde{O}\left(DS\beta^{-\ell d}\right) \leq \tilde{O}\left(DS^2 \beta^{-b(d+1)}\right).$$

Substituting the bounds for V_1 and V_2 into Lemma 5, we obtain a bound for the sum in (12). Thus, with probability $1 - \delta/T^2 - 2\delta/T^3 \geq 1 - 2\delta/T^2$, we have

$$\sum_{k \in B_b} \sum_{t \in I_k} (\rho_k^* - \rho_k)$$

$$\leq 2DV_1 + 2DV_2 + O\left(D\sqrt{N\log(T/\delta)}\right) + 2D|B_b|$$

$$\leq N \cdot 8D\nu_1 \beta^{b-1} + \tilde{O}\left(D^2 S^2 \beta^{-b(d+1)}\right) + O\left(D\sqrt{N\log(T/\delta)}\right) + 2D|B_b|.$$

Finally, by combing this upper bound with the lower bound in (11), we have with probability $1 - 2\delta/T^2$ that

$$N \cdot 9D\nu_1 \beta^{b-1}$$

$$\leq N \cdot 8D\nu_1 \beta^{b-1} + \tilde{O}\left(D^2 S^2 \beta^{-b(d+1)}\right) + O\left(D\sqrt{N\log(N/\delta)}\right) + 2D|B_b|,$$

which implies that $N \leq \tilde{O}\left(DS^2 \beta^{-b(d+2)}\right) + O\left(|B_b|\beta^{-b}\right)$. To complete the proof of Lemma 3, it remains to prove the two propositions, which we do next.

Proof (of Proposition 4). To count the number of steps visiting such a node, let us divide them into two parts: those steps in the last episode visiting the node and those steps before that episode. The number of steps in the first part is at most the length of that episode, which according to Proposition 1 is at most $\tilde{O}(D^2/(D^2\beta^{2b})) \leq \tilde{O}(\beta^{-2b})$ with probability $1 - \delta/T^3$, using the fact that $\rho_k^* - \rho_k \geq \Omega(D\beta^{b-1})$ for any episode $k \in B_b$. To bound the number of steps in the second part, note that as the node has not been split, this number must be at most $\tilde{O}(S\beta^{-2\ell}) \leq \tilde{O}(S\beta^{-2b})$ according to the algorithm's discretization rule in (5). Combining these two bounds together, we have the proposition.   □

Proof (of Proposition 5). When any such a node is visited during an episode $k \in B_b$, it is part of the policy π_k with $\rho^* - \rho(\pi_k) \leq \rho_k^* - \rho_k \leq 9D\nu_1 \beta^{b-2} \leq c\epsilon$, with $c = 9D(\nu_1/\nu_2)\beta^{-2}$ and $\epsilon = \nu_2\beta^{\ell}$. That is, such a policy π_k is in the $c\epsilon$-optimal region. As all such nodes are disjoint, each of which contains a ball of radius $\nu_2\beta^{\ell} = \epsilon$ by Assumption 3, the number of such nodes is at most $\sum_{s \in S} \mathcal{N}(\Pi_{ce}^s, \epsilon)$. According to Definition 1, each $\mathcal{N}(\Pi_{ce}^s, \epsilon)$ is at most $O(c\epsilon^{-d}) \leq \tilde{O}(D\beta^{-\ell d})$, if $\epsilon \leq \epsilon_0$, for the constant ϵ_0 associated with d, and at most $\mathcal{N}(\mathcal{X}, \epsilon_0) \leq \tilde{O}(1)$ otherwise. Thus, the the number of such nodes is at most $\tilde{O}(DS\beta^{-\ell d})$.   □

4.2   Proof of Lemma 4

Recall from Lemma 3 that for any b, the total number of steps over episodes in B_b is at most $\tilde{O}(DS^2\beta^{-b(d+2)}) + O(|B_b|\beta^{-b})$ with probability $1 - 2\delta/T^2$. On the other hand, recall from Proposition 1 that the number of steps in any such an episode is at least $\Omega(\beta^{-2b}\log(T/\delta))$ with probability $1 - \delta/T^3$. Combining these two bounds together, we have with probability $1 - 3\delta/T^2$ that

$$|B_b| \leq \tilde{O}\left(DS^2\beta^{-bd}\right) + o(|B_b|) \leq \tilde{O}\left(DS^2\beta^{-bd}\right).$$

The above bound is good for a small b but may not be so when b is large. For all the B_b's with $b > H$, for some threshold H to be determined in a moment, we can bound the number of all such episodes by

$$\sum_{b>H} |B_b| \leq \frac{T}{\Omega\left(\beta^{-2H}\log(T/\delta)\right)} \leq O\left(T\beta^{2H}\right)$$

with probability $1 - \delta/T^2$, because the total number of steps in these episodes is at most T while each such episode takes at least $\Omega(\beta^{-2H}\log(T/\delta))$ steps with probability $1 - \delta/T^3$. As a result, the total number of episodes is

$$m \leq O\left(T\beta^{2H}\right) + \sum_{b\leq H} \tilde{O}\left(DS^2\beta^{-bd}\right) \leq O\left(T\beta^{2H}\right) + \tilde{O}\left(DS^2\beta^{-Hd}\right)$$

with probability $1 - \delta/T^2 - 3\delta/T \geq 1 - \delta/4$ by recalling that $T \geq 4DS^2 \geq 16$. Then by choosing H such that $\beta^H = (DS^2/T)^{\frac{1}{d+2}}$, we have with probability $1 - \delta/4$ that

$$m \leq \tilde{O}\left((DS^2)^{\frac{2}{d+2}}T^{\frac{d}{d+2}}\right) \leq \tilde{O}\left((DS^2)^{\frac{1}{d+2}}T^{\frac{d+1}{d+2}}\right).$$

References

1. Auer, P., Cesa-Bianchi, N., Schapire, R.: Finite-Time Analysis of the Multi-Armed Bandit Problem. SIAM Journal on Computing **32**(1), 48–77 (2002)
2. Bubeck, S., Munos, R., Stoltz, G., Szepesvári, C.: $\mathcal{X}$-Armed Bandits. Journal of Machine Learning Research **12** (2011)
3. Jaksch, T., Ortner, R., Auer, P.: Near-Optimal Regret Bounds for Reinforcement Learning. Journal of Machine Learning Research **11** (2010)
4. Kleinberg, R., Slivkins, A., Upfal, E.: Multi-armed bandits in metric spaces. In: ACM Symp. on Theory of Computing (STOC) (2008)
5. Kober, J., Bagnell, J.A., Peters J.: Reinforcement Learning in Robotics: A Survey. International Journal of Robotic Research **32** (2013)
6. Munos, R.: From Bandits to Monte-Carlo Tree Search: The Optimistic Principle Applied to Optimization and Planning. Foundations and Trends in Machine Learning **7**(1), 1–130 (2014)
7. Ortner, R., Ryabko, D.: Online Regret Bounds for Undiscounted Continuous Reinforcement Learning. Advances in Neural Information Processing Systems **25**, 1772–1780 (2012)

Adaptive Sampling for Incremental Optimization Using Stochastic Gradient Descent

Guillaume Papa, Pascal Bianchi$^{(\boxtimes)}$, and Stephan Clémençon

Institut Mines Telecom - Telecom ParisTech,
LTCI Telecom ParisTech and CNRS No. 5141, Paris, France
{guillaume.papa,pascal.bianchi,stephan.clemencon}@telecom-paristech.fr

Abstract. A wide collection of popular statistical learning methods, ranging from K-means to Support Vector Machines through Neural Networks, can be formulated as a stochastic gradient descent (SGD) algorithm in a specific setup. In practice, the main limitation of this incremental optimization technique is due to the stochastic noise induced by the choice at random of the data involved in the gradient estimate computation at each iteration. It is the purpose of this paper to introduce a novel implementation of the SGD algorithm, where the data subset used at a given step is not picked uniformly at random among all possible subsets but drawn from a specific adaptive sampling scheme, depending on the past iterations in a Markovian manner, in order to refine the current statistical estimation of the gradient. Beyond an algorithmic description of the approach we propose, rate bounds are established and illustrative numerical results are displayed in order to provide theoretical and empirical evidence of its statistical performance, compared to more "naive" SGD implementations. Computational issues are also discussed at length, revealing the practical advantages of the method promoted.

1 Introduction

In this paper, we consider the generic minimization problem

$$\min_{\theta \in \Theta} f(\theta) = \min_{\theta \in \Theta} \frac{1}{N} \sum_{i=1}^{N} f_i(\theta), \tag{1}$$

where Θ is a Euclidean space, typically $\mathbb{R}^d$ with $d \geq 1$, and f_1, ..., f_N form a collection of real-valued convex continuously differentiable functions on Θ. Such an optimization problem typically arises in a broad variety of statistical learning problems, in particular supervised tasks, where the goal pursued is to learn a predictive model, fully determined by a parameter θ, in order to predict a random variable Y (the response/output) from an input observation X taking its values in a feature space $\mathcal{X}$. The performance of the predictive function defined by θ is measured by the expectation $R(\theta) = \mathbb{E}[\ell(\theta; (X, Y))]$, referred to as the *risk*, where ℓ is a *loss function* assumed convex w.r.t. θ. The distribution (X, Y) being unknown in practice, the risk functional is replaced by its statistical counterpart, the *empirical risk* namely, given by

© Springer International Publishing Switzerland 2015
K. Chaudhuri et al. (Eds.): ALT 2015, LNAI 9355, pp. 317–331, 2015.
DOI: 10.1007/978-3-319-24486-0_21

$$\widehat{R}_N(\theta) = \frac{1}{N} \sum_{i=1}^{N} \ell(\theta; (X_i, Y_i)), \tag{2}$$

based on $N \geq 1$ supposedly available independent training examples (X_1, Y_1), ..., (X_N, Y_N), copies of the random pair (X, Y). The minimization problem (2) can be solved incrementally, by means of variants of the *stochastic approximation* method originally introduced in the seminal contribution of [7]. This consists in computing successive estimates of a minimizer of (2) using the recursive equation

$$\theta_{t+1} = \theta_t - \gamma_t \widehat{r}_t(\theta_t) \tag{3}$$

from a preliminarily picked initial value $\theta_0 \in \Theta$, where $\widehat{r}_t$ denotes an estimate of the gradient $\nabla_\theta \widehat{R}_N$ and γ_t is the *learning rate* or *stepsize*. In contrast to the *batch approach*, where all the data are used to estimate the gradient at each iteration (*i.e.* $\widehat{r}_t(\theta) = \nabla_\theta \widehat{R}_N(\theta)$ for all $t \geq 0$ and $\theta \in \Theta$), subsets of the data sample only are involved in the gradient estimation steps of *sampled incremental algorithms*, with the aim to reduce computational cost when N is large. In the most commonly used implementation of the *stochastic gradient descent* (SGD) algorithm, the gradient estimate is computed from a subset of reduced size $S \leq N$ uniformly drawn without replacement among all possible subsets of the dataset of size S at each step $t \geq 0$.

It is the major goal of the present paper to introduce a specific variant of the SGD algorithm with an *adaptive* sampling scheme, in the sense that it may possibly be different from *sampling without replacement* (SWOR) and vary with t, depending on the past iterations. Rate bounds and limit theorems guaranteeing the theoretical validity of the methodology we propose are established at length. In addition, the Markovian dynamics governing the evolution of the instrumental sampling distribution is shown to offer crucial advantages regarding computational efficiency. Finally, very encouraging experimental results are displayed, supporting the relevance of our method, in comparison to the usual mini-batch SGD implementation or alternative SGD techniques standing for natural competitors.

The paper is structured as follows. A short review of the SGD methods documented in the literature, those based on non SWOR sampling schemes in particular, can be found in section 2. A description of the specific variant we propose in this paper is given in section 3, together with a detailed discussion about the computational cost inherent to its implementation. The analysis assessing the validity of the estimate output by the algorithm proposed is carried out in section 4, whereas illustrative experiments are presented in section 5. Finally, some concluding remarks are collected in section 6.

2   Non Uniform Sampling (NUS) - State of the Art

We start off with a brief review of sampled incremental optimization algorithms, whose archetype is the celebrated SGD algorithm ([7]). Although relevant references in this area are much too numerous to be listed in an exhaustive manner, we point out that significant advances have been recently made in the design of efficient incremental methods, see for instance [10],[13],[16],[14] or [2] that achieve better performances than the traditional SGD method (for instance, by having the variance of the estimator going to 0 as in [13], [2]). In order to study adaptative sampling scheme, we only considered the classical framework of SGD and did not compare to theses methods. In the original implementation of the SGD algorithms, a single observation (*i.e.* $S = 1$), indexed by i_{t+1} say, is chosen uniformly at random in $\{1, \ldots, N\}$ at each iteration $t + 1$ to form the gradient estimate $\nabla f_{i_{t+1}}(\theta_t)$, its convergence following then from basic stochastic approximation theory. More recently, several papers have shown the possible gain from the use of non-uniform sampling, that is, choosing i_{t+1} according to a non-trivial distribution p well-suited to the specific optimization problem considered: this approach boils down to finding an optimal distribution, in the sense that it minimizes an upper bound on the convergence rate of the estimator [18],[4],[15]. As far as NUS is concerned, it is natural to ask what relevant choice of the probability distribution must be chosen in order to achieve the smallest expected risk: for instance, the sampling scheme may depend on the Lipschitz constant of the gradient as proposed in [4] or on upperbounds for the norm of the gradient, see [18]. Despite these recent contributions, some questions remain open. 1) In general, usual analyses of NUS algorithms do not fully grasp the impact of the sampling scheme on the performance. For instance, [2] establish performance bounds which prove the convergence of NUS scheme for SAGA. Although the attractivity of NUS is demonstrated in the simulations, the bound itself does not fully reveal the performance gain w.r.t. uniform sampling. 2) As opposed to uniform sampling, the choice of an index i_{t+1} according to a non-trivial distribution on $\{1,\ldots,N\}$ is obviously more demanding in terms of computational time. The question of an efficient sampling implementation remains posed. In particular, it is important to quantify the increased complexity caused by NUS. 3) Proposed rules for choosing the sampling distribution p depend on global properties of the functions f_i (namely the Lipschitz constants L_i). They do not build upon the amount of information gathered on the optimization problem as the algorithm proceeds. An alternative is to use adaptive sampling, updating the sampling distribution $p = p_t$ at each iteration t in a Markovian fashion, as in the algorithm described below.

As shall be seen in the subsequent analysis, the NUS approach proposed in this paper has theoretical and practical advantages regarding all these aspects. Whereas rate bound analysis by means of standard tools is poorly informative in general in the present setting, the asymptotic viewpoint developed in this paper clearly highlights the benefit to using the specific NUS method we promote.

3   Adaptive Sampling SGD (AS-SGD)

We now turn to the description of the variant of SGD method considered in this paper. The main novelty arises from the use of a specific instrumental sampling distribution evolving at each iteration. We also provide some insight into the gain one may expect from such a method and discuss the computational issues related to its implementation. Here and throughout, if $I = (i_1, \ldots, i_S)$ is a S-uplet on $\{1, \ldots, N\}$ and h is a function on $\{1, \ldots, N\}$, we use the (slightly abusive) notation $\sum_{i \in I} h(i)$ to represent the sum $\sum_{n=1}^{S} h(i_n)$.

3.1 The Algorithmic Principle

Let $S \leq N$ be fixed. At each iteration $t \geq 1$, the generic AS-SGD is implemented in three steps as follows:

1. Compute the instrumental probability distribution p_t on $\{1, \ldots, N\}$ from the information available at iteration t.
2. Form a random sequence of S indexes $I_{t+1} = (i_{t+1}^{(1)}, \ldots, i_{t+1}^{(S)})$ by sampling independently S times according to distribution p_t.
3. Update the estimate using the equation

$$\theta_{t+1} = \Pi_{\mathcal{K}}\left(\theta_t - \frac{\gamma_t}{S} \sum_{i \in I_{t+1}} \frac{\nabla f_i(\theta_t)}{N p_{t,i}}\right). \tag{4}$$

where $\mathcal{K}$ is a compact convex set. Let $\mathcal{F}_t = \sigma(I_1, \ldots, I_t)$ be the σ-algebra generated by the past variables up to time t. Conditioned upon $\mathcal{F}_t$, we have: $i_t^{(1)}, \ldots, i_t^{(S)} \overset{i.i.d.}{\sim} p_t$. Set $p_{t,j} = \mathbb{P}(i_{t+1}^{(1)} = j | \mathcal{F}_t)$ for $j = 1, \ldots, N$. Equipped with these notations, observe that the original SGD algorithm corresponds to the case where $S = 1$ and $p_{t,j} \equiv 1/N$. When $S > 1$, notice that, in contrast to the mini-batch SGD (based on the SWOR scheme), one samples with replacement here and, due to the fact that the $p_{t,j}$'s are not equal in general, it is necessary to normalize the individual gradients by $N p_{t,i}$ in order to guarantee the unbiasedness condition of the increment, which is classically required to ensure proper convergence of the algorithm, see [7], [4], [18], [15]:

$$\mathbb{E}\left(\frac{1}{S} \sum_{i \in I_{t+1}} \frac{\nabla f_i(\theta_t)}{N p_{t,i}} \,\middle|\, \mathcal{F}_t\right) = \sum_{j=1}^{N} p_{t,j} \frac{\nabla f_j(\theta_t)}{N p_{t,j}} = \nabla f(\theta_t). \tag{5}$$

3.2 Ideal Sampling Distribution

In order to provide some insight into the specific dynamics we propose to build the successive sampling distributions p_t, we first evaluate a bound on the amount of decrease of the functional. We assume that hypothesis below is fulfilled.

Assumption 1. *For all* $i \in \{1, \ldots, N\}$, *the function* f_i *is convex, continuously differentiable and its gradient* ∇f_i *is* L_i-*Lipschitz continuous with* $L_i < +\infty$.

Let θ_t be the sequence defined by (4) and $\mathcal{K} = \Theta$. By virtue of Assumption 1 and Theorem 2.1.5 in [11], we have $f(\theta_{t+1}) \leq f(\theta_t) + \langle \nabla f(\theta_t), \theta_{t+1} - \theta_t \rangle + \frac{\bar{L}}{2} \|\theta_{t+1} - \theta_t\|^2$ where $\bar{L} = \frac{1}{N} \sum_i L_i$. Taking the conditional expectation, we obtain that

$$\mathbb{E}(f(\theta_{t+1}) | \mathcal{F}_t) \leq f(\theta_t) - \gamma_t \|\nabla f(\theta_t)\|^2 + \frac{\bar{L} \gamma_t^2 \Delta(\theta_t, p_t)}{2 N^2},$$

where

$$\Delta(\theta_t, p_t) = \mathbb{E}\left(\left\|\frac{1}{S}\sum_{i \in I_{t+1}}\frac{\nabla f_i(\theta_t)}{p_{t,i}}\right\|^2 \middle| \mathcal{F}_t\right)$$

$$= \frac{1}{S}\sum_{i=1}^{N}\frac{\|\nabla f_i(\theta_t)\|^2}{p_{t,i}} + \frac{N^2(S-1)\|\nabla f(\theta_t)\|^2}{S}.$$

At iteration t, the probability $p_t^* = p_t^*(\theta_t)$ ensuring the steepest descent on the above bound is clearly given by $p_t^* = \arg\min_p \Delta(\theta_t, p)$ or equivalently by $p_{t,i}^* = \|\nabla f_i(\theta_t)\|/\sum_{j=1}^{N}\|\nabla f_j(\theta_t)\|$ for $i \in \{1, \ldots, N\}$, as mentionned in [18] and [15]. Unfortunately, practical implementation of the above sampling scheme is prohibitively complex, as it would require to evaluate all gradients to calculate the norms $\|\nabla f_1(\theta_t)\|, \ldots, \|\nabla f_N(\theta_t)\|$ at each iteration which is precisely what we try to avoid. The crucial point is therefore to propose a sampling scheme approximating p_t^* without requiring any additional gradient evaluations.

3.3   A Practical Sampling Distribution - Our Proposal

The main idea is to replace each unknown gradient norm $\|\nabla f_i(\theta_t)\|$ by a (possibly outdated) norm $g_{t,i} = \|\nabla f_i(\theta_k)\|$ at some former instant $k = k(i,t)$ corresponding to the last time $k \le t$ when the ith component was picked. More formally, we define the random sequence g_t as

$$g_{t+1,i} = \begin{cases} \|\nabla f_i(\theta_{t+1})\| & \text{if } i \in \{i_{t+1}^{(1)}, \ldots, i_{t+1}^{(S)}\} \\ g_{t,i} & \text{otherwise.} \end{cases} \tag{6}$$

Then, a natural way to approximate p_t^* is to set for each i

$$\bar{p}_{t,i} = \frac{g_{t,i}}{\sum_{j=1}^{N} g_{t,j}}. \tag{7}$$

It turns out that convergence cannot be guaranteed with the choice (7), because a certain component $\bar{p}_{t,i}$ can get arbitrarily close to zero, so that the i-th index is too rarely, or even never, picked[1]. A possible remedy is to enforce a greedy sampling scheme or, as we will refer to it, a Doeblin-like condition on the transition kernel of the underlying Markov chain, see [17]:

$$\forall i \in \{1, \ldots, N\}, \ p_{t,i} = \rho\nu_i + (1-\rho)\bar{p}_{t,i}, \tag{8}$$

where $\nu = (\nu_1, \ldots, \nu_N)$ is an arbitrary probability distribution satisfying $\nu_i > 0$ for $1 \le i \le N$, and $0 < \rho \le 1$. This condition has the following interpretation: p_t is a mixture between two laws of probability and one of this law (ν) is independent from the past. The AS-SGD is summarized in Algorithm 1 below.

---

[1] Consider for instance the case $N = 2$, $\mathcal{X} = \mathbb{R}$, $f_1(\theta) = \theta^2$, $f_2(\theta) = (\theta - 1)^2$ and $\theta_0 = 0$.

Algorithm 1. AS-SGD

Input: θ_0, ρ, T, S, $(\gamma_t)_{t=0}^{T-1}$, ν
Initialization:
for $i = 1$ **to** N **do**
   Set $g_{0,t} = \|\nabla f_i(\theta_0)\|$
end for
$\mathcal{A}_0 = \texttt{buildtree}(g_0)$
for $t = 0$ **to** $T - 1$ **do**
   Define $\bar{p}_t$ and p_t according to (7) and (8)
   $I_{t+1} = \texttt{sample}(\bar{p}_t, S, \rho, \nu, \mathcal{A}_t)$
   $\theta_{t+1} = \Pi_{\mathcal{K}}(\theta_t - \frac{\gamma_t}{NS}\sum_{i \in I_{t+1}} \frac{\nabla f_i(\theta_t)}{p_{t,i}})$
   Update g_{t+1} according to (6)
   $\mathcal{A}_{t+1} = \texttt{updatetree}(\mathcal{A}_t, I_{t+1}, g_{t+1})$
end for
Return θ_T

3.4   Computationally Efficient Sampling

We point out that there is an additional computational price to pay when implementing NUS instead of uniform sampling. Given a non uniform distribution $p = (p_1, ..., p_N)$ on $\{1, \ldots, N\}$, the simulation time needed to generate a r.v. with distribution p is larger than in the case of uniform distribution. Indeed, one may resort to the inversion method (see [3] for instance), which boils down to inserting of an element in the sorted vector $\tilde{p} = (p_1, p_1 + p_2, ..., p_1 + ...p_N)$ and requires $\lceil \log_2(N) \rceil$ operations. Unfortunately, changing the i-th component of p (just like in the algorithm we propose) changes $N - i$ components in the vector $\tilde{p}$. Our approach based on the notion of *binary research tree* is inspired from [3] and overcome this difficulty.

Building/Updating a Tree. For simplicity, assume that N is even. Define a tree $\mathcal{A}_t$ with N terminal leaves in correspondence with the indexes in $\{1, \ldots, N\}$ the weight $g_{t,i}$ is assigned to the leaf No. i. Each pair $(2k+1, 2(k+1))$ of adjacent terminal leaves, $k \in \{0, \ldots, N/2\}$, admits a common ancestor, which the weight $g_{t,i} + g_{t,i+1}$ is assigned to. Continuing this way in a "bottom-up" fashion, the weight $S_t = \sum_i g_{t,i}$ is assigned to the root node. The function $\texttt{buildtree}$ used in Algorithm 1 generates such a tree from scratch and is used at the initialization $t = 0$. At step $t+1$, g_{t+1} is essentially identical to g_t except for a few S elements which have been updated. The tree $\mathcal{A}_{t+1}$ being close to $\mathcal{A}_t$, it does *not* have to be rebuilt from scratch. The routine $\texttt{updatetree}$ given in the Appendix provides a computationally efficient way to update the tree $\mathcal{A}_{t+1}$ from $\mathcal{A}_t$.

Simulating a Random Index. Suppose that we seek to generate a r.v., say $i_{t+1}^{(1)}$, according to distribution $\bar{p}_t$. Using a r.v. U generated according to the uniform distribution on $[0, 1]$, a path is generated from the root to one of the leaves by comparing U to successive thresholds. The generated variable $i_{t+1}^{(1)}$ is defined as the index of the obtained leaf. The procedure is detailed in the subroutines $\texttt{sample}$ and $\texttt{sample\_tree}$ (see the Appendix). Therefore updating

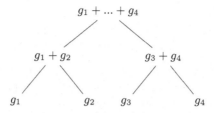

Fig. 1. A binary tree for $N = 4$.

an element i of the distribution is equivalent to update the path from i to the root of the tree and takes $\lceil \log_2(N) \rceil$ operations, while sampling from our distribution is equivalent to follow a path from the root of the tree to one of its leaves. Table 1 summarizes the iteration complexity of the proposed method.

Table 1. Comparison of iteration complexities of AS-SGD and SGD with uniform sampling: c = complexity of pointwise gradient computation, S = sample size.

	SGD	$AS - SGD$
Complexity	Sc	$S(c + (2 - \rho)\lceil \log_2(N) \rceil)$

4   Performance Analysis

We start off by giving a rate bound analysis of the algorithm we proposed and then study the asymptotic behavior of the estimator θ_t produced by our algorithms.

4.1   Preliminary Results

In our analysis, we assume that

Assumption 2. *i) The function f is α-strongly convex, ii) The minimizer θ^* of f belongs to the interior of $\mathcal{K}$.*

The lemma stated below provides a bound on the MSE $a_t = \mathbb{E}(\|\theta_t - \theta^*\|^2)$, where θ_t is generated by Algorithm 1. Its proof is strongly inspired by [5] and [1], where similar bounds are provided.

Lemma 1. *Let Assumptions 1 and 2 hold true. Set $\gamma_t = \gamma_1 t^{-\beta}$ where $\beta \in (0, 1]$ and assume $\gamma_1 > \beta/(2\alpha)$. For each $t \in \mathbb{N}^*$,*

$$a_t \leq C\gamma_t/\rho, \tag{9}$$

where $C = \max(\frac{2B_\nu^2 \gamma_1}{2\alpha\gamma_1 - 1}, \frac{a_1}{\gamma_1})$ when $\beta = 1$ and $C = \max(\frac{B_\nu^2 \gamma_1}{2\alpha}, \frac{a_1}{\gamma_1})$ otherwise, with

$$B_\nu = \frac{1}{SN^2} \sum_{i=1}^{N} \nu_i^{-1} \sup_{\theta \in \mathcal{K}} \|\nabla f_i(\theta)\|^2.$$

Remark 1. *One mights easily check that choosing ν so as to minimize B_ν would lead to take $\nu_i \sim \sup_{\theta \in \mathcal{K}} \|\nabla f_i(\theta)\|$ which is the sampling distribution proposed in [18].*

We emphasize the fact that sharper bounds on a_t can be obtained. One could for instance easily generalize the approach of [5] which provides bounds on a_t which are not only tighter but also valid under weaker assumptions on the step size. This would come at the price of a rather tedious bound in (9). As explained below, such an involved bound would actually be unnecessary for our purpose, and Lemma 1 is in fact sufficient to derive the main result of the next paragraph. It also admits a simpler proof provided in Appendix B. Before skipping to the main result, we first discuss the bound in (9). Lemma 1 establishes that the adaptive sampling scheme preserves the convergence rate in $O(t^{-\beta})$ obtained in the uniform sampling case. Nevertheless, Lemma 1 is merely a sanity check, because unfortunately, the bound does **not** suggest that adaptive non-uniform sampling generates a performance improvement: the minimum of the RHS in (9) is attained for $\rho = 1$ which boils down to selecting a constant sampling probability. This is in contradiction with numerical results, which suggest on the opposite that strong benefits can be obtained by adaptively selecting the sampling probability. In order to obtain results that fully grasp the benefits of our adaptive sampling strategy, we investigate from now on the asymptotic regime $t \to \infty$. The following Lemma provides an estimate of the value $b_{t,i} = \mathbb{E}[\|g_t^i - \nabla f_i(\theta_{t-1})\|^2]$, which quantifies the mean square gap between the current (unobserved) gradients and the outdated gradients used to generate the next samples.

Lemma 2. *Suppose that the assumptions of Lemma 1 hold true. For any $t \in \mathbb{N}^*$,*

$$b_{t,i} \leqslant \frac{(2L_i)^2 2^\beta}{1 - (1 - \rho\nu_i)^S} \frac{C}{\rho t^\beta} + o(t^{-\beta}). \tag{10}$$

4.2   Main Results

We now turn to the analysis of the asymptotic behavior and prove how our algorithm improve on SGD.

Assumption 3. *The function f is twice differentiable in a neighborhood of θ^*.*

We introduce the sampling probability given by $\pi^* = \rho\nu + (1 - \rho)\bar{\pi}^*$, where,

$$\bar{\pi}_i^* = \frac{\|\nabla f_i(\theta^*)\|}{\sum_{j=1}^N \|\nabla f_j(\theta^*)\|} \tag{11}$$

for any $i = 1, \ldots, N$. We define $Q^* = \sum_{i=1}^N \nabla f_i(\theta^*)\nabla f_i(\theta^*)^T/(SN^2\pi_i^*)$ and denote by $H = \nabla^2 f(\theta^*)$ the Hessian at point θ^*.

Theorem 1. *Suppose that Assumptions 1, 2 and 3 hold true and that the stepsize satisfies the condition of Lemma 1. Then the sequence $(\theta_t - \theta^*)/\sqrt{\gamma_t}$*

converges in distribution to a zero-mean Gaussian variable whose covariance matrix $\Sigma = \Sigma(\rho, \nu)$ is the solution to the following Lyapunov equation

$$\Sigma H + H\Sigma = Q^* \quad (if\ \beta < 1)$$
$$\Sigma(I_d + 2\gamma_1 H) + (I_d + 2\gamma_1 H)\Sigma = 2\gamma_1 Q^* \quad (if\ \beta = 1).$$

The proof is provided in Appendix D. The following Corollary is directly obtained by use of the second order delta-method [12]. We denote by $\mathrm{tr}(A)$ the trace of any square matrix A.

Corollary 1. *Under the assumptions of Theorem 1, $\gamma_t^{-1}(f(\theta_t) - f(\theta^*))$ converges in distribution to the r.v. $V = (1/2)Z^T\Sigma(\rho,\nu)^{1/2}H\Sigma(\rho,\nu)^{1/2}Z$ where Z is a Gaussian vector $\mathcal{N}(0, I_d)$. In addition, we have $\mathbb{E}(V) = \mathrm{tr}(H\Sigma(\rho,\nu))/2$.*

We now use Corollary 1 to compare our method with the best possible *fixed* choice of a sampling distribution. Note that the search for an optimal fixed distribution is also discussed in [15].

When the distribution is fixed, say to p, the asymptotic covariance of the normalized error is given by $\Sigma(1, p)$ as defined in Theorem 1. Motivated by Corollary 1, we refer to the optimal fixed sampling distribution as the distribution p minimizing $\mathrm{tr}(H\Sigma(1, p))$. It is straightforward to show that

$$\arg\min_p \mathrm{tr}(H\Sigma(1, p)) = \bar{\pi}^*,$$

where $\bar{\pi}^*$ is defined in (11). We also set $\sigma_*^2 = \mathrm{tr}(H\Sigma(1, \bar{\pi}^*))$. The following proposition follows from standard algebra and its proof is omitted due to the lack of space.

Proposition 1. *Let $\Sigma(\rho, \nu)$ be the asymptotic covariance matrix defined in Theorem 1. Then,*

$$\sigma_*^2 \leq \mathrm{tr}(H\Sigma(\rho, \nu)) \leq \sigma_*^2(1 + S\rho/(1-\rho)).$$

Proposition 1 implies that the asymptotic performance of the proposed AS-SGA can be made arbitrarily closed to the one associated with the best sampling distribution provided that ρ is chosen closed to zero. It is of course tempting to set $\rho = 0$ in (8) however in this case, the statement of Theorem 1 would be no longer valid.

5   Numerical Experiments

We consider the l_2-regularized logistic regression problem. Denoting by N the number of observations and by d the number of features, the optimization problem can be formulated as follows:

$$\min_{\theta \in \mathbb{R}^d} \frac{1}{T} \sum_{i=1}^{N} f(y_i, x_i, \theta) + \frac{\lambda}{2} \|\theta\|^2, \tag{12}$$

where $f(x, y, \theta) = \log(1 + \exp(-yx^T\theta))$ the $(y_i)_{i=1}^{N}$ are in $\{-1, +1\}$, the $(x_i)_{i=1}^{N}$ are in $\mathbb{R}^d$ and $\lambda > 0$ is a scalar. Note that for this problem one mights easily have access to the quantities L_i and $B_i = \sup_{\theta \in \mathbb{R}^d} \|\nabla f_i(\theta)\|$ .We used the benchmark dataset *covtype* with $N = 581012$, $d = 54$, $\lambda = \frac{1}{\sqrt{N}}$ and $\gamma_t = \frac{\gamma_1}{1+\gamma_1 \lambda t}$ as proposed in [9] , where γ_1 is determined using a small sample of the training set. We considered the cases $\nu_i = \frac{1}{N}$ (ASGD), $\nu_i \sim L_i$ (ASGD-Lip) and $\nu_i \sim B_i$ (ASG-B), ran the algorithm for different values of the parameter ρ and compared it to the usual stochastic gradient descent with uniform sampling (SGD), lipschitz sampling (SGD-Lip) and upper-bound sampling (SGD-B) for the same parameters. In this scenario, $\mathcal{C} \sim d$, the computational times related to the SGD and the AS-SGD are comparable (see Table 1). Experiments suggest that choosing $\nu_i \sim L_i$ leads to better performances and that using a small value of ρ leads to poor (respectively good) performance when θ_1 is far (respectively close) from θ^* . This suggests that a strategy could consist in running a classical SGD to get closer to θ^* and then run AS-SGD. One could also use the AS-SGD with a decreasing step-size policy. We dit not study these policies due to space limitations.

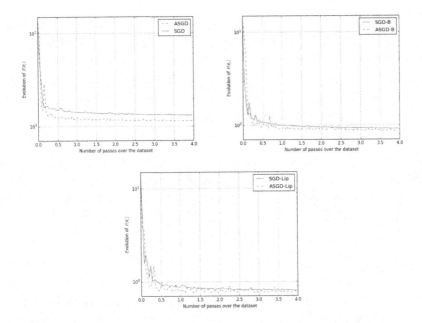

Fig. 2. Evolution of $f(\theta_t)$ with $S = 10$ and $\rho = 0.7$

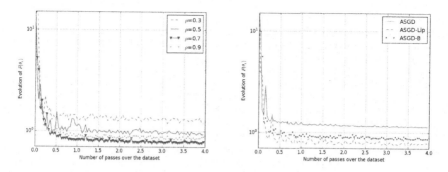

Fig. 3. Evolution of $f(\theta_t)$ with different values of ρ ($\nu_i \sim L_i$, $S = 10$) (left) and different sampling strategies ($\rho = 0.7, S = 10$) (right)

6 Conclusion

Motivated by recent work on SGD with non uniform probability, we introduced a novel implementation of the SGD algorithm with an adaptative sampling. We proposed a specific adaptive sampling scheme, depending on the past iterations in a Markovian manner that achieves low simulation cost. We also proposed a rigorous analysis to justify our approach and gave sufficient conditions to obtain theoretical guarantees.

Acknowledgments. This work has been supported by Chair Machine Learning for Big Data at Telecom ParisTech and by the Orange/Telecom ParisTech think tank phi-TAB.

A Algorithms for Efficient NUS

In this Appendix, we provide the main procedures called by Algorithm 1 to efficiently generate a collection of S iid random indexes on $\{1, \ldots, N\}$. If $\mathcal{A}$ is a tree with N leaves and e is a node, we denote by $w(e)$ the weight of node e. The root of $\mathcal{A}$ is denoted by root(e). Father(e) is the father of a node e (and the empty set if $e =$root($\mathcal{A}$). Son(e) is the list of sons of e (and the emptyset if e is a leaf) and the elements of the list are refered to as son(e)[1], son(e)[2]. The functions isroot(e) and isleaf(e) return boolean values equal to one if e is the root or a leaf respectively. Finally, if e is a leaf, label(e) returns the index of the leaf e in $\{1, \ldots, N\}$. The procedure `buildtree` is omitted but discussed in Section 3.4. The algorithm `Sample` simply consists in writing the probability distribution as a mixture and is also omitted.

Algorithm 2. Sample_tree	**Algorithm 3.** Update_tree
Input: $\mathcal{A}$	**Input:** $\mathcal{A}, I, g$
$e = \text{root}(\mathcal{A})$	**for** $i \in I$ **do**
Draw $U \sim \text{Uniform}([0, w(e)])$	$\quad e = \text{leaf}(\mathcal{A}, i)$
repeat	$\quad \delta = g_i - w(e)$
$\quad$ **if** $U < w(\text{son}(e)[1])$ **then**	$\quad w(e) \leftarrow g_i$
$\quad\quad e \leftarrow \text{son}(e)[1])$	$\quad$ **repeat**
$\quad$ **else**	$\quad\quad e \leftarrow \text{father}(e)$
$\quad\quad U \leftarrow U - w(\text{son}(e)[1])$	$\quad\quad w(e) \leftarrow w(e) + \delta$
$\quad\quad e \leftarrow \text{son}(e)[2]$	$\quad$ **until** isroot(e)
$\quad$ **end if**	**end for**
until isleaf(e)	**Return** $\mathcal{A}$
Return label(e)	

B Proof of Lemma 1

Using the non expansiveness of the projection, the strong convexity and the definition of B_ν, we get $a_{t+1} \leqslant (1 - 2\alpha\gamma_t)a_t + \gamma_t^2 \frac{B_\nu^2}{\rho}$. We will now prove the lemma by induction. The property is checked for $t = 1$ by defintion of C. Assume the result holds true for a_t then, we have $a_{t+1} \leqslant ((1 - 2\alpha\gamma_t)\gamma_t C + \gamma_t^2 B^2)/\rho$ and it is sufficient to show that $(1 - 2\alpha\gamma_t)\gamma_t C + \gamma_t^2 B^2 \leqslant \gamma_{t+1} C$ which is equivalent to $\gamma_t^2 B_\nu^2 \leqslant C(\gamma_{t+1} - \gamma_t + 2\alpha\gamma_t^2)$. If $\beta = 1$, then, using $2\alpha\gamma_1 > 1$, we get: $\gamma_{t+1} - \gamma_t + 2\alpha\gamma_t^2 \geqslant \frac{(2\alpha\gamma_1 - 1)\gamma_1}{t(t+1))} > 0$ and therefore

$$\frac{B_\nu^2 \gamma_t^2}{\gamma_{t+1} - \gamma_t + 2\alpha\gamma_t^2} \leqslant \frac{B_\nu^2 \gamma_1}{t^2} \frac{t(t+1)}{(2\alpha\gamma_1 - 1)} \leqslant \frac{2B_\nu^2 \gamma_1}{2\alpha\gamma_1 - 1}, \tag{13}$$

which gives the result. If $0 < \beta < 1$,

$$\gamma_{t+1} - \gamma_t + 2\alpha\gamma_t^2 = \gamma_1 \int_t^{t+1} \frac{-\beta}{u^{1+\beta}} + 2\alpha\frac{\gamma_1^2}{t^{2\beta}} \geqslant 2\alpha\frac{\gamma_1^2}{t^{2\beta}} - \frac{\gamma_1\beta}{t^{\beta+1}}$$

$$\geqslant \frac{\gamma_1(2\alpha\gamma_1 t^{1-\beta} - \beta)}{t^{\beta+1}} > 0$$

since $2\alpha\gamma_1 > \beta$. We get

$$\frac{B^2\gamma_t^2}{\gamma_{t+1} - \gamma_t + 2\alpha\gamma_t^2} \leqslant \frac{B^2\gamma_1}{t^{2\beta}} \frac{t^{\beta+1}}{(2\alpha\gamma_1 t^{1-\beta} - \beta))} < \frac{B^2}{2\alpha},$$

which concludes the proof.

C Proof of Lemma 2

We consider a stepsize γ_t such that

$$a_t \leqslant \frac{C\gamma_{t+1}}{\rho}, \tag{14}$$

and we denote by $\mathbb{I}_A$ the indicator function of any event A *i.e.*, the r.v. equal to one on this event and to 0 elsewhere. Consider any index i and instant t and let $A^i_{k,t}$ the event *"the index i has not been picked since instant k"*. Using the Doeblin Condition we have:

$$\mathbb{P}(A^i_{k,t}) = \mathbb{E}[\mathbb{I}_{\epsilon^i_t=0}...\mathbb{I}_{\epsilon^i_{k+1}=0}\mathbb{I}_{\epsilon^i_k \neq 0}] \leqslant \mathbb{E}[\mathbb{E}[\mathbb{I}_{\epsilon^i_t=0}|\mathcal{F}_{t-1}]...\mathbb{I}_{\epsilon^i_{k+1}=0}]$$

$$\leqslant (1-\rho\nu_i)^S \mathbb{E}[\mathbb{I}_{\epsilon^i_{t-1}=0}...\mathbb{I}_{\epsilon^i_{k+1}=0}] \leqslant (1-\rho\nu_i)^{S(t-k)}$$

Conditionaly upon $A^i_{k,t}$, we have $g_{t,i} = \nabla f_i(\theta_{k-1})$. Since $(A^i_{k,t})_{k \leqslant t}$ is a partition of the state space, the law of total probability combined with Assumption 1 and the independence induced by the Doeblin Condition yields :
$\mathbb{E}(\|g_{t,i} - \nabla f_i(\theta_{t-1})\|^2) \leqslant L_i^2 \sum_{k=1}^t \mathbb{E}(\|\theta_{k-1} - \theta_{t-1}\|^2)(1 - \rho\nu_i)^{S(t-k)}$. Using $\mathbb{E}(\|\theta_{k-1} - \theta_{t-1}\|^2) \leqslant 2a_{t-1} + 2a_{k-1} \leqslant 4C\gamma_k/\rho$ leads to:

$$\mathbb{E}(\|g_{t,i} - \nabla f_i(\theta_{t-1})\|^2) \leqslant 4(L_i)^2 \frac{C}{\rho} \sum_{k=1}^t \gamma_k (1-\rho\nu_i)^{S(t-k)}$$

$$= 4(L_i)^2 \frac{C}{\rho} \sum_{k=0}^{t-1} \gamma_{t-k}(1-\rho\nu_i)^{Sk}.$$

For all $1 < t_0 < t$, we have by splitting the sum in two terms:

$$\sum_{k=0}^{t-1} \gamma_{t-k}(1-\rho\nu_i)^{Sk} \leqslant \frac{\gamma_{t-t_0}}{1-(1-\rho\nu_i)^S} + (1-\rho\nu_i)^{St_0} \sum_{k=t_0}^{t-1} \gamma_{t-k}.$$

Taking $2t_0 \sim t$ for instance and using the classical integral test for convergence gives the result.

D   Proof of Theorem 1

The proof is prefaced by the following Lemma, whose proof is given at the end of this section.

Lemma 3. *Under the Assumptions of Theorem 1, the sequence (θ_t, p_t) converges to (θ^*, π^*) with probability one.*

We now prove the main result. We use the decomposition $\theta_{t+1} = \theta_t - \gamma_t \nabla f(\theta_t) + \gamma_t e_{t+1} + \gamma_t \eta_{t+1}$, where we set $D_{t+1} = (1/S)\sum_{i \in I_{t+1}} \nabla f_i(\theta_t)/(Np_{t,i})$, $e_{t+1} = \nabla f(\theta_t) - D_{t+1}$, $\eta_{t+1} = (\Pi_K(\phi_{t+1}) - \phi_{t+1})/\gamma_t$, $\phi_{t+1} = \theta_t - \gamma_t D_{t+1}$.

We next check Conditions **C1** to **C4** in [6]. Conditions **C1** and **C4** are immediate consequences of Assumptions 2 and 3. We check that e_t satisfies Condition **C2**. First, by virtue of (5), (e_t) is a martingale increment sequence adapted to $\mathcal{F}_t$ *i.e.*, $\mathbb{E}(e_{t+1}|\mathcal{F}_t) = 0$. Second, for all $t \in \mathbb{N}^*$ and $i = 1,\ldots,N$, $p_{t,i} \geqslant \rho\nu_i > 0$ with probability one. Therefore, it is straightforward to check that $\|e_t\| \leqslant M$ a.s. for some constant M, which only depends on ρ,

ν and $B_1, \ldots, B_N$. Third, we analyze the asymptotic behavior of the conditional covariance $Q_t = \mathbb{E}(e_{t+1}e_{t+1}^T | \mathcal{F}_t)$. After some algebra, we obtain that $Q_t = (1/N^2) \sum_{i=1}^N (1/(Sp_{t,i}))(\nabla f_i(\theta_t) - \nabla f(\theta_t))(\nabla f_i(\theta_t) - \nabla f(\theta_t))^T$. Using Lemma 3 along with the continuity of ∇f_i for each i, we directly obtain that Q_t tends to Q^* with probability one. Condition **C2** in [6] is thus fulfilled. Turning to Condition **C3**, we shall prove that $\gamma_t^{-1/2} \eta_{t+1}$ converges to 0 in L_1. Using Cauchy-Schwarz inequality :

$$\mathbb{E}(\|\gamma_t^{-1/2}\eta_{t+1}\|) = \mathbb{E}(\|\gamma_t^{-1/2}\eta_{t+1}\| \mathbb{I}_{\eta_{t+1} \neq 0})$$
$$\leq \mathbb{E}(\|\eta_{t+1}\|^2)^{1/2} \left(\gamma_t^{-1}\mathbb{P}(\eta_{t+1} \neq 0)\right)^{1/2}.$$

Defining $u_t = \mathbb{E}(\|\eta_{t+1}\|^2)$ and $v_t = \mathbb{P}(\eta_{t+1} \neq 0)$, the above inequality reads

$$\mathbb{E}(\|\gamma_t^{-1/2}\eta_{t+1}\|) \leq \sqrt{u_t}\sqrt{\frac{v_t}{\gamma_t}}. \tag{15}$$

We first analyze u_t. Observe that

$$\|\eta_{t+1}\|^2 = \gamma_t^{-2}\|\Pi_{\mathcal{K}}(\phi_{t+1}) - \phi_{t+1}\|^2 \leq 2\gamma_t^{-2}(\|\Pi_{\mathcal{K}}(\phi_{t+1}) - \theta_t\|^2 + \|\phi_{t+1} - \theta_t\|^2)$$
$$\leq 4\gamma_t^{-2}\|\phi_{t+1} - \theta_t\|^2 = 4\|D_{t+1}\|^2,$$

where the last inequality is due to the non-expansiveness of the projection operator. Therefore Since $p_{t,i}$ admits a fixed deterministic minorant for each i and since the gradients ∇f_i are bounded on $\mathcal{K}$, there exists a deterministic constant M' such that $\|D_{t+1}\|^2 \leq M'$. In particular, the sequence $\|\eta_{t+1}\|^2$ is uniformly integrable. We now prove that $\eta_{t+1} \to 0$ almost surely. Consider an $\epsilon > 0$ such that the ball $B(\theta^*, 2\epsilon)$ of center θ^* and radius ϵ is contained in $\mathcal{K}$. By Lemma 3, for all ω on a set of probability one, there exists $N_0(\omega)$ such that $\|\theta_{t+1}(\omega) - \theta^*\| \leq \epsilon$ for all $t \geq N_0(\omega)$. Using again that $\|D_{t+1}(\omega)\|$ is a bounded sequence and $\gamma_t \to 0$, it is clear that $\gamma_t\|D_{t+1}(\omega)\| < \epsilon$ for t large enough. Thus, for t large enough, $\phi_{t+1}(\omega) \in \mathcal{K}$ which implies that $\eta_{t+1}(\omega) = 0$. Almost surely, the sequence η_{t+1} converges to zero. Putting all pieces together, $\|\eta_{t+1}\|^2$ is a uniformly integrable sequence which tends a.s. to zero. As a consequence, $u_t = \mathbb{E}(\|\eta_{t+1}\|^2)$ tends to zero as $t \to \infty$. We now analyze $v_t = \mathbb{P}(\eta_{t+1} \neq 0)$. For $\epsilon > 0$ be defined as above, note that the event $\{\eta_{t+1} \neq 0\}$ is included in the event $\{\|\phi_{t+1} - \theta^*\| \geq \epsilon\}$. By Markov inequality,

$$v_t \leq \epsilon^{-2}\mathbb{E}(\|\phi_{t+1} - \theta^*\|^2) \leq 2\epsilon^{-2}(\mathbb{E}(\|\phi_{t+1} - \theta_t\|^2) + \mathbb{E}(\|\theta_t - \theta^*\|^2))$$
$$\leq 2\epsilon^{-2}\left(\gamma_t^2\mathbb{E}(\|D_{t+1}\|^2) + \frac{C\gamma_t}{\rho}\right),$$

where we used Lemma 1 to obtain the last inequality. Recalling that D_{t+1} is bounded, it is clear that v_t/γ_t is a bounded sequence. Finally, by inequality (15), we conclude that $\mathbb{E}(\|\gamma_t^{-1/2}\eta_{t+1}\|)$ tends to zero. Condition **C3** in [6] is satisfied. This completes the proof of Theorem 1.

Proof of Lemma 3

Almost sure convergence of θ_t to θ^* directly follows from the Robbins-Siegmund Lemma and follows the same line of reasoning than [8]: it is therefore omitted. It remains to show that for each $j = 1, \ldots, N$, $\bar{p}_{t,j} \rightarrow \bar{\pi}_j^*$ a.s., where $\bar{\pi}_j^* = \|\nabla f_j(\theta^*)\| / \sum_j \|\nabla f_j(\theta^*)\|$. Let A_j denote the event that index j is picked infinitely often (*i.e.*, there exists an infinite sequence $(t_k, n_k)_{k \in \mathbb{N}}$ on $\mathbb{N}^* \times \{1, \ldots, S\}$ such that $i_{t_k}^{(n_k)} = j$). As can be easily checked, the Doeblin condition $p_{t,j} \geq \rho \nu_i$ ensures that A_j has probability one. For a fixed $\omega \in A_j$ and using the continuity of ∇f_j, we obtain that $g_{t,j}(\omega)$ converges to $\|\nabla f_j(\theta^*)\|$. As a consequence, $\bar{p}_{t,j}(\omega) \rightarrow \bar{\pi}_j^*(\omega)$ and the result follows.

References

1. Nemirovski, A., Juditsky, A., Lan, G., Shapiro, A.: Robust stochastic approximation approach to stochastic programming. SIAM J. Optim. (2009)
2. Defazio, A., Bach, F., Lacoste-Julien, S.: SAGA: A Fast Incremental Gradient Method With Support for Non-Strongly Convex Composite Objectives (2014)
3. Devroye, L.: Non-uniform random variate generation (1986)
4. Needell, D., Srebro, N., Ward, R.: Stochastic gradient descent, weighted sampling and the randomized kaczmarz algorithm
5. Bach, F., Moulines, E.: Non-asymptotic analysis of stochastic approximation algorithms for machine learning. In: NIPS, pp. 451–459 (2011)
6. Fort, G.: Central limit theorems for stochastic approximation with controlled Markov Chain. EsaimPS (2014)
7. Robbins, H., Monro, S.: A stochastic approximation method. Ann. Math. Statist. **22** (1951)
8. Bottou, L.: Online algorithms and stochastic approximations. In: Online Learning and Neural Networks
9. Bottou, L.: Stochastic gradient tricks. In: Neural Networks, Tricks of the Trade, Reloaded (2012)
10. Mairal, J.: Incremental Majorization-Minimization Optimization with Application to Large-Scale Machine Learning (2014)
11. Nesterov, Y., Nesterov, I.U.E.: Introductory Lectures on Convex Optimization: A Basic Course. Applied Optimization. Springer (2004)
12. Pelletier, M.: Weak convergence rates for stochastic approximation with application to multiple targets and simulated annealing. Ann. Appl. Prob (1998)
13. Johnson, R., Zhang, T.: Accelerating stochastic gradient descent using predictive variance reduction. In: NIPS, pp. 315–323 (2013)
14. Schmidt, M.W., Le Roux, N., Bach, F.: Minimizing finite sums with the stochastic average gradient. CoRR (2013)
15. Clemencon, S., Bertail, P., Chautru, E.: Scaling up m-estimation via sampling designs: The horvitz-thompson stochastic gradient descent. In: 2014 IEEE Big Data (2014)
16. Shalev-Shwartz, S., Zhang, T.: Stochastic Dual Coordinate Ascent Methods for Regularized Loss Minimization (2012)
17. Meyn, S., Tweedie, R.L.: Markov Chains and Stochastic Stability (2009)
18. Zhao, P., Zhang, T.: Stochastic Optimization with Importance Sampling (2014)

Online Linear Optimization for Job Scheduling Under Precedence Constraints

Takahiro Fujita, Kohei Hatano[(✉)], Shuji Kijima, and Eiji Takimoto

Department of Informatics, Kyushu University, Fukuoka, Japan
{takahiro.fujita,hatano,kijima,eiji}@inf.kyushu-u.ac.jp

Abstract. We consider an online job scheduling problem on a single machine with precedence constraints under uncertainty. In this problem, for each trial $t = 1, \ldots, T$, the player chooses a total order (permutation) of n fixed jobs satisfying some prefixed precedence constraints. Then, the adversary determines the processing time for each job, 9 and the player incurs as loss the sum of the processing time and the waiting time. The goal of the player is to perform as well as the best fixed total order of jobs in hindsight. We formulate the problem as an online linear optimization problem over the permutahedron (the convex hull of permutation vectors) with specific linear constraints, in which the underlying decision space is written with exponentially many linear constraints. We propose a polynomial time online linear optimization algorithm; it predicts almost as well as the state-of-the-art offline approximation algorithms do in hindsight.

1 Introduction

Job scheduling is a fundamental problem in the field of computer science and operations research, and it has been studied extensively for decades. It has broad applications in operating systems, assignments of tasks to workers, manufacturing systems, and many other areas.

Studies of how to schedule jobs that use a single machine under precedence constraints is well studied in the mathematical programming literature. More precisely, assume that there are n fixed jobs and a single processor. Let $[n] = \{1, \ldots, n\}$ be the set of jobs. Each job i needs *processing time* ℓ_i to be completed by the processor. A schedule is a permutation over n jobs, and the processor does the jobs sequentially according to the schedule. For a given schedule, the *completion time* of job i is the sum of the waiting time (the sum of the processing times of the jobs finished before completing job i) and the processing time of job i. There are precedence constraints over n jobs. For example, job 1 needs to be completed before job 3, and job 2 needs to be completed before job 5. The constraints are represented as a set of binary relations $\mathcal{A} \subset [n] \times [n]$, e.g., $\mathcal{A} = \{(1,3),(2,5)\}$. Given the processing time of n jobs and the precedence constraints $\mathcal{A}$, the typical goal is to find a schedule that minimizes the sum of the (weighted) completion times of the n jobs, subject to the constraints $\mathcal{A}$. This problem is categorized as $1|prec| \sum_j C_j$ in the literature[1]. It is known that this problem is NP-hard [18,19].

---

[1] The weighted version is known as $1|prec| \sum_j w_j C_j$.

© Springer International Publishing Switzerland 2015
K. Chaudhuri et al. (Eds.): ALT 2015, LNAI 9355, pp. 332–346, 2015.
DOI: 10.1007/978-3-319-24486-0_22

For further developments, see, e.g., [2,9]. Several 2-approximation algorithms have been proposed for the offline setting [6,7,14,22,25] and a stochastic setting [24,26]. In this paper, we consider a different scenario for the classical problem. What if the processing time of each job is unknown when we determine the schedule? This question is quite natural in many application areas that cope with uncertainty and in which the processing time of each job is uncertain or varies in time. It is impossible to solve this problem directly without knowing the processing time, so we consider an iterative scenario. Each day $t = 1, \ldots, T$, we determine a total order of n fixed jobs satisfying some prefixed precedence constraints. Then, after processing all n jobs according to the schedule, the processing time of each job is revealed. The goal is to minimize *the sum of the completion times* over all jobs and all T days under fixed precedence constraints, where the completion time of job i at day t is the sum of processing times of all jobs prior to i and the processing time of job i.

Now, let us formulate the problem in a formal way. A permutation σ is a bijection from $[n]$ to $[n]$. Another representation of a permutation σ over the set $[n]$ is a vector in $[n]^n$, defined as $\sigma = (\sigma(1), \ldots, \sigma(n))$, which corresponds to σ. For example, $(3, 2, 1, 4)$ is a representation of a permutation for $n = 4$. The vector representation of permutations is convenient, since the sum of the completion times of the jobs according to some permutation σ is expressed as the inner product $\sigma \cdot \ell$, where ℓ is the vector consisting of the processing times of the jobs. For example, there are 4 jobs to be processed according to the order $4, 1, 2, 3$. Each processing time is given as $\ell = (\ell_1, \ell_2, \ell_3, \ell_4)$. The completion times of jobs $i = 4, 1, 2, 3$ are ℓ_4, $\ell_4 + \ell_1$, $\ell_4 + \ell_1 + \ell_2$, and $\ell_4 + \ell_1 + \ell_2 + \ell_3$, respectively, and the sum of completion times is $4\ell_4 + 3\ell_1 + 2\ell_2 + \ell_3$. Note that the completion time exactly corresponds to $\sigma \cdot \ell$, the inner product of ℓ and the permutation vector $\sigma = (3, 2, 1, 4)$. Here, component σ_i of the permutation σ represents the priority of job i.

Let S_n be the set of all permutations over $[n]$, i.e., $S_n = \{\sigma \in [n]^n \mid \sigma$ is a permutation over $[n]\}$. In particular, the convex hull of all permutations is called the permutahedron, denoted as P_n. The set $\mathcal{A}$ of precedence constraints is given as $\mathcal{A} = \{(i_k, j_k) \in [n] \times [n] \mid i_k \neq j_k, k = 1, \ldots, m\}$, meaning that object i_k is preferred to object j_k. The set $\mathcal{A}$ induces the set defined by the linear constraints $\mathrm{Precons}(\mathcal{A}) = \{p \in \mathbb{R}_+^n \mid p_i \geq p_j$ for $(i, j) \in \mathcal{A}\}$. We further assume that there exists a linear ordering consistent with $\mathcal{A}$. In other words, we assume there to exist a permutation $\sigma \in S_n \cap \mathrm{Precons}(\mathcal{A})$.

The online job scheduling problem can be formulated as the following online linear optimization problem over $S_n \cap \mathrm{Precons}(\mathcal{A})$. For each trial $t = 1, \ldots, T$, (i) the player predicts a permutation $\sigma_t \in S_n \cap \mathrm{Precons}(\mathcal{A})$, (ii) the adversary returns a loss vector $\ell_t \in [0, 1]^n$, and (iii) the player incurs loss $\sigma_t \cdot \ell_t$. The goal of the player is to minimize the α-regret for some small $\alpha \geq 1$:

$$\alpha\text{-Regret} = \sum_{t=1}^{T} \sigma_t \cdot \ell_t - \alpha \min_{\sigma \in S_n \cap \mathrm{Precons}(\mathcal{A})} \sum_{t=1}^{T} \sigma \cdot \ell_t.$$

In this paper, we propose an online linear optimization algorithm over $P_n \cap$ Precons($\mathcal{A}$) whose the α-regret is $O(n^2\sqrt{T})$ for $\alpha = 2 - 2/(n+1)$. For each trial, our algorithm runs in polynomial time in n and m. More precisely, the running time at each trial is $O(n^4)$. Further, we show that a lower bound of the 1-regret is $\Omega(n^2\sqrt{T})$.

In addition, we prove that there is no polynomial time algorithm with α-regret $poly(n, m)\sqrt{T}$ with $\alpha < 2 - 2/(n+1)$ unless there exists a randomized approximation algorithm with approximation $\alpha < 2 - 2/(n+1)$ for the corresponding offline problem. Thus far, the state-of-the-art approximation algorithms have an approximation ratio $2 - 2/(n+1)$, and it is a longstanding open problem to find an approximation algorithm with a better ratio [30]. It has been determined that there is no polynomial-time $(1 + \varepsilon)$-approximation algorithm (PTAS) for any constant $\varepsilon > 0$ under some standard assumption of the complexity theory [3]. Therefore, the regret bound is optimal among any polynomial algorithms unless there exists a better approximation algorithm for the offline problem.

Note that our online algorithm is deterministic, so some reader might worry that the algorithm works without any randomization. We show that our problem can be reduced to online problem over continuous space and rounding problem. For the online algorithm, some deterministic algorithms are known and can achieve a good regret bound. Therefore, there is no reason that our algorithm is stochastic.

2   Related Research

There has been related research on the online prediction of permutations. The earliest approach was to directly design online prediction algorithms for permutations. Helmbold and Warmuth [15] were the first to do this, and in their setting, a permutation is given as a permutation matrix, which is a more generic expression than a permutation vector (i.e., permutation matrices can encode permutation vectors). Thus, their algorithm can be used for our problem without precedence constraints. Yasutake et al. [31] proposed an online linear optimization algorithm over the permutahedron when there are no precedence constraints. Ailon proposed another online optimization algorithm with improved regret bound and time complexity [1]. Suehiro et al. [28] extended the result of Yasutake et al. [31] to the submodular base polyhedron; this can be used not only for permutations, but also for other combinatorial objects, such as spanning trees. These algorithms, however, are not designed for precedence-constrained problems.

The second approach is to transform an offline algorithm to an online optimization algorithm. By using the conversion method of Kakade et al. [16] or that of Fujita et al. [13], we can construct online optimization algorithms with α-regret that are similar to ours.

However, with the method of Kakade et al. [16], the resulting time complexity per trial is linear in T, which is not desirable. For the method of Fujita et al., the α-regret is proved to be $\alpha = 2 - 2/(n-1) + \varepsilon$, which is slightly inferior to that of our method. The running time per trial is $poly(n, 1/\varepsilon)$, which is independent of T but depends on $1/\varepsilon$. Also, Fujita et al. showed that if an offline algorithm use an LP-relaxation and a metarounding, then FPL with the algorithm can achieve good bounds on the α-regret. But, the LP-relaxation based algorithm of previous work cannot be directly applied to our problem, because its rounding algorithm is not metarounding.

Our approach is completely different from previous offline algorithms. The known offline approximation algorithms rely on formulations that use completion-time variables or linear-ordering variables. In the first formulation, n variables indicate the time at which the job is completed. In this case, the relaxed problem is formulated as a linear program with exponentially many constraints (see, e.g., [14,25]). The problem can be approximately solved in polynomial time by using the ellipsoid method. In the second case, there are $\binom{n}{2}$ variables, which represent relative comparisons between pairs of jobs. The relaxed problem is also formulated as a linear program with $O(n^2)$ variables and $O(n^3)$ linear constraints (e.g., [7]). In both formulations, the set of linear constraints and associated rounding methods require knowledge of the processing times of the jobs, and these are not available in the online setting. Our approach uses some geometric properties of the permutahedron, and thus it is totally different from the previous approaches. As a result, our rounding algorithm does not require knowledge of the processing times of the jobs. Thus, our approach is suitable for the online problem. On the other hand, a shortcoming of our approach is that it is only able to deal with the online problem of minimizing the unweighted total sum of the completion times.

Online learning approaches for job scheduling problems are not new. In particular, Even-Dar et al. [10] considered an online optimization problem with global functions, and its applications schedule jobs for several machines in order to minimize the makespan (the time at which the final job is completed). Both the objectives and the techniques of Even-Dar et al. [10] are different from ours. For online multi-task learning, Lugosi et al. considered some class of constraints [20]. For some natural class of constraints, they showed that the online task can be reduced to the shortest path problem, which is efficiently solvable. However, the constraints in our setting are much complicated and their method does not seem to be applicable to our problem.

3   Online Linear Optimization Algorithm over the Permutations

In this section, we propose our algorithm PermLearnPrec and prove its α-regret bound. We will use the notion of the permutahedron. The permutahedron P_n is the convex hull of the set of permutations S_n. It is known that P_n can be represented as the set of points $\boldsymbol{p} \in \mathbb{R}^n_+$ satisfying $\sum_{i \in S} p_i \leq \sum_{i=1}^{|S|} (n + 1 - i)$ for any $S \subset [n]$, and $\sum_{i=1}^{n} p_i = n(n + 1)/2$. For further discussion of the permutahedron, see, e.g., [12,32].

3.1   Main Structure

Our algorithm PermLearnPrec is shown as Algorithm 1. The algorithm maintains a weight vector $\boldsymbol{p}_t$ in $\mathbb{R}^n_+$, which represents a mixture of the permutations in S_n. At each trial t, it "rounds" (see next section for description) a vector $\boldsymbol{p}_t$ into a permutation $\boldsymbol{\sigma}_t$ so that $\boldsymbol{\sigma}_t \leq \alpha \boldsymbol{p}_t$ for some $\alpha \geq 1$. After the loss vector $\boldsymbol{\ell}_t$ is given, PermLearnPrec updates the weight vector $\boldsymbol{p}_t$ in an additive way and successively projects it onto the set of linear constraints representing the precedence constraints $\mathrm{Precons}(\mathcal{A})$ and the intersection of the permutahedron P_n and $\mathrm{Precons}(\mathcal{A})$.

The main structure of our algorithm is built on a standard online convex optimization algorithm known as online gradient descent (OGD) [33]. OGD consists of an additive update of the weight vectors, followed by projection to some convex set of interest. In our case, the convex set is $P_n \cap \mathrm{Precons}(\mathcal{A})$. Using these procedures, the regret bound of OGD can be proved to be $O(n^2\sqrt{T})$. Thus, the successive projections are apparently redundant, and only one projection to $P_n \cap \mathrm{Precons}(\mathcal{A})$ would suffice. However, the projection onto $P_n \cap \mathrm{Precons}(\mathcal{A})$ is not known to be tractable, and it contains exponentially many linear constraints. Thus, we take a different approach. Instead of performing the projection directly, we use successive projections onto $\mathrm{Precons}(\mathcal{A})$ and $P_n \cap \mathrm{Precons}(\mathcal{A})$. Below, we will show that these successive projections are the key to an efficient implementation of our algorithm. First, we will prove an α-regret bound of the proposed algorithm, and then we will show that our algorithm can be efficiently realized.

Algorithm 1. PermLearnPrec

Input: parameter $\eta > 0$.

1. Let $\boldsymbol{p}_1 = ((n+1)/2, \ldots, (n+1)/2) \in [0, n]^n$.
2. For $t = 1, \ldots, T$
   (a) (Rounding) Run **Rounding$(\boldsymbol{p}_t)$** and get $\boldsymbol{\sigma}_t \in S_n$ such that $\boldsymbol{\sigma}_t \leq (2 - 2/(n+1))\boldsymbol{p}_t$.
   (b) Incur a loss $\boldsymbol{\sigma}_t \cdot \boldsymbol{\ell}_t$.
   (c) Update $\boldsymbol{p}_{t+\frac{1}{3}}$ as $\boldsymbol{p}_{t+\frac{1}{3}} = \boldsymbol{p}_t - \eta\boldsymbol{\ell}_t$.
   (d) (1st projection) Let $\boldsymbol{p}_{t+\frac{2}{3}}$ be the Euclidean projection onto the set $\mathrm{Precons}(\mathcal{A})$, .i.e.,
   $$\boldsymbol{p}_{t+\frac{2}{3}} = \arg\min_{\boldsymbol{p} \in \mathrm{Precons}(\mathcal{A})} \|\boldsymbol{p} - \boldsymbol{p}_{t+\frac{1}{3}}\|_2^2.$$
   (e) (2nd projection) Let $\boldsymbol{p}_{t+1}$ be the projection of $\boldsymbol{p}_{t+\frac{2}{3}}$ onto the set $P_n \cap \mathrm{Precons}(\mathcal{A})$, that is,
   $$\boldsymbol{p}_{t+1} = \arg\min_{\boldsymbol{p} \in P_n \cap \mathrm{Precons}(\mathcal{A})} \|\boldsymbol{p} - \boldsymbol{p}_{t+\frac{2}{3}}\|_2^2.$$

We begin our analysis of PermLearnPrec with the following lemma. The lemma guarantees the progression of $\boldsymbol{p}_t$ towards any vector in $P_n \cap \mathrm{Precons}(\mathcal{A})$, as measured by the Euclidean norm squared.

Lemma 1. *For any $q \in P_n \cap \mathrm{Precons}(\mathcal{A})$ and for any $t \geq 1$,*

$$\|q - p_t\|_2^2 - \|q - p_{t+1}\|_2^2 \geq 2\eta(q - p_t) \cdot \ell_t - \eta^2 \|\ell_t\|_2^2.$$

Proof. By using the generalized Pythagorean theorem (e.g., [5]),

$$\|q - p_{t+\frac{2}{3}}\|_2^2 \geq \|q - p_{t+1}\|_2^2 + \|p_{t+1} - p_{t+\frac{2}{3}}\|_2^2$$

and

$$\|q - p_{t+\frac{1}{3}}\|_2^2 \geq \|q - p_{t+\frac{2}{3}}\|_2^2 + \|p_{t+\frac{2}{3}} - p_{t+\frac{1}{3}}\|_2^2.$$

By combining these, we obtain

$$
\begin{aligned}
&\|q - p_t\|_2^2 - \|q - p_{t+1}\|_2^2 \\
&\geq \|q - p_t\|_2^2 - \|q - p_{t+\frac{1}{3}}\|_2^2 + \|p_{t+1} - p_{t+\frac{2}{3}}\|_2^2 + \|p_{t+\frac{2}{3}} - p_{t+\frac{1}{3}}\|_2^2 \\
&\geq \|q - p_t\|_2^2 - \|q - p_{t+\frac{1}{3}}\|_2^2,
\end{aligned}
\tag{1}
$$

where the last inequality follows since Euclidean distance is nonnegative. Then, because $p_{t+\frac{1}{3}} = p_t - \eta \ell_t$, the right-hand side of inequality (1) is

$$\|q - p_t\|_2^2 - \|q - p_{t+\frac{1}{3}}\|_2^2 = 2\eta(q - p_t) \cdot \ell_t - \eta^2 \|\ell_t\|_2^2. \tag{2}$$

By combining (1) and (2), we complete the proof.

Lemma 2 (Cf. Zinkevich [33]). *For any $T \geq 1$ and $\eta = (n+1)/(2\sqrt{T})$,*

$$\sum_{t=1}^{T} p_t \cdot \ell_t \leq \min_{p \in P_n \cap \mathrm{Precons}(\mathcal{A})} \sum_{t=1}^{T} p \cdot \ell_t + \frac{n(n+1)}{2}\sqrt{T}.$$

Proof. By Lemma 1, summing the inequality from $t = 1$ to T and rearranging, we obtain that for any $q \in P_n \cap \mathrm{Precons}(\mathcal{A})$,

$$
\begin{aligned}
\sum_{t=1}^{T}(p_t - q) \cdot \ell_t &\leq \frac{1}{2\eta}\sum_{t=1}^{T}(\|q - p_t\|_2^2 - \|q - p_{t+1}\|_2^2) + \frac{\eta}{2}\sum_{t=1}^{T}\|\ell_t\|_2^2 \\
&= \frac{1}{2\eta}(\|q - p_1\|_2^2 - \|q - p_T\|_2^2) + \frac{\eta}{2}\sum_{t=1}^{T}\|\ell_t\|_2^2 \\
&\leq \frac{1}{2\eta}n(\frac{n+1}{2})^2 + \frac{\eta}{2}nT,
\end{aligned}
$$

where the last inequality holds since for any $i \in [n]$, $(q_i - p_{i,1})^2$ is at most $p_{1,i}^2 = (\frac{n+1}{2})^2$, and $\ell_t \in [0,1]^n$. By setting $\eta = (n+1)/(2\sqrt{T})$, we have the cumulative loss bound as desired.

4   Efficient Implementations of Projection and Rounding

In this section, we propose efficient algorithms for successive projections onto $\mathrm{Precons}(\mathcal{A})$ and $P_n \cap \mathrm{Precons}(\mathcal{A})$. We then show an implementation of the procedure Rounding.

4.1 Projection onto the Set Precons($\mathcal{A}$) of the Precedence Constraints

The problem of projection onto Precons($\mathcal{A}$) is described as follows:

$$\min_{\boldsymbol{p} \in \mathbb{R}^n} \|\boldsymbol{p} - \boldsymbol{q}\|_2^2$$

$$\text{sub.to:} \quad p_i \geq p_j, \quad \text{for } (i,j) \in \mathcal{A}.$$

This problem is known as the isotonic regression problem [21,23,27]. Previously known algorithms for the isotonic regression run in $O(mn^2 \log n)$ or $O(n^4)$ time (see [21] for details), where $m = |\mathcal{A}|$.

4.2 Projection of a Point in Precons($\mathcal{A}$) onto $P_n \cap$ Precons($\mathcal{A}$)

In this subsection, we show an efficient algorithm, which we will call Projection, for computing the projection a point in Precons($\mathcal{A}$) onto the intersection of the permutahedron P_n and the set Precons($\mathcal{A}$) of the precedence constraints. In fact, we will show that the problem can be reduced to projection onto P_n only, and thus we can use the algorithm of Suehiro et al. [28] for finding the projection onto P_n. This is shown as Algorithm 2.

Formally, the problem is stated as follows:

$$\min_{\boldsymbol{p} \in \mathbb{R}^n} \|\boldsymbol{p} - \boldsymbol{q}\|_2^2$$

$$\text{sub. to:} \sum_{j \in S} p_j \leq \sum_{j=1}^{|S|} (n + 1 - j), \quad \text{for any } S \subset [n],$$

$$\sum_{j=1}^{n} p_j = \frac{n(n+1)}{2},$$

$$p_i \geq p_j, \quad \text{for } (i,j) \in \mathcal{A}.$$

Without loss of generality, we may assume that elements in $\boldsymbol{q}$ are sorted in descending order, i.e., $q_1 \geq q_2 \geq \cdots \geq q_n$. This can be achieved in time $O(n \log n)$ by sorting $\boldsymbol{q}$. First, we show that this projection preserves the order in $\boldsymbol{q}$.

Lemma 3 (Order Preserving Lemma [28]). *Let $\boldsymbol{p}^*$ be the projection of $\boldsymbol{q}$ onto P_n s.t. $q_1 \geq q_2 \geq \cdots \geq q_n$. Then, the projection $\boldsymbol{p}^*$ also satisfies $p_1^* \geq p_2^* \geq \cdots \geq p_n^*$.*

Furthermore, we need to show that the projection onto P_n preserves equality, and this is guaranteed by the following lemma.

Lemma 4 (Equality Preserving Lemma). *Let $\boldsymbol{p}^*$ be the projection of $\boldsymbol{q}$ onto P_n. Then, the projection $\boldsymbol{p}^*$ satisfies $p_i = p_j$ if $q_i = q_j$.*

Proof. Assume that the lemma is false. Then, there exists a pair i and j such that $q_i = q_j$ and $p_i^* < p_j^*$. Let $\boldsymbol{p}'$ be the vector obtained by letting $p_i' = p_j' = (p_i^* + p_j^*)/2$ and $p_k' = p_k^*$ for $k \neq i, j$. It can be easily verified that $\boldsymbol{p}'s \in P_n$. Now, observe that

$$\|\boldsymbol{p}^* - \boldsymbol{q}\|_2^2 - \|\boldsymbol{p}' - \boldsymbol{q}\|_2^2 = p_i^{*2} + p_j^{*2} - p_i'^2 - p_j'^2 + 2\boldsymbol{p}' \cdot \boldsymbol{q} - 2\boldsymbol{p}^* \cdot \boldsymbol{q}$$
$$= p_i^{*2} + p_j^{*2} - (p_i^* + p_j^*)^2/2 + 2(p_i' - p_i^*)q_i + 2(p_j' - p_j^*)q_j$$
$$= \frac{1}{2}(p_i^* - p_j^*)^2 + 2(p_i' + p_j' - p_i^* - p_j^*)q_i$$
$$= \frac{1}{2}(p_i^* - p_j^*)^2 > 0,$$

which contradicts the fact that $\boldsymbol{p}^*$ is the projection.

Now we are ready to show one of our main technical lemmas.

Lemma 5. *For any $\boldsymbol{q} \in \mathrm{Precons}(\mathcal{A})$,*

$$\arg\min_{\boldsymbol{p} \in P_n} \|\boldsymbol{p} - \boldsymbol{q}\| = \arg\min_{\boldsymbol{p} \in P_n \cap \mathrm{Precons}(\mathcal{A})} \|\boldsymbol{p} - \boldsymbol{q}\|.$$

Proof. Let $\boldsymbol{p}^* = \arg\min_{\boldsymbol{p} \in P_n} \|\boldsymbol{p} - \boldsymbol{q}\|$. By definition of the projection, for any $\boldsymbol{p} \in P_n \cap \mathrm{Precons}(\mathcal{A}) \subseteq P_n$, $\|\boldsymbol{p} - \boldsymbol{q}\| \geq \|\boldsymbol{p}^* - \boldsymbol{q}\|$. Further, by Lemmas 3 and 4, $\boldsymbol{p}^*$ preserves the order and equality in $\boldsymbol{q}$. That is, $\boldsymbol{p}^*$ also satisfies the constraints defined by $\mathrm{Precons}(\mathcal{A})$. Therefore, we have $\boldsymbol{p}^* \in \mathrm{Precons}(\mathcal{A})$. These facts imply that $\boldsymbol{p}^*$ is indeed the projection of $\boldsymbol{q}$ onto $P_n \cap \mathrm{Precons}(\mathcal{A})$.

So, by Lemma 2, when a vector $\boldsymbol{q} \in \mathrm{Precons}(\mathcal{A})$ is given, we can compute the projection of $\boldsymbol{q}$ onto $P_n \cap \mathrm{Precons}(\mathcal{A})$ by computing the projection of $\boldsymbol{q}$ onto P_n only. By applying the projection algorithm of Suehiro et al. [28] for the base polyhedron (which generalizes the permutahedron), we obtain the following result.

Theorem 1. *There exists an algorithm with input $\boldsymbol{q} \in \mathrm{Precons}(\mathcal{A})$ that outputs the projection of $\boldsymbol{q}$ onto $P_n \cap \mathrm{Precons}(\mathcal{A})$ in time $O(n^2)$ and space $O(n)$.*

4.3 Rounding

Algorithm 3 is Rounding. The algorithm is simple. Roughly speaking, if the input $\boldsymbol{p} \in P_n \cap \mathrm{Precons}(\mathcal{A})$ is sorted as $p_1 \geq \cdots \geq p_n$, the algorithm outputs $\boldsymbol{\sigma}$ such that $\sigma_1 \geq \cdots \geq \sigma_n$, i.e., $\boldsymbol{\sigma} = (n, n-1, \ldots, 1)$. Note that we need to break ties in $\boldsymbol{p}$ to construct $\boldsymbol{\sigma}$. Let $\mathcal{A}^*$ be the transitive closure of $\mathcal{A}$. Then, given an equivalence set $\{j \mid p_i = p_j\}$, we break ties so that if $(i, j) \in \mathcal{A}^*$, $\sigma_i \geq \sigma_j$. This can be done by, e.g., quicksort. First, we will show that Rounding guarantees that for each $i \in [n]$, $\sigma_i \leq (2 - 2/(n+1))p_i$, and then discuss its time complexity.

We prove the following lemma for Rounding.

Algorithm 2. Projection onto $P_n \cap \mathrm{Precons}(\mathcal{A})$

Input: $q \in \mathrm{Precons}(\mathcal{A})$ s.t. $q_1 \geq q_2 \geq \cdots \geq q_n$.
Output: projection x of q onto P_n.

1. Let $i_0 = 0$.
2. **For** $k = 1, \ldots,$
   (a) Let $C^k(i) = \frac{g(i) - g(i_{k-1}) - \sum_{j=i_{k-1}+1}^{i} q_j}{i - i_{k-1}}$,
       where $g(i) = \sum_{j=1}^{i}(n + 1 - j)$,
       and $i_k = \arg\min_{i:i_{k-1}+1 \leq i \leq n} C^k(i)$;
       if there are multiple minimizers, choose the largest one as i_k.
   (b) Set $x_i = q_i + C^t(i_k)$ (for $i_{k-1} + 1 \leq i \leq i_k$).
   (c) **If** $i_k = n$, **then** break.
3. **Output** x.

Algorithm 3. Rounding

Input: $p \in P_n \cap \mathrm{Precons}(\mathcal{A})$ satisfying $p_1 \geq p_2 \geq \cdots \geq p_n$ and the transitive closure $\mathcal{A}^*$ of $\mathcal{A}$
Output: Permutation $\sigma \in S_n \cap \mathrm{Precons}(\mathcal{A})$

1. Sort elements of p in the descending order, where for elements i, j such that $p_i = p_j$, i is larger than j if $(i, j) \in \mathcal{A}^*$, otherwise beak the tie arbitrarily.
2. Output the permutation σ s.t. $\sigma_i = (n + 1) - r_i$, where r_i is the ordinal of i in the above order.

Lemma 6. *For any $p \in P_n \cap \mathrm{Precons}(\mathcal{A})$ s.t. $p_1 \geq \cdots \geq p_n$, given p, the output σ of Rounding satisfies that for each $i \in [n]$, $\sigma_i \leq (2 - 2/(n+1))p_i$.*

Proof. For each $i \in [n]$, by definition of the permutahedron, we have

$$\sum_{j=1}^{i} p_j \leq \sum_{j=1}^{i-1} j = \frac{i(i-1)}{2}. \tag{3}$$

By the assumption that $p_1 \geq \cdots \geq p_n$, the average of $p_i + p_{i+1} + \cdots + p_n$ is not larger than p_i. Thus, we have

$$p_i \geq \frac{\sum_{j=i}^{n} p_j}{n + 1 - i} = \frac{\sum_{j=1}^{n} p_j - \sum_{j=1}^{i-1} p_j}{n + 1 - i} \geq \frac{(n + i)(n + 1 - i)}{2(n + 1 - i)} = \frac{n + i}{2},$$

where the second inequality follows from (3). Thus, for each $i \in [n]$,

$$\frac{\sigma_i}{p_i} \leq \frac{n - i + 1}{\frac{1}{2}(n + i)} = \frac{2(n + i - 1)}{n + i} = 2 - \frac{4i - 2}{n + i}.$$

Here, the second term $\frac{4i-2}{n+i}$ is minimized when $i = 1$. Therefore, $\sigma_i/p_i \leq 2 - 2/(n+1)$, as claimed.

For computing Rounding, we need to construct the transitive closure $\mathcal{A}^*$ of $\mathcal{A}$ before the protocol begins. It is well known that a transitive closure can be computed by using algorithms for all-pairs shortest paths. For this problem, the Floyd-Warshall algorithm can be used; it runs in time $O(n^3)$ and space $O(n^2)$ (see, e.g., [8]). When $\mathcal{A}$ is small, for example, $m << n^2$, we can use Johnson's algorithm, which runs in time $O(n^2 \log n + nm)$ and space $O(m^2)$.

The time complexity of Rounding is $O(n^2)$, which is due to the sorting. The space complexity is $O(n^2)$, if we use the Floyd-Warshall algorithm with an adjacency matrix. The space complexity can be reduced to $O(m^2)$ if we employ Johnson's algorithm, which uses an adjacency list. On the other hand, we need an extra $O(\log m)$ factor in the time complexity since we need $O(\log m)$ time to check if $(i, j) \in \mathcal{A}^*$ when $\mathcal{A}^*$ is given as an adjacency list.

4.4   Main Result

We are now ready to prove the main result. From Lemma 2, Lemma 6, Theorem 1 and the fact that for any $x \in \mathbb{R}_+^n$, $\min_{p \in P_n \cap \text{Precons}(\mathcal{A})} p \cdot x \leq \min_{\sigma \in S_n \cap \text{Precons}(\mathcal{A})} \sigma \cdot x$, we immediately get the following theorem.

Theorem 2. *There exists an online linear optimization algorithm over $P_n \cap$ Precons$(\mathcal{A})$ such that*

1. *its $(2 - 2/(n + 1))$-regret is $O(n^2 \sqrt{T})$, and*
2. *its per-trial running time is $O(n^4)$.*

5   Lower Bound

In this section, we derive a lower bound for the regret for our online prediction problem over the permutahedron P_n. Here, we consider the special case of no precedence constraint being given.

Theorem 3. *For our prediction problem over the permutahedron P_n, the 1-regret is $\Omega(n^2 \sqrt{T})$.*

Proof. We consider an adversary who makes random choices. More precisely, at each trial t, the adversary randomly chooses a loss vector ℓ_t from ℓ^0, ℓ^1, where ℓ^0 (ℓ^1) is the loss vector in which the first $\frac{n}{2}$ elements are 0s (1s) and the remaining elements are 1s (0s). Then, for any online optimization algorithm that outputs $\sigma_t \in S_t$ at trial t,

$$E[\sum_{t=1}^{T} \sigma_t \cdot \ell_t] = \frac{n(n+1)T}{4}.$$

Now, let us consider the best fixed permutation. Let $\sigma^0 = (n, n-1, n-2, \ldots, 1)$ and $\sigma^1 = (1, 2, 3, 4, \ldots, n)$. Suppose that ℓ^0 appears more frequently than ℓ^1 by k. The best permutation is σ^0, and its cumulative loss is

$$\sum_{i=1}^{\frac{n}{2}} i \left(\frac{T}{2} + \frac{k}{2} \right) + \sum_{i=\frac{n}{2}+1}^{n} i \left(\frac{T}{2} - \frac{k}{2} \right)$$

$$= \frac{n(n+1)T}{4} + \frac{k}{2} \left(2 \frac{\frac{n}{2}(\frac{n}{2}+1)}{2} - \frac{n(n+1)}{2} \right) \frac{k}{2}$$

$$= \frac{n(n+1)T}{4} - \frac{k}{2} n \left(\frac{n+1}{2} - \frac{\frac{n}{2}+1}{2} \right)$$

$$= \frac{n(n+1)T}{4} - \frac{k}{2} \frac{n^2}{4}.$$

The same argument follows for the opposite case, where ℓ^1 is more frequent by k. In fact, k can be expressed as $k = \sum_{t=1}^{T} \delta_t$, where each δ_t is a discrete uniform random variable that takes values of ± 1. Then, the expected regret of any online optimization algorithm is at least $\frac{n^2}{8} E\left[\left| \sum_{t=1}^{T} \delta_t \right| \right]$. By the central limit theorem, the distribution of $\sum_{t=1}^{T} \delta_t$ converges to a Gaussian distribution with mean 0 and variance $\sqrt{T}$. Thus, for sufficiently large T, $\Pr[|\sum_{t=1}^{T} \delta_t| \geq \sqrt{T}]$ is a constant: c $(0 < c < 1)$. Therefore, the expected regret bound has a lower bound of $\frac{n^2}{8} c\sqrt{T}$. This implies that there exists a sequence of loss vectors that enforces any online optimization algorithm to incur regret that is at least $\Omega(n^2 \sqrt{T})$.

In general, this lower bound on 1-regret is tight, since there are online algorithms that achieve a 1-regret with $O(n^2 \sqrt{T})$ ([1, 28]).

It is natural to ask if the $(2 - 2/(n+1))$-regret $O(n^2 \sqrt{T})$ is tight under precedence constraints. We do not yet have a lower bound for this case, but we will show that our algorithm is optimal unless there is an offline algorithm with an approximation ratio $\alpha < 2$.

Theorem 4. *If there exists a polynomial-time online linear optimization algorithm with an α-regret of $poly(n, m)\sqrt{T}$, then there also exists a randomized polynomial-time algorithm for the offline problem with an approximation ratio α.*

Proof. The proof is based on standard online-to-offline conversion methods that can be found in the online learning literature (see, e.g., [11]). Let A be such an online linear optimization algorithm, and let its output at each trial t be denoted as σ_t. Let $\ell \in [0, 1]^n$ be the loss vector in the offline problem. We consider an adversary who returns $\ell_t = \ell$ at each trial t. Then, the cumulative loss of A divided by T is bounded as follows:

$$\frac{1}{T} \sum_{t=1}^{T} \sigma_t \cdot \ell \leq \alpha \min_{\sigma \in S_n \cap \text{Precons}(\mathcal{A})} \sigma \cdot \ell + \frac{poly(n, m)}{T}.$$

Now, let $\widehat{\sigma}$ be a uniformly and randomly chosen permutation from $\{\sigma_1, \ldots, \sigma_T\}$. Then,

$$E[\widehat{\sigma} \cdot \ell] \le \alpha \min_{\sigma \in S_n \cap \mathrm{Precons}(\mathcal{A})} \sigma \cdot \ell + \frac{poly(n, m)}{T}.$$

By setting $T = poly(n, m)$, the expected cumulative loss of $\widehat{\sigma}$ is at most α times the cumulative loss of the best permutation (with a constant additive term), which completes the proof.

6   Experiments

In this section, we show preliminary experiments with artificial data sets in order to compare the performance of our algorithm with other methods. The experiments were performed on a server with four cores of Intel Xeon CPU X5560 2.80 GHz and a memory of 198 GB. We implemented the programs using Matlab with its Optimization Toolbox. To generate the loss vector at each trial t, we independently and randomly specified each element $\ell_{t,i}$ of the loss vector ℓ_t as follows: Let $\ell_{t,i} = 1$ with probability r_i and $\ell_{t,i} = 0$, otherwise. We set $r_i = i/n$ so that $E[\ell_t] = (1/n, 2/n, ..., 1)$. We constructed random acyclic precedence constraints on n jobs in the following way. First, we constructed a random total order over n jobs (vertices). Then, we constructed an acyclic directed graph over n vertices by adding $\binom{n}{2}$ directed edges according to the total order. Finally, we kept each edge (i, j) alive with probability $\pi = 0.2$, and otherwise, we removed the edge. The resulting directed graph represented the set of precedence constraints.

Using the above method, for each fixed n and T, we constructed three random sequences of loss vectors and three random sets of precedence constraints. The results (cumulative loss or computation time) were then averaged.

We compared our algorithm PermLearnPrec (PLP) to the following algorithms. We used the offline-to-online conversion techniques of Kakade et al. [16] (KKL) and the metarounding technique of Fujita et al. [13] combined with (FPL) ([17]; FPLM). For the metarounding of Fujita et al., we set $\epsilon = 0.01$ to guarantee $(\alpha + \varepsilon)$-regret when using an α-approximation offline algorithm. We used the linear programming (LP) relaxation-based scheduling algorithm of Chudak and Hochbaum [7] as the offline algorithm. This algorithm solves a minimum-cut problem on a network with $O(n^2)$ nodes and $O(n^3)$ arcs. We used the maxflow algorithm of Boykov and Kolmogorov [4] to solve the minimum-cut problem. In our PLP algorithm, we solved the isotonic regression by using the standard quadratic programming (QP) solver in Matlab.

In Figure 1, we summarize the cumulative losses and total computation times, both averaged over three random data sets. As can be seen, the cumulative losses of all of the algorithms are quite similar. KKL performed slightly better than the other two algorithms, which is not surprising since they have almost identical α-regret bounds. On the other hand, if we consider the computation times of the algorithms, there is a very large difference. Our algorithm runs roughly 20 to 30 times faster than the other methods. The reason for this is that since the data are

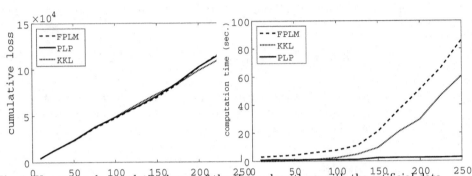

Fig. 1. Upper panel: cumulative losses of the three algorithms with the artificial data set for $n = 50$ and $T = 10, 20, 50, 100, 200, 250$; lower panel: total computation times of the three algorithms with $T = 50$ with $n = 10, 20, 50, 100, 200, 250$.

relatively "easy," the best permutation might not change frequently over time. Thus, in many trials, the projections onto the set of precedence constraints are already satisfied, and if this is the case, our algorithm can skip this step, whereas the other methods must compute the precedence constraints for every trial.

7   Conclusion

In this paper, we propose a polynomial-time online linear optimization algorithm over the permutahedron under precedence constraints. Our algorithm achieves a $(2 - 2/(n+1))$-regret bound $O(n^2\sqrt{T})$, which means that it can predict as well as the state-of-the art offline approximation algorithms in hindsight. The approximation algorithm for which the approximation ratio is strictly less than $2 - 2/(n+1)$.

An interesting open question is how our online framework can be extended to minimize the sum of weighted completion times. We note that Woeginger [29] showed that the offline problem of minimizing the sum of weighted completion time can be reduced to that of minimizing the unweighted sum. This reduction might be useful for designing an online version.

Acknowledgments. We thank anonymous reviewers for useful comments. Hatano is grateful to the supports from JSPS KAKENHI Grant Number 25330261. Takimoto is grateful to the supports from JSPS KAKENHI Grant Number 15H02667. In addition, the authors acknowledge the support from MEXT KAKENHI Grant Number 24106010 (the ELC project).

References

1. Ailon, N.: Improved bounds for online learning over the permutahedron and other ranking polytopes. In: Proceedings of 17th International Conference on Artificial Intelligence and Statistics (AISTAT 2014), pp. 29–37 (2014)

2. Ambühl, C., Mastrolilli, M., Mutsanas, N., Svensson, O.: On the Approximability of Single-Machine Scheduling with Precedence Constraints Christoph Ambühl. Mathematics of Operations Research **36**(4), 653–669 (2011)
3. Ambühl, C., Mastrolilli, M., Swensson, O.: Inapproximability results for sparsest cut, optimal linear arrangement, and precedence constrained scheduling. In: Proceedings of the 48th Annual IEEE Symposium on Foundations of Computer Science (FOCS 2007), pp. 329–337 (2007)
4. Boykov, Y., Kolmogorov, V.: An Experimental Comparison of Min-Cut/Max-Flow Algorithms for Energy Minimization in Computer Vision. IEEE Transactions on Pattern Analysis and Machine Intelligence **26**(9), 1124–1137 (2004)
5. Cesa-Bianchi, N., Lugosi, G.: Prediction, Learning, and Games. Cambridge University Press (2006)
6. Chekuri, C., Motwani, R.: Precedence constrained scheduling to minimize sum of weighted completion times on a single machine. Discrete Applied Mathematics **98**(1–2), 29–38 (1999)
7. Chudak, F.A., Hochbaum, D.S.: A half-integral linear programming relaxation for scheduling precedence-constrained jobs on a single machine. Operations Research Letters **25**, 199–204 (1999)
8. Cormen, T.H., Leiserson, C.E., Rivest, R.L., Clifford, S.: Introduction to Algorithms, 3rd edn. The MIT Press (2009)
9. Correa, J.R., Schulz, A.S.: Single-machine scheduling with precedence constraints. Mathematics of Operations Research **30**(4), 1005–1021 (2005)
10. Even-Dar, E., Kleinberg, R., Mannor, S., Mansour, Y.: Online learning for global cost functions. In: Proceedings of the 22nd Conference on Learning Theory (COLT 2009) (2009)
11. Freund, Y., Schapire, R.E.: Large Margin Classification Using the Perceptron Algorithm. Machine Learning **37**(3), 277–299 (1999)
12. Fujishige, S.: Submodular functions and optimization, 2nd edn. Elsevier Science (2005)
13. Fujita, T., Hatano, K., Takimoto, E.: Combinatorial online prediction via metarounding. In: Jain, S., Munos, R., Stephan, F., Zeugmann, T. (eds.) ALT 2013. LNCS, vol. 8139, pp. 68–82. Springer, Heidelberg (2013)
14. Hall, L.A., Schulz, A.S., Shmoys, D.B., Wein, J.: Scheduling to Minimize Average Completion Time: Off-Line and On-Line Approximation Algorithms. Mathematics of Operations Research **22**(3), 513–544 (1997)
15. Helmbold, D.P., Warmuth, M.K.: Learning Permutations with Exponential Weights. Journal of Machine Learning Research **10**, 1705–1736 (2009)
16. Kakade, S., Kalai, A.T., Ligett, L.: Playing games with approximation algorithms. SIAM Journal on Computing **39**(3), 1018–1106 (2009)
17. Kalai, A., Vempala, S.: Efficient algorithms for online decision problems. Journal of Computer and System Sciences **71**(3), 291–307 (2005)
18. Lawler, E.L.: On Sequencing jobs to minimize weighted completion time subject to precedence constraints. Annals of Discrete Mathematics **2**(2), 75–90 (1978)
19. Lenstra, J.K., Kan, A.H.G.R.: Complexity of scheduling under precedence constraints. Operations Research **26**, 22–35 (1978)
20. Lugosi, G., Papaspiliopoulos, O., Stoltz, G.: Online multi-task learning with hard constraints. In: Proceedings of the 22nd Conference on Learning Theory (COLT 2009) (2009)
21. Luss, R., Rosset, S., Shahar, M.: Efficient regularized isotonic regression with application to gene-gene interaction search. Annals of Applied Statistics **6**(1) (2012)

22. Margot, F., Queyranne, M., Wang, Y.: Decompositions, Network Flows, and a Precedence Constrained Single-Machine Scheduling ProblemDecompositions, Network Flows, and a Precedence Constrained Single-Machine Scheduling Problem. Operations Research **51**(6), 981–992 (2003)
23. Maxwell, W., Muckstadt, J.: Establishing consistent and realistic reorder intervals in production-distribution systems. Operations Research **33**, 1316–1341 (1985)
24. Mohring, H.R., Schulz, A.S., Uetz, M.: Approximation in Stochastic Scheduling: The Power of LP-based Priority Policies. Journal of the ACM **46**(6), 924–942 (1999)
25. Schulz, A.S.: Scheduling to minimize total weighted completion time: performance guarantees of LP-based heuristics and lower bounds. In: Cunningham, W.H., Queyranne, M., McCormick, S.T. (eds.) IPCO 1996. LNCS, vol. 1084, pp. 301–315. Springer, Heidelberg (1996)
26. Skutella, M., Uetz, M.: Stochastic Machine Scheduling with Precedence Constraints. SIAM Journal on Computing **34**(4), 788–802 (2005)
27. Spouge, J., Wan, H., Wilbur, W.: Least squares isotonic regression in two dimensions. J. Optimization Theory and Apps. **117**, 585–605 (2003)
28. Suehiro, D., Hatano, K., Kijima, S., Takimoto, E., Nagano, K.: Online prediction under submodular constraints. In: Bshouty, N.H., Stoltz, G., Vayatis, N., Zeugmann, T. (eds.) ALT 2012. LNCS, vol. 7568, pp. 260–274. Springer, Heidelberg (2012)
29. Woeginger, G.J.: On the approximability of average completion time scheduling under precedence constraints. In: Orejas, F., Spirakis, P.G., van Leeuwen, J. (eds.) ICALP 2001. LNCS, vol. 2076, pp. 887–897. Springer, Heidelberg (2001)
30. Woeginger, G.J., Schuurman, P.: Polynomial time approximation algorithms for machine scheduling: Ten open problems. Journal of Scheduling **2**, 203–213 (1999)
31. Yasutake, S., Hatano, K., Kijima, S., Takimoto, E., Takeda, M.: Online linear optimization over permutations. In: Asano, T., Nakano, S., Okamoto, Y., Watanabe, O. (eds.) ISAAC 2011. LNCS, vol. 7074, pp. 534–543. Springer, Heidelberg (2011)
32. Ziegler, G.M.: Lectures on Polytopes. Graduate Texts in Mathematics, vol. 152. Springer-Verlag (1995)
33. Zinkevich, M.: Online convex programming and generalized infinitesimal gradient ascent. In: Proceedings of the Twentieth International Conference on Machine Learning (ICML 2003), pp. 928–936 (2003)

Kolmogorov Complexity, Algorithmic Information Theory

Solomonoff Induction Violates Nicod's Criterion

Jan Leike$^{(\boxtimes)}$ and Marcus Hutter

Australian National University, Canberra, Australia
{jan.leike,marcus.hutter}@anu.edu.au

Abstract. *Nicod's criterion* states that observing a black raven is evidence for the hypothesis H that all ravens are black. We show that Solomonoff induction does not satisfy Nicod's criterion: there are time steps in which observing black ravens *decreases* the belief in H. Moreover, while observing any computable infinite string compatible with H, the belief in H decreases infinitely often when using the unnormalized Solomonoff prior, but only finitely often when using the normalized Solomonoff prior. We argue that the fault is not with Solomonoff induction; instead we should reject Nicod's criterion.

Keywords: Bayesian reasoning · Confirmation · Disconfirmation · Hempel's paradox · Equivalence condition · Solomonoff normalization

1 Introduction

Inductive inference, how to generalize from examples, is the cornerstone of scientific investigation. But we cannot justify the use of induction on the grounds that it has reliably worked before, because this argument presupposes induction. Instead, we need to give *deductive* (logical) arguments for the use of induction. Today we know a formal solution to the problem of induction: Solomonoff's theory of learning [16,17], also known as *universal induction* or *Solomonoff induction*. It is a method of induction based on Bayesian inference [9] and algorithmic probability [11]. Because it is solidly founded in abstract mathematics, it can be justified purely deductively.

Solomonoff defines a prior probability distribution M that assigns to a string x the probability that a universal monotone Turing machine prints something starting with x when fed with fair coin flips: Solomonoff's prior encompasses *Ockham's razor* by favoring simple explanations over complex ones: algorithmically simple strings have short programs and are thus assigned higher probability than complex strings that do not have short programs. Moreover, Solomonoff's prior respects *Epicurus' principle* of multiple explanation by never discarding possible explanations: any possible program that explains the string contributes to the probability [8].

For data drawn from a computable probability distribution μ, Solomonoff induction will converge to the correct belief about any hypothesis [1]. Moreover, this can be used to produce reliable predictions extremely fast: Solomonoff induction will make a total of at most $E+O(\sqrt{E})$ errors when predicting the next data

© Springer International Publishing Switzerland 2015
K. Chaudhuri et al. (Eds.): ALT 2015, LNAI 9355, pp. 349–363, 2015.
DOI: 10.1007/978-3-319-24486-0_23

points, where E is the number of errors of the informed predictor that knows μ [7]. In this sense, Solomonoff induction solves the induction problem [15]. It is incomputable, hence it can only serve as an ideal that any practical learning algorithm should strive to approximate.

But does Solomonoff induction live up to this ideal? Suppose we entertain the hypothesis H that all ravens are black. Since this is a universally quantified statement, it is refuted by observing one counterexample: a non-black raven. But at any time step, we have observed only a finite number of the potentially infinite number of possible cases. Nevertheless, Solomonoff induction maximally confirms the hypothesis H asymptotically.

This paper is motivated by a problem of inductive inference extensively discussed in the literature: the *paradox of confirmation*, also known as *Hempel's paradox* [5]. It relies on the following three principles.

- *Nicod's criterion* [14, p.67]: observing an F that is a G increases our belief in the hypothesis that all Fs are Gs.
- *The equivalence condition*: logically equivalent hypothesis are confirmed or disconfirmed by the same evidence.
- *The paradoxical conclusion*: a green apple confirms H.

The argument goes as follows. The hypothesis H is logically equivalent to the hypothesis H' that all non-black objects are non-ravens. According to Nicod's criterion, any non-black non-raven, such as a green apple, confirms H'. But then the equivalence condition entails the paradoxical conclusion.

The paradox of confirmation has been discussed extensively in the literature on the philosophy of science [3–6,12,13,19]; see [18] for a survey. Support for Nicod's criterion is not uncommon [6,12,13] and no consensus is in sight.

Using results from algorithmic information theory we show that Solomonoff induction avoids the paradoxical conclusion because it does not fulfill Nicod's criterion. There are time steps when (counterfactually) observing a black raven disconfirms the hypothesis that all ravens are black (Theorem 7 and Corollary 12). In the deterministic setting Nicod's criterion is even violated infinitely often (Theorem 8 and Corollary 13). However, if we *normalize* Solomonoff's prior and observe a deterministic computable infinite string, Nicod's criterion is violated at most finitely many times (Theorem 11). Our results are independent of the choice of the universal Turing machine. A list of notation can be found on page 363.

2   Preliminaries

Let $\mathcal{X}$ be some finite set called *alphabet*. The set $\mathcal{X}^* := \bigcup_{n=0}^{\infty} \mathcal{X}^n$ is the set of all finite strings over the alphabet $\mathcal{X}$, and the set $\mathcal{X}^\infty$ is the set of all infinite strings over the alphabet $\mathcal{X}$. The empty string is denoted by ϵ, not to be confused with the small positive rational number ε. Given a string $x \in \mathcal{X}^*$, we denote its length by $|x|$. For a (finite or infinite) string x of length $\geq k$, we denote with $x_{1:k}$ the first k characters of x, and with $x_{<k}$ the first $k-1$ characters of x. The notation

$x_{1:\infty}$ stresses that x is an infinite string. We write $x \sqsubseteq y$ iff x is a prefix of y, i.e., $x = y_{1:|x|}$.

A *semimeasure* over the alphabet $\mathcal{X}$ is a probability measure on the probability space $\mathcal{X}^{\sharp} := \mathcal{X}^* \cup X^{\infty}$ whose σ-algebra is generated by the *cylinder sets* $\Gamma_x := \{xz \mid z \in \mathcal{X}^{\sharp}\}$ [11, Ch.4.2]. If a semimeasure assigns zero probability to every finite string, then it is called a *measure*. Measures and semimeasures are uniquely defined by their values on cylinder sets. For convenience we identify a string $x \in \mathcal{X}^*$ with its cylinder set Γ_x.

For two functions $f, g : \mathcal{X}^* \to \mathbb{R}$ we use the notation $f \overset{\times}{\geq} g$ iff there is a constant $c > 0$ such that $f(x) \geq cg(x)$ for all $x \in \mathcal{X}^*$. Moreover, we define $f \overset{\times}{\leq} g$ iff $g \overset{\times}{\geq} f$ and we define $f \overset{\times}{=} g$ iff $f \overset{\times}{\leq} g$ and $f \overset{\times}{\geq} g$. Note that $f \overset{\times}{=} g$ does *not* imply that there is a constant c such that $f(x) = cg(x)$ for all x.

Let U denote some universal Turing machine. The *Kolmogorov complexity* $K(x)$ of a string x is the length of the shortest program on U that prints x and then halts. A string x is *incompressible* iff $K(x) \geq |x|$. We define $m(t) := \min_{n \geq t} K(n)$, the *monotone lower bound on K*. Note that m grows slower than any unbounded computable function. (Its inverse is a version of the *busy beaver* function.) We also use the same machine U as a monotone Turing machine by ignoring the halting state and using a write-only output tape. The *monotone Kolmogorov complexity* $Km(x)$ denotes the length of the shortest program on the monotone machine U that prints a string starting with x. Since monotone complexity does not require the machine to halt, there is a constant c such that $Km(x) \leq K(x) + c$ for all $x \in X^*$.

Solomonoff's prior M [16] is defined as the probability that the universal monotone Turing machine computes a string when fed with fair coin flips in the input tape. Formally,

$$M(x) := \sum_{p:\, x \sqsubseteq U(p)} 2^{-|p|}.$$

Equivalently, the Solomonoff prior M can be defined as a mixture over all lower semicomputable semimeasures [20].

The function M is a lower semicomputable semimeasure, but not computable and not a measure [11, Lem.4.5.3]. It can be turned into a measure M_{norm} using *Solomonoff normalization* [11, Sec.4.5.3]: $M_{\text{norm}}(\epsilon) := 1$ and for all $x \in \mathcal{X}^*$ and $a \in \mathcal{X}$,

$$M_{\text{norm}}(xa) := M_{\text{norm}}(x) \frac{M(xa)}{\sum_{b \in \mathcal{X}} M(xb)} \tag{1}$$

since $M(x) > 0$ for all $x \in \mathcal{X}^*$.

Every program contributes to M, so we have that $M(x) \geq 2^{-Km(x)}$. However, the upper bound $M(x) \overset{\times}{\leq} 2^{-Km(x)}$ is generally false [2]. Instead, the following weaker statement holds.

Lemma 1 ([10] as cited in [2, p.75]). *Let $E \subset \mathcal{X}^*$ be a recursively enumerable and prefix-free set. Then there is a constant $c_E \in \mathbb{N}$ such that $M(x) \leq 2^{-Km(x)+c_E}$ for all $x \in E$.*

Proof. Define

$$\nu(x) := \begin{cases} M(x), & \text{if } x \in E, \text{ and} \\ 0, & \text{otherwise.} \end{cases}$$

The semimeasure ν is lower semicomputable because E is recursively enumerable. Furthermore, $\sum_{x \in \mathcal{X}^*} \nu(x) \le 1$ because M is a semimeasure and E is prefix-free. Therefore ν is a discrete semimeasure. Hence there are constant c and c' such that $Km(x) \le K(x) + c \le -\log \nu(x) + c + c' = -\log M(x) + c + c'$ [11, Cor.4.3.1]. $\square$

Lemma 2 ([11, Sec.4.5.7]). *For any computable measure μ the set of μ-Martin-Löf-random sequences has μ-probability one:*

$$\mu(\{x \in \mathcal{X}^\infty \mid \exists c \forall t.\ M(x_{1:t}) \le c\mu(x_{1:t})\}) = 1.$$

3   Solomonoff and the Black Ravens

Setup. In order to formalize the black raven problem (in line with [15, Sec.7.4]), we define two predicates: blackness B and ravenness R. There are four possible observations: a black raven BR, a non-black raven $\overline{B}R$, a black non-raven $B\overline{R}$, and a non-black non-raven $\overline{B}\overline{R}$. Therefore our alphabet consists of four symbols corresponding to each of the possible observations, $\mathcal{X} := \{BR, \overline{B}R, B\overline{R}, \overline{B}\overline{R}\}$. We will not make the formal distinction between observations and the symbols that represent them, and simply use both interchangeably.

We are interested in the hypothesis 'all ravens are black'. Formally, it corresponds to the set

$$H := \{x \in \mathcal{X}^\sharp \mid x_t \ne \overline{B}R\ \forall t\} = \{BR, B\overline{R}, \overline{B}\overline{R}\}^\sharp, \tag{2}$$

the set of all finite and infinite strings in which the symbol $\overline{B}R$ does not occur. Let $H^c := \mathcal{X}^\sharp \setminus H$ be the complement hypothesis 'there is at least one non-black raven'. We fix the definition of H and H^c for the rest of this paper.

Using Solomonoff induction, our prior belief in the hypothesis H is

$$M(H) = \sum_{p:\, U(p) \in H} 2^{-|p|},$$

the cumulative weight of all programs that do not print any non-black ravens. In each time step t, we make one observation $x_t \in \mathcal{X}$. Our *history* $x_{<t} = x_1 x_2 \ldots x_{t-1}$ is the sequence of all previous observations. We update our belief with Bayes' rule in accordance with the Bayesian framework for learning [9]: our *posterior belief* in the hypothesis H is

$$M(H \mid x_{1:t}) = \frac{M(H \cap x_{1:t})}{M(x_{1:t})}.$$

We say that the observation x_t *confirms* the hypothesis H iff $M(H \mid x_{1:t}) > M(H \mid x_{<t})$ (the belief in H increases), and we say that the observation x_t *disconfirms* the hypothesis H iff $M(H \mid x_{1:t}) < M(H \mid x_{<t})$ (the belief in H decreases). If $M(H \mid x_{1:t}) = 0$, we say that H is *refuted*, and if $M(H \mid x_{1:t}) \to 1$ as $t \to \infty$, we say that H is *(maximally) confirmed asymptotically.*

Confirmation and Refutation. Let the sequence $x_{1:\infty}$ be sampled from a computable measure μ, the *true environment*. If we observe a non-black raven, $x_t = \overline{B}R$, the hypothesis H is refuted since $H \cap x_{1:t} = \emptyset$ and this implies $M(H \mid x_{1:t}) = 0$. In this case, our enquiry regarding H is settled. For the rest of this paper, we focus on the interesting case: we assume our hypothesis H is in fact true in μ ($\mu(H) = 1$), i.e., μ does not generate any non-black ravens. Since Solomonoff's prior M dominates all computable measures, there is a constant w_μ such that

$$\forall x \in \mathcal{X}^* \quad M(x) \geq w_\mu \mu(x). \tag{3}$$

Thus Blackwell and Dubins' famous merging of opinions theorem [1] implies

$$M(H \mid x_{1:t}) \to 1 \text{ as } t \to \infty \text{ with } \mu\text{-probability one.}[1] \tag{4}$$

Therefore our hypothesis H is confirmed asymptotically [15, Sec.7.4]. However, convergence to 1 is extremely slow, slower than any unbounded computable function, since $1 - M(H \mid x_{1:t}) \overset{\times}{\geq} 2^{-m(t)}$ for all t.

In our setup, the equivalence condition holds trivially: a logically equivalent way of formulating a hypothesis yields the same set of infinite strings, therefore in our formalization it constitutes the same hypothesis. The central question of this paper is Nicod's criterion, which refers to the assertion that BR and $\overline{B}R$ confirm H, i.e., $M(H \mid x_{1:t}BR) > M(H \mid x_{<t})$ and $M(H \mid x_{1:t}\overline{B}R) > M(H \mid x_{<t})$ for all strings $x_{<t}$.

4   Disconfirming H

We first illustrate the violation of Nicod's criterion by defining a particular universal Turing machine.

Example 3 (Black Raven Disconfirms). The observation of a black raven can falsify a short program that supported the hypothesis H. Let $\varepsilon > 0$ be a small rational number. We define a semimeasure ρ as follows.

$$\rho(\overline{B}R^\infty) := \tfrac{1}{2} \quad \rho(BR^\infty) := \tfrac{1}{4} \quad \rho(BR\,\overline{B}R^\infty) := \tfrac{1}{4} - \varepsilon \quad \rho(x) := 0 \text{ otherwise.}$$

To get a universally dominant semimeasure ξ, we mix ρ with the universally dominant semimeasure M.

$$\xi(x) := \rho(x) + \varepsilon M(x).$$

For computable ε, the mixture ξ is a lower semicomputable semimeasure. Hence there is a universal monotone Turing machine whose Solomonoff prior is equal to ξ [20, Lem.13]. Our a priori belief in H at time $t = 0$ is

$$\xi(H \mid \epsilon) = \xi(H) \geq \rho(\overline{B}R^\infty) + \rho(BR^\infty) = 75\%,$$

---

[1] Blackwell-Dubins' theorem refers to (probability) measures, but technically M is a semimeasure. However, we can view M as a measure by introducing an extra symbol to our alphabet [11, p.264]. This preserves dominance (3), and hence absolute continuity, which is the precondition for Blackwell-Dubins' theorem.

$M(\cdot)$	H	H^c
$\bigcup_{a \neq x_t} \Gamma_{x_{<t}a}$	A	B
$\Gamma_{x_{1:t}}$	C	D
$\{x_{<t}\}$	E	0

$$A := \sum_{a \neq x_t} M(x_{<t}a \cap H)$$

$$B := \sum_{a \neq x_t} M(x_{<t}a \cap H^c)$$

$$C := M(x_{1:t} \cap H)$$

$$D := M(x_{1:t} \cap H^c)$$

$$E := M(x_{<t}) - \sum_{a \in \mathcal{X}} M(x_{<t}a)$$

Fig. 1. The definitions of the values A, B, C, D, and E. Note that by assumption, $x_{<t}$ does not contain non-black ravens, therefore $M(\{x_{<t}\} \cap H^c) = M(\emptyset) = 0$.

while our a posteriori belief in H after seeing a black raven is

$$\xi(H \mid BR) = \frac{\xi(H \cap BR)}{\xi(BR)} \leq \frac{\rho(BR^\infty) + \varepsilon}{\rho(BR^\infty) + \rho(BR\overline{BR}^\infty)} = \frac{\frac{1}{4} + \varepsilon}{\frac{1}{2} - \varepsilon} < 75\%$$

for $\varepsilon \leq 7\%$. Hence observing a black raven in the first time step disconfirms the hypothesis H.   ◇

The rest of this section is dedicated to show that this effect occurs independent of the universal Turing machine U and on all computable infinite strings.

4.1   Setup

At time step t, we have seen the history $x_{<t}$ and now update our belief using the new symbol x_t. To understand what happens, we split all possible programs into five categories.

(a) Programs that *never* print non-black ravens (compatible with H), but become falsified at time step t because they print a symbol other than x_t.
(b) Programs that eventually print a non-black raven (contradict H), but become falsified at time step t because they print a symbol other than x_t.
(c) Programs that *never* print non-black ravens (compatible with H), and predict x_t correctly.
(d) Programs that eventually print a non-black raven (contradict H), and predict x_t correctly.
(e) Programs that do not print additional symbols after printing $x_{<t}$ (because they go into an infinite loop).

Let A, B, C, D, and E denote the cumulative contributions of these five categories of programs to M. A formal definition is given in Figure 1, and implicitly depends on the current time step t and the observed string $x_{1:t}$. The values of A,

B, C, D, and E are in the interval $[0,1]$ since they are probabilities. Moreover, the following holds.

$$M(x_{<t}) = A + B + C + D + E \qquad\qquad M(x_{1:t}) = C + D \qquad (5)$$

$$M(x_{<t} \cap H) = A + C + E \qquad\qquad M(x_{1:t} \cap H) = C \qquad (6)$$

$$M(H \mid x_{<t}) = \frac{A+C+E}{A+B+C+D+E} \qquad\qquad M(H \mid x_{1:t}) = \frac{C}{C+D} \qquad (7)$$

We use results from algorithmic information theory to derive bounds on A, B, C, D, and E. This lets us apply the following lemma which states a necessary and sufficient condition for confirmation/disconfirmation at time step t.

Lemma 4 (Confirmation Criterion). *Observing x_t confirms (disconfirms) the hypothesis H if and only if $AD + DE < BC$ $(AD + DE > BC)$.*

Proof. The hypothesis H is confirmed if and only if

$$M(H \mid x_{1:t}) - M(H \mid x_{<t}) \overset{(7)}{=} \frac{C}{C+D} - \frac{A+C+E}{A+B+C+D+E} = \frac{BC-AD-DE}{(A+B+C+D+E)(C+D)}$$

is positive. Since the denominator is positive, this is equivalent to $BC > AD + DE$. $\square$

Example 5 (Confirmation Criterion Applied to Example 3). In Example 3 we picked a particular universal prior and $x_1 = BR$. In this case, the values for A, B, C, D, and E are

$$A \in [\tfrac{1}{2}, \tfrac{1}{2} + \varepsilon] \quad B \in [0, \varepsilon] \quad C \in [\tfrac{1}{4}, \tfrac{1}{4} + \varepsilon] \quad D \in [\tfrac{1}{4} - \varepsilon, \tfrac{1}{4}] \quad E \in [0, \varepsilon].$$

We invoke Lemma 4 with $\varepsilon := 7\%$ to get that $x_1 = BR$ disconfirms H:

$$AD + DE \geq \tfrac{1}{8} - \tfrac{\varepsilon}{2} = 0.09 > 0.0224 = \tfrac{1}{4} + \varepsilon^2 \geq BC. \qquad \Diamond$$

Lemma 6 (Bounds on $ABCDE$). *Let $x_{1:\infty} \in H$ be some computable infinite string. The following statements hold for every time step t.*

(i) $0 < A, B, C, D, E < 1$

(ii) $A + B \overset{\times}{\leq} 2^{-K(t)}$

(iii) $A, B \overset{\times}{\geq} 2^{-K(t)}$

(iv) $C \overset{\times}{\geq} 1$

(v) $D \overset{\times}{\geq} 2^{-m(t)}$

(vi) $D \to 0$ *as* $t \to \infty$

(vii) $E \to 0$ *as* $t \to \infty$

Proof. Let p be a program that computes the infinite string $x_{1:\infty}$.

(i) Each of A, B, C, D, E is a probability value and hence bounded between 0 and 1. These bounds are strict because for any finite string there is a program that prints that string.

(ii) A proof is given in the appendix of [8]. Let $a \neq x_t$ and let q be the shortest program for the string $x_{<t}a$, i.e., $|q| = Km(x_{<t}a)$. We can reconstruct t by running p and q in parallel and counting the number of characters printed until their output differs. Therefore there is a constant c independent of t such that $K(t) \leq |p| + |q| + c = |p| + Km(x_{<t}a) + c$. Hence

$$2^{-Km(x_{<t}a)} \leq 2^{-K(t)+|p|+c} \tag{8}$$

The set $E := \{x_{<t}a \mid t \in \mathbb{N}, a \neq x_t\}$ is recursively enumerable and prefix-free, so Lemma 1 yields a constant c_E such that

$$M(x_{<t}a) \leq 2^{-Km(x_{<t}a)+c_E} \overset{(8)}{\leq} 2^{-K(t)+|p|+c+c_E}.$$

With $A + B \leq (\#\mathcal{X} - 1) \max_{a \neq x_t} M(x_{<t}a)$ follows the claim.

(iii) Let $a \neq x_t$ and let q be the shortest program to compute t, i.e., $|q| = K(t)$. We can construct a program that prints $x_{<t}a\overline{BR}$ by first running q to get t and then running p until it has produced a string of length $t - 1$, and then printing $a\overline{BR}$. Hence there is a constant c independent of t such that $Km(x_{<t}a\overline{BR}) \leq |q| + |p| + c = K(t) + |p| + c$. Therefore

$$M(x_{<t}a \cap H^c) \geq M(x_{<t}a\overline{BR}) \geq 2^{-Km(x_{<t}a\overline{BR})} \geq 2^{-K(t)-|p|-c}.$$

For the bound on $M(x_{<t}a \cap H)$ we proceed analogously except that instead of printing $\overline{BR}$ the program goes into an infinite loop.

(iv) Since by assumption the program p computes $x_{1:\infty} \in H$, we have that $M(x_{1:t} \cap H) \geq 2^{-|p|}$.

(v) Let n be an integer such that $K(n) = m(t)$. We proceed analogously to (iii) with a program q that prints n such that $|q| = m(t)$. Next, we write a program that produces the output $x_{1:n}\overline{BR}$, which yields a constant c independent of t such that

$$M(x_{1:t} \cap H^c) \geq M(x_{1:n}\overline{BR}) \geq 2^{-Km(x_{1:n}\overline{BR})} \geq 2^{-|q|-|p|-c} = 2^{-m(t)-|p|-c}.$$

(vi) This follows from Blackwell and Dubins' result (4):

$$D = (C + D)\left(1 - \frac{C}{C+D}\right) \leq (1 + 1)(1 - M(H \mid x_{1:t})) \to 0 \text{ as } t \to \infty.$$

(vii) $\sum_{t=1}^{\infty} M(\{x_{<t}\}) = M(\{x_{<t} \mid t \in \mathbb{N}\}) \leq 1$, thus $E = M(\{x_{<t}\}) \to 0$.   □

Lemma 6 states the bounds that illustrate the ideas to our results informally: From $A \overset{\times}{=} B \overset{\times}{=} 2^{-K(t)}$ (iii,iii) and $C \overset{\times}{=} 1$ (iv) we get

$$AD \overset{\times}{=} 2^{-K(t)}D, \qquad\qquad BC \overset{\times}{=} 2^{-K(t)}.$$

According to Lemma 4, the sign of $AD + DE - BC$ tells us whether our belief in H increases (negative) or decreases (positive).

Since $D \to 0$ (vi), the term $AD \overset{\times}{=} 2^{-K(t)}D$ will eventually be smaller than $BC \overset{\times}{=} 2^{-K(t)}$. Therefore it is crucial how fast $E \to 0$ (vii). If we use M, then $E \to 0$ slower than $D \to 0$ (v), therefore $AD+DE-BC$ is positive infinitely often (Theorem 8). If we use M_{norm} instead of M, then $E = 0$ and hence $AD + DE - BC = AD - BC$ is negative except for a finite number of steps (Theorem 11).

4.2   Unnormalized Solomonoff Prior

Theorem 7 (Counterfactual Black Raven Disconfirms H). *Let $x_{1:\infty}$ be a computable infinite string such that $x_{1:\infty} \in H$ ($x_{1:\infty}$ does not contain any non-black ravens) and $x_t \neq BR$ infinitely often. Then there is a time step $t \in \mathbb{N}$ (with $x_t \neq BR$) such that $M(H \mid x_{<t}BR) < M(H \mid x_{<t})$.*

Proof. Let t be time step such that $x_t \neq BR$. From the proof of Lemma 6 (iii) we get $M(H^c \cap x_{\dot{<}t}BR) \geq 2^{-K(t)-c}$ and thus

$$M(H \mid x_{<t}BR) \leq \frac{M(H \cap x_{<t}BR) + M(H^c \cap x_{<t}BR) - 2^{-K(t)-c}}{M(x_{<t}BR)}$$

$$= 1 - \frac{2^{-K(t)-c}}{M(x_{<t}BR)} \leq 1 - \frac{2^{-K(t)-c}}{A+B} \stackrel{(iii)}{\leq} 1 - 2^{-c-c'}.$$

From (4) there is a t_0 such that for all $t \geq t_0$ we have $M(H \mid x_{<t}) > 1 - 2^{-c-c'} \geq M(H \mid x_{<t}BR)$. Since $x_t \neq BR$ infinitely often according to the assumption, there is a $x_t \neq BR$ for $t \geq t_0$.   □

Note that the black raven in Theorem 7 that we observe at time t is *counterfactual*, i.e., not part of the sequence $x_{1:\infty}$. If we picked the binary alphabet $\{BR, \overline{BR}\}$ and denoted only observations of ravens, then Theorem 7 would not apply: the only infinite string in H is BR^∞ and the only counterfactual observation is $\overline{BR}$, which immediately falsifies the hypothesis H. The following theorem gives an on-sequence result.

Theorem 8 (Disconfirmation Infinitely Often for M). *Let $x_{1:\infty}$ be a computable infinite string such that $x_{1:\infty} \in H$ ($x_{1:\infty}$ does not contain any non-black ravens). Then $M(H \mid x_{1:t}) < M(H \mid x_{<t})$ for infinitely many time steps $t \in \mathbb{N}$.*

Proof. We show that there are infinitely many $n \in \mathbb{N}$ such that for each n there is a time step $t > n$ where the belief in H decreases. The ns are picked to have low Kolmogorov complexity, while the ts are incompressible. The crucial insight is that a program that goes into an infinite loop at time t only needs to know n and not t, thus making this program much smaller than $K(t) \geq \log t$.

Let q_n be a program that starting with $t = n + 1$ incrementally outputs $x_{1:t}$ as long as $K(t) < \log t$. Formally, let $\phi(y, k)$ be a computable function such that $\phi(y, k+1) \leq \phi(y, k)$ and $\lim_{k \to \infty} \phi(y, k) = K(y)$.

```
program qn:
    t  :=  n + 1
    output x<t
    while true:
        k  :=  0
        while φ(t, k) ≥ log t:
            k  :=  k + 1
        output xt
        t  :=  t + 1
```

The program q_n only needs to know p and n, so we have that $|q_n| \leq K(n) + c$ for some constant c independent of n and t. For the smallest $t > n$ with $K(t) \geq \log t$, the program q_n will go into an infinite loop and thus fail to print a t-th character. Therefore

$$E = M(\{x_{<t}\}) \geq 2^{-|q_n|} \geq 2^{-K(n)-c}. \tag{9}$$

Incompressible numbers are very dense, and a simple counting argument shows that there must be one between n and $4n$ [11, Thm.3.3.1(i)]. Furthermore, we can assume that n is large enough such that $m(4n) \leq m(n) + 1$ (since m grows slower than the logarithm). Then

$$m(t) \leq m(4n) \leq m(n) + 1 \leq K(n) + 1. \tag{10}$$

Since the function m grows slower than any unbounded computable function, we find infinitely many n such that

$$K(n) \leq \tfrac{1}{2}(\log n - c - c' - c'' - 1), \tag{11}$$

where c' and c'' are the constants from Lemma 6 (iii,v). For each such n, there is a $t > n$ with $K(t) \geq \log t$, as discussed above. This entails

$$m(t) + K(n) + c + c'' \overset{(10)}{\leq} 2K(n) + 1 + c + c'' \overset{(11)}{\leq} \log n - c' \leq \log t - c' \leq K(t) - c'. \tag{12}$$

From Lemma 6 we get

$$AD + DE \overset{(i)}{>} DE \overset{(9),(v)}{\geq} 2^{-m(t)-c-K(n)-c''} \overset{(12)}{\geq} 2^{-K(t)+c'} \overset{(i,iii)}{\geq} BC.$$

With Lemma 4 we conclude that x_t disconfirms H. $\square$

To get that M violates Nicod's criterion infinitely often, we apply Theorem 8 to the computable infinite string BR^∞.

4.3 Normalized Solomonoff Prior

In this section we show that for computable infinite strings, our belief in the hypothesis H is non-increasing at most finitely many times if we normalize M.

For this section we define A', B', C', D', and E' analogous to A, B, C, D, and E as given in Figure 1 with M_{norm} instead of M.

Lemma 9 ($M_{\mathrm{norm}} \geq M$). $M_{\mathrm{norm}}(x) \geq M(x)$ for all $x \in \mathcal{X}^*$.

Proof. We use induction on the length of x: $M_{\mathrm{norm}}(\epsilon) = 1 = M(\epsilon)$ and

$$M_{\mathrm{norm}}(xa) = \frac{M_{\mathrm{norm}}(x)M(xa)}{\sum_{b \in \mathcal{X}} M(xb)} \geq \frac{M(x)M(xa)}{\sum_{b \in \mathcal{X}} M(xb)} \geq \frac{M(x)M(xa)}{M(x)} = M(xa).$$

The first inequality holds by induction hypothesis and the second inequality uses the fact that M is a semimeasure. $\square$

The following lemma states the same bounds for M_{norm} as given in Lemma 6 except for (i) and (vii).

Lemma 10 (Bounds on $A'B'C'D'E'$). *Let $x_{1:\infty} \in H$ be some infinite string computed by program p. The following statements hold for all time steps t.*

(i) $A \leq A'$, $B \leq B'$,
$\quad C \leq C'$, $D \leq D'$
(ii) $A' + B' \overset{\times}{\leq} 2^{-K(t)}$
(iii) $A', B' \overset{\times}{\geq} 2^{-K(t)}$

(iv) $C' \overset{\times}{\geq} 1$
(v) $D' \overset{\times}{\geq} 2^{-m(t)}$
(vi) $D' \to 0$ as $t \to \infty$
(vii) $E' = 0$

Proof.   (i) Follows from Lemma 9.

(ii) Let $a \neq x_t$. From Lemma 6 (iii) we have $M(x_{<t}a) \overset{\times}{\leq} 2^{-K(t)}$. Thus

$$M_{\mathrm{norm}}(x_{<t}a) \overset{(1)}{=} \frac{M_{\mathrm{norm}}(x_{<t})M(x_{<t}a)}{\sum_{b \in \mathcal{X}} M(x_{<t}b)} \overset{\times}{\leq} \frac{M_{\mathrm{norm}}(x_{<t})2^{-K(t)}}{\sum_{b \in \mathcal{X}} M(x_{<t}b)} \overset{\times}{\leq} 2^{-K(t)}.$$

The last inequality follows from $\sum_{b \in \mathcal{X}} M(x_{<t}b) \geq M(x_{1:t}) \overset{\times}{\geq} 1$ (Lemma 6 (iv)) and $M_{\mathrm{norm}}(x_{<t}) \leq 1$.

(iii-v) This is a consequence of (i) and Lemma 6 (iii-v).

(vi) Blackwell and Dubins' result also applies to M_{norm}, therefore the proof of Lemma 6 (vi) goes through unchanged.

(vii) Since M_{norm} is a measure, it assigns zero probability to finite strings, i.e., $M_{\mathrm{norm}}(\{x_{<t}\}) = 0$, hence $E' = 0$.   □

Theorem 11 (Disconfirmation Finitely Often for M_{norm}). *Let $x_{1:\infty}$ be a computable infinite string such that $x_{1:\infty} \in H$ ($x_{1:\infty}$ does not contain any non-black ravens). Then there is a time step t_0 such that $M_{\mathrm{norm}}(H \mid x_{1:t}) > M_{\mathrm{norm}}(H \mid x_{<t})$ for all $t \geq t_0$.*

Intuitively, at time step t_0, M_{norm} has learned that it is observing the infinite string $x_{1:\infty}$ and there are no short programs remaining that support the hypothesis H but predict something other than $x_{1:\infty}$.

Proof. We use Lemma 10 (iii,iii,iv,vii) to conclude

$$A'D' + D'E' - B'C' \leq 2^{-K(t)+c}D' + 0 - 2^{-K(t)-c'-c''} \leq 2^{-K(t)+c}(D' - 2^{-c-c'-c''}).$$

From Lemma 10 (vi) we have that $D' \to 0$, so there is a t_0 such that for all $t \geq t_0$ we have $D' < 2^{-c-c'-c''}$. Thus $A'D' + D'E' - B'C'$ is negative for $t \geq t_0$. Now Lemma 4 entails that the belief in H increases.   □

Interestingly, Theorem 11 does not hold for M since that would contradict Theorem 8. The reason is that there are quite short programs that produce $x_{<t}$, but do not halt after that. However, from p and $x_{<t}$ we cannot reconstruct t, hence a program for $x_{<t}$ does not give us a bound on $K(t)$.

Since we get the same bounds for M_{norm} as in Lemma 6, the result of Theorem 7 transfers to M_{norm}:

Corollary 12 (Counterfactual Black Raven Disconfirms H). *Let $x_{1:\infty}$ be a computable infinite string such that $x_{1:\infty} \in H$ ($x_{1:\infty}$ does not contain any non-black ravens) and $x_t \neq BR$ infinitely often. Then there is a time step $t \in \mathbb{N}$ (with $x_t \neq BR$) such that $M_{\mathrm{norm}}(H \mid x_{<t}BR) < M_{\mathrm{norm}}(H \mid x_{<t})$.*

For incomputable infinite strings the belief in H can decrease infinitely often:

Corollary 13 (Disconfirmation Infinitely Often for M_{norm}). *There is an (incomputable) infinite string $x_{1:\infty} \in H$ such that $M_{\mathrm{norm}}(H \mid x_{1:t}) < M_{\mathrm{norm}}(H \mid x_{<t})$ infinitely often as $t \to \infty$.*

Proof. We iterate Corollary 12: starting with $\overline{BR}^{\infty}$, we get a time step t_1 such that observing BR at time t_1 disconfirms H. We set $x_{1:t_1} := \overline{BR}^{t_1-1} BR$ and apply Corollary 12 to $x_{1:t_1}\overline{BR}^{\infty}$ to get a time step t_2 such that observing BR at time t_2 disconfirms H. Then we set $x_{1:t_2} := x_{1:t_1}\overline{BR}^{t_2-t_1-1} BR$, and so on. $\square$

4.4   Stochastically Sampled Strings

The proof techniques from the previous subsections do not generalize to strings that are sampled stochastically. The main obstacle is the complexity of counterfactual observations $x_{<t}a$ with $a \neq x_t$: for deterministic strings $Km(x_{<t}a) \to 0$, while for stochastically sampled strings $Km(x_{<t}a) \nrightarrow 0$. Consider the following example.

Example 14 (Uniform IID Observations). Let λ_H be a measure that generates uniform i.i.d. symbols from $\{BR, B\overline{R}, \overline{BR}\}$. Formally,

$$\lambda_H(x) := \begin{cases} 0 & \text{if } \overline{BR} \in x, \text{ and} \\ 3^{-|x|} & \text{otherwise.} \end{cases}$$

By construction, $\lambda_H(H) = 1$. By Lemma 2 we have $A, C, E \overset{\times}{=} 3^{-t}$ and $B, D \overset{\times}{=} 3^{-t}2^{-m(t)}$ with λ_H-probability one. According to Lemma 4, the sign of $AD + DE - BC$ is indicative for the change in belief in H. But this is inconclusive both for M and M_{norm} since each of the summands AD, BC, and DE (in case $E \neq 0$) go to zero at the same rate:

$$AD \overset{\times}{=} DE \overset{\times}{=} BC \overset{\times}{=} 3^{-2t}2^{-m(t)}.$$

Whether H gets confirmed or disconfirmed thus depends on the universal Turing machine and/or the probabilistic outcome of the string drawn from λ_H. $\Diamond$

5   Discussion

We chose to present our results in the setting of the black raven problem to make them more accessible to intuition and more relatable to existing literature. But these results hold more generally: our proofs follow from the bounds on A, B, C, D, and E given in Lemma 6 and Lemma 10. These bounds rely on the fact that we are observing a computable infinite string and that at any time step t there are programs consistent with the observation history that contradict the hypothesis and there are programs consistent with the observation history that are compatible with the hypothesis. No further assumptions on the alphabet, the hypothesis H, or the universal Turing machine are necessary.

In our formalization of the raven problem given in Section 3, we used an alphabet with four symbols. Each symbol indicates one of four possible types of observations according to the two binary predicates blackness and ravenness. One could object that this formalization discards important structure from the problem: BR and $\overline{B}R$ have more in common than BR and $\overline{BR}$, yet as symbols they are all the same. Instead, we could use the latin alphabet and spell out 'black', 'non-black', 'raven', and 'non-raven'. The results given in this paper would still apply analogously.

Our result that Solomonoff induction does not satisfy Nicod's criterion is not true for every time step, only for some of them. Generally, whether Nicod's criterion should be adhered to depends on whether the paradoxical conclusion is acceptable. A different Bayesian reasoner might be tempted to argue that a green apple *does* confirm the hypothesis H, but only to a small degree, since there are vastly more non-black objects than ravens [3]. This leads to the acceptance of the paradoxical conclusion, and this solution to the confirmation paradox is known as the *standard Bayesian solution*. It is equivalent to the assertion that blackness is equally probable regardless of whether H holds: $P(\text{black}|H) \approx P(\text{black})$ [19]. Whether or not this holds depends on our prior beliefs.

The following is a very concise example against the standard Bayesian solution [4]: There are two possible worlds, the first has 100 black ravens and a million other birds, while the second has 1000 black ravens, one white raven, and a million other birds. Now we draw a bird uniformly at random, and it turns out to be a black raven. Contrary to what Nicod's criterion claims, this is strong evidence that we are in fact in the second world, and in this world non-black ravens exist.

For another, more intuitive example: Suppose you do not know anything about ravens and you have a friend who collects atypical objects. If you see a black raven in her collection, surely this would not increase your belief in the hypothesis that all ravens are black.

We must conclude that violating Nicod's criterion is not a fault of Solomonoff induction. Instead, we should accept that for Bayesian reasoning Nicod's criterion, in its generality, is false! Quoting the great Bayesian master E. T. Jaynes [9, p.144]:

> In the literature there are perhaps 100 'paradoxes' and controversies which are like this, in that they arise from faulty intuition rather than faulty mathematics. Someone asserts a general principle that seems to him intuitively right. Then, when probability analysis reveals the error, instead of taking this opportunity to educate his intuition, he reacts by rejecting the probability analysis.

Acknowledgments. This work was supported by ARC grant DP150104590.

References

1. Blackwell, D., Dubins, L.: Merging of opinions with increasing information. The Annals of Mathematical Statistics, 882–886 (1962)
2. Gács, P.: On the relation between descriptional complexity and algorithmic probability. Theoretical Computer Science **22**(1–2), 71–93 (1983)
3. Good, I.J.: The paradox of confirmation. British Journal for the Philosophy of Science, 145–149 (1960)
4. Good, I.J.: The white shoe is a red herring. The British Journal for the Philosophy of Science **17**(4), 322–322 (1967)
5. Hempel, C.G.: Studies in the logic of confirmation (I.). Mind, 1–26 (1945)
6. Hempel, C.G.: The white shoe: No red herring. The British Journal for the Philosophy of Science **18**(3), 239–240 (1967)
7. Hutter, M.: New error bounds for Solomonoff prediction. Journal of Computer and System Sciences **62**(4), 653–667 (2001)
8. Hutter, M.: On universal prediction and Bayesian confirmation. Theoretical Computer Science **384**(1), 33–48 (2007)
9. Jaynes, E.T.: Probability Theory: The Logic of Science. Cambridge University Press (2003)
10. Levin, L.A.: Laws of information conservation (nongrowth) and aspects of the foundation of probability theory. Problemy Peredachi Informatsii **10**(3), 30–35 (1974)
11. Li, M., Vitányi, P.M.B.: An Introduction to Kolmogorov Complexity and Its Applications. Texts in Computer Science, 3rd edn. Springer (2008)
12. Mackie, J.L.: The paradox of confirmation. British Journal for the Philosophy of Science, 265–277 (1963)
13. Maher, P.: Inductive logic and the ravens paradox. Philosophy of Science, 50–70 (1999)
14. Nicod, J.: Le Problème Logique de L'Induction. Presses Universitaires de France (1961)
15. Rathmanner, S., Hutter, M.: A philosophical treatise of universal induction. Entropy **13**(6), 1076–1136 (2011)
16. Solomonoff, R.: A formal theory of inductive inference. Parts 1 and 2. Information and Control **7**(1), 1–22 and 224–254 (1964)
17. Solomonoff, R.: Complexity-based induction systems: Comparisons and convergence theorems. IEEE Transactions on Information Theory **24**(4), 422–432 (1978)
18. Swinburne, R.G.: The paradoxes of confirmation: A survey. American Philosophical Quarterly, 318–330 (1971)
19. Vranas, P.B.: Hempel's raven paradox: A lacuna in the standard Bayesian solution. The British Journal for the Philosophy of Science **55**(3), 545–560 (2004)
20. Wood, I., Sunehag, P., Hutter, M.: (Non-)equivalence of universal priors. In: Dowe, D.L. (ed.) Solomonoff Festschrift. LNCS, vol. 7070, pp. 417–425. Springer, Heidelberg (2013)

List of Notation

$:=$	defined to be equal
$\#A$	the cardinality of the set A, i.e., the number of elements
$\mathcal{X}$	a finite alphabet
$\mathcal{X}^*$	the set of all finite strings over the alphabet $\mathcal{X}$
$\mathcal{X}^\infty$	the set of all infinite strings over the alphabet $\mathcal{X}$
$\mathcal{X}^\sharp$	$\mathcal{X}^\sharp := \mathcal{X}^* \cup \mathcal{X}^\infty$, the set of all finite and infinite strings over the alphabet $\mathcal{X}$
Γ_x	the set of all finite and infinite strings that start with x
x, y	finite or infinite strings, $x, y \in \mathcal{X}^\sharp$
$x \sqsubseteq y$	the string x is a prefix of the string y
ϵ	the empty string
ε	a small positive rational number
t	(current) time step
n	natural number
$K(x)$	Kolmogorov complexity of the string x: the length of the shortest program that prints x and halts
$m(t)$	the monotone lower bound on K, formally $m(t) := \min_{n \geq t} K(n)$
$Km(x)$	monotone Kolmogorov complexity of the string x: the length of the shortest program on the monotone universal Turing machine that prints something starting with x
BR	a symbol corresponding to the observation of a black raven
$\overline{BR}$	a symbol corresponding to the observation of a non-black raven
$B\overline{R}$	a symbol corresponding to the observation of a black non-raven
$\overline{B\overline{R}}$	a symbol corresponding to the observation of a non-black non-raven
H	the hypothesis 'all ravens are black', formally defined in (2)
U	the universal (monotone) Turing machine
M	the Solomonoff prior
M_{norm}	the normalized Solomonoff prior, defined according to (1)
p, q	programs on the universal (monotone) Turing machine

On the Computability of Solomonoff Induction
and Knowledge-Seeking

Jan Leike[✉] and Marcus Hutter

Australian National University, Canberra, Australia
{jan.leike,marcus.hutter}@anu.edu.au

Abstract. Solomonoff induction is held as a gold standard for learning, but it is known to be incomputable. We quantify its incomputability by placing various flavors of Solomonoff's prior M in the arithmetical hierarchy. We also derive computability bounds for knowledge-seeking agents, and give a limit-computable weakly asymptotically optimal reinforcement learning agent.

Keywords: Solomonoff induction · Exploration · Knowledge-seeking agents · General reinforcement learning · Asymptotic optimality · Computability · Complexity · Arithmetical hierarchy · Universal turing machine · AIXI · BayesExp

1 Introduction

Solomonoff's theory of learning [11,19,20], commonly called *Solomonoff induction*, arguably solves the induction problem [18]: for data drawn from any computable measure μ, Solomonoff induction will converge to the correct belief about any hypothesis [1]. Moreover, convergence is extremely fast in the sense that the expected number of prediction errors is $E + O(\sqrt{E})$ compared to the number of errors E made by the informed predictor that knows μ [4].

In *reinforcement learning* an agent repeatedly takes actions and receives observations and rewards. The goal is to maximize cumulative (discounted) reward. Solomonoff's ideas can be extended to reinforcement learning, leading to the Bayesian agent AIXI [3,5]. However, AIXI's trade-off between exploration and exploitation includes insufficient exploration to get rid of the prior's bias [9], which is why the universal agent AIXI does not achieve asymptotic optimality [13,15].

For extra exploration, we can resort to Orseau's *knowledge-seeking agents*. Instead of rewards, knowledge-seeking agents maximize entropy gain [14,16] or expected information gain [17]. These agents are apt explorers, and asymptotically they learn their environment perfectly [16,17].

A reinforcement learning agent is *weakly asymptotically optimal* if the value of its policy converges to the optimal value in Cesàro mean [7]. Weak asymptotic optimality stands out because it currently is the only known nontrivial objective notion of optimality for general reinforcement learners [7,9,15]. Lattimore defines

© Springer International Publishing Switzerland 2015
K. Chaudhuri et al. (Eds.): ALT 2015, LNAI 9355, pp. 364–378, 2015.
DOI: 10.1007/978-3-319-24486-0_24

Table 1. The computability results on M, M_{norm}, $\overline{M}$, and $\overline{M}_{\mathrm{norm}}$ proved in Section 3. Lower bounds on the complexity of $\overline{M}$ and $\overline{M}_{\mathrm{norm}}$ are given only for specific universal Turing machines.

P	$\{(x,q) \in \mathcal{X}^* \times \mathbb{Q} \mid P(x) > q\}$	$\{(x,y,q) \in \mathcal{X}^* \times \mathcal{X}^* \times \mathbb{Q} \mid P(xy \mid x) > q\}$
M	$\Sigma_1^0 \setminus \Delta_1^0$	$\Delta_2^0 \setminus (\Sigma_1^0 \cup \Pi_1^0)$
M_{norm}	$\Delta_2^0 \setminus (\Sigma_1^0 \cup \Pi_1^0)$	$\Delta_2^0 \setminus (\Sigma_1^0 \cup \Pi_1^0)$
$\overline{M}$	$\Pi_2^0 \setminus \Delta_2^0$	$\Delta_3^0 \setminus (\Sigma_2^0 \cup \Pi_2^0)$
$\overline{M}_{\mathrm{norm}}$	$\Delta_3^0 \setminus (\Sigma_2^0 \cup \Pi_2^0)$	$\Delta_3^0 \setminus (\Sigma_2^0 \cup \Pi_2^0)$

the agent BayesExp by grafting a knowledge-seeking component on top of AIXI and shows that BayesExp is a weakly asymptotically optimal agent in the class of all stochastically computable environments [6, Ch. 5].

The purpose of models such as Solomonoff induction, AIXI, and knowledge-seeking agents is to answer the question of how to solve (reinforcement) learning *in theory*. These answers are useless if they cannot be approximated in practice, i.e., by a regular Turing machine. Therefore we posit that any ideal model must at least be *limit computable* (Δ_2^0).

Limit computable functions are the functions that admit an *anytime algorithm*. More generally, the *arithmetical hierarchy* specifies different levels of computability based on *oracle machines*: each level in the arithmetical hierarchy is computed by a Turing machine which may query a halting oracle for the respective lower level.

In previous work [10] we established that AIXI is limit computable if restricted to ε-optimal policies, and placed various versions of AIXI, AINU, and AIMU in the arithmetical hierarchy. In this paper we investigate the (in-)computability of Solomonoff induction and knowledge-seeking. The universal prior M is lower semicomputable and hence its conditional is limit computable. But M is a semimeasure: it assigns positive probability that the observed string has only finite length. This can be circumvented by normalizing M. Solomonoff's normalization M_{norm} preserves the ratio $M(x1)/M(x0)$ and is limit computable. If we remove the contribution of programs that compute only finite strings, we get a semimeasure $\overline{M}$, which can be normalized to $\overline{M}_{\mathrm{norm}}$ by multiplication with a constant. We show that both $\overline{M}$ and $\overline{M}_{\mathrm{norm}}$ are *not* limit computable. Our results on the computability of Solomonoff induction are stated in Table 1 and proved in Section 3. In Section 4 we show that for finite horizons both the entropy-seeking and the information-seeking agent are Δ_3^0-computable and have limit-computable ε-optimal policies. The weakly asymptotically optimal agent BayesExp relies on optimal policies that are generally not limit computable [10, Thm.16]. In Section 5 we give a weakly asymptotically optimal agent based on BayesExp that is limit computable. A list of notation can be found on page 377.

2   Preliminaries

We use the setup and notation from [10].

2.1   The Arithmetical Hierarchy

A set $A \subseteq \mathbb{N}$ *is* Σ_n^0 iff there is a computable relation S such that

$$k \in A \iff \exists k_1 \forall k_2 \ldots Q_n k_n\ S(k, k_1, \ldots, k_n) \tag{1}$$

where $Q_n = \forall$ if n is even, $Q_n = \exists$ if n is odd [12, Def. 1.4.10]. A set $A \subseteq \mathbb{N}$ *is* Π_n^0 iff its complement $\mathbb{N} \setminus A$ is Σ_n^0. We call the formula on the right hand side of (1) a Σ_n^0-*formula*, its negation is called Π_n^0-*formula*. It can be shown that we can add any bounded quantifiers and duplicate quantifiers of the same type without changing the classification of A. The set A *is* Δ_n^0 iff A is Σ_n^0 and A is Π_n^0. We get that Σ_1^0 as the class of recursively enumerable sets, Π_1^0 as the class of co-recursively enumerable sets and Δ_1^0 as the class of recursive sets.

We say the set $A \subseteq \mathbb{N}$ is Σ_n^0-*hard* (Π_n^0-*hard*, Δ_n^0-*hard*) iff for any set $B \in \Sigma_n^0$ ($B \in \Pi_n^0$, $B \in \Delta_n^0$), B is many-one reducible to A, i.e., there is a computable function f such that $k \in B \leftrightarrow f(k) \in A$ [12, Def. 1.2.1]. We get $\Sigma_n^0 \subset \Delta_{n+1}^0 \subset \Sigma_{n+1}^0 \subset \ldots$ and $\Pi_n^0 \subset \Delta_{n+1}^0 \subset \Pi_{n+1}^0 \subset \ldots$. This hierarchy of subsets of natural numbers is known as the *arithmetical hierarchy*.

By Post's Theorem [12, Thm. 1.4.13], a set is Σ_n^0 if and only if it is recursively enumerable on an oracle machine with an oracle for a Σ_{n-1}^0-hard set.

2.2   Strings

Let $\mathcal{X}$ be some finite set called *alphabet*. The set $\mathcal{X}^* := \bigcup_{n=0}^{\infty} \mathcal{X}^n$ is the set of all finite strings over the alphabet $\mathcal{X}$, the set $\mathcal{X}^\infty$ is the set of all infinite strings over the alphabet $\mathcal{X}$, and the set $\mathcal{X}^\sharp := \mathcal{X}^* \cup \mathcal{X}^\infty$ is their union. The empty string is denoted by ϵ, not to be confused with the small positive real number ε. Given a string $x \in \mathcal{X}^*$, we denote its length by $|x|$. For a (finite or infinite) string x of length $\geq k$, we denote with $x_{1:k}$ the first k characters of x, and with $x_{<k}$ the first $k-1$ characters of x. The notation $x_{1:\infty}$ stresses that x is an infinite string. We write $x \sqsubseteq y$ iff x is a prefix of y, i.e., $x = y_{1:|x|}$.

2.3   Computability of Real-Valued Functions

We fix some encoding of rational numbers into binary strings and an encoding of binary strings into natural numbers. From now on, this encoding will be done implicitly wherever necessary.

Definition 1 (Σ_n^0-, Π_n^0-, Δ_n^0-**computable**). *A function* $f : \mathcal{X}^* \to \mathbb{R}$ *is called* Σ_n^0-computable (Π_n^0-computable, Δ_n^0-computable) *iff the set* $\{(x, q) \in \mathcal{X}^* \times \mathbb{Q} \mid f(x) > q\}$ *is* Σ_n^0 (Π_n^0, Δ_n^0).

Table 2. Connection between the computability of real-valued functions and the arithmetical hierarchy.

	$\{(x,q) \mid f(x) > q\}$	$\{(x,q) \mid f(x) < q\}$
f is computable	Δ_1^0	Δ_1^0
f is lower semicomputable	Σ_1^0	Π_1^0
f is upper semicomputable	Π_1^0	Σ_1^0
f is limit computable	Δ_2^0	Δ_2^0
f is Δ_n^0-computable	Δ_n^0	Δ_n^0
f is Σ_n^0-computable	Σ_n^0	Π_n^0
f is Π_n^0-computable	Π_n^0	Σ_n^0

A Δ_1^0-computable function is called *computable*, a Σ_1^0-computable function is called *lower semicomputable*, and a Π_1^0-computable function is called *upper semicomputable*. A Δ_2^0-computable function f is called *limit computable*, because there is a computable function ϕ such that

$$\lim_{k \to \infty} \phi(x, k) = f(x).$$

The program ϕ that limit computes f can be thought of as an *anytime algorithm* for f: we can stop ϕ at any time k and get a preliminary answer. If the program ϕ ran long enough (which we do not know), this preliminary answer will be close to the correct one.

Limit-computable sets are the highest level in the arithmetical hierarchy that can be approached by a regular Turing machine. Above limit-computable sets we necessarily need some form of halting oracle. See Table 2 for the definition of lower/upper semicomputable and limit-computable functions in terms of the arithmetical hierarchy.

Lemma 2 (Computability of Arithmetical Operations). *Let $n > 0$ and let $f, g : \mathcal{X}^* \to \mathbb{R}$ be two Δ_n^0-computable functions. Then*

(i) $\{(x,y) \mid f(x) > g(y)\}$ is Σ_n^0,
(ii) $\{(x,y) \mid f(x) \leq g(y)\}$ is Π_n^0,
(iii) $f + g$, $f - g$, and $f \cdot g$ are Δ_n^0-computable,
(iv) f/g is Δ_n^0-computable if $g(x) \neq 0$ for all x, and
(v) $\log f$ is Δ_n^0-computable if $f(x) > 0$ for all x.

3   The Complexity of Solomonoff Induction

A *semimeasure* over the alphabet $\mathcal{X}$ is a function $\nu : \mathcal{X}^* \to [0,1]$ such that (i) $\nu(\epsilon) \leq 1$, and (ii) $\nu(x) \geq \sum_{a \in \mathcal{X}} \nu(xa)$ for all $x \in \mathcal{X}^*$. A semimeasure is called *(probability) measure* iff for all x equalities hold in (i) and (ii).

$$M(xy \mid x) > q \iff \forall \ell \exists k \, \frac{\phi(xy, k)}{\phi(x, \ell)} > q \iff \exists k \exists \ell_0 \forall \ell \geq \ell_0 \, \frac{\phi(xy, k)}{\phi(x, \ell)} > q$$

Fig. 1. A Π_2^0-formula and an equivalent Σ_2^0-formula defining conditional M. Here $\phi(x, k)$ denotes a computable function that lower semicomputes $M(x)$.

Solomonoff's prior M [19] assigns to a string x the probability that the reference universal monotone Turing machine U [11, Ch. 4.5.2] computes a string starting with x when fed with uniformly random bits as input. Formally,

$$M(x) := \sum_{p: \, x \sqsubseteq U(p)} 2^{-|p|}. \tag{2}$$

The function M is a lower semicomputable semimeasure, but not computable and not a measure [11, Lem. 4.5.3]. A semimeasure ν can be turned into a measure ν_{norm} using *Solomonoff normalization*: $\nu_{\mathrm{norm}}(\epsilon) := 1$ and for all $x \in \mathcal{X}^*$ and $a \in \mathcal{X}$,

$$\nu_{\mathrm{norm}}(xa) := \nu_{\mathrm{norm}}(x) \frac{\nu(xa)}{\sum_{b \in \mathcal{X}} \nu(xb)}. \tag{3}$$

By definition, M_{norm} and $\overline{M}_{\mathrm{norm}}$ are measures [11, Sec. 4.5.3]. Moreover, since $M_{\mathrm{norm}} \geq M$, normalization preserves universal dominance. Hence Solomonoff's theorem implies that M_{norm} predicts just as well as M.

The *measure mixture* $\overline{M}$ [2, p. 74] is defined as

$$\overline{M}(x) := \lim_{n \to \infty} \sum_{y \in \mathcal{X}^n} M(xy). \tag{4}$$

The measure mixture $\overline{M}$ is the same as M except that the contributions by programs that do not produce infinite strings are removed: for any such program p, let k denote the length of the finite string generated by p. Then for $|xy| > k$, the program p does not contribute to $M(xy)$, hence it is excluded from $\overline{M}(x)$.

Similarly to M, the measure mixture $\overline{M}$ is not a (probability) measure since $\overline{M}(\varepsilon) < 1$, but in this case normalization (3) is just multiplication with the constant $1/\overline{M}(\epsilon)$, leading to the *normalized measure mixture* $\overline{M}_{\mathrm{norm}}$. When using the Solomonoff prior M (or one of its sisters M_{norm}, $\overline{M}$, or $\overline{M}_{\mathrm{norm}}$) for sequence prediction, we need to compute the conditional probability $M(xy \mid x) := M(xy)/M(x)$ for finite strings $x, y \in \mathcal{X}^*$. Because $M(x) > 0$ for all finite strings $x \in \mathcal{X}^*$, this quotient is well-defined.

Theorem 3 (Complexity of M, M_{norm}, $\overline{M}$, and $\overline{M}_{\mathrm{norm}}$).

(i) $M(x)$ *is lower semicomputable*

(ii) $M(xy \mid x)$ *is limit computable*

(iii) $M_{\mathrm{norm}}(x)$ *is limit computable*

(iv) $M_{\mathrm{norm}}(xy \mid x)$ *is limit computable*

(v) $\overline{M}(x)$ *is Π_2^0-computable*

(vi) $\overline{M}(xy \mid x)$ *is Δ_3^0-computable*

(vii) $\overline{M}_{\mathrm{norm}}(x)$ *is Δ_3^0-computable*

(viii) $\overline{M}_{\mathrm{norm}}(xy \mid x)$ *is Δ_3^0-computable*

Proof. (i) By [11, Thm. 4.5.2]. Intuitively, we can run all programs in parallel and get monotonely increasing lower bounds for $M(x)$ by adding $2^{-|p|}$ every time a program p has completed outputting x.

(ii) From (i) and Lemma 2 (iv), since $M(x) > 0$ (see also Figure 1).

(iii) By Lemma 2 (iii,iv) and $M(x) > 0$.

(iv) By (iii) and Lemma 2 (iv), since $M_{\mathrm{norm}}(x) \geq M(x) > 0$.

(v) Let ϕ be a computable function that lower semicomputes M. Since M is a semimeasure, $M(xy) \geq \sum_z M(xyz)$, hence $\sum_{y \in \mathcal{X}^n} M(xy)$ is nonincreasing in n and thus $\overline{M}(x) > q$ iff $\forall n \exists k \sum_{y \in \mathcal{X}^n} \phi(xy, k) > q$.

(vi) From (v) and Lemma 2 (iv), since $\overline{M}(x) > 0$.

(vii) From (v) and Lemma 2 (iv).

(viii) From (vi) and Lemma 2 (iv), since $\overline{M}_{\mathrm{norm}}(x) \geq \overline{M}(x) > 0$. □

We proceed to show that these bounds are in fact the best possible ones. If M were Δ_1^0-computable, then so would be the conditional semimeasure $M(\cdot \mid \cdot)$. Thus we could compute the M-adversarial sequence $z_1 z_2 \ldots$ defined by

$$z_t := \begin{cases} 0 & \text{if } M(1 \mid z_{<t}) > \frac{1}{2}, \\ 1 & \text{otherwise.} \end{cases}$$

The sequence $z_1 z_2 \ldots$ corresponds to a computable deterministic measure μ. However, we have $M(z_{1:t}) \leq 2^{-t}$ by construction, so dominance $M(x) \geq w_\mu \mu(x)$ with $w_\mu > 0$ yields a contradiction with $t \to \infty$:

$$2^{-t} \geq M(z_{1:t}) \geq w_\mu \mu(z_{1:t}) = w_\mu > 0$$

By the same argument, the normalized Solomonoff prior M_{norm} cannot be Δ_1^0-computable. However, since it is a measure, Σ_1^0- or Π_1^0-computability would entail Δ_1^0-computability.

For $\overline{M}$ and $\overline{M}_{\mathrm{norm}}$ we prove the following two lower bounds for specific universal Turing machines.

Theorem 4 ($\overline{M}$ is not Limit Computable). *There is a universal Turing machine U' such that the set $\{(x, q) \mid \overline{M}_{U'}(x) > q\}$ is not in Δ_2^0.*

Proof. Assume the contrary, let A be Π_2^0 but not Δ_2^0, and let S be a computable relation such that

$$n \in A \iff \forall k \exists i\, S(n, k, i). \tag{5}$$

For each $n \in \mathbb{N}$, we define the program p_n as follows.

```
output 1^{n+1}0
k := 0
while true:
    i := 0
    while not S(n, k, i):
        i := i + 1
    k := k + 1
    output 0
```

Each program p_n always outputs $1^{n+1}0$. Furthermore, the program p_n outputs the infinite string $1^{n+1}0^\infty$ if and only if $n \in A$ by (5). We define U' as follows using our reference machine U.

- $U'(1^{n+1}0)$: Run p_n.
- $U'(00p)$: Run $U(p)$.
- $U'(01p)$: Run $U(p)$ and bitwise invert its output.

By construction, U' is a universal Turing machine. No p_n outputs a string starting with $0^{n+1}1$, therefore $\overline{M}_{U'}(0^{n+1}1) = \frac{1}{4}\big(\overline{M}_U(0^{n+1}1) + \overline{M}_U(1^{n+1}0)\big)$. Hence

$$\overline{M}_{U'}(1^{n+1}0) = 2^{-n-2}\mathbb{1}_A(n) + \tfrac{1}{4}\overline{M}_U(1^{n+1}0) + \tfrac{1}{4}\overline{M}_U(0^{n+1}1)$$
$$= 2^{-n-2}\mathbb{1}_A(n) + \overline{M}_{U'}(0^{n+1}1)$$

If $n \notin A$, then $\overline{M}_{U'}(1^{n+1}0) = \overline{M}_{U'}(0^{n+1}1)$. Otherwise, we have $|\overline{M}_{U'}(1^{n+1}0) - \overline{M}_{U'}(0^{n+1}1)| = 2^{-n-2}$.

Now we assume that $\overline{M}_{U'}$ is limit computable, i.e., there is a computable function $\phi : \mathcal{X}^* \times \mathbb{N} \to \mathbb{Q}$ such that $\lim_{k\to\infty} \phi(x,k) = \overline{M}_{U'}(x)$. We get that

$$n \in A \iff \lim_{k\to\infty} \phi(0^{n+1}1, k) - \phi(1^{n+1}0, k) > 2^{-n-3},$$

thus A is limit computable, a contradiction. □

Corollary 5 ($\overline{M}_{\mathrm{norm}}$ is not Σ_2^0- or Π_2^0-computable). *There is a universal Turing machine U' such that $\{(x,q) \mid \overline{M}_{\mathrm{norm}U'}(x) > q\}$ is not in Σ_2^0 or Π_2^0.*

Proof. Since $\overline{M}_{\mathrm{norm}} = c \cdot \overline{M}$, there exists a $k \in \mathbb{N}$ such that $2^{-k} < c$ (even if we do not know the value of k). We can show that the set $\{(x,q) \mid \overline{M}_{\mathrm{norm}U'}(x) > q\}$ is not in Δ_2^0 analogously to the proof of Theorem 4, using

$$n \in A \iff \lim_{k\to\infty} \phi(0^{n+1}1, k) - \phi(1^{n+1}0, k) > 2^{-k-n-3}.$$

If $\overline{M}_{\mathrm{norm}}$ were Σ_2^0-computable or Π_2^0-computable, this would imply that $\overline{M}_{\mathrm{norm}}$ is Δ_2^0-computable since $\overline{M}_{\mathrm{norm}}$ is a measure, a contradiction. □

Since $M(\epsilon) = 1$, we have $M(x \mid \epsilon) = M(x)$, so the conditional probability $M(xy \mid x)$ has at least the same complexity as M. Analogously for M_{norm} and $\overline{M}_{\mathrm{norm}}$ since they are measures. For $\overline{M}$, we have that $\overline{M}(x \mid \epsilon) = \overline{M}_{\mathrm{norm}}(x)$, so Corollary 5 applies. All that remains to prove is that conditional M is not lower semicomputable.

Theorem 6 (Conditional M is not Lower Semicomputable). *The set $\{(x, xy, q) \mid M(xy \mid x) > q\}$ is not recursively enumerable.*

Proof. Assume to the contrary that $M(xy \mid x)$ is lower semicomputable. According to [8, Thm. 12] there is an infinite string $z_{1:\infty}$ such that $z_{2t} = z_{2t-1}$ for all $t > 0$ and

$$\liminf_{t\to\infty} M(z_{1:2t} \mid z_{<2t}) < 1. \tag{6}$$

Define the semimeasure

$$
\nu(x_{1:t}) := \begin{cases} \prod_{k=1}^{\lceil t/2 \rceil} M(x_{<2k} \mid x_{<2k-1}) & \text{if } \forall 0 < 2k \le t \; x_{2k} = x_{2k-1} \\ 0 & \text{otherwise.} \end{cases}
$$

Since we assume $M(x_{<2k} \mid x_{<2k-1})$ to be lower semicomputable, ν is lower semicomputable. Therefore there is a constant $c > 0$ such that $M(x) \ge c\nu(x)$ for all $x \in \mathcal{X}^*$. With the chain rule we get for even-lengthed x with $x_{2k} = x_{2k-1}$

$$
c \le \frac{M(x)}{\nu(x)} = \frac{\prod_{i=1}^{t} M(x_{1:i} \mid x_{<i})}{\prod_{k=1}^{t/2} M(x_{<2k} \mid x_{<2k-1})} = \prod_{k=1}^{t/2} M(x_{1:2k} \mid x_{<2k}).
$$

Plugging in the sequence $z_{1:\infty}$, we get a contradiction with (6):

$$
0 < c \le \prod_{k=1}^{t} M(z_{1:2k} \mid z_{<2k}) \xrightarrow{t \to \infty} 0 \qquad \qquad \square
$$

4 The Complexity of Knowledge-Seeking

In general reinforcement learning the agent interacts with an environment in cycles: at time step t the agent chooses an *action* $a_t \in \mathcal{A}$ and receives a *percept* $e_t = (o_t, r_t) \in \mathcal{E}$ consisting of an *observation* $o_t \in \mathcal{O}$ and a real-valued *reward* $r_t \in \mathbb{R}$; the cycle then repeats for $t + 1$. A *history* is an element of $(\mathcal{A} \times \mathcal{E})^*$. We use $\ae \in \mathcal{A} \times \mathcal{E}$ to denote one interaction cycle, and $\ae_{1:t}$ to denote a history of length t. A *policy* is a function $\pi : (\mathcal{A} \times \mathcal{E})^* \to \mathcal{A}$ mapping each history to the action taken after seeing this history. We assume $\mathcal{A}$ and $\mathcal{E}$ to be finite.

The environment can be stochastic, but is assumed to be semicomputable. In accordance with the AIXI literature [5], we model environments as lower semi-computable *chronological conditional semimeasures* (LSCCCSs). The class of of all LSCCCSs is denoted with $\mathcal{M}$. A *conditional semimeasure* ν takes a sequence of actions $a_{1:t}$ as input and returns a semimeasure $\nu(\cdot \parallel a_{1:t})$ over $\mathcal{E}^\sharp$. A conditional semimeasure ν is *chronological* iff percepts at time t do not depend on future actions, i.e., $\nu(e_{1:t} \parallel a_{1:k}) = \nu(e_{1:t} \parallel a_{1:t})$ for all $k > t$. Despite their name, conditional semimeasures do *not* specify conditional probabilities; the environment ν is *not* a joint probability distribution on actions and percepts. Here we only care about the computability of the environment ν; for our purposes, chronological conditional semimeasures behave just like semimeasures.

Equivalently to (2), the Solomonoff prior M can be defined as a mixture over all lower semicomputable semimeasures using a lower semicomputable *universal prior* [21]. We generalize this representation to chronological conditional semimeasures: we fix the lower semicomputable universal prior $(w_\nu)_{\nu \in \mathcal{M}}$ with $w_\nu > 0$ for all $\nu \in \mathcal{M}$ and $\sum_{\nu \in \mathcal{M}} w_\nu \le 1$, given by the reference machine U according to $w_\nu := 2^{-K_U(\nu)}$ [5, Sec. 5.1.2]. The universal prior w gives rise to the *universal mixture* ξ, which is a convex combination of all LSCCCSs $\mathcal{M}$:

$$
\xi(e_{<t} \parallel a_{<t}) := \sum_{\nu \in \mathcal{M}} w_\nu \nu(e_{<t} \parallel a_{<t})
$$

The universal mixture ξ is analogous to the Solomonoff prior M but defined for reactive environments. Analogously to Theorem 3 (i), the universal mixture ξ is lower semicomputable [5, Sec. 5.10]. Moreover, we have $\xi_{\text{norm}} \geq \xi$, preserving universal dominance analogously to M.

4.1   Knowledge-Seeking Agents

We discuss two variants of knowledge-seeking agents: entropy-seeking agents (Shannon-KSA) [14,16] and information-seeking agents (KL-KSA) [17]. The entropy-seeking agent maximizes the Shannon entropy gain, while the information-seeking agent maximizes the expected Bayesian information gain (KL-divergence) in the universal mixture ξ. These quantities are expressed in the *value function*.

In this section we use a finite lifetime m (possibly dependent on time step t): the knowledge-seeking agent maximizes entropy/information received up to and including time step m. We assume that the function m (of t) is computable.

Definition 7 (Entropy-Seeking Value Function [16, Sec. 6]). *The entropy-seeking value of a policy π given history $æ_{<t}$ is*

$$V_H^\pi(æ_{<t}) := \sum_{e_{t:m}} -\xi_{\text{norm}}(e_{1:m} \mid e_{<t} \,\|\, a_{1:m}) \log_2 \xi_{\text{norm}}(e_{1:m} \mid e_{<t} \,\|\, a_{1:m})$$

where $a_i := \pi(e_{<i})$ for all $i \geq t$.

Definition 8 (Information-Seeking Value Function [17, Def. 1]). *The information-seeking value of a policy π given history $æ_{<t}$ is*

$$V_I^\pi(æ_{<t}) := \sum_{e_{t:m}} \sum_{\nu \in \mathcal{M}} w_\nu \frac{\nu(e_{1:m} \,\|\, a_{1:m})}{\xi_{\text{norm}}(e_{<t} \,\|\, a_{<t})} \log_2 \frac{\nu(e_{1:m} \mid e_{<t} \,\|\, a_{1:m})}{\xi_{\text{norm}}(e_{1:m} \mid e_{<t} \,\|\, a_{1:m})}$$

where $a_i := \pi(e_{<i})$ for all $i \geq t$.

We use V^π in places where either of the entropy-seeking or the information-seeking value function can be substituted.

Definition 9 ((ε-)Optimal Policy). *The optimal value function V^* is defined as $V^*(æ_{<t}) := \sup_\pi V^\pi(æ_{<t})$. A policy π is optimal iff $V^\pi(æ_{<t}) = V^*(æ_{<t})$ for all histories $æ_{<t} \in (\mathcal{A} \times \mathcal{E})^*$. A policy π is ε-optimal iff $V^*(æ_{<t}) - V^\pi(æ_{<t}) < \varepsilon$ for all histories $æ_{<t} \in (\mathcal{A} \times \mathcal{E})^*$.*

An entropy-seeking agent is defined as an optimal policy for the value function V_H^* and an information-seeking agent is defined as an optimal policy for the value function V_I^*.

The entropy-seeking agent does not work well in stochastic environments because it gets distracted by noise in the environment rather than trying to distinguish environments [17]. Moreover, the unnormalized knowledge-seeking agents may fail to seek knowledge in deterministic semimeasures as the following example demonstrates.

Example 10 (Unnormalized Entropy-Seeking). Suppose we use ξ instead of ξ_{norm} in Definition 7. Fix $\mathcal{A} := \{\alpha, \beta\}$, $\mathcal{E} := \{0, 1\}$, and $m := 1$ (we only care about the entropy of the next percept). We illustrate the problem on a simple class of environments $\{\nu_1, \nu_2\}$:

$$\alpha/0/0.1 \; \circlearrowleft (\nu_1) \circlearrowright \; \beta/0/0.5 \qquad \alpha/1/0.1 \; \circlearrowleft (\nu_2) \circlearrowright \; \beta/0/0.5$$

where transitions are labeled with action/percept/probability. Both ν_1 and ν_2 return a percept deterministically or nothing at all (the environment ends). Only action α distinguishes between the environments. With the prior $w_{\nu_1} := w_{\nu_2} := 1/2$, we get a mixture ξ for the entropy-seeking value function V_H^π. Then $V_H^*(\alpha) \approx 0.432 < 0.5 = V_H^*(\beta)$, hence action β is preferred over α by the entropy-seeking agent. But taking action β yields percept 0 (if any), hence nothing is learned about the environment. $\Diamond$

Solomonoff's prior is extremely good at learning: with this prior a Bayesian agent learns the value of its own policy asymptotically (on-policy value convergence) [5, Thm. 5.36]. However, generally it does not learn the result of counterfactual actions that it does not take. Knowledge-seeking agents learn the environment more effectively, because they focus on exploration. Both the entropy-seeking agent and the information-seeking agent are *strongly asymptotically optimal* in the class of all deterministic computable environments [16, 17, Thm. 5]: the value of their policy converges to the optimal value in the sense that $V^\pi \to V^*$ almost surely. Moreover, the information-seeking agent also learns to predict the result of counterfactual actions [17, Thm. 7].

4.2 Knowledge-Seeking is Limit Computable

We proceed to show that ε-optimal knowledge-seeking agents are limit computable, and optimal knowledge-seeking agents are in Δ_3^0.

Theorem 11 (Computability of Knowledge-Seeking). *There are limit-computable ε-optimal policies and Δ_3^0-computable optimal policies for entropy-seeking and information-seeking agents.*

Proof. Since ξ, ν, and w_ν are lower semicomputable, the value functions V_H^* and V_I^* are Δ_2^0-computable according to Lemma 2 (iii-v). The claim now follows from the following lemma. $\square$

Lemma 12 (Complexity of (ε-)Optimal Policies [10, Thm. 8 & 11]). *If the optimal value function V^* is Δ_n^0-computable, then there is an optimal policy π^* that is in Δ_{n+1}^0, and there is an ε-optimal policy π^ε that is in Δ_n^0.*

5 A Weakly Asymptotically Optimal Agent in Δ_2^0

In reinforcement learning we are interested in *reward-seeking* policies. Rewards are provided by the environment as part of each percept $e_t = (o_t, r_t)$ where

$o_t \in \mathcal{O}$ is the *observation* and $r_t \in [0, 1]$ is the *reward*. In this section we fix a computable discount function $\gamma : \mathbb{N} \to \mathbb{R}$ with $\gamma(t) \geq 0$ and $\sum_{t=1}^{\infty} \gamma(t) < \infty$. The *discount normalization factor* is defined as $\Gamma_t := \sum_{i=t}^{\infty} \gamma(i)$. The *effective horizon* $H_t(\varepsilon)$ is a horizon that is long enough to encompass all but an ε of the discount function's mass:

$$H_t(\varepsilon) := \min\{k \mid \Gamma_{t+k}/\Gamma_t \leq \varepsilon\}.$$

Definition 13 (Reward-Seeking Value Function [10, Def. 20]). *The re-ward-seeking value of a policy π in environment ν given history $æ_{<t}$ is*

$$V_\nu^\pi(æ_{<t}) := \frac{1}{\Gamma_t} \sum_{m=t}^{\infty} \sum_{e_{t:m}} \gamma(m) r_m \nu(e_{1:m} \mid e_{<t} \parallel a_{1:m})$$

if $\Gamma_t > 0$ and $V_\nu^\pi(æ_{<t}) := 0$ if $\Gamma_t = 0$ where $a_i := \pi(e_{<i})$ for all $i \geq t$.

Definition 14 (Weak Asymptotic Optimality [7, Def. 7]). *A policy π is weakly asymptotically optimal in the class of environments $\mathcal{M}$ iff the reward-seeking value converges to the optimal value on-policy in Cesàro mean, i.e.,*

$$\frac{1}{t} \sum_{k=1}^{t} \left(V_\nu^*(æ_{<k}) - V_\nu^\pi(æ_{<k}) \right) \xrightarrow{t \to \infty} 0 \quad \nu\text{-almost surely for all } \nu \in \mathcal{M}.$$

Not all discount functions admit weakly asymptotically optimal policies [7, Thm. 8]; a necessary condition is that the effective horizon grows sublinearly [6, Thm. 5.5]. This is satisfied by geometric discounting, but not by harmonic or power discounting [5, Tab. 5.41].

This condition is also sufficient [6, Thm. 5.6]: Lattimore defines a weakly asymptotically optimal agent called *BayesExp* [6, Ch. 5]. BayesExp alternates between phases of exploration and phases of exploitation: if the optimal in-formation-seeking value is larger than ε_t, then BayesExp starts an exploration phase, otherwise it starts an exploitation phase. During an exploration phase, BayesExp follows an optimal information-seeking policy for $H_t(\varepsilon_t)$ steps. During an exploitation phase, BayesExp follows an ξ-optimal reward-seeking policy for one step [6, Alg. 2].

Generally, optimal reward-seeking policies are Π_2^0-hard [10, Thm. 16], and for optimal knowledge-seeking policies we only proved that they are Δ_3^0. Therefore we do not know BayesExp to be limit computable, and we expect it not to be. However, we can approximate it using ε-optimal policies preserving weak asymptotic optimality.

Theorem 15 (A Limit-Computable Weakly Asymptotically Optimal Agent). *If there is a nonincreasing computable sequence of positive reals $(\varepsilon_t)_{t\in\mathbb{N}}$ such that $\varepsilon_t \to 0$ and $H_t(\varepsilon_t)/(t\varepsilon_t) \to 0$ as $t \to \infty$, then there is a limit-computable policy that is weakly asymptotically optimal in the class of all com-putable stochastic environments.*

Proof. Analogously to Theorem 3 (i) we get that ξ is lower semicomputable, and hence the optimal reward-seeking value function V_ν^* is limit computable [10, Lem. 21]. Hence by Lemma 12, there is a limit-computable 2^{-t}-optimal reward-seeking policy π_ξ for the universal mixture ξ [10, Cor. 22]. By Theorem 11 there are limit-computable $\epsilon_t/2$-optimal information-seeking policies π_I^t with lifetime $t + H_t(\varepsilon_t)$. We define a policy π analogously to BayesExp with π_I^t and π_ξ instead of the optimal policies:

If $V_I^*(\ae_{<t}) > \varepsilon_t$ for lifetime $t + H_t(\varepsilon_t)$, then follow π_I^t for $H_t(\varepsilon_t)$ steps.

Otherwise, follow π_ξ for one step.

Since V_I^*, π_I, and π_ξ are limit computable, the policy π is limit computable. Furthermore, π_ξ is 2^{-t}-optimal and $2^{-t} \to 0$, so $V_\xi^{\pi_\xi}(\ae_{<t}) \to V_\xi^*(\ae_{<t})$ as $t \to \infty$.

Now we can proceed analogously to the proof of [6, Thm. 5.6], which consists of three parts. First, it is shown that the value of the ξ-optimal reward-seeking policy π_ξ^* converges to the optimal value for exploitation time steps (second branch in the definition of π) in the sense that $V_\mu^{\pi_\xi^*} \to V_\mu^*$. This carries over to the 2^{-t}-optimal policy π_ξ, since the key property is that on exploitation steps, $V_I^* < \varepsilon_t$; i.e., π only exploits if potential knowledge-seeking value is low. In short, we get for exploitation steps

$$V_\xi^{\pi_\xi}(\ae_{<t}) \to V_\xi^{\pi_\xi^*}(\ae_{<t}) \to V_\mu^{\pi_\xi^*}(\ae_{<t}) \to V_\mu^*(\ae_{<t}) \text{ as } t \to \infty.$$

Second, it is shown that the density of exploration steps vanishes. This result carries over since the condition $V_I^*(\ae_{<t}) > \varepsilon_t$ that determines exploration steps is exactly the same as for BayesExp and π_I^t is $\varepsilon_t/2$-optimal.

Third, the results of part one and two are used to conclude that π is weakly asymptotically optimal. This part carries over to our proof.   □

6 Summary

When using Solomonoff's prior for induction, we need to evaluate conditional probabilities. We showed that conditional M and M_{norm} are limit computable (Theorem 3), and that $\overline{M}$ and $\overline{M}_{\text{norm}}$ are not limit computable (Theorem 4 and Corollary 5); see Table 1 on page 365. This result implies that we can approximate M or M_{norm} for prediction, but not the measure mixture $\overline{M}$ or $\overline{M}_{\text{norm}}$.

In some cases, normalized priors have advantages. As illustrated in Example 10, unnormalized priors can make the entropy-seeking agent mistake the entropy gained from the probability assigned to finite strings for knowledge. From $M_{\text{norm}} \geq M$ we get that M_{norm} predicts just as well as M, and by Theorem 3 we can use M_{norm} without losing limit computability.

Any method that tries to tackle the reinforcement learning problem has to balance between exploration and exploitation. AIXI strikes this balance in the Bayesian way. However, this does not lead to enough exploration [9,15]. Our agent cares more about the present than the future—hence an investment in form of exploration is discouraged. To counteract this, we can add a knowledge-seeking component to the agent. In Section 4 we discussed two variants of knowledge-seeking agents: entropy-seekers [16] and information-seekers [17]. We showed that ε-optimal knowledge-seeking agents are limit computable and optimal knowledge-seeking agents are Δ_3^0 (Theorem 11).

We set out with the goal of finding a perfect reinforcement learning agent that is limit computable. Weakly asymptotically optimal agents can be considered a suitable candidate, since they are currently the only known general reinforcement learning agents which are optimal in an objective sense [9]. We discussed Lattimore's BayesExp [6, Ch. 5], which relies on Solomonoff induction to learn its environment and on a knowledge-seeking component for extra exploration. Our results culminated in a limit-computable weakly asymptotically optimal agent (Theorem 15). based on Lattimore's BayesExp. In this sense our goal has been achieved.

Acknowledgments. This work was supported by ARC grant DP150104590. We thank Tom Sterkenburg for feedback on the proof of Theorem 6.

References

1. Blackwell, D., Dubins, L.: Merging of opinions with increasing information. The Annals of Mathematical Statistics, 882–886 (1962)
2. Gács, P.: On the relation between descriptional complexity and algorithmic probability. Theoretical Computer Science **22**(1–2), 71–93 (1983)
3. Hutter, M.: A theory of universal artificial intelligence based on algorithmic complexity. Technical Report cs.AI/0004001 (2000). http://arxiv.org/abs/cs.AI/0004001
4. Hutter, M.: New error bounds for Solomonoff prediction. Journal of Computer and System Sciences **62**(4), 653–667 (2001)
5. Hutter, M.: Universal Artificial Intelligence: Sequential Decisions Based on Algorithmic Probability. Springer (2005)
6. Lattimore, T.: Theory of General Reinforcement Learning. PhD thesis, Australian National University (2013)
7. Lattimore, T., Hutter, M.: Asymptotically optimal agents. In: Kivinen, J., Szepesvári, C., Ukkonen, E., Zeugmann, T. (eds.) ALT 2011. LNCS, vol. 6925, pp. 368–382. Springer, Heidelberg (2011)
8. Lattimore, T., Hutter, M., Gavane, V.: Universal prediction of selected bits. In: Kivinen, J., Szepesvári, C., Ukkonen, E., Zeugmann, T. (eds.) ALT 2011. LNCS, vol. 6925, pp. 262–276. Springer, Heidelberg (2011)
9. Leike, J., Hutter, M.: Bad universal priors and notions of optimality. In: Conference on Learning Theory (2015)
10. Leike, J., Hutter, M.: On the computability of AIXI. In: Uncertainty in Artificial Intelligence (2015)
11. Li, M., Vitányi, P.M.B.: An Introduction to Kolmogorov Complexity and Its Applications. Texts in Computer Science, 3rd edn. Springer (2008)
12. Nies, A.: Computability and Randomness. Oxford University Press (2009)
13. Orseau, L.: Optimality issues of universal greedy agents with static priors. In: Hutter, M., Stephan, F., Vovk, V., Zeugmann, T. (eds.) Algorithmic Learning Theory. LNCS, vol. 6331, pp. 345–359. Springer, Heidelberg (2010)

14. Orseau, L.: Universal knowledge-seeking agents. In: Kivinen, J., Szepesvári, C., Ukkonen, E., Zeugmann, T. (eds.) ALT 2011. LNCS, vol. 6925, pp. 353–367. Springer, Heidelberg (2011)
15. Orseau, L.: Asymptotic non-learnability of universal agents with computable horizon functions. Theoretical Computer Science **473**, 149–156 (2013)
16. Orseau, L.: Universal knowledge-seeking agents. Theoretical Computer Science **519**, 127–139 (2014)
17. Orseau, L., Lattimore, T., Hutter, M.: Universal knowledge-seeking agents for stochastic environments. In: Jain, S., Munos, R., Stephan, F., Zeugmann, T. (eds.) ALT 2013. LNCS, vol. 8139, pp. 158–172. Springer, Heidelberg (2013)
18. Rathmanner, S., Hutter, M.: A philosophical treatise of universal induction. Entropy **13**(6), 1076–1136 (2011)
19. Solomonoff, R.: A formal theory of inductive inference. Parts 1 and 2. Information and Control **7**(1), 1–22 and 224–254 (1964)
20. Solomonoff, R.: Complexity-based induction systems: Comparisons and convergence theorems. IEEE Transactions on Information Theory **24**(4), 422–432 (1978)
21. Wood, I., Sunehag, P., Hutter, M.: (Non-)equivalence of universal priors. In: Dowe, D.L. (ed.) Solomonoff Festschrift. LNCS, vol. 7070, pp. 417–425. Springer, Heidelberg (2013)

List of Notation

$:=$ defined to be equal

$\mathbb{N}$ the natural numbers, starting with 0

A, B sets of natural numbers

$\mathbb{1}_A$ the characteristic function that is 1 if its argument is an element of the set A and 0 otherwise

$\mathcal{X}^*$ the set of all finite strings over the alphabet $\mathcal{X}$

$\mathcal{X}^\infty$ the set of all infinite strings over the alphabet $\mathcal{X}$

$\mathcal{X}^\sharp$ $\mathcal{X}^\sharp := \mathcal{X}^* \cup \mathcal{X}^\infty$, the set of all finite and infinite strings over the alphabet $\mathcal{X}$

x, y finite or infinite strings, $x, y \in \mathcal{X}^\sharp$

$x \sqsubseteq y$ the string x is a prefix of the string y

ϵ the empty string, the history of length 0

ε a small positive real number

$\mathcal{A}$ the (finite) set of possible actions

$\mathcal{O}$ the (finite) set of possible observations

$\mathcal{E}$ the (finite) set of possible percepts, $\mathcal{E} \subset \mathcal{O} \times \mathbb{R}$

M Solomonoff's prior defined in (2)

$\overline{M}$ the measure mixture defined in (4)

ν_{norm} Solomonoff normalization of the semimeasure ν defined in (3)

α, β two different actions, $\alpha, \beta \in \mathcal{A}$

a_t the action in time step t

e_t the percept in time step t

o_t the observation in time step t

r_t the reward in time step t, bounded between 0 and 1

$æ_{<t}$ the first $t-1$ interactions, $a_1 e_1 a_2 e_2 \ldots a_{t-1} e_{t-1}$

γ the discount function $\gamma : \mathbb{N} \to \mathbb{R}_{\geq 0}$

Γ_t a discount normalization factor, $\Gamma_t := \sum_{i=t}^{\infty} \gamma(i)$

$H_t(\varepsilon)$ the effective horizon, $H_t(\varepsilon) = \min\{H \mid \Gamma_{t+H}/\Gamma_t \leq \varepsilon\}$

π a policy, i.e., a function $\pi : (\mathcal{A} \times \mathcal{E})^* \to \mathcal{A}$

V_H^π the entropy-seeking value of the policy π (see Theorem 7)

V_I^π the information-seeking value of the policy π (see Theorem 8)

V_ν^π the reward-seeking value of policy π in environment ν (see Theorem 13)

V^π the entropy-seeking/information-seeking/reward-seeking value of policy π

V^* the optimal entropy-seeking/information-seeking/reward-seeking value

ϕ a computable function

S a computable relation over natural numbers

n, k, i natural numbers

t (current) time step

m lifetime of the agent (a function of the current time step t)

$\mathcal{M}$ the class of all lower semicomputable chronological conditional semimeasures; our environment class

ν lower semicomputable semimeasure

μ computable measure, the true environment

ξ the universal mixture over all environments in $\mathcal{M}$

Open Questions

1. Can the upper bound of Δ_3^0 for knowledge-seeking policies be improved?
2. Is BayesExp limit computable?
3. Does the lower given in Theorem 4 and Theorem 5 hold for any universal Turing machine?

We expect the answers to questions 1 and 2 to be negative and the answer to question 3 to be positive.

Two Problems for Sophistication

Peter Bloem[(⊠)] , Steven de Rooij, and Pieter Adriaans

System and Network Engineering Group, University of Amsterdam,
Amsterdam, The Netherlands
uva@peterbloem.nl, steven.de.rooij@gmail.com, p.w.adriaans@uva.nl

Abstract. Kolmogorov complexity measures the amount of information in data, but does not distinguish structure from noise. Kolmogorov's definition of the *structure function* was the first attempt to measure only the structural information in data, by measuring the complexity of the smallest model that allows for optimal compression of the data. Since then, many variations of this idea have been proposed, for which we use *sophistication* as an umbrella term. We describe two fundamental problems with existing proposals, showing many of them to be unsound. Consequently, we put forward the view that the problem is fundamental: it may be impossible to objectively quantify the sophistication.

1 Introduction

Kolmogorov complexity gives us a sound definition of the amount of information contained in a binary string. It does not, however, capture what most people would consider complexity. For example, a sequence of a million coin flips will almost certainly have maximal Kolmogorov complexity, even though there is nothing complex about flipping a coin repeatedly. Many scholars have defined additional measures in the spirit of Kolmogorov complexity, aimed at quantifying not *all* information in a binary string, but only the *meaningful*. While this concept has been given many names, we use *sophistication* as an umbrella term. In this paper, we investigate two serious problems with sophistication. We conclude with two arguments suggesting the problems are fundamental, explaining our belief that sophistication cannot be defined in a satisfactory manner.

The Kolmogorov complexity $C(x)$ of a binary string x is, informally, the length of the shortest computer program to print x. This length depends on the choice of programming language, but, by the invariance theorem [15, Section 2.1], only by a constant, independent of x. For sufficiently complex objects, the choice of programming language becomes irrelevant and Kolmogorov complexity becomes an *objective* measure. A definition of sophistication $S(x)$ in the spirit of $C(x)$ should have similar guarantees:

1. $S(x)$ should count the bits required for an effective description of the structural properties of a binary string.
2. An analogue of invariance should hold: there must be strict limits on how much sophistication can be affected by a change in programming language.

© Springer International Publishing Switzerland 2015
K. Chaudhuri et al. (Eds.): ALT 2015, LNAI 9355, pp. 379–394, 2015.
DOI: 10.1007/978-3-319-24486-0_25

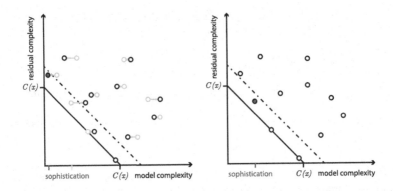

Fig. 1. (left) Two-part representations of x by the two components of their code. The Kolmogorov complexity $C(x)$, appearing as a black diagonal, provides a lower bound on the total codelength. We consider only representations that are close to this optimum—called *candidates*—with the threshold represented by a dashed line. The size of the smallest model below the threshold is the sophistication of the data. (right) The same image, after a constant perturbation in the model complexity caused by a change in numbering.

3. There should be no constant c such that $S(x) \leq c$ for every input x. If sophistication is bounded, then knowing its value under one programming language provides no constraints on its value under another language (except it is also bounded).

4. Similarly, there should be no constant c such that $|C(x) - S(x)| \leq c$ for all x, because then sophistication would be equivalent to Kolmogorov complexity.

There have been many proposals for such a measure, all based on a *two-part code*: we encode a *model* in the first part of the code, which is interpreted as a representation of x's structural properties. The model does not fully specify x, but when combined with the second part of the code, which specifies the noise, the original string becomes fully determined.[1]

For any string x, there may be many different two-part codes. The total length can never be less than the Kolmogorov complexity, but it can come close. Figure 1 illustrates the principle. The key to sophistication is to take the representations that come close to the Kolmogorov complexity, the *candidates*, and define the sophistication as the size of the smallest model in this set. However, for most definitions, we can prove that they fail one of the conditions above. For others, we cannot *prove* they conflict with our requirements, but we show these methods only assign substantial sophistication to strings that require an enormous amount of processing to construct.

A valid definition of $S(x)$ must contend with two important issues. First, the details of the way the model is encoded are important. There are two technically distinct approaches; in one of these one has to deal with the so-called

---

[1] Some variants deviate from the two-part coding format, see Section 4.3.

"nickname problem" that strangely remains unresolved in several publications. These definitions yield a sophistication that is highly dependent on the chosen programming language, unless special care is taken, as discussed in Section 3.

The second issue is that of striking the right balance between under- and overfitting, which we consider in Section 4. Overfitting is a common problem in statistics, that refers to the tendency to choose a complex model that provides a very good fit to the observed data, but does not generalise well to unseen data. In the case of sophistication, overfitting occurs if the model that determines the sophistication contains much or even all of the noise. In statistics, overfitting is often addressed by penalising complex models. In sophistication, however, such penalties tend to break the balance between structural information and noise, and lead to the opposite problem: underfitting.

Underfitting occurs when the selected model is simple, but fails to capture all structure in the data. This is also a problem for sophistication because the models under consideration are so powerful. In particular, in any programming language, there are programs that implement an interpreter for *another* language. Such *universal models* are *simple*, since they can be described with a relatively small number of bits, yet are able to represent any data using a code within a constant from the Kolmogorov complexity. Such a two-part representation essentially encodes all information as noise. If complex models are penalized, then the problem becomes to make sure that universal models are not *always* preferred for complex data. The usual workaround is to restrict the set of allowed models, for instance to total functions. While this excludes universal models, it is questionable whether it adequately solves the problem of underfitting in general.

Finally, in Section 5 we argue that while two-part coding can yield useful insights into the structure of the data and identifies some models as poor representations, it is probably not possible to objectively separate structure from noise and identify a *single* model as "best": many models of different complexities may be reasonable representations. Rather than doggedly trying to "fix" this property of algorithmic statistics, we propose embracing the idea that the data allows for multiple, equivalent interpretations of which information is structured, and which is random, and that there is no such thing as sophistication.

2 Notation

The following notation allows us to generalize across all definitions and variants, save the occasional exception which we will highlight individually.

Let $\mathbb{B} = \{0,1\}^*$. We deal with partial computable functions $f : \mathbb{B} \times \mathbb{B} \to \mathbb{B}$, which we also call *models*. f is called *prefix* if $\mathrm{dom}_z(f) = \{y : f(y, z) \neq \infty\}$, is a prefix free set for all z, i.e. no string in $\mathrm{dom}_z(f)$ is a prefix of another. A function f is *total* if $\forall_z \mathrm{dom}_z(f) = \mathbb{B}$. In most cases, we do not use the second argument, and let $f(x) = f(x, \epsilon)$.

A *numbering* is an enumeration of the partial computable functions, denoted by $\psi_1, \psi_2, \ldots$ or simply ψ. We fix one canonical numbering ϕ, chosen to be *effective*: ie. given i and y, we can effectively compute $\phi_i(y)$. We call a numbering

ψ *acceptable* if there exist total, computable functions $a, b : \mathbb{N} \to \mathbb{N}$ with $\forall : i$, $\phi_i = \psi_{b(i)}$ and $\psi_i = \phi_{a(i)}$.

A *model class* is a set of indices in a numbering ψ. We define four classes:

- The indices of the partial computable functions $\mathcal{C} = \mathbb{N}$.
- The total functions $\mathcal{T} = \{i : \psi_i \text{ is total}\}$. Note that $\mathcal{T}$ is not computably enumerable.
- $\mathcal{K}$ is an enumerable set such that $\{\psi_i : i \in \mathcal{K}\}$ is the set of all partial computable prefix functions.
- The finite sets: $\mathcal{F}$ is an enumerable set such that $\{\psi_i : i \in \mathcal{F}\}$ is the set of uniform codes for all finite sets.[2]

Let $\bar{x}$ denote the prefix-encoded representation for x. We require that the mapping satisfies $|\bar{x}| = |x| + O(\log|x|)$ (see eg. [15, Section 1.4]). To simplify notation, we will sometimes conflate natural numbers and binary strings, implicitly using the ordering $(0, \epsilon), (1, 0), (2, 1), (3, 00), (4, 01), \ldots$

For technical reasons, we deviate slightly from the traditional notation of Kolmogorov complexity: let $\mathcal{M}$ be a model class and ψ an acceptable numbering, then let $C^{\mathcal{M},\psi}(x \mid z) = \min\{|\bar{i}y| : \psi_i(y, z) = x, i \in \mathcal{M}\}$, with $C^{\mathcal{M},\psi}(x) = C^{\mathcal{M},\psi}(x \mid \epsilon)$. We omit the numbering when the distinction is not relevant. $C^{\mathcal{C}}(x)$ corresponds to the plain Kolmogorov complexity $C(x)$ and $C^{\mathcal{K}}(x)$ corresponds to the prefix-free version $K(x)$. Note that the notation $C^{\{i\},\psi}(x)$ represents the smallest two-part description of x using model ψ_i.

We use the principle of a numbering for the purpose normally served by the universal Turing machine. We prefer to work with numberings as it highlights an important issue: while Kolmogorov complexity is invariant to the choice of numbering this property does not immediately carry over to sophistication: for some treatments, the result is highly dependent on the chosen numbering, as we will see in the next section.

3   Inefficient Indices

The simplest approach to sophistication would be to 'open up' the Kolmogorov complexity and to see which program achieves the smallest description length: the program that *witnesses* the Kolmogorov complexity. This witness is a two-part coding; it consists of a model and an input.

Definition 1 (Index sophistication). *Let ψ be an acceptable numbering. Let $\mathcal{M}$ be the model class from which candidates are chosen, and let $\mathcal{N}$ be the model class that determines the minimum achievable complexity. Let c be a fixed constant. The* index sophistication *is:*

$$S_{\text{index}}^{\mathcal{M},\mathcal{N},\psi,c}(x) = \min\left\{|i| : C^{\{i\},\psi}(x) \leq C^{\mathcal{N},\psi}(x) + c, \ i \in \mathcal{M}\right\}.$$

When $\mathcal{M} = \mathcal{N}$, we will use $S_{\text{index}}^{\mathcal{M},\psi,c}$. If the set over which the minimum is taken is empty, the sophistication is undefined.

---

[2] A uniform code for a set F is a surjective prefix function $f : \{0,1\}^{\lceil \log|F|\rceil} \to F$.

Koppel and Atlan's treatment [13,14], where the name *sophistication* originates, follows this basic logic, although it contains idiosyncracies like the use of monotonic models, and an extension to infinite strings. As the subsequent history of sophistication has discarded these, we will not discuss them here.

In [4,3] Koppel's principle is limited to finite strings, with $\mathcal{J}$ as a model class. The definition is similar to $S_{\text{index}}^{\mathcal{J},\mathcal{C},\psi,c}$, except the total complexity of a witness (i, y) is measured as $|i| + |y|$ without the cost of delimiting the two. This difference is not relevant to the current discussion. The restriction to $\mathcal{J}$ is a common approach, which avoids underfitting, as discussed in the next section.

Lemma 1. *Let S_{index}^{ψ} denote any index sophistication with respect to numbering ψ (with any choice for $\mathcal{M}$, $\mathcal{N}$ and c). There are acceptable numberings ψ and ξ such that for all x: $|S_{\text{index}}^{\psi}(x) - S_{\text{index}}^{\xi}(x)| \geq \frac{1}{2}\min\{S_{\text{index}}^{\psi}(x), S_{\text{index}}^{\xi}(x)\}$.*

Proof. Let $z_i \in \mathbb{B}$ consist of $2^i - 1$ zeroes followed by a one. Define ψ, ξ such that $\psi_j(x) = \phi_i(x)$ for $j = z_{2i}$ and $\xi_j(x) = \phi_i(x)$ for $j = z_{2i+1}$, with all other functions returning ∞ for all inputs. Choose any x and assume w.l.o.g. that $S_{\text{index}}^{\psi}(x) \leq S_{\text{index}}^{\xi}(x)$. By construction, we have $2S_{\text{index}}^{\psi}(x) \leq S_{\text{index}}^{\xi}(x)$. □

Thus, the length of the index is a very poor indicator of model complexity. For a robust measure, we define the complexity of a function f as in [11,20] by

$$C^{\mathcal{M},\psi}(f) = \min\{C^{\mathcal{M},\psi}(i) : \psi_i = f\}. \tag{1}$$

Lemma 3 in the appendix shows that $C^{\mathcal{C}}(f)$ and $C^{\mathcal{K}}(f)$ are invariant. Note that the perversely inefficient numberings of Lemma 1 are no issue for Kolmogorov complexity: we can use a UTM with a more efficient numbering as a model at only a constant penalty. For sophistication, however, the numbering is crucial.

There are two ways to use $C^{\mathcal{M}}(f)$ for more robust attempts to define sophistication. Confusingly, both are used in the literature. First, we can measure the complexity of the *model* ϕ_i as $C^{\mathcal{K}}(\phi_i)$, which is then the size of the first part of a two-part code describing the data. This approach is used in [7,8,20,9].

Second, we can stick to using the length of the index as the measure of sophistication, but restrict the allowed numberings to those that can represent a given function *efficiently*. This approach is taken by Adriaans in [1], who defines *facticity* as $S_{\text{index}}^{\mathcal{C},\psi,0}$, but only allows *faithful* numberings. Formally, a faithful numbering has the property that $\forall i \exists j : \psi_i = \psi_j, |j| \leq C^{\mathcal{C}}(\psi_j) + c$, for some constant c. Essentially, this means that a faithful numbering can represent a function f with an index the same length as the Kolmogorov complexity $C^{\mathcal{C}}(f)$.

We prove in the appendix that—contrary to Adriaans' suggestion—there do exist faithful, acceptable numberings (Lemma 4). However, even choosing a faithful numbering is not enough. The Kolmogorov complexity uses representations of the form $\bar{i}y$, with $\psi_i(y) = x$, where the bar denotes some straightforward prefix encoding to delimit the model description i from its input y. If we define a second prefix encoding $\tilde{i}$, with $|\bar{i}| - |\tilde{i}|$ unbounded, we can define a second representation $\bar{u}\tilde{i}y$, with $\psi_u(\tilde{i}y) = \psi_i(y)$, at a constant overhead $|\bar{u}|$, and gain more than $|\bar{u}|$ for sufficiently complex strings, resulting again in a bounded sophistication.

We continue with a sophistication that avoids the issues of inefficient indices and of inefficient prefix encodings. We change the definition of index sophistication so that its two-part representations use $C^{\mathcal{K}}(\phi_i)$ bits for the representation of the model. We first introduce the following notation for the $\mathcal{M}$-Kolmogorov complexity using such compact two-part representations:

$$C_{\text{comp}}^{\mathcal{M},\psi}(x) = \min\{C^{\mathcal{K},\psi}(\psi_i) + |y| : \psi_i(y) = x, i \in \mathcal{M}\}.$$

For model classes $\mathcal{K}$ and $\mathcal{C}$ this is equivalent to the existing definition and invariant to the numbering. Note that again, we use $C_{\text{comp}}^{\{i\},\psi}(x)$ to represent the smallest two-part code using model ψ_i.

Definition 2 (Sophistication).

$$S^{\mathcal{M},\mathcal{N},\psi,c}(x) = \min\left\{ C^{\mathcal{K}}(\phi_i) : C_{\text{comp}}^{\{i\},\psi} \leq C_{\text{comp}}^{\mathcal{N},\psi}(x) + c, i \in \mathcal{M} \right\}.$$

4   Balancing Under- and Overfitting

In the last section, we began to see the delicate balance between the two code components. We will study this balance, starting with the variant $S^{\mathcal{K},\psi,c}$, which is not used in the literature, but helps to illustrate the issues we wish to discuss.

$\mathcal{K}$ has optimal representations with all but a constant part of the information in the input and it has optimal representations with all information in the model. The downside to this balance is that it becomes easy to show a lack of invariance. We can tweak the numbering so that models in a specific subset $\mathcal{M}' \subset \mathcal{K}$ become cheaper to represent by an arbitrary amount relative to others: we can ensure that a model in $\mathcal{M}'$ always determines the sophistication. For instance, if we let $\mathcal{M}'$ contain only a universal model we get a bounded sophistication.

Theorem 1 (Underfitting). *Let $\mathcal{M}, \mathcal{N}$ be model classes with $\mathcal{M} \subseteq \mathcal{N}$ and let $\mathcal{M}$ contain a universal model ϕ_u, with the property that $\exists c \forall i \in \mathcal{N}, x \in \mathbb{B} : C_{\text{comp}}^{\{u\},\phi}(x) \leq C_{\text{comp}}^{\mathcal{N},\phi}(x) + c$. Then, for some numbering ψ, $S^{\mathcal{M},\mathcal{N},\psi,c}$ is bounded.*

This problem is well known and many treatments avoid it by restricting the model class. Less well known, perhaps, is that the same holds in the other direction: if $\mathcal{M}'$ is the set of *singleton* models—those models that output a single x for an empty input—we get a sophistication equal to the Kolmogorov complexity.

Theorem 2 (Overfitting). *Let $\mathcal{X} \subseteq \mathbb{B}$. Let $\mathcal{M} \subseteq \mathcal{N} \subseteq \mathcal{K}$ be model classes where for every $x \in \mathcal{X}$ there is a singleton model $i \in \mathcal{M}$ with $\phi_i(\epsilon) = x$. Then there is a numbering ψ, and a constant c, such that for all $x \in \mathcal{X}$ we have $C^{\mathcal{K}}(x) - S^{\mathcal{M},\mathcal{N},\psi,c}(x) \leq c$.*

The proofs of both theorems rely on a simple principle: there exist numberings which have the effect of penalizing $C^{\mathcal{K}}(\phi_i)$ for any model outside $\mathcal{M}'$ by an arbitrary constant amount. We can use this to effectively 'push' these models outside of the range of candidates, ensuring that, under this numbering, a model in $\mathcal{M}'$ always determines the sophistication. The requirements for $\mathcal{M}'$ are somewhat complex. The following lemma gives a set of sufficient conditions.

Lemma 2. *Let $\mathcal{M}$ and $\mathcal{N}$ be any model class, let $\mathcal{X}$ be any set of binary strings and let $D : \mathbb{B} \to \mathbb{N}$ be a partial computable decoding function with a prefix-free domain that maps function descriptions to their indices in ϕ. Let $\mathcal{M}' = \mathrm{range}(D)$. Further assume there is a constant c such that:*

(a) $\forall_{m \in \mathcal{M}'} : \min\{|p| : \phi_{D(p)} = \phi_m\} \leq C^{\mathcal{K},\phi}(\phi_m) + c$

(b) $\forall_{x \in \mathcal{X}} : C_{\mathrm{comp}}^{\mathcal{M}',\phi}(x) - C_{\mathrm{comp}}^{\mathcal{N},\phi}(x) \leq c$.

Then, there is a ψ such that if $S^{\mathcal{M},\mathcal{N},\psi,k}(x)$ is defined, then $S^{\mathcal{M},\mathcal{N},\psi,k}(x) = S^{\mathcal{M}',\mathcal{N},\psi,k}(x)$ up to a constant.

Proof. Pick any $x \in \mathcal{X}$. Let f and g be ϕ-indices such that $f \in \mathcal{M}'$ and $g \notin \mathcal{M}'$ nor is ϕ_g equivalent to any function indexed by $\mathcal{M}'$. Furthermore let $C_{\mathrm{comp}}^{\{f\},\phi}(x)$ and $C_{\mathrm{comp}}^{\{g\},\phi}(x)$ both be within a constant q of $C_{\mathrm{comp}}^{\mathcal{N},\phi}(x)$. Assumption (b) ensures that $\mathcal{M}'$ always provides such an f.

We will show that for every integer r, there is a numbering ψ such that $C_{\mathrm{comp}}^{\{g'\},\psi}(x) - C_{\mathrm{comp}}^{\{f'\},\psi}(x) \geq r$ for all $x \in \mathcal{X}$, where f' and g' are the ψ-indices equivalent to f and g. Thus, for large enough r, ϕ_g is eliminated as a candidate model, while ϕ_f remains in place. Thus, under ψ, a member of $\mathcal{M}'$ determines the sophistication, or the sophistication is undefined.

Let d be a positive constant. We will show later how to choose it to achieve the required result. We define ψ as follows:

$$\psi_0(p) = 0^d 1 D(p), \quad \psi_{0^d 1 i}(p) = \phi_i(p), \quad \psi_j(\cdot) = \infty \text{ if } j \neq 0 \text{ and } j \neq 0^d 1 \ldots$$

The key to the proof is the way that the function complexity $C^{\mathcal{K}}(\cdot)$ changes when we change the numbering from ϕ to ψ. For f, the value increases by no more than a fixed constant, but for g, it increases by a constant that we can arbitrarily increase by increasing d.

We will first show that for f, the value does not increase by more than a constant c_f. Assume w.l.o.g. that $0 \in \mathcal{K}$.

$$
\begin{aligned}
C^{\mathcal{K},\psi}(\phi_f) &= \min\left\{|\bar{j}q| : \psi_{\psi_j(q)} = \phi_f, j \in \mathcal{K}\right\} && \text{rewriting (1)} \\
&\leq \min\left\{|\bar{0}q| : \psi_{\psi_0(q)} = \phi_f\right\} && \text{choose } j = 0 \\
&= \min\left\{|q| : \phi_{D(q)} = \phi_f\right\} + |\bar{0}| \\
&\leq C^{\mathcal{K},\phi}(\phi_f) + c_f && \text{by assumption (a).}
\end{aligned}
$$

In order to show that for g, we can increase the difference by an arbitrary constant, we first show that, for any z not in the range of ψ_0, the Kolmogorov complexity itself increases by at least d when we switch from ϕ to ψ:

$$C^{\mathcal{K},\psi}(z) = \min\left\{|\bar{i}y| : \psi_i(y) = z, i \in \mathcal{K}\right\} \qquad\qquad \text{by definition}$$

$$= \min\left\{|\overline{0^d 1j}y| : \psi_{0^d 1j}(y) = z\right\} \qquad \text{since } z \notin \text{range}(\psi_0)$$

$$\geq \min\left\{|\bar{j}y| : \phi_j(y) = z\right\} + d$$

$$= C^{\mathcal{K},\phi}(z) + d. \qquad\qquad\qquad\qquad (2)$$

We now show the increase in model complexity for g. First, assume $\phi_g \neq \psi_0$:

$$C^{\mathcal{K},\psi}(\phi_g) = \min\left\{C^{\mathcal{K},\psi}(i) : \psi_i = \phi_g\right\}$$

$$= \min\left\{C^{\mathcal{K},\psi}(0^d 1j) : \phi_j = \phi_g\right\}$$

$$= C^{\mathcal{K},\psi}(0^d 1j)$$

$$\geq C^{\mathcal{K},\phi}(0^d 1j) + d \qquad\qquad\qquad \text{by (2)}$$

$$\geq C^{\mathcal{K},\phi}(j) - c_g + d \qquad\qquad \text{since } C^{\mathcal{K}}(j) \leq C^{\mathcal{K}}(0^d 1j) + c_0$$

$$\geq C^{\mathcal{K},\phi}(\phi_g) - c_g + d.$$

Now assume $\phi_g = \psi_0$. We have $C^{\mathcal{K},\psi}(\phi_g) = \min\left\{C^{\mathcal{K},\psi}(i) : \psi_i = \psi_0\right\} \geq d$. This follows from the fact that the minimum is achieved either at $i = 0$ or at $i = 0^d 1m$ with $m \notin \mathcal{M}'$. Neither have a representation using a function with a ψ-index without the $0^d 1$ prefix.

Choosing $d \geq r + \max\left\{C^{\mathcal{K},\phi}(\psi_0), c_g\right\} + c_f + 2q$ ensures that for both cases, we have $C^{\mathcal{K},\psi}(\phi_g) \geq C^{\mathcal{K},\phi}(\phi_g) + r + c_f + 2q$. While $C^{\mathcal{K}}(\psi_0)$ depends on the choice of d, we have $C^{\mathcal{K}}(\psi_0) \leq C^{\mathcal{K}}(d) + C^{\mathcal{K}}(D)$, up to a constant, which is in $O(\log d)$, so we can choose d to satisfy the inequality.

Finally, we can show the result:

$$C_{\text{comp}}^{\{g'\},\psi}(x) - C_{\text{comp}}^{\{f'\},\psi}(x)$$

$$= C^{\mathcal{K},\psi}(\phi_g) + \min\{|y| : \phi_g(y) = x\} - C^{\mathcal{K},\psi}(\phi_f) - \min\{|y| : \phi_f(y) = x\}$$

$$\geq C^{\mathcal{K},\phi}(\phi_g) + r + c_f + 2q + \min\{|y| : \phi_g(y) = x\}$$

$$\quad - C^{\mathcal{K},\phi}(\phi_f) - c_f - \min\{|y| : \phi_f(y) = x\}$$

$$= C_{\text{comp}}^{\{g\},\phi}(x) - C_{\text{comp}}^{\{f\},\phi}(x) + r + 2q \geq r. \qquad\qquad \square$$

Theorems 1 and 2 follow as corollaries. For Theorem 1:

Proof. Let D be a prefix function as in Lemma 2 that returns the index of u for the argument ϵ and ∞ for any other argument. That is, $\mathcal{M}' = \{u\}$. This construction satisfies the conditions 1 and 2 from Lemma 2. Invoking it, we find that there exists an acceptable numbering ψ for which $S^{\mathcal{M},\mathcal{N},\psi,k}(x) = S^{\mathcal{M}',\mathcal{N},\phi,k}(x) + c$. Since $\mathcal{M}'$ contains only a single model, $S^{\mathcal{M}',\mathcal{N},\phi,c}(x)$ is constant. $\qquad \square$

And for Theorem 2:

Proof. Let x be any string. Given a description of x, we construct some index i such that $\phi_i(\epsilon) = x$ (a singleton for x). Thus, $C^{\mathcal{K},\psi}(\phi_i) \leq C^{\mathcal{K},\psi}(x)$ up to a constant. Likewise, given ϕ we can produce x, so that $|C^{\mathcal{K},\phi}(\phi_i) - C^{\mathcal{K},\phi}(x)| \leq c$ for some constant c.

We now define a computable function D by $D(\bar{i}y) = j$ where $\phi_j(\epsilon) = \phi_i(y)$ and $i \in \mathcal{K}$, and let $\mathcal{M}'$ be its range. We will show that the two conditions of Lemma 2 hold for the prefix function D.

(a) Let $f \in \mathcal{M}'$ with $\phi_f(\epsilon) = x$. Then $\min\{|p| : \psi_{D(p)} = \phi_f\} = \min\{|\bar{i}q| : \phi_i(q) = x\} = C^{\mathcal{K}}(x) \leq C^{\mathcal{K}}(\phi_f) + c$. (b) On the one hand $C^{\mathcal{M}',\psi}_{\text{comp}}(x) \leq C^{\mathcal{K}}(\phi_f) + |\epsilon| \leq C^{\mathcal{K}}(x) + c$. On the other hand, the witness to $C^{\mathcal{M},\psi}_{\text{comp}}(x)$ is an effective description of x, so $C^{\mathcal{K}}(x)$ is at most a constant larger.

Now, by Lemma 2 there is a numbering ψ such that we have $S^{\mathcal{M},\mathcal{N},\psi,k}(x) = S^{\mathcal{M}',\mathcal{N},\psi,k}(x) + c$. We observed that $|C^{\mathcal{K}}(\phi_i) - C^{\mathcal{K}}(x)| \leq c_0$ for all singletons, so $S^{\mathcal{M}',\mathcal{N},\psi,k}(x) \geq C^{\mathcal{K},\psi}(x) - c_0$. This proves the theorem.   □

Thus, in this balanced sophistication, there is no invariance: all information can be seen as structure, or as noise, depending on the numbering. To avoid these issues, existing proposals upset the balance to exclude or penalize the universal models, and possibly the singleton models.

4.1 Overfitting

We will now review the treatments in the literature that show overfitting. The first is the structure function, proposed by Kolmogorov, most likely the first attempt at separating structure from noise in an objective manner. Kolmogorov defined the following function, using the finite sets $\mathcal{F}$ as models:

$$h_x(\alpha) = \min\left\{\log|F| : x \in F, C^{\mathcal{K}}(F) \leq \alpha\right\}$$

and suggested that the smallest set for which $C^{\mathcal{K}}(F) + \log|F| \leq C^{\mathcal{K}}(x) + c$ holds for some pre-chosen constant c, can be seen as capturing all the structure in x [7]. This is equivalent to the sophistication $S^{\mathcal{F},\mathcal{K},\psi,c}(x)$. Theorem 2 shows there are numberings for which this sophistication is always equal to $C^{\mathcal{K}}(x)$. Thus, either this is true for all numberings, or this sophistication is not invariant.

In [8] the structure function is extended to an *algorithmic sufficient statistic*. This is, again, essentially the witness to the sophistication $S^{\mathcal{F},\mathcal{K},\psi,c}(x)$. A probabilistic version is also introduced, which uses the model class $\mathcal{P}$, which indexes the set of functions that compute computable probability semimeasures up to a multiplicative constant error, yielding $S^{\mathcal{P},\mathcal{K},\psi,c}(x)$. For both, Lemma 2 gives us a numbering such that the singleton is always the minimal sufficient statistic.

It may be argued that the slack parameter c in the sophistication, which determines the allowed gap between a candidate representation and the complexity, should depend on the numbering, but this dependence has not been mentioned in the literature and there is no obvious method to choose this constant for a given numbering.

In traditional statistics, overfitting is often addressed by a penalty on complex models. As we have seen, a strong penalty, such as the one imposed by an inefficient prefix encoding of the model, will cause underfitting. A more subtle approach is to allow descriptions that are not self-delimiting. The gap between the smallest self-delimiting description and the smallest non self-delimiting description grows without bound [15, Section 4.5.5], so that some information ends up in the noise, since placing all information in the model results in a self-delimiting, and thus non-optimal description. This eliminates the singletons as viable candidates. This approach is taken by Vitányi [20] and by Adriaans [1]. Such measures reduce the overfitting problem, but they only increase the tendency to underfit. We also pay the price that the models can no longer be equated with probability measures, weakening the link to traditional statistics.

4.2   Underfitting

Universal models are a widely acknowledged problem for sophistication, and most proposals avoid them by limiting the allowed models to exclude them. It is known that there are strings x for which $S^{\mathcal{F},\mathcal{K},\psi,c}(x)$, $S^{\mathcal{T},\psi,c}(x)$ and $S^{\mathcal{T},\mathcal{K},\psi,c}(x)$ are close to $|x|$ (up to a logarithmic term). Proofs can be found in [8], [4] and [20] respectively. These are the *absolutely non-stochastic strings* [17]. The existence of these strings is independent of the numbering.

However, the problem of the singletons remains. Only one model class eliminates both the singletons and the universal model: $\mathcal{T}$. The only proposal we are aware of that uses an efficient model representation *and* excludes the universal models *and* excludes the singletons is: $S^{\mathcal{T},\mathcal{K},\psi,c}$, from [20]. While this avoids our proofs of boundedness, there is no evidence that $S^{\mathcal{T},\mathcal{K},\psi,c}$ is actually invariant.

While high sophistication strings exist for $S^{\mathcal{T},\mathcal{K},\psi,c}$, they may not conform to sophistication's motivating intuition. To show this, we use the concept of depth:

Definition 3 (Depth[5,2]). *Let U be some universal Turing machine, so that $U(\bar{\imath}y) = \phi_i(y)$. Let U^t be a simulation of this machine, which is allowed to run for at most t steps, and returns 0 if it has not yet finished at that point. Let $C_t^{\mathcal{M}}(x) = \min\{|\bar{\imath}y| : U^t(\bar{\imath}y) = x, \phi_i \in \mathcal{M}\}$. The c-depth is $d^{\mathcal{M},c}(x) = \min\left\{t : C_t^{\mathcal{M}}(x) - C^{\mathcal{M}}(x) \le c\right\}$.*

Deep strings are those that can only be optimally compressed with a great investment of time. We note that it is exceedingly unlikely that a deep string is sampled from a shallow distribution [6,5].

Theorem 3. *Let $A(n)$ be the single-argument Ackermann function and c_d some constant. For all k, there is a numbering ψ such that for all strings with depth $d^{\mathcal{C},c_d}(x) \le A(C^{\mathcal{C}}(x))$ the sophistication $S^{\mathcal{T},\mathcal{K},\psi,k}(x)$ is bounded.*

Proof. Let $U(\bar{\imath}y)$ be some universal Turing machine, and let $U^A(\bar{\imath}y)$ be a simulation of that machine which outputs 0 if the number of steps taken exceeds $A(|\bar{\imath}y|)$. Let u be the index of the function U^A in the standard enumeration.

Let $D(\epsilon) = u$. We can instantiate Lemma 2 with D, $\mathcal{M}' = \{u\}$ and $\mathcal{X} = \{x : d^{\mathcal{C},c_d}(x) \leq A(C^{\mathcal{C}}(x))\}$. This tells us that there exists a numbering ψ for which $S^{\mathcal{J},\mathcal{K},\psi,k}(x) = S^{\mathcal{M}',\mathcal{K},\psi,k}(x) + |\bar{0}| \leq c$ for all $x \in \mathcal{X}$.   □

This shows that while high-sophistication strings exist, they do not behave as expected. Consider a string that is typical for a shallow model, say some elaborate probabilistic automaton. Under $S^{\mathcal{J},\mathcal{K},\psi,c}$, no matter how high the complexity of the automaton, the sophistication is bounded. We could encode the collected works of Shakespeare in its transition graph, and this information would be counted as noise. Any structure simple enough to be exploited within the time bound of the Ackermann function will not be seen as 'meaningful information'. Only structure so deep that it would take beyond the lifetime of the universe to decompress would count towards sophistication. In the remainder we will refer to strings x with $d^{\mathcal{C},c}(x) \leq A(C^{\mathcal{C}}(x))$ as *shallow* strings. Note that any string whose shortest program can be run in any time bound represented by a primitive recursive function is shallow.

The relation between $S(x)$ and $d(x)$ is also investigated in [3], where it is shown that within a logarithmic error term on the sophistication and the slack, they are identical. Our point is not the similarity between the two, but that for all practical strings, the sophistication is bounded. This contradicts the intuition that sophistication measures structure, as it seems to suggest that all strings we can possibly hope to understand or generate contain no structure, save a constant amount. The alternative is that under other numberings these strings *do* have structure, but then the sophistication is not invariant.

As for the strings with high sophistication, they have the property that they can be compressed far better with partial functions than with total: they are non-typical for the model class $\mathcal{J}$. This suggests that the 'non-stochastic' property of strings with high sophistication [17,19] says more about depth and totality than it does about structure and noise.

4.3   Other Variants

By moving away from the idea of two-part coding, the mechanics of lemma 2 can be avoided. In [16], the *naive sophistication* is introduced. We will define a generic version, parametrized by model class. Let $C_{\psi_i}(x) = \min\{|y| : \psi_i(y) = x\}$. Then we define the *naive sophistication* as:

$$S_{\text{naive}}^{\mathcal{M},\psi,c}(x) = \min\{C^{\mathcal{K},\psi}(\psi_i) : C_{\psi_i}(x) - C^{\mathcal{K},\psi}(x \mid i) \leq c, i \in \mathcal{M}\}.$$

The condition now is not that the two-part code length is minimal, but that the *randomness deficiency* $C_{\psi_i}(x) - C^{\mathcal{K},\psi}(x \mid i)$ is less than a constant. $S_{\text{naive}}^{\mathcal{J},\psi,c}(x)$ corresponds to the version in [16]. The switch to the randomness deficiency avoids Theorem 2, but we end up with the same problem as in Theorem 3: for shallow strings $S_{\text{naive}}^{\mathcal{J},\psi,c}$ is defined by the model U^A, and thus bounded.

We cannot show that $S_{\text{naive}}^{\mathcal{J},\psi,c}(x)$ is bounded for shallow strings, but this is only a consequence of the use of sets, not of the switch to randomness deficiency as a condition. *Any* set sophistication is necessarily lower-bounded by the function $\text{set}(x) = \min\left\{C^{\mathcal{K}}(F) : x \in F\right\}$ and if this function were bounded, it would suggest that a finite amount of finite sets contained all strings.

Theorem 4. *Let ψ be any acceptable numbering. Then for all shallow x and large enough c, $S_{\text{naive}}^{\mathcal{J},\mathcal{K},\psi,c}(x)$ is bounded and for come constant c_F, we have:*

$$\text{set}(x) \leq S_{\text{naive}}^{\mathcal{F},\mathcal{K},\psi,c}(x) \leq C^{\mathcal{K},\psi}(C^{\mathcal{K},\psi}(x)) + c_F .$$

Proof. Let ψ be any acceptable numbering. Let the Turing machine U be defined as $U(\bar{\imath}y) = \psi_i(y)$ if $i \in \mathcal{K}$, and $U(\bar{\imath}y) = \infty$ otherwise. Let U^A be derived from U as in section in Section 4.2 and let ϕ_u compute U^A.

For the first part we have $C_{\psi_u}(x) - C^{\mathcal{K},\psi}(x \mid u) \leq c_0$ for some c_0, thus for large enough c, $S_{\text{naive}}^{\mathcal{J},\psi,c}(x) \leq C^K(\psi_u)$. For the second part, let $C^{\mathcal{K},\psi}(x) = k$ and $F_k^A = \{x \mid \exists p : U^A(p) = x, |p| = k\}$. $|F_k^A| \leq |\{p : |p| = k\}|$, so that $\log|F_k^A| \leq k$, which gives us $\log|F_k^A| - C^{K,\psi}(x \mid F_k^A) \leq c_1$. Thus, for large enough c, $S_{\text{naive}}^{\mathcal{F},\psi,c}(x) \leq C^K(F_k^A)$. From a description of k, we can compute F_k^A with a finite program, so that $C^{\mathcal{K},\psi}(F_k^A) \leq C^{\mathcal{K},\psi}(k) + c_F$, which completes the proof. $\qquad \square$

Note that the constant c only needs to be large enough to ensure that $C^{\mathcal{K},\psi}(x) - C^{\mathcal{K},\psi}(x \mid u) \leq c$ and $C^{\mathcal{K},\psi}(x) - C^{\mathcal{K},\psi}(x \mid F_k^A) \leq c$. Since u and F_k^A are generally of no value in computing x, c is likely very small.

Another approach is the *coarse sophistication* [4], defined in [16] as:

$$S_{\text{coarse}}^{\mathcal{M},\mathcal{N},\psi}(x) = \min_c \left\{S^{\mathcal{M},\mathcal{N},\psi,c}(x) + c\right\} .$$

Again, this variant avoids the pitfalls of Theorem 2. If there are candidates that are as good as the singletons but with smaller size by more than a constant, the constant penalty c will eventually be much less than the gain for the simpler witness, and the singletons will not determine the coarse sophistication. The coarse sophistication is within a logarithmic term of the busy beaver depth [4]. As with the naive sophistication, we can show that for shallow strings, the total function version is bounded, and the set version grows very slowly:

Theorem 5. *Let ψ be any acceptable numbering. For all shallow x, $S_{\text{coarse}}^{\mathcal{J},\mathcal{K},\psi}(x)$ is bounded and there is a constant c_F such that:*

$$\text{set}(x) \leq s_{\text{coarse}}^{\mathcal{F},\mathcal{K},\psi}(x) \leq 2C^{\mathcal{K},\psi}(C^{\mathcal{K},\psi}(x)) + c_F .$$

Proof. Let ψ be any acceptable numbering and define U^A, ϕ_u and F_k^A as in the proof of Theorem 4. For the first part, we know that for some constant c_0, $C_{\text{comp}}^{\{u\},\psi}(x) \leq C_{\text{comp}}^{\mathcal{K},\psi}(x) + c_0$ so that $S^{\mathcal{J},\mathcal{K},\psi,c_0}(x) \leq C^{\mathcal{K},\psi}(\phi_u)$, thus $S_{\text{coarse}}^{\mathcal{J},\mathcal{K},\psi}(x) \leq C^{\mathcal{K},\psi}(\phi_u) + c_0$. For the second part, we know that for some c_1, $C^{\mathcal{K},\psi}(F_k^A) + \log|F_k^A| \leq C^{\mathcal{K},\psi}(k) + C^{\mathcal{K},\psi}(x) + c_1$, so that $S^{\mathcal{F},\mathcal{K},\psi,C^{\mathcal{K}}(k)+c_1}(x) \leq C^{\mathcal{K},\psi}(F_k^A)$. Thus, $S_{\text{coarse}}^{\mathcal{F},\mathcal{K},\psi}(x) \leq C^{\mathcal{K},\psi}(F_k^A) + C^{\mathcal{K},\psi}(k) + c_1 = 2C^{\mathcal{K},\psi}(C^{\mathcal{K},\psi}(x)) + c_F.$ $\qquad \square$

In [18], Vereshchagin proposes a *strongly algorithmic sufficient statistic*. Where the regular algorithmic sufficient statistic F from [8] has $C^{\mathcal{K}}(F \mid x)$ constant, the strong variant imposes the stronger requirement that $C^{\mathcal{T}}(F \mid x)$ is also constant. This reduces the problems of underfitting discussed in this section, but since $C^{\mathcal{T}}(\{x\} \mid x)$ is bounded, by Theorem 2, overfitting remains a problem: there exist numberings under which the singletons are the only candidates.

Finally, *effective complexity* [9], proposed by Gell-Man and Lloyd, was formulated from the perspective of physics, but fits the mold of sophistication. The model class consists of all computable probability distributions on finite sets. The complexity of the model is measured by its Kolmogorov complexity, avoiding the problems of Section 3. Theorem 2, however, still applies to effective complexity. Unlike other sophistication measures, it is not the candidate with the smallest model which is chosen, but the one which reproduces the data within the shortest time. Thus, if there are multiple candidates, this approach would likely favor the singletons. In [10], the authors abandon this approach, and note that the choice from the set of candidates is a subjective one, which depends on context, which is in line with the view we express in the next section.

5 Discussion and Conclusion

We have criticized existing measures of sophistication and shown technical problems with all of them. But that does not in itself mean that it should be impossible to come up with a sound measure. The common intuition, starting with the structure function, appears to be that the crucial property is whether a string is typical for a model, and that this typicality can be tested: another random choice from that model should select a string with the same structure. This idea is bold, but not unreasonable. Nevertheless, we offer the opinion that such a clean-cut separation *cannot* be made to work. We provide two arguments.

For the first argument, we take a generative perspective. We can generate data from a model $\phi_i, i \in \mathcal{K}$, by feeding it random bits until it produces an output. We will call the resulting probability distribution p_i. Call a sophistication *consistent* if, for sufficiently large data, it reflects the complexity of the source of the data. Now, let $\phi_u(\bar{i}y) = \phi_i(y)$ and sample from p_u. Then the initial bits will determine the prefix encoded index $\bar{i}$ of the function ϕ_i that ϕ_u will subsequently emulate, and the remaining bits are used as inputs to ϕ_i. We now ask, what should be the sophistication of the resulting data?

Certainly, if we have to judge based only on the data, we cannot exclude the possibility that the data was sampled from p_u: after all, it was. Yet, neither can we deny that it may have came from p_i, as again, it did! Eliminating the universal models does not solve this problem: the same argument holds if ϕ_u indexes, for instance, only those models computable by finite automata. *Any* model that dominates a set of other models creates this kind of ambiguity.

Consider the following metaphor. We are given a a bitmap image of the painting *Impression of a Sunrise*. There are many good models for this string, from very generic to very specific. Sophistication suggests that we can choose one of these as the objective, intrinsic model of the data. The universal model says that it is 'some compressible, finite object'. Another might say that it is 'an image'. Even more specific would be 'a painting', 'a Monet', or specifically 'the painting *Impression of a Sunrise*'. A sound sophistication should be able to select one of these as the proper representation of structure in the data, and disqualify the others as over- or underfitting. But how should we be able to say that the data is intrinsically more of a painting than an image? More of a Monet than a painting? Intuitively, such distinctions require further assumptions, or a second sample from the same distribution.

The second reason we doubt sophistication is more technical. Consider the set of all possible two-part representations of x. When the numbering is changed, the codelength of the model part of all these representations will change. This is illustrated in the second diagram in Figure 1. The invariance theorem expresses that this change is limited by a constant term. However, even this small shift can push some representations out of the acceptable region (indicated by the dashed line), and pull others in. This may lead to a different representation determining the sophistication, one whose *total* codelength is close to what it was before, but whose *model* codelength can be anywhere between 0 and $C^{\mathcal{C}}(x)$. If such jumps can occur, the sophistication is not invariant. And while we cannot *prove* in general that such jumps can always occur, there seems to be no reason to believe that they do not. Indeed, in [3] it is shown that logarithmic changes in the slack parameter can already cause these effects.

So we take a skeptical view of sophistication. Note that part of the theory is fine: there is nothing wrong with evaluating models for the data by comparing their two-part code lengths. In fact, the randomness deficiency $-\log p_i(x) - C^{\mathcal{K}}(x \mid i)$ has a direct statistical interpretation as a measure of counterevidence—under p_i, the probability of a randomness deficiency above k is less than 2^{-k} [6, Lemma 6]. In the Monet example above, this will allow us to disqualify the model expressing that the data is actually, say, a recording of jazz music.

But fundamental problems arise as soon as a hard cut-off is introduced on how far we are allowed to deviate from the minimum determined by the Kolmogorov complexity. In our opinion, a lot of measures taken in the literature, such as restricting the model class or introducing model penalties, complicate the method and make problems harder to analyse, without actually addressing the fundamental issue. This is dangerous: if such ad-hoc fixes result in a theory that is hard to prove either wrong or right, it creates an artificial dead end for a valuable area of research. When the hard cut-off on candidates is avoided, however, all such measures are no longer necessary. What remains is an elegant theory that can be used to sift through all possible models, disproving most while retaining a select number of interesting candidates for our further consideration.

Acknowledgments. This publication was supported by the Dutch national program COMMIT and by the Netherlands eScience center. We thank Tom Sterkenburg for interesting discussions.

References

1. Adriaans, P.: Facticity as the amount of self-descriptive information in a data set (2012). arXiv:1203.2245
2. Antunes, L., Fortnow, L., van Melkebeek, D., Vinodchandran, N.V.: Computational depth: concept and applications. Th. Comp. Sc. **354**(3), 391–404 (2006)
3. Antunes, L.F.C., Bauwens, B., Souto, A., Teixeira, A.: Sophistication vs logical depth (2013). http://arxiv.org/abs/1304.8046
4. Antunes, L.F.C., Fortnow, L.: Sophistication revisited. Theory Comput. Syst. **45**(1), 150–161 (2009)
5. Bennett, C.H.: Logical depth and physical complexity. In: The Universal Turing Machine: A Half-Century Survey. Oxford University Press (1988)
6. Bloem, P., Mota, F., de Rooij, S., Antunes, L., Adriaans, P.: A safe approximation for kolmogorov complexity. In: Auer, P., Clark, A., Zeugmann, T., Zilles, S. (eds.) ALT 2014. LNCS, vol. 8776, pp. 336–350. Springer, Heidelberg (2014)
7. Cover, T.M.: Kolmogorov complexity, data compression, and inference. In: The Impact of Processing Techniques on Communications, pp. 23–33. Springer (1985)
8. Gács, P., Tromp, J., Vitányi, P.M.B.: Algorithmic statistics. IEEE Tr. Inf. Th. **47**(6), 2443–2463 (2001)
9. Gell-Mann, M., Lloyd, S.: Information measures, effective complexity, and total information. Complexity **2**(1), 44–52 (1996)
10. Gell-Mann, M., Lloyd, S.: Effective complexity. Nonextensive Entropy-Interdisciplinary Applications, by Edited by Murray Gell-Mann and C Tsallis, pp. 440. Oxford University Press, Apr 2004. ISBN-10: 0195159764. ISBN-13: 9780195159769, 1 (2004)
11. Grünwald, P., Vitányi, P.M.B.: Shannon information and Kolmogorov complexity (2004). arXiv:cs/0410002
12. Kleene, S.C.: On notation for ordinal numbers. J. Symb. Log., 150–155 (1938)
13. Koppel, M.: Structure. In: The Universal Turing Machine: A Half-Century Survey. Oxford University Press (1988)
14. Koppel, M., Atlan, H.: An almost machine-independent theory of program-length complexity, sophistication, and induction. Inf. Sci. **56**(1–3), 23–33 (1991)
15. Li, M., Vitányi, P.M.B.: An introduction to Kolmogorov complexity and its applications. Springer-Verlag (1993)
16. Mota, F., Aaronson, S., Antunes, L., Souto, A.: Sophistication as randomness deficiency. In: Jurgensen, H., Reis, R. (eds.) DCFS 2013. LNCS, vol. 8031, pp. 172–181. Springer, Heidelberg (2013)
17. Shen, A.K.: The concept of (α, β)-stochasticity in the Kolmogorov sense, and its properties. Soviet Math. Dokl **28**(1), 295–299 (1983)
18. Vereshchagin, N.: Algorithmic minimal sufficient statistics: a new approach. Theory of Computing Systems, 1–19 (2015)
19. Vereshchagin, N.K., Vitányi, P.M.B.: Kolmogorov's structure functions and model selection. IEEE Tr. Inf. Th. **50**(12), 3265–3290 (2004)
20. Vitányi, P.M.B.: Meaningful information. IEEE Tr. Inf. Th. **52**(10) (2004)

A Appendix

Lemma 3 (Invariance of function complexity). *Let ψ and η be any two acceptable numberings Let f be any partial computable function. There exists a constant c independent of f such that*

$$\left| C^{\mathcal{K},\psi}(f) - C^{\mathcal{K},\eta}(f) \right| \le c \text{ and } \left| C^{\mathcal{C},\psi}(f) - C^{\mathcal{C},\eta}(f) \right| \le c.$$

Proof. Let $g(i)$ be the function such that $\psi_i = \eta_{g(i)}$.

$$
\begin{aligned}
C^{\mathcal{C},\psi}(f) &= \min\left\{ C^{\mathcal{C},\psi}(i) : \psi_i = f \right\} \ge \min\left\{ C^{\mathcal{C},\eta}(i) : \psi_i = f \right\} - c \\
&= \min\left\{ C^{\mathcal{C},\eta}(i) : \eta_{g(i)} = f \right\} - c = \min\left\{ C^{\mathcal{C},\eta}(g(i)) : \eta_{g(i)} = f \right\} - c' \\
&\ge \min\left\{ C^{\mathcal{C},\eta}(j) : \eta_j = f \right\} - c' = C^{\mathcal{C},\eta}(f).
\end{aligned}
$$

Reverse ψ and η for the opposite inequality. The same proof holds for $C^{\mathcal{K}}$.   □

Lemma 4. *There are faithful acceptable numberings.*

Proof. Let $d \in \mathbb{N}$ be an index such that $\phi_d(y) = \infty$ for all y. Define

$$
\psi_q = \begin{cases}
\phi_{\phi_i(p)} & \text{if } q \text{ can be written as } \bar{i}p \text{ and } \phi_i(p) < \infty, \\
\phi_d & \text{otherwise.}
\end{cases}
$$

It may seem that the second line requires a test whether $\phi_i(p)$ halts, for ψ to be acceptable, but as we will show below, this is not the case.

To show that ψ is faithful, pick any function f. Then

$$
\begin{aligned}
C^{\mathcal{C},\phi}(f) &= \min\{ C^{\mathcal{C},\phi}(i) : \phi_i = f \} = \min\{ \min\{ |\bar{a}b| : \phi_a(b) = i \} : \phi_i = f \} \\
&= \min\{ |\bar{a}b| : \phi_{\phi_a(b)} = f \} = \min\{ |\bar{a}b| : \psi_{\bar{a}b} = f \}.
\end{aligned}
$$

This shows there is a sufficiently small ψ index.

To show that ψ is acceptable, let $\phi_j(z) = z$. Then a ϕ-index i can be mapped to a ψ-index with $r(i) = \bar{j}i$, so that $\psi_{r(i)}(y) = \psi_{\bar{j}i}(y) = \phi_i(y)$. For the reverse, define $\phi_v(\bar{i}p, y) = \phi_{\phi_i(p)}(y)$. For fixed $\bar{i}p$, the s^n_m-theorem [12] states that we can compute the h such that $\phi_h(y) = \phi_v(\bar{i}p, y)$. Let $h(\bar{i}p)$ denote this index as a function of the program; further define $h(q) = d$ if q cannot be expressed as $\bar{i}p$. By construction h is total and computable. To check that the mapping returns the correct function, rewrite $\phi_{h(\bar{i}p)}(y) = \phi_v(\bar{i}p, y) = \phi_{\phi_i(p)}(y) = \psi_{\bar{i}p}(y)$. Note that if q can be written as $\bar{i}p$, but $\phi_i(p)$ diverges, $h(q)$ will still return a function, but one which doesn't halt, making it equivalent to ϕ_d as required.   □

Author Index

Printed in the United States
By Bookmasters